Glacial and Periglacial
Geomorphology

To the Memory of

W. Vaughan Lewis 1907-1961

who inspired in us a lasting interest
in glaciers and glacial geomorphology

Glacial and Periglacial Geomorphology

Clifford Embleton

Reader in Geography, University of London King's College

Cuchlaine A. M. King

Professor of Geography, University of Nottingham

Edward Arnold

First published 1968 by
Edward Arnold (Publishers) Ltd.
25 Hill Street, London WIX 8LL

Reprinted 1969
Reprinted 1971
Reprinted 1974

ISBN: 0 7131 5377 6

Printed in Great Britain
by Unwin Brothers Limited
The Gresham Press, Old Woking, Surrey
A member of the Staples Printing Group

Preface

The ice age has left a direct imprint on a large part of the world's land surface. Ice is also still active in many areas today, modifying the landscape and creating distinctive landforms. This book deals with glacial and periglacial landscapes in which the unifying theme is provided by ice. The emphasis throughout is on the landforms and processes of erosion and deposition. Many aspects of glaciology, such as the study of sea ice or glacier surface features, are either omitted or only touched on briefly, nor is it the authors' intention to provide any comprehensive account of Pleistocene geology, stratigraphy, and chronology. The indirect effects of glaciation, such as changes in land and sea level, were important and widespread, but limitations of space prohibit more than an outline of these aspects. Ten years ago, two major works in the English language, both largely concerned with Pleistocene geology and stratigraphy, were published: the second edition of R. F. Flint's *Glacial and Pleistocene Geology* and the two-volume survey of *The Quaternary Era* by J. K. Charlesworth. Although both contain sections dealing with glacial and periglacial geomorphology, there is need for a fuller treatment of these subjects, taking into account the many important advances and discoveries made in the last ten years.

The writing of this book, commenced in 1964, was inspired by W. Vaughan Lewis, whose untimely death in 1961 meant such a tragic loss to glaciology and geomorphology. Under Vaughan Lewis, to whose memory this book is dedicated, we both received our first introduction to glaciers and glacial landforms, especially in the several expeditions which he organized or initiated to study glaciers such as Vesl-Skautbreen, Veslgjuv-breen, and Austerdalsbreen in Norway. His infectious enthusiasm and unfailing cheerfulness, his generosity and helpfulness to his students, will always be remembered.

We have attempted to concentrate on the more modern contributions to the subject. The literature available is of gigantic proportions and is accumulating at an ever-increasing rate. The need to collate and summarize the results of research, from time to time, is just as important as the need for research itself. The text was completed in March 1967. References to the older literature are necessarily very selective, and those who seek a fuller treatment of historical aspects and discarded theories are referred to Professor Charlesworth's masterly summary. In compiling material for certain parts of the book, his survey of the Quaternary period has proved invaluable. We also acknowledge the help in collecting the most recent material provided by *Geomorphological Abstracts*, edited by Professor K. M. Clayton.

References are listed at the end of each Chapter. There is inevitably some duplication of references for this reason. Abbreviations of the titles of journals are in accordance with the recommendations given in the *World List of Scientific Periodicals* (4th edition). All measurements used in the book are given in the metric system.

The authors make no apology for the occasional overlap of material between one Chapter and another. Cross references are given where this occurs. Such overlap is unavoidable when a particular topic is relevant to more than one Chapter, and is essential for clarity of presentation.

The authors express their gratitude to the numerous persons who have assisted in various ways during the preparation of this book. Professor K. C. Edwards of the University of Nottingham and Professor J. C. Pugh of the University of London King's College have generously placed the facilities of their respective departments of geography at our disposal. The librarians at the Geological Survey Museum, the Science Museum, and the Royal Geographical Society have given every possible assistance. For help in typing parts of the manuscript, the authors are indebted to Miss D. M. Hooper and Miss C. Obert. Most of the illustrations have been prepared in the drawing office of the Department of Geography, King's College, London, and the authors express their gratitude particularly to Mr J. Davie for his meticulous work. Mr G. Zalavary gave expert assistance with photographic work.

CLIFFORD EMBLETON
CUCHLAINE A. M. KING

July 1967

Contents

PART I.   BASIC CONCEPTS OF GLACIATION
AND GLACIER BEHAVIOUR

PART II.   GLACIAL AND FLUVIOGLACIAL EROSION

PART III.   GLACIAL AND FLUVIOGLACIAL DEPOSITION

PART IV.   PERIGLACIAL GEOMORPHOLOGY

List of Plates

Figures in the Text

For the moving of large masses of rock, the most powerful engines without doubt which nature employs are the glaciers. . . . These great masses are in perpetual motion, . . . impelled down the declivities on which they rest by their own enormous weight, together with that of the innumerable fragments of rock with which they are loaded. These fragments they gradually transport to their utmost boundaries. . . . In this manner, before the valleys were cut out in the form they are now . . . huge fragments of rock may have been carried to a great distance; and it is not wonderful, if these same masses, greatly diminished in size, and reduced to gravel or sand, have reached the shores, or even the bottom of the ocean.

(J. PLAYFAIR, 1802)

Introduction

Geographers and geologists today have grown up with the idea that glaciers and ice-sheets once extended much farther south in Europe and North America, so that it is difficult to appreciate the doubts and difficulties that attended the establishment of the glacial theory. The complex of forms associated with mountain glaciation, the U-shaped valleys, rock-basin lakes, cirques and other features, are so well known that they are among the most easily recognized of landforms. They are, however, by no means easy to explain in detail, and many less striking but equally important features of glaciation may even be entirely overlooked.

The extent and character of the effects of the ice-age only became known gradually (F. J. North, 1943). W. Buckland, who was born in 1784, was aware of the presence of large erratics that could not have been carried to their present positions by rivers. He explained these and many other features, however, as the result of the Noachian flood. It was this idea that led to the introduction of the term 'drift'. Later, Buckland was converted to the glacial view by Louis Agassiz. The glacial theory had its beginning in the Alps, whose glaciers were larger in the late eighteenth and early nineteenth centuries, when the glacial theory was being slowly developed, than they are now.

As early as 1723, J. J. Scheuchzer (1672-1733) put forward a theory of glacier movement. He suggested that water entered crevasses in the ice and froze, causing the ice to move downhill. One of the chief initiators of the glacial theory was H. B. de Saussure (1740-99). His views were a mixture of modern and medieval. He appreciated the formation, movement and some of the effects of glaciers, especially their ability to transport large boulders. De Saussure greatly influenced the views of J. Hutton on glaciation. In 1815, J. P. Perraudin, a Swiss guide, suggested that glaciers had formerly been more extensive, a view which was supported by J. Venetz, an engineer, in 1821. Another early advocate of the glacial theory was J. de Charpentier (1786-1855). It is interesting to note that de Charpentier credited J. W. von Goethe, the poet, with the discovery of the ice-age around 1829-30. A. Bernhardi also published a paper on the glacial theory in 1832. The term 'ice age' was first coined in 1837 by K. Schimper.

The work of these pioneers and others, such as that of H. T. de la Beche in 1842, was largely ignored until the second half of the nineteenth century. It was not until Louis Agassiz turned his attention to the glacial theory that much notice was taken of the new ideas by the most influential scientists of the day. Agassiz was born in 1807. He was convinced by de Charpentier of the validity of the glacial theory. Agassiz himself, however, did not in fact add much new to the theory, but his well-known name gave greater weight to the new views. He published his views in 1840 under the title *Études sur les glaciers*. Agassiz converted Buckland and C. Lyell to the glacial theory when he visited Britain in 1840. After travelling to various parts of the British Isles, Agassiz came to the conclusion that all of this area had once been beneath an ice-sheet. Although a great weight of evidence was presented to members of the Geological Society of London in 1840 by Agassiz, Buckland and Lyell, there was still considerable hesitation in accepting the implications of the glacial theory. In 1847 Agassiz left for North America where he also stimulated interest in the glacial theory.

In Britain, the glacial theory did not receive much further support until A. C. Ramsay started his work on glaciation. In 1860, he published a study of the glacial features of Switzerland and North Wales. It was mainly Ramsay who eventually overcame the iceberg theory, to which Lyell had reverted by this time and which was widely held. He worked with T. F. Jamieson, who first put forward the true explanation of the parallel roads of Glen Roy. Meanwhile, many sound glaciological observations were being made in Switzerland by J. D. Forbes, J. Tyndall and others, who were working on the mechanics of glacier movement. These early workers have been followed by countless others, through all of whose efforts light has gradually been shed on the complexities of glacial processes and on the landscapes which they produce.

The Pleistocene ice-sheets and glaciers moved over a landscape that had previously been fashioned by subaerial processes and especially river action. The extent to which the ice modified these earlier landscapes varied with many factors. These factors included the nature of the pre-glacial relief, the amount of ice accumulation and the ease with which it could flow outward. Thus in many areas, the landscape now shows an assemblage of forms that owe their character to both glacial and non-glacial processes. The fact that so many landscapes are polygenetic greatly complicates any analysis of their evolution. The situation is rendered even more complex by the geologically rapid waxing and waning of the ice-sheets during the Pleistocene period, which is now thought to have covered at least the last one and a half million years.

The appearances of glacial landforms belonging to different periods of the ice age are also significantly different. There is a very striking difference between the character of the glacial landforms in the areas covered by the Newer Drift and the Older Drift in Britain. The Older Drift has subdued relief and the deposits are deeply weathered. They occur mainly on the interfluves as erosion has removed them from the valley floors. These characteristics contrast sharply with the fresh appearance of the Newer Drift, with its constructional forms and its frequent occurrence in the valley bottoms. This contrast has been referred to by K. M. Clayton (1957) as the 'attitude' of the drift.

The features produced by glacial action also develop and change with time from immature, youthful forms to mature forms. The ice age, however, was short from the viewpoint of geological time. It is unlikely, therefore, that glacial forms have developed beyond a fairly young stage in most areas. This applies particularly to those areas that lay near the limits of glaciation. The most advanced forms of glacial action described by D. L. Linton (1963) occur in the Antarctic, where they still remain partially buried beneath a thick cover of ice. Thus when glacial forms are examined, it must always be remembered that the time they have had to develop has been relatively short. The glacial processes have modified a pre-existing landscape, and the resultant landforms have also suffered modification under different conditions since their emergence from beneath the ice.

In view of the short duration of the ice age, it is apparent that in many circumstances the action of ice is very effective in producing a new landscape, which bears unmistakable evidence of its formative agent. Active glaciers are one of the most effective of all geomorphological agents, acting both destructively and constructively. Nevertheless there are also conditions under which ice can protect the land surface on which it rests.

There is still much to be learnt about the initiation of the glacial period and the factors on which the fluctuations of the ice-sheets and glaciers depend. It is not known whether the ice masses are now decaying for the last time or whether the present time is an interglacial period. The present situation cannot, however, be prolonged indefinitely. Either the ice will all melt eventually, with the result that sea-level would rise by at least 50 m and flood many major cities; alternatively, the northern ice-sheets might re-establish themselves and creep southward once more over northern Europe, Britain and North America. At present, the snow-line lies just above the top of Ben Nevis in Scotland, so that only a small lowering of the temperature would allow snow to remain throughout the summer and consolidate into small cirque glaciers.

There are few parts of the world which have not been affected directly or indirectly by the Pleistocene ice age. A great deal of work has been done and much has been written concerning the effects of the ice age. Much of this work has been devoted to a study of the chronology of glaciation, as revealed in the stratigraphy of glacial deposits on land and in the sedimentary record in the sea. This aspect is not, however stressed in this book, which is concerned essentially with the direct effects of glacial action and periglacial action on the landscape. The whole of the British Isles, much of Europe and North

America as well as large parts of Asia, South America and smaller parts of Australia show evidence in their landforms of the influence of glacial or periglacial processes. The areas beyond the outer limits of the major Pleistocene ice-sheets were modified by the cold climates that extended well beyond the glacial limits.

In the large area covered either by the ice-sheets or by periglacial conditions, a very wide range of relief types, rock types and climatic conditions occurs. The range of these variables leads in turn to a great diversity of glacial and periglacial forms that have been recognized and studied. This variety means that many different approaches to the problem of understanding the genesis of glacial and periglacial landforms have been adopted, and many different techniques have been applied in these studies.

Present-day glaciers provide a valuable laboratory in which to examine glacial processes and glacial landforms in the making. The surface movement of glaciers can easily be observed, but it is along their sides and at their bases that erosion and most deposition occurs. For this reason, much effort has been expended in drilling boreholes through glaciers and digging tunnels into them to examine their movement relative to their beds and to study the ice-rock interface. Instruments have been developed to measure the rate of movement of the ice, its temperature, its crystal structure and other characteristics. Observations such as these have provided important evidence on the manner in which ice moves and moulds its bed. The field-work has been supplemented by laboratory studies on ice and by theoretical studies of its behaviour under specified conditions. The theoretical studies and laboratory results can then be checked in the field by direct observation. In this way, valuable results have been obtained concerning the processes of glacial erosion and deposition in terms of the flow characteristics of the ice.

Glacial deposits and depositional forms are less easy to observe directly because they usually form beneath moving ice. Some marginal features, however, can be observed in the process of formation: for example, the formation of eskers has been observed on glacier margins. In depositional forms, the character of the deposit itself gives the most valuable evidence concerning their method of formation. For this reason, much work has been devoted to the study of till fabrics and the internal structure of the depositional features.

In periglacial activity also, the study of the deposits throws much light on periglacial processes. Some deposits, for example, can be shown to be frost-susceptible while others are not. Direct observations of periglacial features in the field can be usefully supplemented by laboratory experiments and by theoretical calculations. A combination of these methods has given a much clearer idea of how the patterns that are characteristic of the periglacial environment have been formed as well as many other landforms of peri-glacial origin.

In order to understand as fully as possible the glacial landforms that have been revealed by the withdrawal of the ice from areas previously covered by the ice-sheets, it is necessary to consider the studies that have been made of the existing ice masses. A study of the processes acting at present can throw light on the landforms that were created by similar processes in the past. Thus, the first section of the book is devoted to an account of the character-istics of the glaciers and ice-sheets that still cover substantial portions of the earth's surface. This section is followed by studies of the forms produced by glacial and fluvioglacial erosion (Part II) and by glacial and fluvioglacial deposition (Part III). Part IV is devoted to the periglacial environment.

REFERENCES

L. AGASSIZ (1840), Études sur les glaciers (Neuchâtel)

A. BERNHARDI (1832), 'Wie kamen die aus dem Norden stammenden Felsbruchsücke. und Geschiebe . . . an ihre gegenwärtigen Fundorte?' Heidelb. Jb. Minert. Geogn. Petrefaktenk. 3, 257-67

W. BUCKLAND (1823), Reliquiae Diluvianae (London)

J. DE CHARPENTIER (1835), 'Notice sur la cause probable du transport des Blocs erratiques de la Suisse', Annls Mines, Paris 3, 8, 219-36

K. M. CLAYTON (1957), 'The differentiation of the glacial drifts of the East Midlands', E. Midld Geogr. 7, 31-40

H. T. DE LA BECHE (1832), Geological Manual (London)

J. D. FORBES (1843), Travels through the Alps of Savoy (Edinburgh)

J. HUTTON (1795), Theory of the Earth (Edinburgh)

T. F. JAMIESON (1862), 'On the ice-worn rocks of Scotland', Q.J. geol. Soc. Lond. 18, 164-84

D. L. LINTON (1963), 'Some contrasts in landscapes in British Antarctic Territory', Geogr. J. 129, 274-82

F. J. NORTH (1943), 'Centenary of the Glacial Theory', Proc. Geol. Ass. 46, 1-28

J. P. PERRAUDIN (1815), quoted in J. VENETZ (1821)

A. C. RAMSAY (1860), The old glaciers of Switzerland and North Wales (London)

A. C. RAMSAY (1864), 'On the erosion of valleys and lakes', Phil. Mag. 4, 28, 293-311

H. B. DE SAUSSURE (1779-96), Voyages dans les Alpes (Paris), 4 vols

J. J. SCHEUCHZER (1723), Itinera per Helvetiae Regiones Alpinas (Leyden)

K. SCHIMPER (1837), 'Über die Eiszeit', Mém. Soc. helv. sci. nat. (1837), 5, 38-51

J. TYNDALL (1858), 'On some physical properties of ice', Phil. Trans. R. Soc. 148, 211-29

J. TYNDALL (1860), Glaciers of the Alps (London)

J. VENETZ (1833), 'Mémoire sur les variations de la température dans les Alpes de la Suisse', Mém. Soc. helv. sci. nat. (1833), 1, pt. 2

PART I

Basic Concepts of Glaciation and Glacier Behaviour

Glacier ice is able to produce the phenomena for which we seek to account; and that in every place where those phenomena occur. (M. H. CLOSE, 1866)

Chapter 1

Ice Ages and World Glaciation

Finally, two or three hitherto silent guests called to their aid a period of intense cold, with glaciers descending from the highest mountain ranges, far into the low country. . . .
(J. W. VON GOETHE, *Wilhelm Meister*, 1829)

Throughout the geological history of the earth, long periods of warm climate have alternated with shorter phases of generally cold climate during some of which ice-sheets and glaciers have appeared. There is good evidence that glaciation occurred more than once in the Pre-Cambrian and again in the Permo-Carboniferous period, as well as in the Pleistocene. Pre-Cambrian tillites and boulder-beds are known from many widely separated areas of the world, such as Scotland (C. Kilburn, W. S. Pitcher, and R. M. Shackleton, 1965) or northern Michigan (U.S.A.). There is even clearer evidence of the Permo-Carboniferous ice age. In South Africa, for instance, the Dwyka tillite at the base of the Karroo System contains scratched and faceted stones, and rests in many places on a striated rock surface; there are also varve-like sediments. The approximate world extent of the Permo-Carboniferous glaciation is known, but the same cannot be said for any of the Pre-Cambrian glaciations where the evidence is very much more fragmentary. In neither case is it known how long glaciation lasted, nor whether there was a series of glacial and interglacial episodes within one major glaciation.

The Permo-Carboniferous glaciation was followed by a lengthy span of time, from the Triassic to the Tertiary, when world temperatures were higher than those of today, and when ice-sheets and glaciers were absent. In Eocene times, the mean annual temperature for central Europe may have been 21-22°C. (P. Woldstedt, 1954), but during succeeding phases of the Tertiary, temperatures fell gradually, the total drop in temperature through the Tertiary being assessed at about 10°C. Around 1·5 to 2 million years B.P., the onset of the Pleistocene refrigeration was heralded by a relatively sudden further fall of temperature.

The Pleistocene Period

a) General Introduction

'Pleistocene' was the name given by C. Lyell (1839) to the 'most recent' epoch of geological history.* Repeated glaciation of certain areas was the outstanding feature of this period; at times, ice spread out over an area totalling 46 million km², or 3·1 times the present ice cover of the world (Figs. 1-1 and 1-2). Agreement has still to be reached on the number of glacial periods, but four or five principal ones are generally accepted (Table 1). Each of these glacial periods may have included as many as three or four separate cold phases with milder intervals or 'interstadials'.

TABLE 1   *Subdivisions of the Pleistocene in the Main Glaciated Areas*

Provisional Numerical Order	Great Britain	Alps	Northern Europe	European Russia	North America
Last Glaciation	Newer Drift	Würm	Weichselian	Valdai	Wisconsinan
Last Interglacial	*Ipswichian*		*Eemian*		*Sangamonian*
Fourth Glaciation	Gipping	Riss	Saale	Dnepr	Illinoian
Third Interglacial	*Hoxnian*		*Holstein*		*Yarmouthian*
Third Glaciation	Lowestoft	Mindel	Elster	Likhvin	Kansan
Second Interglacial	*Cromerian*				*Aftonian*
Second Glaciation		Gunz	Pre-Elster	Pre-Likhvin	Nebraskan
First Glaciation		Donau			Pre-Nebraskan

Note: Named interglacials are shown in italics

The interglacials were probably longer than the glacial periods, and certainly lasted many times longer than the Post-glacial period. Correlations suggested between various parts of Europe in Table 1 are reasonably certain, but the links with North America, apart from the correlation of the Wisconsinan with the Würm/Weichselian, are more tenuous. Some recent correlations even equate the Nebraskan, Kansan, Illinoian with the Donau, Günz and Mindel respectively, assigning the Riss to the earliest Wisconsinan-Sangamonian stage. It is also possible that the sequences are only the last of a longer series, in which records of still older glaciations have not survived, though most geologists now regard the sequence Donau through Würm as spanning practically the whole of the Pleistocene. The distribution of ice was not precisely the same in each glacial period. The Riss/Saale glaciation in Europe and the Illinoian in North America probably represented the maximum spread of ice, though in some areas, as in west-central North America, the preceding Kansan (Mindel) ice extended even farther (Fig. 1-2). The last (Würm, Wisconsinan, etc.) glaciation was somewhat less extensive. On a

---

* Most geologists today use the term Quaternary for the whole of the period since the Tertiary, and divide the Quaternary into Pleistocene and Recent. The Recent period (sometimes called the Post-glacial), is variously taken as beginning around 6000 B.P. (at the end of the Flandrian transgression), at 8800 B.P. (by De Geer), or around 11,000 B.P. (when data from a number of sources indicate a sudden amelioration of climate).

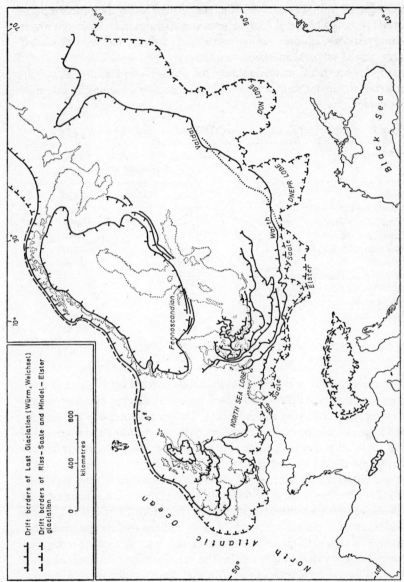

Fig. 1-1 European Pleistocene ice limits (R. F. Flint, *Glacial and Pleistocene Geology*, John Wiley & Sons, New York 1957—with minor amendments)

world-wide view, however, these differences are of relatively minor importance, and it should be noted that the several glaciations in fact covered much
the same territory, adopted similar patterns of ice-flow, and were governed
by very similar physical and climatic controls.

The following table, mainly abbreviated from W. L. Donn et al. (1962),
summarizes the major distributions of Pleistocene ice according to the most
recent evidence:

TABLE 2   *Areas Covered by Ice in the Last Glaciation and at the Pleistocene Maximum*

Area	Maximum extent	Extent in last glaciation
	million of km²	
Antarctica	13·20	13·20
Laurentide ice-sheet	13·79	12·74
Scandinavian ice-sheet	6·67	4·09
Siberian ice-sheet	3·73	1·56
N. American Cordilleran ice	2·50	2·20
Greenland	2·16	2·16
All others in N. Hemisphere	4·07	3·45
Southern hemisphere excluding Antarctica and including all South America	1·02	0·90
	47·14	40·30

cf. Present-day ice-covered area: 14·97

Both the Antarctic and Greenland ice-sheets differed little in extent from
their present sizes, for they were limited as now by calving into deep water.
They were, however, very much thicker. The great Laurentide and Scandinavian ice-sheets are now represented by quite insignificant and fragmented
masses: over 99 per cent of their bulk has melted away. Relatively little is
known about the thicknesses of ice in the Pleistocene, except for a few small
areas, and consequently estimates of the volume of ice vary widely. H. Valentin (1952) computed a figure of 60·9 million km³; a more recent assessment
by W. L. Donn and others (1962) puts the figure of the Riss/Illinoian
glaciation at 84 to 99 million, and 71 to 84 million for the last glaciation.
These figures represent about three times as much ice as remains today.

The climatology of the Pleistocene is also very speculative and inadequately
understood: some hypotheses to explain the changing climate will be summarized later. There is an enormous literature concerned with probable
climatic conditions at different stages, yet opinions are so divergent (except
in regard to the last 15,000 years) that it is unsafe to generalize beyond such
statements that the average world temperature in the glacials was possibly

5-7°C. cooler than at present while the interglacials may have been somewhat warmer than at present in middle to high latitudes. The snow-line in the glacials was depressed everywhere, but not uniformly. In the Alps, it was reduced in elevation by as much as 1300 m, in Britain by up to 1200 m, and in the Western Cordillera of North America by about 800 m.

b) Stratigraphy and Correlation

Because of the relative youthfulness and therefore often well-preserved nature of its record in terms of deposits, landforms, and fossil plants and animals, the Pleistocene is the best documented of all geological periods. This very wealth of data, and the multiplicity of changes within a short time span have led to great difficulties of interpretation and wide differences of opinion concerning sequences of events and their correlation. Pleistocene stratigraphy is complicated at the outset by the enormous range of environments in which deposits accumulated—marine, fluviatile, lacustrine, glacial, arid, and volcanic sequences are included. Even more formidable from the point of view of correlation and attempts to establish successions is the fact that the great bulk of the Pleistocene sediments now visible consist of terrestrial and particularly glacial accumulations, characterized by rapid lateral and vertical variation, often poorly understood mode of deposition, and lack of significant fossil content. Exposures, moreover, are often inadequate, and lack clarity because of the unconsolidated nature of the sediments. Nor was the Pleistocene period one of crustal stability; not only were large areas affected by glacio-isostatic deformation, but tectonic disturbances resulting in faulting, tilting, and warping were widespread in certain areas (see, for example, P. B. King, 1965).

The principal bases of correlation of Pleistocene successions may be summarized as follows:

i) *Fossils* The stratigraphy of the Pleistocene, unlike that of the older geological periods (whose time span was individually vastly longer), does not depend to any great extent on fossil zonation, whether by plants, molluscs, or vertebrates. This is fundamentally because of the short duration of the Pleistocene, the frequent absence or rarity of fossil remains, and the time-transgressive nature of the fossils that do exist. The fossil content of deep-sea cores has, however, been of value in estimating Pleistocene changes of temperature.

ii) *Lithological units* The tracing and correlation of lithological units is a basic tool (and one of the oldest) for elucidating Pleistocene successions. The mapping of particular drift sheets, solifluction sheets, loess horizons, and spreads of fluvioglacial gravels (including terraces) has been the foundation of Pleistocene stratigraphy in North America and Europe. The method is not without its hazards, especially the result of rapid lateral variations in the unit being used for correlation.

iii) *Palaeosols and weathering zones*   Weathered layers and fossil soils have long been used stratigraphically to separate Pleistocene deposits, especially in the Mid-West of North America (R. V. Ruhe, 1965). Here, they have been the primary basis for separation of the four principal drift sheets; they have also been used to correlate pluvial events in the Great Basin with glacial successions in the Sierra Nevada (R. B. Morrison, 1964) and elsewhere.

iv) *Prehistoric cultures*   Archaeological investigations have helped (though in some cases also served to hinder) the unravelling of Pleistocene successions and their approximate dating. Such methods are now recognized as rather unreliable, because of the way in which the cultures are strongly time-transgressive. J. K. Charlesworth (1957, pp. 1028-39) gives a detailed summary of some relevant literature.

v) *Pollen stratigraphy*   The study of pollen sequences has proved to be one of the most useful of all methods in elucidating the climatic changes of the last 15,000 years (see, for example, H. Godwin, 1956). A series of stratigraphical units and pollen zones has been established in the British Isles and northern Europe, and is also being worked out in North America. Table 3 gives the principal time divisions now widely adopted in Europe, though there are minor differences between the schemes formulated independently in the several countries. The names in column 2 are based on the classical work of Blytt and Sernander in the late nineteenth and early twentieth centuries.

For each zone in a particular region, a characteristic flora is reflected in the pollen assemblage. There are, of course, variations from one region to

TABLE 3   *Pollen Zones of the Late- and Post-glacial Periods in North-western Europe*

Period		Zone	Time, radiocarbon years B.C.
Post-Glacial	IX	Sub-Atlantic	Post-500
	VIII	Sub-Boreal	500-2500
	VII	Atlantic	2500-5500
	VI	Late Boreal	5500-6500
	V	Early Boreal	6500-7500
	IV	Pre-Boreal	7500-8200
Late-Glacial	III	Younger Dryas	8200-9000
	II	Allerød	9000-10,200
	I c	Older Dryas	
	b	Bölling	10,200-12,800
	a	Oldest Dryas	

Note: In North America, Zone III corresponds to the Valders advance, Zone II to the Two Creeks interval, Zone I to the Mankato advance.

another within any one zone and, taking this into account, it is possible broadly to reconstruct vegetation maps for each period which throw light on the climatic conditions prevailing, and to learn about the distribution of glacial ice in particular periods. Zone III, for instance, witnessed the last phase in Britain when glaciers reappeared.

c) Pleistocene Chronology

The last twenty years have seen the development and application of entirely new methods to the elucidation of stratigraphical problems and correlation in the Pleistocene. Prior to about 1950, Pleistocene chronology rested on very uncertain foundations or on techniques of very limited application; it is well, however, to outline these older techniques first, before considering the new methods.

i) *Rates of operation of geomorphological processes*   Such assessments were the basis of the earliest attempts to devise a time-scale. A well-known example is the measurement by various workers of the rate and distance of recession of Niagara Falls; other workers have investigated rates of delta-building since an area was uncovered by ice, rates of weathering of drift sheets, or rates of limestone solution around erratics deposited originally on the limestone surface but now found resting on a limestone pedestal (for a summary, see Charlesworth, 1957, pp. 1517-22). All such measurements were subject to grave difficulties of interpretation, and many were later shown to be invalid. Extrapolation of rates of denudation backwards in time should not rest on the assumption that the rates have been constant.

ii) *Dendrochronology*   The study of annual growth rings of trees can provide an accurate clock within particular regions. The method is limited for obvious reasons to the last 3000-4000 years at the most, but within this limit, tree-ring studies yield reliable climatic and hydrological data (H. C. Fritts, 1965).

iii) *Rhythmite studies*   The use of rhythmites or varves in glacial lake deposits for chronological purposes was pioneered by De Geer and his associates in Sweden in the early part of this century, and later applied by E. Antevs in eastern North America. The method is discussed in Chapter 19. Until the advent of radiometric dating, this was the most reliable basis of absolute dating of late Pleistocene events in Scandinavia and North America. Much valuable information was accumulated, though the validity of some correlations has been questioned.

iv) *Fluorine-phosphate ratios in fossil bones*   The fluorine content of bones affected by percolating ground water containing minute traces of this element increases with time. The method excited public interest with the exposure of the Piltdown forgery, but the finding of suitable naturally emplaced bones in Pleistocene successions is obviously a rare event.

v) *Dating by radioactive isotopes*   The introduction of these techniques has revolutionized the whole subject of Pleistocene chronology. For the first time, relatively accurate dating of horizons in Pleistocene successions is becoming possible, and the next decades are likely to witness greater progress in the unravelling of Pleistocene stratigraphy and the correlation of events than has been made in the whole of the last century or so of investigation. Space does not permit more than a brief indication of the methods so far established and their potentialities: for a fuller recent discussion, see W. S. Broecker (1965). The first of the methods to be explored was based on the isotope carbon 14 (W. F. Libby, 1955), and can be applied to organic and carbonate materials. At first, the method was restricted to dates within the last 35,000 years, but with pre-enrichment and sensitive detectors, a range of 70,000 years is now feasible, which is found to cover most if not all of the last glaciation (Würm/Wisconsinan). The method is not without its problems and sources of error, which cannot be elaborated here; but the possibility of contamination by younger carbon means that ages of organic materials over 40,000 years and of carbonate materials over 20,000 years must be considered as minima.

The second method depends on the decay of the isotope potassium 40 to argon 40. It can yield reliable ages for certain minerals in which the argon produced has not been able to escape and where there has been no likelihood of contamination by atmospheric argon. It has so far been applied only to volcanic rocks in the Pleistocene, and the main problem is thus of finding suitable rocks in critical stratigraphical positions. The method is rather uncertain for dates younger than 20,000 years, but even this gives a possible overlap with radiocarbon dating. A most interesting result recently obtained is the dating of the Bishop Tuff in western North America at 700,000 years, for the Bishop Tuff overlies glacial till, possibly the earliest in North America.

Thirdly, there are the uranium series methods based on uranium 234, thorium 230, and protactinium 231 which have recently been applied to some Pleistocene marine and other carbonate sediments, but this work is still in an early stage. Provisional Pa^{231}/Th^{230} dating of the Sangamon (Riss–Würm) interglacial suggests an age around 100,000–150,000 years B.P.

The methods so far mentioned are summarized in Table 4. Others, not yet properly evaluated, are based on the isotopes helium 4, chlorine 36, and beryllium 10. With the aid of radioactive isotopes, a reliable chronology for the last 150,000 years is being steadily built up, and there is considerable promise that its extension farther back in the Pleistocene will be feasible.

vi) *Tephrochronology*   Introduced by S. Thorarinsson (1944, 1949) in Iceland, and developed by S. Kaizuka (1961) and others in Japan, this method of dating and correlation depends on the fact that a mantle of volcanic ash,

TABLE 4 *Principal Radioactive Isotopes used in the Dating of Pleistocene Events*

	Half-life (years)	Approximate range of method (years)
C^{14}	5730 ± 40	0–70,000
K^{40}	$1 \cdot 3 \times 10^9$	20,000+
U^{234}	250,000	50,000–1,000,000
Th^{230}	75,000	0–400,000
Pa^{231}	32,000	5000–120,000

deposited over a wide region in a brief moment of geological time and later buried, forms an ideal stratigraphical marker. Its age may be ascertained by the K^{40} method, or by C^{14} dating of associated carbonaceous material, if any. An example is the Pearlette ash layer of the central-western U.S.A., emanating from a still unlocated source. It forms an important link between Kansan glacial deposits in the Great Plains, the Cedar Ridge glaciation in the Rockies, an ancient subaerial stage of Lake Bonneville, and alluvial deposits in the unglaciated part of the Great Plains. Its detailed distribution has yet to be systematically studied, and its dating is awaited with keen interest (see R. E. Wilcox, 1965).

vii) *Palaeomagnetic studies*   Palaeomagnetic correlation of igneous and some sedimentary rocks is possible because the direction of the earth's magnetic field is thought to have suddenly and periodically reversed at certain points in geological time, and because certain rocks become permanently magnetized in the earth's field at the time they are formed. The reversals are necessarily world-wide and synchronous, and are now broadly dated by the K^{40} method. Consequently, correlations and further dates can now be obtained by simple palaeomagnetic measurements. Four epochs are recognized:

TABLE 5 *Late Cenozoic Epochs of Normal and Reversed Magnetization*

Age (10^6 years)	Name	Magnetization
0–1	Brunhes epoch	Normal
1–2·5	Matuyama epoch	Reversed
2·5–3·4	Gauss epoch	Normal
>3·4	Gilbert epoch	Reversed

Glaciation in North America began at least in the early part of the Brunhes epoch; in Iceland, glacial deposits are associated with the Matuyama and possibly even earlier (M. G. Rutten and H. Wensink, 1960). With such a broad time scale of world-wide significance established, dating of early Pleistocene and pre-Pleistocene events and deposits is in progress. Already, it seems likely that the Villafranchian sequence in Europe terminated within the Matuyama epoch. The main problem hindering further progress is that, in common with the K^{40} method, of finding juxtapositions of rocks suitable for palaeomagnetic dating, and glacial or fossiliferous deposits (A. Cox, R. R. Doell, and G. B. Dalrymple, 1965).

viii) *Studies of deep-sea sediments*   Development of deep-sea core-sampling techniques has made possible the stratigraphical study of Pleistocene deep-sea sediments. Changes in the type and abundance of fossils in various layers can be used to reveal changes of sea-water temperature, which to some extent must reflect changes in Pleistocene climate. C. Emiliani (1955) developed an additional technique based on oxygen isotope ratios for pelagic foraminiferal shells, enabling actual temperature curves to be deduced with some degree of confidence, and suggesting that Pleistocene temperatures of tropical sea-surfaces fluctuated within a range 23-29°C., and that the lowest temperatures of presumed glacial maxima were much the same, as were the highest temperatures of the interglacials. Having obtained temperature curves, the problem remains of dating the events shown on them. If the most recent layers of a core are dated by such means as carbon 14, extrapolations of age can be made to the older layers (more than 30,000 years old) by assuming uniform rates of sediment accumulation. This is, of course, very uncertain, and herein lies the weakness of the system until absolute dates for the older layers are available to act as independent checks. Meanwhile, Emiliani suggests that the maxima of the Riss-Saale glaciation occurred at about 110,000 B.P., of the Mindel-Elster at about 180,000 B.P., and of the Gunz at about 280,000 B.P.; W. S. Broecker and others (1958) extended Emiliani's time scale by about 50 per cent, by allowing for variable rates of sedimentation. Completely different correlations between the glacial sequence and the deep-sea sediment record have been made by D. B. Ericson, M. Ewing, and G. Wollin (1964). According to them, for instance, the first glacial period (? Gunz) is dated at about 1,500,000 years, and there is only agreement with Emiliani over the last 120,000 years. A. Holmes (1966) also favours a 'long' time-scale placing the first (Donau) glaciation between 1·8 and 1·37 million years, and the Gunz between 1·15 and 0·9 million years.

d) The Duration of the Pleistocene

A fundamental and still controversial matter in this respect is the definition and location of the Plio-Pleistocene boundary, a subject too large to enter into here, beyond observing that geologists in England now place the boundary at the base of the Red Crag, and in Italy at the base of the Calabrian. These horizons mark a distinct deterioration in climate and significant faunal changes.

The duration of the Pleistocene has been a constant subject of discussion and disagreement. Early estimates were given boldly and confidently;

caution has now supervened. Estimates based on such vague measures as the degrees of weathering of different drifts (for example, F. Leverett, 1930, arrived at 1 million years since the Nebraskan) or on hypotheses such as that of M. Milankovitch (1930; insolation curves said to support a figure of 700,000-750,000 years for the Donau glaciation) are not now regarded as of any real value. The indications of a few K[40] dates and palaeomagnetic studies are that the beginning of the Pleistocene is likely to have been not less than 800,000 years ago and possibly as far back as 2·5 million years. D. B. Ericson and G. Wollin (1964) prefer a figure of 1·5 million years, and postulate four main glacial periods since that date. K. P. Oakley (1964) adopts the figure of 1·5 million years but proposes five main glacials beginning with the Donau. In North America, the Sherwin till of California-Nevada is older than 700,000 years (see p. 16 for the dating of the Bishop Tuff above it); in Iceland, the record of glaciation exceeds 2·5 million years (see p. 17) but this may extend back into the Pliocene.

Present-day Glacierization

About 15 million km², or 10 per cent of the earth's land area, are ice-covered at the present day. Table 6 shows the principal glacierized regions, ranked in order of size. It should be noted that the two largest, Antarctica and Greenland, account for 96 per cent of the total ice-covered area. The figures are those quoted by R. F. Flint (1957), with minor amendments, and they should be regarded, like all measurements of large areas, as approximations

TABLE 6   *Present-day Ice-covered Areas (in km²)*

Antarctic	12,653,000
Greenland	1,802,600
North-east Canada	153,200
Central Asian ranges	124,500
Spitsbergen group	58,000
Other Arctic islands	54,000
Alaska	51,500
South American ranges	25,000
West Canadian ranges	24,900
Iceland	11,800
Scandinavia	5,000
Alps	3,600
Caucasus	2,000
New Zealand	1,000
U.S.A.	650
Others	about 800

(Total volume of present ice: 28 to 35 million km³)

only. Many ice-covered regions are still imperfectly mapped, as for instance in Central Asia.

The table demonstrates how varied are the climatic circumstances under which glacier ice exists in the world today. By far the greatest part of the world's ice, in the Antarctic and northern Greenland especially, is found in high latitudes characterized by very low winter temperatures (the lowest temperature yet recorded is $-88°C$. at Vostok near to the south geomagnetic pole at 3500 m elevation), low summer temperatures, small annual precipitation, and minimal ablation. These ice masses as a result are relatively inactive and stable. Greater activity is noticeable in areas of less extreme cold and moderate precipitation, as in Southern Greenland for example. The most active glaciers, generally speaking, are those in regions where winter temperatures are low but summer temperatures high, where there is abundant precipitation in the form of snow, but at the same time where ablation losses are often heavy. Such conditions prevail in the Alps, in New Zealand, and in many other regions of middle latitudes.

The vertical distribution of ice, now as in the past, is controlled primarily by the altitude of the firn-line on the ice, for this affects the size of the accumulation area above it (see Chapter 2). The snow-line in ice-free areas varies considerably in altitude both in any given area and seasonally; it is often highly irregular, and strongly influenced by such considerations as shade and snow-drifting as well as by precipitation and ablation. For this reason, it is more useful usually to deal with the average 'regional' rather than the 'local' snow-line, the regional snow-line* being defined as the average lower limit of snow persisting throughout the year. The height of the regional snow-line varies at present from sea-level, to over 6000 m in the Central Andes. In northern Britain, its theoretical height is put at about 1500 m. Its height is controlled both by snowfall and by summer temperatures, which determine the extent to which winter snow can survive throughout the summer season. Any rain falling in summer, too, is a potent instrument aiding the dissipation of winter snow.

The distribution of ice in the world is a constantly changing one. Glacier fluctuations are obvious in terms of the retreat or advance of glacier snouts over periods of a few years or decades, and are readily measurable by comparisons of maps of different dates, or more precisely by ground survey or photogrammetry. Less obvious, less easily measured, but of even greater significance are changes in the thickness of glaciers, which in turn cause changes in area or length. Trim-lines, representing abrupt changes in vegetation, sometimes mark the former higher margin of a glacier (Chapter 15). It is well known that since the mid- or late-nineteenth century, the world's

* Sometimes termed the 'climatic' snow-line.

glaciers have in general been receding—the tongue of the Rhône glacier for instance now hangs high above the village of Gletsch, to which it closely approached in 1920. Many glaciers this century have stagnated; some, as in the Pyrenees, have vanished entirely. Alpine glaciers overall have lost more than one-third of their area since the 1870s, though there have been minor interruptions in this recession. A few glaciers in the world, exceptionally, have advanced in this period (see Chapter 2): one such is the Taku glacier in Alaska. Data on glacier fluctuations relating mainly to the period 1850-1950 are extensively summarized in Charlesworth (1957, pp. 142-4).

Glaciers vary greatly in their rates of response to climatic fluctuations (see Chapter 2). The least sensitive are the largest and most slowly moving ice-sheets, lying in high-altitude regions of severe climate. Much better indicators of climatic change are glaciers of middle latitudes with a critical mass balance, where slight changes in rates of accumulation or ablation may have very profound consequences. It is quite possible that only small changes of climatic régime were necessary to induce the rapid expansion of middle-latitude ice-sheets in the Pleistocene.

Causes of Ice Ages

Any modern theory of the causes of ice ages must be able to satisfy the following requirements, most of which are now well established:

1. There were Ice Ages at various geological epochs before the Pleistocene.
2. There is evidence of a slow cooling of the earth's atmosphere through the Tertiary period.
3. Pleistocene glacial periods were cold climatically, and temperatures over the whole earth were less than those now experienced.
4. There were interglacial periods in the Pleistocene, some of which at least were relatively warm.
5. The difference of mean annual air temperature between glacial and inter-glacial periods was of the order of 6°C. in the Tropics (as shown by deep-sea sediment studies) and possibly 8° or 9°C. in higher latitudes.
6. Precipitation in and around ice-covered areas in the glacial periods was slightly less than in these areas today (G. Manley, 1951, for instance, suggests 80 per cent of present precipitation for British Highlands), but more of it fell as snow.
7. The snow-line in the Pleistocene rose and fell through a vertical interval of as much as 1300 m in places.
8. There were minor fluctuations of climate superimposed on the general rhythm of glacial and interglacial episodes.

9. The climatic fluctuations in the northern and southern hemispheres were probably synchronous (see, for example, J. T. Hollin, 1962): this is definite for the last 10,000-15,000 years, but for the period prior to this, has been disputed, for instance, by A. T. Wilson (1966).

10. Land and water had no radically different distribution from that of today.

11. The earth's poles were in much the same positions as they are today.

The last two conditions apply to the Pleistocene, but not to the pre-Pleistocene glacial epochs.

The past century and a half has seen innumerable theories expressed on the cause of Ice Ages ranging 'from the remotely possible to the mutually contradictory and the palpably inadequate' (Charlesworth, 1957, p. 1532). Few, indeed, can claim to be able to meet more than a small number of the conditions given above. The older theories are adequately reviewed elsewhere (C. E. P. Brooks, 1949, and J. K. Charlesworth, 1957, for instance), and the chief purpose of this section is to classify, as concisely as possible, the main groups of the more plausible theories. Attention will be focussed on those that have emerged in the last twenty years, and both primary and secondary mechanisms will be considered.

A. Geophysical Theories

a) Continental Drift   As a hypothesis which has returned to favour amongst geologists in recent years as a result of palaeomagnetic and other studies, it has been used as a means of bringing various regions into polar positions at various times. Its principal application is in substantially helping to solve the riddle of the Permo-Carboniferous glaciation, but there is no evidence for any significant differential movement of the continents in the $1\frac{1}{2}$ million years or so of the Pleistocene: on the contrary, the fact that successive Pleistocene glaciations covered mainly similar areas is strong evidence against such movement. It is also a mistakenly-held belief that polar positions provide the optimum conditions for glaciation, whereas in fact much more favourable positions are those of middle-latitudè highlands experiencing heavy snowfall and cool summers.

b) Polar wandering   This is another means of achieving the same aim as that of supposed Pleistocene continental drift, and it encounters the same problems, especially the evidence against any significant polar wandering in the Pleistocene. The idea, however, has been recently used (M. Ewing and W. L. Donn, 1956-9) as a means of explaining the onset of glaciation at the beginning of the Pleistocene, it being postulated that at this time the earth's poles first took up roughly their present positions. Other factors were then brought into play to account for the glacial-interglacial sequence.

B. *Land-water Changes*

a) Oceanic changes   Since some major ocean drifts and currents serve to transfer considerable quantities of heat from tropical to polar latitudes, changes in these water movements could provoke changes in polar climates. The most critical point in regard to the Arctic is the Faroe-Icelandic sill, shallowing of which could result in less tropical water entering the Arctic Ocean. Glacio-eustatic lowering of sea-level (Chapter 5) could in itself produce this effect and it thereby becomes a secondary mechanism of promoting cooling and ice extension in the Arctic once a glacial period is under way. Ewing and Donn (1956-9) utilize this idea. Earlier, C. E. P. Brooks (1949) also postulated rearrangements of oceanic circulation caused by tectonically-created land bridges to cause climatic cooling of certain areas. J. T. Hollin (1962) also points out that sea-level lowering in the glacial periods would result in northward displacement of the 'grounding-line' in Antarctica—the line along which the ice-shelves begin to float. This was an important factor aiding extension of the Antarctic ice and the implications are discussed more fully in Chapter 2.

b) Changes of land-mass altitude   The effect of elevating the land is to bring larger areas above the snow-line. This was a view long held to account for Pleistocene glaciation following on the widespread orogenic activity of Tertiary time, until the fact of the glacial-interglacial succession became established. Although there is no evidence for rapidly alternating uplifts and subsidences to explain climatic fluctuations within the Pleistocene, the basic idea of the Pleistocene following (and including) a period of tectonic uplift is a sound one. There is abundant geological and geomorphological evidence of late-Tertiary and Pleistocene uplift in many areas. The Alps, for instance, are thought to have been raised some 2000 m from the Pliocene to the mid-Pleistocene; the Sierra Nevada of California may have risen 2000 m or more in the Pleistocene (see P. B. King, 1965); the Himalayas were possibly uplifted 3000 m in the Pliocene-early Pleistocene. It is well known that quite small altitudinal changes can have a disproportionately great effect in increasing snowfall and lowering temperatures, and the widespread nature of Cenozoic uplift undoubtedly helped to promote Pleistocene refrigeration, even though it cannot explain the fact of multiple glaciation. Some authorities believe that the gradual increase in mean land altitude throughout the Tertiary was responsible, partly at least, for the gradual lowering of temperatures in this time (see p. 9). The idea is also applicable, it should be noted, to the Permo-Carboniferous glaciation which also succeeded an epoch of major orogenesis in many parts of the world.

A further point in connection with land-mass altitude and glaciation concerns glacial isostasy. Once ice on a land area reaches a certain thickness, isostatic depression of the crust commences, and may eventually amount to one-quarter to one-third of the ice thickness. Thus the accumulation of great thicknesses of ice is encouraged, and this in turn leads to more prolonged survival of the ice.

C. *Atmospheric Changes*

a) *Variations in carbon dioxide content*   By reducing the amount of CO_2 in the atmosphere, the rate of absorption by the latter of long-wave radiation will also be reduced, and thus the atmosphere cooled. The essence of this theory was stated as long ago as 1899 by T. C. Chamberlin. Modern calculations suggest that even to lower surface air temperatures by 3°C. would necessitate halving the present atmospheric CO_2 content, and it is practically impossible to devise a way of upsetting the earth's carbon cycle to such an extent. However, it may be an important secondary factor in the glacial-interglacial alternation, inasmuch as with the formation of ice-sheets (containing little carbonate matter), the atmosphere's content of CO_2 would be indirectly increased, causing some warming of the atmosphere and helping to bring glaciation to an end.

b) *Variations in the amount of suspended volcanic dust*   Volcanic (or other) dust in the atmosphere serves to cut out some solar radiation, but the screening effect is essentially regional rather than world-wide, and would have to be maintained for long periods. Evidence of volcanic activity on a sufficient scale throughout the Pleistocene is lacking.

c) *Other atmospheric changes* involving ozone and water vapour content, for example, have been considered, but we do not know what might cause such changes, nor how the glacial-interglacial succession might thereby be explained.

D. *Meteorological Theories*

These consist essentially of secondary mechanisms for promoting glaciation once started, or for bringing a glacial period temporarily to an end. The basic idea is that quite small changes in climate, or in the atmospheric or oceanic circulatory systems, might have far-reaching effects, especially on ice-masses in a critical condition. C. E. P. Brooks (1949) argued, for instance, that quite a small drop in temperature of an ocean already near freezing-point would serve to initiate an ice-sheet, and that factors such as the increased albedo of the ice compared with the ocean would then generate lower temperatures. This would encourage expansion of the ice, further cooling, and so on, until widespread glaciation resulted. The arguments are valid, though exaggerated, but the problems are that such processes might be slow, and that there is no explanation of how the ice is to be later removed and glaciation terminated. Moreover the cause of the initial lowering of temperature remains unknown.

E. *Changes in the Earth's Orbit and Axis*

There are three main variables in the geometrical relationships between earth and sun:

i) The precession of the equinoxes, with a periodicity of about 21,000 years. The distribution of solar radiation over the earth's surface varies slightly according to whether the equinox, for instance, occurs in aphelion or perihelion.

ii) The obliquity of the plane of the ecliptic varies from 21° 58' to 24° 36', with a periodicity of about 40,400 years.

iii) The eccentricity of the earth's orbit; periodicity about 91,800 years.

All these, singly and in combination, affect the receipt of solar radiation and its distribution over the globe, and are indisputable. Their overall effect on earth temperatures has been studied and discussed for many years, by none more so than M. Milankovitch (1920 *et seq.*). The main problem is not their reality but the smallness of their probable effect, which is likely to be no more than 1° or 2°C., yet on the other hand, the periodicities can be fitted to certain known Pleistocene fluctuations. The idea has been used recently by C. Emiliani and J. Geiss (1957) in a complex theory involving also changes in solar radiation and topographical-climatic changes. The chief problem in using planetary changes alone is that they offer no explanation for the non-occurrence of glacial-interglacial periods between the Permian and the Pleistocene, and more specifically, for the gradual cooling throughout the Tertiary. It seems certain that the planetary changes must be superimposed on changes resulting from other causes.

F. *Changes in Solar Emission*

In our present state of knowledge, this seems to be the most likely first cause of major climatic fluctuations, with certain other changes (especially E and B(b)) superimposed as additional mechanisms. The sun is the basic energy source for all atmospheric processes. Changes in the quality of radiation, particularly in the ultra-violet and cosmic components, are known to occur; and there may also be short-term changes in the quantity of radiation associated with sunspot cycles of 11, 22, 44 years, etc. Over the last few decades, however, measurements have failed to reveal any sizeable changes in the so-called 'solar constant', and the often-suggested correlations between sunspot activity and climatic change over the last 200 years may have little statistical significance.

More speculative, but possibly of vital importance to the solution of the Ice Age problem, are longer-term changes in solar emission, such as Sir George Simpson (1934, 1957) has postulated, with a periodicity of, for instance, 200,000 or 400,000 years. Certain aspects of Simpson's theory are unacceptable in the light of recent evidence showing colder conditions in the Tropics during glacial periods, but the basic mechanism is still available. It was used by R. F. Flint (1957) in his 'Solar-topographic concept', which appealed to changes in solar emission in the first place, and secondly to Cenozoic mountain-building and uplift. Flint contended that solar variations had occurred throughout geological time, but that they were only effective in causing a glacial-interglacial succession on land-areas provided with high-altitude elements: highlands were said to be the pre-requisite for extensive glaciation. Substantially similar is the theory offered by R. P. Beckinsale (1965). This postulates long-term changes in land-mass altitude through the Pleistocene: a considerable increase in mean land altitude is held to be essential for the formation of an iceage. Superimposed on this are shorter-term oscillations in the receipt of solar radiation due to changes in planetary geometry, and variations in solar emission. The answer to the question of the causation of ice ages seems likely to reside in compound theories such as these.

Rates of Formation and Dissipation of Pleistocene Ice-Sheets

Recently, some attention has been given to the question of the rates of growth or shrinkage of the Pleistocene ice-sheets. J. Weertman (1964) has analysed the time required to build up or destroy an ice-sheet of given size, using the model of perfect plasticity for ice-flow. If net accumulation averaged 20 to 60 cm a year, the time required to form an ice-age ice-sheet would be of the order of 15,000 to 30,000 years, possibly the length of a glacial sub-stage. On the other hand, if during deglaciation the average ablation rate over large areas were substantially larger than the average accumulation rate, and ice movement ceased, the ice-sheets would disappear comparatively rapidly.

The time required for the build-up of the ice-sheets can be considerably lessened, however, by bringing in another factor, that of changes in the temperature of the basal ice (A. T. Wilson, 1964; J. Weertman, 1966). If we begin with a cold ice-cap frozen to bedrock, outward movement will be slow (see Chapter 4). As increased accumulation causes it to thicken, there will be a gradual warming of the basal ice layer by geothermal heat, owing to the blanketing effect of the ice. Eventually, the basal ice layer will reach pressure melting-point, and substantial sliding over bedrock will be possible, facilitated by basal meltwater. Spreading of the ice will thus result, but at the same time, as Weertman (1966) argues, these changes may in fact reduce the equilibrium thickness of the ice-sheet by a factor of 1·4 to 2 times, since the basal shear stress will be very much less. Thus after a phase of relatively rapid expansion, shrinkage may set in until the ice-sheet either disappears or approaches a new equilibrium size. These suggestions considerably help with one of the most difficult problems that many ice-age theories encounter— that of bringing glaciation to an end once it has started. Weertman's theory considerably reduces the size of any change in climate required to make an ice-sheet disappear.

Conclusion

Major ice ages have occurred in the Pre-Cambrian, Permo-Carboniferous, and Pleistocene periods of earth history. About 1·5-2·0 million years ago, the Pleistocene period was ushered in by relatively sudden falls of world temperature. The maximum extent of ice in the Pleistocene was about three times its present spread and volume, and was associated not with the last (Würm, Weichsel, Wisconsinan) glaciation, but with earlier glacial periods, possibly the Riss (Saale, Illinoian).

Pleistocene stratigraphy is based on the fossil content of marine and some terrestrial formations, on the correlation of lithological units such as till sheets, on the existence of weathering zones or palaeosols, on archaeological evidence, and on palynology. The absolute chronology of the period is now founded mainly on radiometric methods, using particularly the isotopes carbon 14 and potassium 40, on palaeomagnetic, tephrochronological, and deep-sea sediment studies, and to a less extent on glacial lake rhythmites and dendrochronology. The earliest glaciation in the Pleistocene is now dated at more than 700,000 years in North America and 1-2 million years in Iceland.

Glacier ice today covers about 10 per cent of the world's land surface, and exists in a great variety of climatic and topographical conditions. Fluctuations in its extent, vertically and areally, are often well documented for the last 200 years, and fairly well established in the Late- and Post-glacial periods.

The causes of ice-ages are still the subject of speculation and controversy, which it is outside the scope of this book to examine in any detail. Most probably, a number of different factors is involved, and the most plausible theories are those which are based on a combination of changes in land-mass altitude, short-term changes in the receipts of solar radiation resulting from the varying geometrical relationships of earth and sun, and longer-term changes in the quantity and quality of solar emission.

REFERENCES

R. P. BECKINSALE (1965), 'Climatic change: a critique of modern theories' in *Essays in Geography for Austin Miller* (ed. J. B. WHITTOW and P. D. WOOD), 1-38

W. S. BROECKER (1965), 'Isotope geochemistry and the Pleistocene climatic record' in *The Quaternary of the United States* (ed. H. E. WRIGHT and D. G. FREY), 737-54

W. S. BROECKER, K. K. TUREKIAN, and B. C. HEEZEN (1958), 'The relation of deep-sea sedimentation rates to variations in climate', *Am. J. Sci.* **256**, 503-17

C. E. P. BROOKS (1949), *Climate through the Ages*

T. C. CHAMBERLIN (1899), 'An attempt to frame a working hypothesis of the cause of glacial periods on an atmospheric basis', *J. Geol.* **7**, 545-84

J. K. CHARLESWORTH (1957), *The Quaternary Era*

A. COX, R. R. DOELL, and G. B. DALRYMPLE (1965), 'Quaternary paleomagnetic stratigraphy' in *The Quaternary of the United States* (ed. H. E. WRIGHT and D. G. FREY), 817-30

W. L. DONN, W. R. FARRAND, and M. EWING (1962), 'Pleistocene ice volumes and sealevel lowering', *J. Geol.* **70**, 206-14

D. B. ERICSON, M. EWING, and G. WOLLIN (1964), 'The Pleistocene epoch in deep-sea sediments', *Science* **146**, 723-32

D. B. ERICSON and G. WOLLIN (1964), *The Deep and the Past*

C. EMILIANI (1955), 'Pleistocene temperatures', *J. Geol.* **63**, 538-78

C. EMILIANI and J. GEISS (1957), 'On glaciations and their causes', *Geol. Rdsch.* **46**, 576-601

M. EWING and W. L. DONN (1956-59), 'A theory of Ice Ages', *Science* **123**, 1061-6; **126**, 1157-62; **129**, 463-5

R. F. FLINT (1957), *Glacial and Pleistocene geology*

H. C. FRITTS (1965), 'Dendrochronology' in *The Quaternary of the United States* (ed. H. E. WRIGHT and D. G. FREY), 871-80

J. GENTILLI (1948), 'Present-day volcanicity and climatic change', *Geol. Mag.* **85**, 172-5

H. GODWIN (1956), *The history of the British flora*

J. T. HOLLIN (1962), 'On the glacial history of Antarctica', *J. Glaciol.* **4**, 173-95

S. KAIZUKA (1961), 'Geochronology based on volcanic ejecta and its contributions to archeology in Japan', *Asian Perspective (Far east. prehist. Ass.)* **5**, 193-5

C. KILBURN, W. S. PITCHER, and R. M. SHACKLETON (1965), 'The stratigraphy and origin of the Portaskaig Boulder Bed Series (Dalradian)', *Geol. J.* **4**, 343-60

P. B. KING (1965), 'Tectonics of Quaternary time in middle North America' in *The Quaternary of the United States* (ed. H. E. WRIGHT and D. G. FREY), 831-70

H. H. LAMB and A. I. JOHNSON (1959), 'Climatic variation and observed changes in the general circulation', *Geogr. Annlr* **41**, 94-134

F. LEVERETT (1930), 'Relative length of Pleistocene glacial and interglacial stages', *Science* **72**, 193-5

W. F. LIBBY (1955), *Radiocarbon dating* (Chicago, 2nd Ed.)

C. LYELL (1839), *Elements of Geology* (French translation), 621. See also *Ann. Mag. nat. Hist.* **3** (1839), 323

G. MANLEY (1951), 'The range of variation of the British climate', *Geogr. J.* **117**, 43-68

M. MILANKOVITCH (1920), *Théorie mathématique des phénomènes thermiques produits par la radiation solair* (Paris)

M. MILANKOVITCH (1930), 'Mathematische Klimalehre' in *Handbuch der Klimatologie* by W. KÖPPEN and R. GEIGER (Berlin)

R. B. MORRISON (1964), 'Lake Lahontan—geology of southern Carson desert, Nevada', *U.S. geol. Surv. Prof. Pap.* **401**

K. P. OAKLEY (1964), *Frameworks for dating fossil man*

R. V. RUHE (1965), 'Quaternary paleopedology' in *The Quaternary of the United States* (ed. H. E. WRIGHT and D. G. FREY), 755-64

M. G. RUTTEN and H. WENSINK (1960), 'Paleomagnetic dating, glaciations and the chronology of the Plio-Pleistocene in Iceland', *Rep. 21st int. geol. Congr. (Copenhagen)* **4**, 62-70

G. C. SIMPSON (1934), 'World climate during the Quaternary period', *Q.J.R. met. Soc.* **60**, 425-78

G. C. SIMPSON (1957), 'Further studies in world climate', *Q.J.R. met. Soc.* **83**, 459-81

G. C. SIMPSON (1959), 'World temperatures during the Pleistocene', *Q.J.R. met. Soc.* **85**, 332-49

S. THORARINSSON (1944), 'Tefrokronologiska studier på Island', *Geogr. Annlr* **26**, 1-203

S. THORARINSSON (1949), 'Some tephrochronological contributions to the vulcanology and glaciology of Iceland', *Geogr. Annlr* **31**, 239-56

H. VALENTIN (1952), 'Die Küsten der Erde', *Petermanns Mitt., Ergänz.* **246**, 118 p.

J. WEERTMAN (1964), 'Rate of growth or shrinkage of non-equilibrium ice sheets', *J. Glaciol.* **5**, 145-58

J. WEERTMAN (1966), 'Effect of a basal water layer on the dimensions of ice-sheets', *J. Glaciol.* **6**, 191-207

R. E. WILCOX (1965), 'Volcanic-ash chronology' in *The Quaternary of the United States* (ed. H. E. WRIGHT and D. G. FREY), 807-16

A. T. WILSON (1964), 'Origin of ice ages: an ice shelf theory for Pleistocene glaciation', *Nature, Lond.* **201**, 147-9

A. T. WILSON (1966), 'Variation in solar insolation to the South Polar region as a trigger which induces instability in the Antarctic Ice Sheet', *Nature, Lond.* **210**, 477-8

P. WOLDSTEDT (1954), 'Die Klimakurve des Tertiärs und Quartärs in Mitteleuropa', *Eiszeitalter Gegenw.* **4**, 5-9

Chapter 2

Glacier Régimes

Consider, then, the glaciers as the outlets of the vast reservoirs of snow ... —as icy streams moving downwards, and continually supplying their own waste in the lower valleys, into which they intrude themselves like unwelcome guests, in the midst of vegetation, and to the very threshold of habitations. ... (J. D. FORBES, 1843)

Glacier régimes are concerned with the loss and gain of snow in a glacier. When the glacier or ice-sheet has a positive régime, it is gaining ice and will advance and thicken, while the reverse applies when the régime is negative. Glacier régimes are intimately related to climate and climatic changes, and therefore to changes in sea-level relative to the land. They also provide important hydrological data. From the point of view of glacial geomorphology, the régimes of glaciers are important in that they affect the behaviour of the ice. They can be directly related to the formation of moraines, which may mark stages of halt and readvance, trim-lines, some meltwater channels and other features.

First, some terms will be defined and the methods of obtaining glacier régimes will be mentioned. Some characteristic glacier régimes in different climates and environments will then be discussed. Then the effect of climatic change on glaciers will be considered in order that these factors may be correctly related.

Definitions   The two essential factors that make up the régime or mass balance of a glacier system are the accumulation and the ablation, the gain and loss of mass respectively. The term *accumulation* includes all those processes by which solid ice and snow are added to a glacier or ice-sheet. *Ablation* refers to all the processes by which ice and snow are lost from the glacier. Accumulation on the glacier is brought about mainly by the precipitation of snow. Rain, if it freezes on impact, can also be included. Refreezing of liquid water and condensation of ice direct from vapour (sublimation) also cause accumulation, and material may be added to the glacier by wind-blown snow or avalanches. The processes that cause ablation include melting, evaporation, calving, wind erosion, and removal of snow or ice by avalanches. One of the problems of measuring accumulation and ablation is that the processes operate on or near the surface as well as within and beneath the ice. Changes deep within and beneath the ice are usually ignored as they are small compared with the surface and near surface changes, except on floating ice shelves or glacier tongues.

The budget year is the most important unit of time when glacier mass balance studies are being considered. The budget year runs from the time when ablation has reached its maximum extent after the summer season of one year until the similar state in the following year. This period need not be an exact calendar year. The accumulation and ablation are recorded in water equivalent (in cm^3 or m^3) over the whole glacier surface, or in cm or m of water equivalent at a point. In order to measure the water equivalent, it is necessary to know the density of the snow or ice.

The terms used in mass balance studies include the following:

1. *Gross annual accumulation* is the total volume of water equivalent added to a glacier during the budget year.

2. The *net annual accumulation* is the amount of water equivalent added to the glacier and still present by the end of the budget year. This material is limited to the accumulation area of the glacier in which there is a surplus at the end of the budget year.

3. The *gross annual ablation* is the total amount of water equivalent of snow and ice consumed during the budget year by all processes of ablation.

4. The *net annual ablation* is the volume of water equivalent actually lost from the glacier during the budget year. The difference between the gross and the net ablation is made up by such processes as refreezing of melt-water on or within the glacier, the melting of temporary accumulation in the ablation area, and the internal storage of water.

In the ablation area all the material gained that year is lost before the end of the budget year. This loss in temperate glaciers especially takes place largely by surface melting. The melting occurs mainly in the lower parts of the glacier during the summer ablation season, although some ablation may go on at low levels throughout the whole of the year in some glaciers. In other areas, however, such as the Antarctic ice-sheet, calving is the most important process of ablation.

The line dividing the accumulation area from the ablation area is called the *equilibrium line* (Fig. 2-1). This line should not be confused with the *firn-line*, or *annual snow-line*. The two lines have sometimes been confused or regarded as identical, although in fact they may not coincide. The firn-line is the uppermost line on the glacier to which the snowfall of the winter season melts during the summer ablation season. It is a clearly marked line, separating hard blue ice below from snow above in many glaciers. However, in some glaciers there may be an accumulation of dense ice between the firn-line and the equilibrium line, which lies at a lower level. This dense ice is also called *superimposed ice* and it is formed by the refreezing of snow-melt water or rain. In high arctic glaciers in particular there is often a complex transition zone between the accumulation and ablation zones.

The problems of this zone are discussed by F. Müller (1962) in connection with a study of the glaciers on Axel Heiberg Island. A number of zones with different characteristics can be identified. The *dry snow zone* is the highest and always has a temperature below 0°C., so that there is no penetration of melt-water. The *percolation zone* lies below this zone, and is divided into two parts.

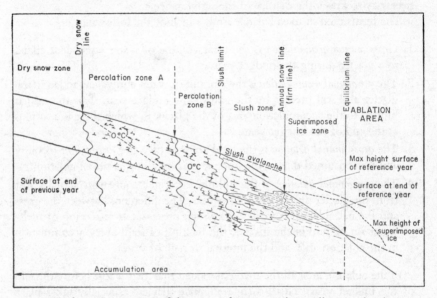

Fig. 2-1 Schematic zonation of the accumulation area (F. Müller, *J. Glaciol.*, 1962, by permission of the Glaciological Society)

In the upper one, *zone A*, percolation does not reach to the previous year's summer surface, although the temperature does rise to 0°C. below the surface and meltwater can penetrate a little way. In *zone B* some of the current season's meltwater penetrates and refreezes to form ice lenses in the earlier accumulation. The 0°C. line sinks below the previous summer surface. Below zone B lies the *slush zone*, from which material can be lost by slush avalanches. These avalanches are caused by the saturation of the snow above the impermeable superimposed ice layers and can cause the removal of material on gentle slopes of only 3 or 4 degrees. The upward limit of this zone is called the *slush limit*. The slush avalanches also carry away some of the superimposed ice and even the older accumulation beneath it below the equilibrium line. The older accumulation of the previous summer can be separated from the accumulation of the current season by the development of *depth hoar* between the two layers. The depth hoar consists of large cup-shaped or bar-like crystals and results from changes of snow character with seasonal weather changes. Some of these zones are shown in Figure 2-1.

Mass Balance Studies

The mass balance or régime of a glacier can be assessed in two ways. It is equal to the gross annual accumulation less the net ablation during the budget year. Alternatively, it is equal to the net annual accumulation in the accumulation area at the end of the budget year less the dense ice lost from the surface exposed at the end of the previous budget year in the ablation zone. The quantities required for the second method are often easier to measure. Details of the measurements necessary to establish the glacier budget are given in the Technical Note No. 1 of the Glaciological Research Sub-Committee of the Glaciological Society.

A detailed study of the problems of establishing the mass balance of a glacier or ice-sheet has been given by M. F. Meier (1962). The budget quantities that he gives are measured in a vertical direction so that they can be readily related to area on a map. The results are given in water equivalent values, which means that it is necessary to observe the density of the ice or snow. In the accumulation area, where the annual accumulation is in the form of snow or firn, a known volume of material must be obtained and its weight gives the density. The water equivalent values can be given for each unit area. Meier uses the term 'specific mass budget' to define the budget at a single point. If these values are integrated over the whole area of the glacier, the total mass budget can be obtained.

Accumulation and ablation vary with both time and space over the whole glacier system. The accumulation reaches a peak at any one point during the early part of the budget year and may fall off to small or zero values during the latter part of the year, during the summer season. The ablation changes, on the other hand, have an almost opposite trend. The difference between the curves for accumulation and ablation shows the specific mass budget at any one point. The value can also be shown cumulatively. This gives the net specific budget throughout the budget year.

The change of the specific budget with altitude is important in calculating the total budget. In the lower part of a mountain glacier, the specific budget will be negative, while in the upper part it will be positive. A direct comparison of these values will not give a true idea of the total budget unless the area of the glacier at the different altitudes is taken into account. Budget values for different altitudes must be related to the areas between the appropriate contours. The total mass budget can be relatively accurately assessed by using these values. An important point on a curve showing the variation of either the specific or total budget with altitude is the point at which the curve crosses from the positive upper part to the negative lower part

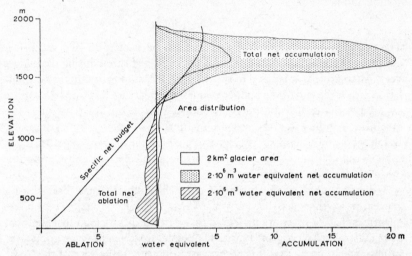

Fig. 2-2 The mass balance of Nigardsbreen for the budget year 1961-2, when the surplus accumulation amounted to 94.9×10^6 m³ water equivalent (G. Østrem, *Norsk geogr. Tidsskr.*, 1961-2)

(Fig. 2-2). This point indicates the position of the equilibrium line, separating the ablation and accumulation areas.

In a glacier system that has a high rate of accumulation and also a high rate of ablation, the rate of flow of the ice must be considerable. This applies for example to the Franz Josef glacier, which flows to within about 200 m of sea-level in a relatively low latitude. Its accumulation area lies in an area of very heavy precipitation on the western side of the Southern Alps of New Zealand, while its snout penetrates down into the coastal rain forests. At the other extreme are some of the Antarctic glaciers that receive a very small quantity of precipitation, owing to the extreme cold, and whose only method of ablation is by calving, as temperatures rarely rise to freezing-point. These glaciers move extremely slowly (Chapter 4). The Ferrar glacier near the Ross Sea, for example, moves at only about 5 cm/day although it is a very large glacier. An improvement of the climate, with higher temperatures, would allow the air to hold more moisture, while still remaining below freezing-point. This would increase the snowfall and thus the glaciers would become more active.

The size of a glacier budget relative to the glacier area gives a good indication of the state of activity of the glacier. If the budget is large relative to the glacier area, the ice will flow fast through it. The reverse will also apply. In both instances if the budget is balanced, the glacier will maintain its volume and position.

MEASUREMENT OF ACCUMULATION

It is generally easier to measure net accumulation than gross accumulation. The former measurement also has the advantage that accumulation of several seasons can be measured at one time. This can be achieved by digging pits at suitable positions in the accumulation area. These give accurate values for the accumulation of several years at the positions and altitudes chosen. Then the total net accumulation can be reasonably accurately calculated from the different specific values obtained from the individual pits.

In analysing the results obtained from pit observations, it is clearly necessary to be able to establish the boundaries between different budget years of accumulation. Sometimes this can be done fairly easily if dirt layers have accumulated on the snow surface at the end of the budget year, but this criterion is not always accurate. The summer snow also usually hardens at the surface where the upper layers of snow have partially melted and refrozen. This gives a thin hard crust that can be identified in the pit profile. The formation of depth hoar in very cold climates has also been mentioned.

Where the accumulation is heavy, pits must be deep to penetrate through a number of budget years of accumulation. For example in Iceland an annual accumulation of 2·38 m was recorded in a pit dug near the southern margin of the Vatnajökull ice-cap at 1200 m. A pit 6 m deep in this area, therefore, can only penetrate through about two complete accumulation layers, especially if the pit is dug early in the ablation season. In this area, the budget years can be fairly readily distinguished by dust layers as the pit was sited fairly close to the rock rim of the ice-cap.

The reverse situation is found in the Antarctic where accumulation is very small. The individual layers are difficult to identify owing to the lack of summer melting. Here the average accumulation is about 30 to 60 cm, but this amount is graually reduced by compression to layers 10 to 20 cm in thickness. A pit 6 m deep in this area would cover somewhere between 20 and 50 years of accumulation.

Another difference between these two areas is the density of the snow in the pits. The density of the Antarctic snow is much less than that in more temperate areas. In the Antarctic, it varies between 0·3 and 0·45 over a period of three years accumulation, the density being least at the beginning of the budget year. In the Iceland pit, on the other hand, the density varied between 0·55 and 0·6 in snow of the previous season. This is caused by the greater pressure of overlying material and much more active melting.

The density within any one budget year does not necessarily vary in the same way with depth in all climates. For example, in the Austrian Alps the density at one point was found to be at a minimum at the bottom and reached a maximum somewhere in the centre of the annual accumulation layer. Changes in crystal size and colour allowed the beginning of the budget year to be identified in this area. One of the characteristics of very mountainous areas of this type is the very uneven nature of the accumulation. Many pits must be dug to obtain a good value for the total net accumulation in these conditions. However, where the accumulation area is broad and even, as in the case of the Vatnajökull plateau, it is only necessary to dig representative pits in one or two positions at each level.

In order to calculate the mass budget of a glacier, the net accumulation is the only accumulation value that it is essential to obtain. Methods of obtaining the gross accumulation on a glacier are more complex and will not be discussed.

ABLATION MEASUREMENTS

The net accumulation must be equated with the ablation of dense ice to obtain the mass balance. There is no need to measure the ablation in the accumulation area in order to obtain the mass balance. However, when the net accumulation is measured in pits during the summer season, the figures obtained will refer to past budget years, while the measure of ablation of solid ice in the ablation area will refer to the current year. This value of ablation must be balanced against the net accumulation found in the following summer season or at the end of the current budget year. The loss of dense ice in the ablation area can be found by continuous measurements of ablation during the whole of the ablation season. These measurements must start while the glacier is still covered with winter snow by drilling stakes down into the solid ice beneath. The depth of snow can then be recorded to indicate the position of the ice surface. However, the level of the ice surface may rise as the season progresses owing to the addition of frozen meltwater beneath the snow. This must be allowed for before the loss of older ice starts and net ablation begins. Some observations of ice density must be made to obtain the water equivalent of the ice lost during the ablation season.

A large number of ablation stakes must be inserted to give an accurate value of the loss of dense ice. The stakes need frequent redrilling during the height of the ablation season when up to 15 cm or more of ice may be lost from the surface of temperate glaciers in one day near their snouts. For example, stakes drilled to a depth of 7·5 m about the middle of the ablation season in 1955 had completely melted out by the early part of the following ablation season near the snout of Austerdalsbreen in Norway. It has been estimated that about 12 m of ice are lost from near the snouts of many Icelandic, Alaskan and Norwegian glaciers during one ablation season. Ablation on Nigardsbreen in Norway exceeds 20 m in one summer.

On the whole, ablation does not vary annually as much as accumulation, although it can vary considerably locally over a glacier surface. It is great near rock walls and where warm, dry katabatic winds blow down the glacier from higher levels. Where there are waves on the glacier surface, the ablation tends to be greater on the crests of the waves, as on Austerdalsbreen, Norway. The amount of ablation often decreases linearly with height and also varies greatly with the season. It rises sharply in the northern hemisphere to a maximum in the first half of July and falls off rapidly again to low values in early September, when the rate resembles that of mid-May.

Another useful observation to make in connection with ablation and mass balance studies is the measurement of the discharge of streams issuing from the glacier. This includes both meltwater and rain falling on the glacier basin during the summer where this does not freeze before reaching the glacier snout. The discharge cannot give the total ablation as ice lost by evaporation is not included in the meltwater, but for hydrological purposes the amount of meltwater is important. It is closely associated both with the glacier mass régime, as discussed by G. Østrem (1963), and with the river régime of the meltwater stream. A measurement of the load of the meltwater stream also gives a useful indication of the effectiveness of glacial erosion by the glacier concerned (Chapter 11).

Glacier Mass Balance

The mass balance of the glacier depends on the balance between accumulation and ablation. Equilibrium is rarely achieved for long and the problem of the relationship between the mass balance and the movement of the glacier is a complex one. It involves changes in the volume, form and flow. Variations in régime may be short-term fluctuations and many areas vary very much from year to year. The glacier does not respond at once to these short-term annual changes as the variations often cancel each other out in succeeding years before the glacier can react to them. Rather more prolonged trends, such as a succession of warm or snowy years, are required before the glacier responds to these climatic variations by a change in form.

The characteristics of the glacier snout can be used to estimate the budget of an unknown glacier. An advancing glacier with a positive budget normally has a steep or vertical front. A glacier in equilibrium has a front of moderate gradient and one that has a negative budget and is retreating generally has a gently sloping front. This gentle slope often becomes buried under morainic debris. When this cover becomes thick, the ice disappears largely by slow down-wasting and the exact end of the glacier may be difficult to determine. This difference of frontal types enables an aerial survey to be used to distinguish glaciers in various states of mass balance in areas where no detailed work has been undertaken. An example of three adjacent glaciers each of which shows clearly one of the three states mentioned is the Three Congruent Glacier of the St. Elias Range in Alaska (E. LaChapelle, 1962, p. 291). As this example shows, there are often interesting variations in response between neighbouring glaciers to similar climatic changes. These depend on the characteristics of the individual glacier basins, such as the ratio of the accumulation area to ablation area and altitude distribution. Before considering the response of glaciers to changes in climate, a few examples of specific mass balance studies in different climatic areas will be considered. The examples are arranged in order of latitude, beginning with the highest latitudes.

ANTARCTIC

The Antarctic continent contains by far the greatest amount of ice and snow in the world. Its ice-sheet covers about 11·5 million km², an area seven times that of Greenland's ice-sheet and one and a half times the area of the United States. The mean thickness of the ice is about 2000 m, or possibly as much as 2500 m. If all the ice were to melt, sea-level would rise by about 61 m. This amount would be reduced to about 40 m if isostatic recovery of the continent were allowed for.

In the last decade, observations of accumulation, ablation and flow in the Antarctic have increased considerably, but there is still some doubt concerning its mass balance. The accumulation over most of the ice-sheet is very low and density only increases slowly with depth in the ice-sheet. Near the surface, the firn has a density of between 0·3 and 0·4. The density increases to 0·82 at a depth between 50 and 100 m. This density increase is accompanied by an increase in the size of the crystals downward, while the enclosed bubbles of air become smaller downward. At a depth of 300 m, it is estimated that in one area the ice is 1600 years old.

A new method has had to be developed to establish the seasonal layers of accumulation both in the Greenland and Antarctic ice-sheets, because in these areas dirt layers are absent and changes in the texture of the ice are small. The method depends upon the proportion of the isotopes oxygen 18 and oxygen 16. The former is greater in importance during the colder season. These measurements can also give some indication of actual temperatures when the snow, now found as ice deeply buried in the ice-sheet, was originally deposited on the surface. Dating of ice layers has also been carried out by C^{14} analysis of the carbon-dioxide bubbles trapped in the ice. One result gives an age of 3100 years for a sample of ice in Greenland.

Accumulation in the Antarctic is difficult to measure owing to the large amount of drifting in the high winds characteristic of the area. Nevertheless there are many estimates on this positive side of the budget. M. Mellor (1959) has discussed the accumulation in the Mawson area. The results suggest very uneven addition of material, varying from a minimum of 2 cm water equivalent a year 160 km from the coast to a maximum of 12 cm water equivalent at 130 and 230 km inland. These are very low values and estimates from other areas give rather higher figures. Values will be given in cm water equivalent per annum. P. A. Shumskiy (in M. Mellor, 1959), for example, gave a value of 85 cm within 60 km from the coast at Mirny, falling to 20 to 30 cm at 70 to 450 km inland and 8 cm at 850 km inland. In other parts of the continent, a value of 15 cm for the area from 100 to 600 km inland has been given. Results of the Norwegian-British-Swedish expedition give 36·5 cm for the shelf, 25 cm for the plateau slopes and 12 cm for the edge of the high plateau of Dronning Maud Land. Ice-shelf accumulation in Wilkes Land is estimated at 35 to 40 cm. An estimate by P. A. Siple (1958) for the South Pole gives 5 to 6 cm for 10 months February through November. A. J. Gow (1965) gives 20 cm /year firn or 7·4 cm water equivalent at the Pole.

J. T. Hollin (1962) has assembled the available data on accumulation in the Antarctic. For the Indian Ocean sector, he suggests a rapid increase in the rate of gross accumulation from zero actually at the coast to a maximum of nearly 30 g/cm^2 near the coast. From this point, the value falls gradually to a latitude of 70°S. and thereafter more rapidly to very low values of less than 10 g/cm^2 around 86°S. in the centre of the continent. This sector of the ice-cap is one where accumulation is reduced by strong and persistent katabatic winds, so that the figures are conservative. F. Loewe (1960) gives a tentative value of accumulation for the area without shelves ranging from a minimum of 10 cm/year to a maximum of 12 cm/year. When the ice-shelves are included, he gives a minimum value of 10·4 cm and a maximum of 13 cm/year. On the shelves, the minimum is 15 cm and the maximum 25 cm/year. The total positive side of the budget using these figures and taking the area of the ice-sheet into account is a minimum without shelves of 125×10^{10} tons and a maximum with shelves of 175×10^{10} tons.

Against these rather variable estimates of the accumulation on the Antarctic continent and its ice-shelves must be set the estimated loss of material by the various processes of ablation. The Antarctic ice-sheet differs from others in that a considerable part of its loss is provided by calving of the large ice-shelves around the coast. The largest is the Ross ice-shelf (Fig. 3-6), and the Filchner ice-shelf in the Weddell Sea is also very extensive. From these floating shelves, pieces break off to form large tabular icebergs. Melting, on the other hand, is one of the least important methods of loss in the Antarctic, where temperatures are low and ablation is restricted to favourable localities. One method by which snow is removed from the land is by wind drifting. Mellor (1959) has estimated that the annual mass transport by the wind at Mawson is 0.22×10^{14} g/km/year. The estimates of Loewe (1960) of the loss of snow by this process range between a minimum value of -2×10^{10} tons and -50×10^{10} tons. Ablation also takes place by evaporation throughout the year. Melting only occurs locally during December and January. Measurement of ablation rates of snow during three years at 60 m elevation in the Mawson area gives figures varying between 53.5 and 79 cm/year. Values fall to lower rates with increasing altitude, for example 22 cm at 365 m. Hollin (1962) suggests that ablation as a whole is extremely small away from the coast, but that at the coast it may be double the value given for accumulation.

The most effective form of loss from the ice-sheet is by flow offshore followed by calving. Against this loss must be set the possible accretion of ice below the shelves. There is, however, doubt as to whether the shelves lose or gain underneath. It seems likely that there is some melting, possibly about 20 cm from the seaward margin of the shelves. F. Debenham (1948) has, however, shown that in a few places the shelves gain by freezing on their lower side. Such a gain would help to account for the headless fish and mirabilite deposits which have been found resting on the surface of the shelf. In these areas of unusual relief conditions, where rock outcrops occur, the surface is melting and the underside of the shelf is freezing.

The removal of mass by iceberg formation, both from the continental ice-sheet and the glaciers, but excluding the shelves, has been estimated by Loewe to be between a minimum of -14 and a maximum of -30×10^{10} tons. The greater part of this loss is supplied by the glaciers, because some of these move with speeds of 300 to 600 m/year, while the ice-sheet only moves at a velocity of about 30 m/year. The marginal ice-shelves, however, supply the greatest amount of loss by this process of calving. Loewe estimates a lost of between -36 and -65×10^{10} tons from the ice-shelves. These move as

300 to 500 m/year and have a thickness of between 190 and 200 m and calving takes place from an estimated length of about 7000 km.

The mass balances produced by various workers vary as much as the constituent ingredients. On the whole, however, most of the estimates indicate that the ice-sheet may have a positive mass balance. Mellor (1959) suggests an annual gain of $1 \cdot 7 \times 10^{18}$ g and a loss of $0 \cdot 69 \times 10^{18}$ g. Loewe's (1960) budget varies between $+ 88 \times 10^{10}$ tons for minimum values of all the different items to $+ 30 \times 10^{10}$ tons for a maximum value for all the items. Of eight values of the mass balance, one is negative, two suggest equilibrium, and the four most recent all show a fairly large positive budget. The average value of these four estimates is $+ 1 \cdot 10 \times 10^{18}$ g/year.

In view of the probable dominance of calving on the negative side of the budget, it is important to note that the movement of the Antarctic ice is probably not as small as was thought at one time. Some velocities have already been mentioned and Hollin (1962) is of the opinion that the mean velocity around the periphery of the ice-sheet is of the order of hundreds of metres a year. This applies to many of the floating shelves.

An important point concerning the fluctuations of the ice-sheet is the grounding line beyond which the ice moving seaward starts to float (p. 23). There is reason to believe that the Antarctic ice-sheet is thinning and retreating slightly at present in phase with the Northern Hemisphere retreat that is occurring in many areas. However, it can reasonably be argued that warmer conditions in the Antarctic are liable to lead to increasing accumulation, but little change in ablation would be expected, and thus it is reasonable to suggest that the advances and retreats of the Northern and Southern Hemispheres should be out of phase. But there is evidence to suggest that the changes are in fact in phase, as Hollin (1962) shows.

He relates this correspondence in phases of retreat and advance to changes in sea-level. When the northern glaciers and ice-sheets fluctuate, which they can do much more readily than the Antarctic owing to their ending on land and their more temperate environment, sea-level fluctuates with them. A fall of sea-level accompanying increasing northern hemisphere glaciation causes the grounding line of the Antarctic ice-sheet to advance seawards. Thus the whole profile of the Antarctic ice-sheet can shift seaward. A considerable change in thickness, amounting to 1230 m, would occur at the original grounding line if the grounding line advanced 90 km. This distance depends on the slope of the sea-floor and the extent of the lowering of sea-level. To obtain the result given above, the slope was taken to be $0 \cdot 1$ degrees and the fall of sea-level 150 m.

The reverse would occur with a rise of sea-level. The profile of the ice-sheet would retreat inland, causing a lowering of ice-level near the margin.

The change in thickness is rapidly reduced inland and depends on the existence of a stable profile. It seems likely that this process keeps the Antarctic in phase with northern hemisphere fluctuations. These changes must be initiated in the north and are then followed in the Antarctic because of the resulting changes of sea-level. Causes of major glacial fluctuations must, therefore, be sought in the Northern Hemisphere.

Hollin considers that the Antarctic has been ice-covered throughout the Pleistocene and that the ice-sheet may well have been initiated during the Tertiary period. There is no prospect of its disappearance in the foreseeable future, owing to the lack of ablation. It is likely that the Antarctic ice-sheet, although frozen to its bed in the peripheral parts, is at the pressure melting-point in the interior. This is partly because of the small accumulation, which prevents rapid downward transference of cold from the surface. There is a possibility that the ice-sheet would reach the pressure melting-point at its base over a wide area if the thickness of the ice-sheet were to increase. The ice-sheet might then suddenly surge forward and extend rapidly as far as 50°S. in extensive shelves according to A. T. Wilson (1964). This would cause a sudden rise of sea-level of about 33 m. The cooling resulting from such an increase of southern hemisphere ice would lead to a renewed northern ice advance, starting a new glacial period. J. Weertman (1966), however, has shown that ice-flow mechanism is not yet sufficiently well known to predict this sequence of events. Weertman suggests that even if the ice-sheet were to spread catastrophically it would do so over a period of 50 to 100 years. This would be long enough for the ice-shelves to break up before reaching the dimensions suggested by Wilson. Weertman, however, considers that a surge of this type is possible though not very likely.

GREENLAND

The mass balance of the Greenland ice-sheet has been worked out by A. Bauer (1955). The total area of the ice-sheet is about $1 \cdot 7264 \times 10^6$ km². The greatest area, 18 per cent, lies between 2440 and 2740 m and 6·5 per cent lies between 3050 and 3390 m. The firn-line is at about 1400 m and 83 per cent of the total area lies in the accumulation zone above this elevation. The estimated budget consists, on the positive side, of accumulation at a mean rate of 0·31 m of water equivalent, giving a total of 446 km³ of water equivalent a year. The ablation is estimated at 1·1 m water equivalent or 315 km³ of water equivalent. The discharge from glaciers represents 240 km³ of ice and is divided between 10 km³ of ice in north Greenland, 90 km³ of ice on the west coast, 120 km³ on the east coast and 20 km³ of ice in Melville Bay. The total deficit is about 100 km³ water equivalent for the whole ice-sheet. The accuracy of these values is, on the whole, low and other estimates suggest a balanced budget.

BARNES ICE-CAP, BAFFIN ISLAND

This ice-cap is an example of a rather unusual arctic type of glacier. The ice-cap has an area of 5900 km² and is elliptical in shape, its maximum length being 150 km and its maximum width 62 km. It rests on a slightly dissected plateau at about 610 m. The centre of the ice-cap forms a broad dome rising to 1128 m at its maximum. The margins are cut by deep surface meltwater streams. The area of the ice-cap is fairly static at present and its movement is very sluggish or negligible. The ice-cap is unusual in that it all lies below the firn-line and has no firn accumulation zone. Its balance is maintained by the addition of superimposed ice.

The total accumulation for the years 1961-2, 1962-3, and 1963-4 is estimated by R. B. Sagar (1966) to be 960×10^6 m³, 1020×10^6 m³, and 1060×10^6 m³ respectively. This yearly increment amounts to about 0·12 per cent of the total mass of ice, which is estimated to be about 910×10^9 m³. The greatest addition of mass occurs in the higher parts of the ice-cap. Except for 1962 when ablation occurred throughout nearly all the ice-cap, ablation was limited to the lower levels of the ice-cap. The accumulation area of the glacier lies above the equilibrium line and has the characteristics of the per-colation zone B and the slush zone. In these zones, percolating water refreezes, adding to the accumulation and densifying the snow at depth.

As melting proceeds in summer, the snow becomes isothermal, slush forms and streams flow. The surface is lowered by these ablation processes at the margin of the ice-cap. The ablation season lasts for about two months. Ice layers form within the snow during this season and these form the super-imposed ice. When this is allowed for, the ablation is about 20 to 21 cm of water equivalent.

The total net budget was found to be a loss of $1·9 \times 10^9$ m³ in 1961-2, which was an abnormally warm and dry year. For 1962-3 and 1963-4, the values were $-0·12 \times 10^9$ m³ and $+0·13 \times 10^9$ m³, giving a total loss for the three seasons of $-0·23$ per cent of the total mass. This caused some wasting of the northern part of the ice-cap.

NORTHERN SWEDEN

The small glacier, Storglaciären, at Kebnekajse in Swedish Lapland has been studied in detail for 16 years since 1946 by V. Schytt (1962). This small glacier has an area of 3·1 km². The average accumulation is 4·0 × 10⁶ m³ of water equivalent and the average ablation is 6·2 × 10⁶ m³ per budget year. Thus the average net loss is about 55 per cent more than the total net accumulation. This results in rapid glacier wastage and retreat. Of the fourteen years of detailed observations, twelve years showed a deficit, one year the budget was balanced, but in only one year, 1948-9, was there a surplus. The years of greatest loss were 1946-7 and 1959-60, both characterized by exceptionally hot dry summers with an excess of ablation. The budget varies very much from year to year so that the mean over a considerable number of years is required to give a good indication of the trend. The glacier will respond to the longer-term trends, but not to the short-term variation from year to year.

ICELAND

The mass balances of many of the outlet glaciers of Vatnajökull vary considerably from year to year. Estimates based on limited observations for the small outlet glacier of Morsárjökull from the years 1951-2, 1952-3, and 1953-4 suggest a total negative balance of approximately 6 × 10⁶ m³ and 26 × 10⁶ m³ of water equivalent for the first two years respectively, while in the third year there was a net gain of about 2 × 10⁶ m³. In general, however, the balance has been negative during this period. This is despite a relatively large proportion of the glacier lying above the equilibrium line. The whole accumulation area, however, lies at a fairly low altitude. The generally negative budget has been operating for several decades and the glacier has retreated about 1 km during the period from 1904 to 1953.

This glacier may be compared with two nearly adjacent glaciers, Skaptafellsjökull and Svinafellsjökull. These two glaciers, and particularly the latter, differ from Morsárjökull in having a smaller accumulation area relative to their ablation area. On the other hand, their accumulation areas attain a considerably greater altitude. This means that winter snowfall is greater and, more important, the loss of snow from the upper part of the accumulation area is much less. Indeed, snow can fall throughout the whole year at the higher levels. The greater net accumulation has resulted in a smaller amount of retreat, particularly in Svinafellsjökull. One part of this glacier, which is fed from the highest part of the accumulation area, has not retreated at all during the 50 years before 1954. Fifty per cent of Svinafellsjökull lies above 1400 m, while the figures for the area above 1400 m of Skaptafellsjökull and Morsárjökull are 22 per cent and zero respectively.

H. W. Ahlmann (1948) has reported the results of work on the budget of Hoffellsjökull, situated a little farther east, for the period 1935-8. The results gave a budget of -364×10^6 m³, $+191 \times 10^6$ m³ and -111×10^6 m³ for the three successive years. These figures also indicate the variability of these outlet glaciers and the generally negative budget at this time is also apparent. The variations in this instance were the result of changes in both accumulation and ablation. Lower ablation accounted for the positive balance in 1936-7. The mass balance had the high total mean value of 1415×10^6 m³ water equivalent for a total glacier area of 312 km². This type of régime is characteristic of a very maritime climate with heavy precipitation of rain and snow, and relatively mild temperatures at all seasons.

SOUTHERN NORWAY

The detailed study by Østrem of Nigardsbreen may be taken as an example of the mass balance of a glacier in southern Norway. This glacier drains from the mountain ice-cap of Jostedalsbreen. The district has a climate rather similar to that of Iceland, and snowfall is high, exceeding 5 m in places. Nigardsbreen has a very large accumulation area on the plateau ice-cap and a narrow outlet glacier tongue. The maximum area occurs between 1600 and 1700 m.

In the year 1961-2, when detailed mass balance studies were made, the specific net balance at different altitudes was such that the maximum accumulation took place between 1700 and 1900 m. The equilibrium line was at 1300 m. Thus the accumulation area occupied the altitude range of the glacier where the area was great. The total net accumulation was very high owing to the near coincidence of the maximum area and the maximum specific accumulation. The very high ablation rate over the narrow glacier trunk at low elevation was not such an important item in the balance because the area involved was small. As a result, in this particular year, there was a strongly positive mass balance. The accumulation was 122×10^6 m³ and the ablation was 27×10^6 m³, giving a balance of $+95 \times 10^6$ m³. This positive budget was caused by unusually low ablation values throughout the whole area as the precipitation was normal (Fig. 2-2).

In this particular year, the mass balance was unusual, because during the last few decades Norwegian glaciers have been retreating rapidly. However, since the mid-1950s there has been some indication of more positive mass balances. This has resulted in the snouts of some of the shorter glaciers advancing during the period 1950 to 1960. Briksdalsbreen, which is a very steep glacier draining from the north side of Jostedalsbreen, has advanced during several years. The larger glaciers, Nigardsbreen and Austerdalsbreen, have continued to retreat until 1962. These two glaciers, however, advanced during the first decade of this century and again during the 1920s. Both these advances were associated with terminal moraine formation. The glacier fluctuations have been correlated with climatic variations. The periods of glacier advance were associated with spells of cooler and wetter conditions. Temperatures rose rapidly and precipitation fell after 1930, giving conditions associated with very rapid glacier recession during the next two decades.

ALASKA

The Alaskan glaciers are of considerable interest as they show some irregularities. Some glaciers undergo rapid advance while neighbouring ones are retreating. The Black Rapids glacier suddenly advanced several kilometres down its valley over a period of 5 months in 1936-7, and then started to retreat again. The advance was probably the result of a succession of very snowy years which affected the glacier after a time lag of 7 years. This type of sudden advance is created by a surge in the glacier, a process that will be considered in more detail at the end of the chapter.

The Malaspina glacier system is one of the larger Alaskan glaciers, covering 4200 km² and containing about 1700 km³ of ice. Its accumulation area lies around the high peaks of the Mount St. Elias range, forming the Upper Seward glacier, which covers 1318 km² at 1520-2130 m. The firn limit is at about 915 m and lies across the outlet glacier, the Lower Seward Glacier, which feeds the Malaspina Glacier. The Lower Seward glacier is 5 to 6·5 km wide and 32 km long, leading down into the piedmont lobe of the Malaspina Glacier. This lobe spreads out over 2640 km² and rises to 610 m. It therefore lies wholly in the ablation area. The outer 8 km are debris-covered and consist of dead ice, on which spruce trees 100 years old are growing.

The annual accumulation of the Upper Seward glacier averages 75 cm of water equivalent, but is variable. It was 65 cm in 1947-8 and 175 cm in 1948-9. The normal budget of the glacier is a deficit of 13×10^6 m³, but during 1948-9, a favourable year for the glacier, the balance was positive. The glacier is, however, slowly wasting away as a negative budget is much more common than the occasional positive one (R. P. Sharp, 1958).

ALPINE GLACIERS

The Hintereisferner has been studied in detail by H. Hoinkes and R. Rudolf (1962) and may be taken as an example of an Alpine glacier. The observations cover the period 1952-61. The main glacier had an area of nearly 10 km² in 1959, its mean height being 2981 m. The mass balance, like that of the other glaciers mentioned, is very variable from year to year. With one exception, however, it was negative. The maximum negative balance of − 98 cm or -9.83×10^6 m³ of water equivalent over the whole glacier occurred in 1957-8. The only year with a positive balance was 1954-5, when the glacier gained 0.77×10^6 m³. The mean value throughout the nine-year period was -3.60×10^6 m³, with an accuracy of $\pm 0.54 \times 10^6$ m³.

These results are similar to those given for the other areas mentioned. They show that strongly negative budgets have been widespread in the last two decades in many of the temperate northern areas. In this Alpine glacier, the accumulation area is about 47 per cent larger than the ablation area, the percentage varying from season to season. Because of the prolonged negative budget, the glacier snout has been retreating on average about 25 m a year, but the retreat has been very variable. At times, about 100 m of dead ice become detached from the snout and gradually melt away.

SOUTHERN ALPS, NEW ZEALAND

The highest mountains of the Southern Alps of New Zealand (43-45°S.) support a number of glaciers. There is an interesting contrast in mass balance between the glaciers on the wetter western slopes and those of the drier eastern side. The longest glacier is the Tasman, draining south-east and south on the east side of the watershed. This glacier flows considerably slower than the Franz Josef on the western side and the former has a much smaller mass balance. The rate of flow of the Tasman falls off from 0.5 m/day at 19 km from the snout, to 0.35 m/day at 10 km from the snout. The ice in this glacier has been wasting away slowly during the last few decades and its lower 8 km are covered with a thick layer of ablation moraine. This down-wasting has resulted in a thinning of the glacier by about 60 m, although the debris-covered snout has not noticeably changed position. The snout lies at a height of about 760 m.

On the western side of the Alps, the Franz Josef and Fox glaciers differ greatly from the Tasman. The Franz Josef glacier descended until about a decade ago to a level of about 200 m above sea-level, reaching into an area of dense rain forest. The glacier moves actively right to its snout, which consists of clean ice with only a thin layer of debris. Records of its velocity at the end of the last century suggest that it flowed at 525 cm/day about 3 km from the snout and 60 cm/day only 0.5 km from the snout. The glacier descends from a very high accumulation area where the precipitation probably exceeds 500 mm/year and is evenly distributed. On the low ground near the glacier snout, the annual precipitation averages between 340 and 286 mm. Records of the movement of the snout show that the glacier retreated slowly from 1910 to 1922, then advanced by a few tens of metres until 1933. A rapid retreat of nearly 1000 m then started and continued until

1946, when a renewed advance of a little over 200 m lasted until 1951. This was followed by renewed and accelerating retreat of 600 m until the early

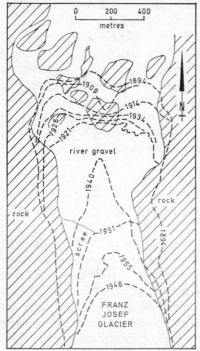

1960s. This phase has come to an end with a sudden and vigorous advance (Fig. 2-3).

Changes in the mass balance of the glacier appear to be related mainly to the precipitation. Variations in the accumulation are, therefore, mainly responsible for the changes in the snout position. There is a close correlation between the measured precipitation near the snout and the movement of the snout. The response of the snout lags five years behind the change in precipitation. The glacier starts to advance five years after the wetter period begins. In general, the periods of heavier precipitation were also periods of lower temperature and less sun. The mass balance was affected both by a rise in accumulation and a fall in ablation.

Fig. 2-3 Variations in the position of the snout of the Franz Josef glacier on the western side of the Southern Alps, New Zealand (R. P. Suggate, *N. Z. J. Sci. Technol*, 1951; Department of Scientific and Industrial Research, Wellington)

EQUATORIAL AFRICA

At the highest levels on the Equatorial mountains of Africa there are still some small glaciers. They are of interest as their régimes depend on different seasonal changes from those of the temperate and high-latitude glaciers already mentioned. On Kilimanjaro, the Penck glacier is the longest. It falls from 5800 m to 4600 m in 2·4 km. No details of the mass balance are available but it seems likely that there are two ablation seasons, the longer occurring from July to September and the shorter in January and February. The total ablation in 1957 at the snout of the glacier probably did not exceed 35 cm and the accumulation was probably of the same order. The main accumulation season is March to June. This glacier differs from those in other climates in that the maximum snowfall probably occurs near the snout of the glacier and decreases upward. The main accumulation zone at present is probably

PLATE I   AERIAL VIEW IN SOUTH BAFFIN ISLAND

Cirques, arêtes and horns rise above a series of glaciers in which the contributory ice streams can be distinguished. The lake on the lower left has been dammed by morainic and fluvioglacial debris, but the meltwater from the glacier turns away from the lake to flow to the west (right). (Surveys and Mapping Branch, Department of Energy, Mines and Resources; *Exposure no.* A16817–102. Canadian Government Copyright)

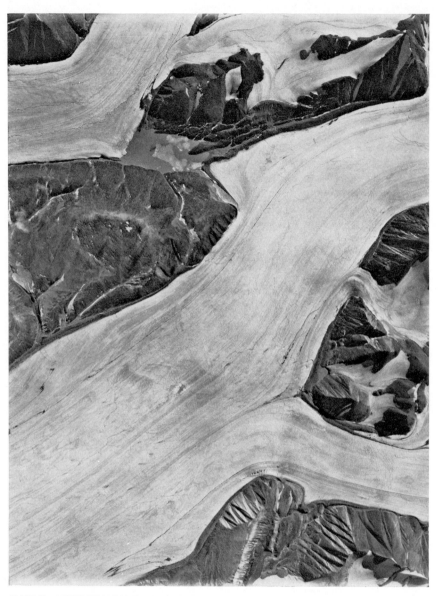

PLATE II   AERIAL VIEW IN NORTH BAFFIN ISLAND.
A former transfluent glacier linking the two main glacier systems is now only represented by a short tongue projecting into the col, where a lake is now impounded. Marginal moraines and supraglacial meltwater channels can be traced, as well as various crevasse patterns. (Surveys and Mapping Branch, Department of Energy, Mines and Resources; *Exposure no.* A.16350–34. Canadian Government Copyright)

in the lower part of the glacier and melting takes place at the top of the mountain. Clearly this régime cannot be stable for long.

On the volcanic peak of Mount Kenya, there are twelve small glaciers of which the Lewis glacier is the largest. This glacier has an area of 0·36 km² and is the lowest, descending to 4480 m. During the period from 1934 to 1958, the glacier receded up to 60 m. A total loss of about 18×10^5 m³ of ice has taken place, though the mass balance of the glacier is not accurately known. Accumulation occurs at two seasons, March to May and November to December, and is very variable. Glaciers can only survive where they are sheltered from the sun and exposed to the precipitation. C. M. Platt (1966) suggests that the annual amounts of accumulation and ablation on this glacier are small compared with those of temperate glaciers.

The Response of Glaciers to Climatic Change; Glacier Surges

Any positive change in the net budget of a glacier arising from climatic or other events will cause corresponding perturbations in glacier thickness, length and rate of flow. The response of glaciers to such events has been considered theoretically by J. F. Nye (1958, 1960, 1963, 1965a, and 1965b), making use of the kinematic wave theory of Lighthill and Whitham. They applied this theory to river floods and traffic flow on roads, both of which are group phenomena. The same theory can be applied to the movement of a wave through a glacier at a speed greater than that of the ice velocity.

Nye's theory of the influence of climatic change on glacier variation considers the effects of seasonal as well as longer-term changes of climate. These changes give rise to variations in accumulation and ablation. His results are based on the assumption that there are no changes in temperature in the ice and that the ice is incompressible. His other basic assumption is that the discharge at any cross-section in the glacier is dependent upon the level of the ice and the slope of the upper glacier surface. Nye's theory is applicable to a valley glacier of varying width, but the simpler case of a valley of uniform rectangular cross-section of unit width may be considered first. Two quantities must be considered, the discharge, q, and the ice thickness, h. The relationship between these two variables is approximately a fourth power one, such that $q \propto h^4$. The term 'kinematic wave' is used to refer to a moving point for which q is constant. The velocity of the moving point is c and this varies down the glacier. The velocity of the kinematic wave under the conditions given is $c = \dfrac{\delta q}{\delta h}$ and is four times that of the ice if the former relationship holds. Because of this, the wave will reach the glacier snout long before any new material could actually be transported that far. Arriving at the snout, the kinematic wave velocity is approximately equal to the ice velocity and gives rise to a sudden forward surge. A wave passing down the Nisqually glacier (Washington) has been closely studied since 1945 (M. F. Meier and A. Johnson, 1962), and its velocity has been found to vary between two and six times the mean ice velocity.

The kinematic wave is the mechanism by which a glacier responds to changes in mass balance. If accumulation and ablation are taken into account, the value of q is no longer constant, but it changes with a, the accumulation (a is negative for ablation). The effects of changes in accumulation and ablation are stable in those parts of the glacier where the flow is extending (that is, where the velocity of ice-flow is increasing down-glacier). But in regions where the ice-flow is compressing, or decelerating down-glacier, any change will tend to be accentuated and the glacier becomes inherently unstable.

It is assumed that the glacier accelerates steadily to a point P (Fig. 2–4) and then decelerates. If there is a uniform and shortlived addition of snow throughout the glacier, the upper part will increase uniformly in thickness. The lower part on the other hand will thicken at an increasing rate as c_0, the steady state kinematic wave velocity, moves down the glacier at a decelerating rate. The kinematic wave is initiated at P, where a step occurs. This step appears because the upper part is falling to its original level while the lower part is still rising. This increase in elevation or step is magnified as it moves down the glacier, so that the effect near the snout is much greater than that elsewhere. In the case of the Nisqually glacier referred to above, the increase in surface level with the arrival of the wave amounted to more than 30 m in places. After the kinematic wave has passed, the glacier surface subsides downstream of P.

The passage of a kinematic wave can sometimes be traced by photogrammetric methods. In front of the wave, there is a region where the ice is compressing and becoming thicker; behind the wave, the ice is in tension, it is intensely crevassed, and its surface level drops as noted. The tension-compression boundary can sometimes be picked out visually on photographs, and successive photogrammetric surveys may enable the down-glacier velocity of the kinematic wave to be determined. A. E. Harrison (1964) in this way found the wave velocity on the Muldrow glacier (Alaska) in 1956 to be 350 m/day.

Many glaciers have experienced spasmodic surges this century, unlike the majority, which are in recession. Some surges have been minor and brief, but others have been major events in the history of a glacier, and for a time have completely transformed the terminal area. The year 1966 has been a remarkable one for surges in Alaska. A dozen major glaciers have surged

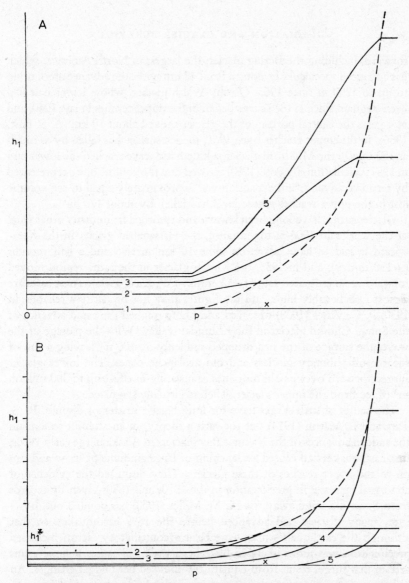

Fig. 2-4 Diagrams to show the effect of a kinematic wave in a glacier. The graph plots h_1, increase in thickness of the ice, against x, distance down glacier. Curves 1 to 5 refer to five successive time intervals, at 0, 5, 10, 20 and 40 years respectively if the strain rate is $+3\%$ a year above P and -3% a year below P, giving a time constant of 10 years. The broken line curves show the progress of the kinematic wave from P. The upper Figure has different initial conditions, with the glacier in a steady state. There is then a sudden, permanent and uniform increase in the rate of accumulation. In the lower Figure, there is only one increase on the glacier (J. F. Nye, *Int. Ass. Scient. Hydrol., Snow and Ice Commission*, **54**, 1961)

forward, including the Bering glacier, the largest in North America, which has advanced spectacularly along a front 42 km wide, the advance amounting to up to 1200 m since 1963. On the Walsh glacier, whose lower half had been stagnant since 1918, a surge began in the upper reaches in late 1960, and in 4 years the central portion of the glacier moved about 10 km (A. S. Post, 1966). In the upper reaches since 1961, the ice surface has fallen by as much as 150 m. On the Muldrow glacier, a kinematic wave reached the lower end in 1957; calculations by Post (1960) showed that the volume of ice represented by terminal advance and expansion was approximately equal to the volume lost in the upper reaches whose level had fallen by about 170 m.

Glacier surges have long been known and recorded from many other parts of the world. In 1904-5, for instance, the Hassanabad glacier in the Karakoram is said to have advanced some 10 km in two and a half months (=130 m/day), and in 1953, the Kutiah glacier in the same region moved forward at an estimated rate of 113 m/day (W. Kick, 1958). These rates are almost unbelievably high, and the figures may not be entirely reliable. In 1930, P. C. Visser (1938) observed and photographed the rapid advance of the Sultan Chusku glacier in the Khumdan valley. With the passage of the wave, the surface of the firn dropped suddenly by 100 m, leaving a rim of sheared-off tributary glaciers and old avalanche cones. The lower glacier surged forward over its old moraines, translating in effect up to 300 million m^3 of ice from the upper glacier. Then it virtually stagnated.

The causes of such surges have for long been a matter of dispute. R. S. Tarr and L. Martin (1914) put forward a theory, in an attempt to explain the sudden advances of the Yakutat Bay glaciers in Alaska in the early 1900s, that earthquakes had caused avalanching of large amounts of snow and firn on to the upper reaches of these glaciers. They dismissed the evidence of abnormal increases in precipitation in the 1870s and 1880s given by records from Sitka 200 miles away. As M. M. Miller (1958) has pointed out, however, many glaciers had advanced before the 1899 earthquakes, so that although the earthquakes may have been a contributory factor, increased precipitation was more likely to have been the basic cause. It now seems certain that surges result from climatically-induced net budget changes. An imbalance between accumulation and ice discharge develops over a period, a critical threshold is reached, and a massive transfer of ice into the ablation zone follows.

The earthquake mechanism must not be entirely abandoned, however, for there is no doubt that the surging glaciers of Alaska, the Karakoram, or Chile lie in seismically active areas. Once a state of imbalance in a glacier has been reached, a seismic disturbance may well trigger off a surge. Another possible relationship between surging glaciers and seismically active areas is that the latter are often regions of abnormally high geothermal heat flow. This may cause a slow build-up of basal meltwater until there is a drastic change in the coupling of the glacier to its bed (Meier, 1965). The important part played by meltwater in facilitating catastrophic advances will be considered further in Chapter 4.

The response time of smaller glaciers to climatic change varies between 3 and 30 years, but it may amount to hundreds or even thousands of years for very large ice masses. For the Antarctic ice-sheet, it is probably about 5000 years. This means that some apparently different ice-sheet advances in the Pleistocene may not be of world-wide chronological significance, but may merely reflect different frequency responses to the same climatic event. The seasonal changes in accumulation and ablation are of high frequency compared with the response time. The resulting thickness changes in the glacier lag about three months behind the greatest seasonal accumulation. For longer-term climatic changes, the response of the glacier comes into phase with the climatic change in its upper parts, but in the lower parts the response is more complex. There are two separate effects. First, there is the direct response, which may not be in phase with the cause. Secondly, there is the delayed, indirect, response caused by the travelling down-glacier of the wave form. The size of the wave varies down-glacier, instability tending to cause an increase while diffusion effects cause a decrease. The direct and indirect effects may cause one glacier to behave differently from its neighbour if it is of different dimensions. Thus, it is possible to explain the great variety of glacier response to changes in accumulation and ablation.

The response of each glacier can be calculated if the relevant data are available. These data include q, the discharge, B, the breadth, h, the height above an arbitrary datum line, a, the rate of accumulation averaged along a transverse line, and α, the slope of the ice surface. These values are required for different points down the glacier. Using these values, the changes in the glacier can be assessed by calculating c, the velocity, and D, the diffusion co-efficient for the kinematic waves, if observed rates of change of accumulation are known. The formulae for c and D are

$$c = \frac{1}{B} \cdot \frac{\delta q}{\delta h} \qquad D = \frac{1}{B} \cdot \frac{\delta q}{\delta a}$$

Nye (1965) has applied his theoretical analysis to actual glaciers, using a computer program to calculate the frequency response of South Cascade glacier (Washington) and Storglaciären (north Sweden). The latter glacier moves at about half the speed of the former and its response time is double that of the Cascade glacier, the values being 55 years and 26 years respectively. The amplitude of change is twice as great in Storglaciären, owing to its lower velocity of flow. Nye (1965) has also shown that it is possible to work back from the known changes in the position of the glacier snout to obtain the glacier budget in the past. In the more remote past, the evidence of glacier fluctuation is largely geomorphological. This evidence can then be more accurately related to climatic change, through calculated values of accumulation and ablation. The annual budget changes are not recorded by snout changes, but the observed changes related to the mean of ten years' budget values agree well with theoretical changes at the snout.

Conclusions

The glacier or ice-sheet régime depends upon the balance between accumulation and ablation. This balance is rarely achieved for a long period so that glaciers are continually fluctuating. The mass balance depends on the difference between accumulation and ablation. But equally important from the point of view of glacier activity are the actual amounts of accumulation and ablation, because these are related to glacier flow. Where the amount of snow added is high the glacier will be active, as in the case of the glaciers of south Iceland and Norway. But where the accumulation is very small and ablation equally low because of extremely low temperatures, glaciers will tend to move more slowly in relation to their dimensions and are geomorphologically less active. From the geomorphological point of view, therefore, both net and total budget values must be taken into account. The net budget determines the movement of the glacier snout and the variations in ice thickness. These changes in turn lead to the formation of terminal moraines, lateral moraines and trim-lines. The total budget determines in part the degree of glacial activity, and therefore is related to the erosive and depositional effects on the landscape.

The application of the kinematic wave theory to glacier fluctuations and surges provides a useful method of linking climatic change with glacier régime and glacier response, and hence with geomorphological features. This link can work in both directions.

REFERENCES

H. W. AHLMANN (1948), 'Glaciological research on the North Atlantic coasts', R. geogr. Soc. Res. Ser. 1, 83 p.

A. BAUER (1955), 'The balance of the Greenland ice-sheet', J. Glaciol. 2, 456-62

C. BULL (1958), 'Snow accumulation in North Greenland', J. Glaciol. 3, 237-48

F. E. CHARNLEY (1959), 'Some observations on the glaciers of Mount Kenya', J. Glaciol. 3, 480-92

F. DEBENHAM (1948), 'The problem of the Great Ross Barrier', *Geogr. J.* 112, 196–218

A. DESIO (1954), 'An exceptional glacier advance in the Karakoram-Ladakh region', *J. Glaciol.* 2, 383–5

A. J. GOW (1965), 'On the accumulation and seasonal stratification of snow at the South Pole', *J. Glaciol.* 5, 467–77

A. E. HARRISON (1964), 'Ice surges on the Muldrow glacier, Alaska', *J. Glaciol.* 5, 365–8

H. HOINKES and R. RUDOLPH (1962), 'Mass balance studies on the Hintereisferner, Ötztal Alps, 1952–61', *J. Glaciol.* 4, 266–80

J. T. HOLLIN (1962), 'On the glacial history of Antarctica', *J. Glaciol.* 4, 173–95

D. W. HUMPHRIES (1959), 'Preliminary notes on the glaciology of Kilimanjaro', *J. Glaciol.* 3, 475–9

W. KICK (1958), 'Exceptional glacier advances in the Karakoram', *J. Glaciol.* 3, 229

C. A. M. KING and J. D. IVES (1955), 'Glaciological observations on some of the outlet glaciers of South-West Vatnajökull, Iceland', *J. Glaciol.* 2, 563–9

E. LACHAPELLE (1962), 'Assessing glacier mass budgets by reconnaissance aerial photography', *J. Glaciol.* 4, 290–7

F. LOEWE (1960), 'Notes concerning the mass budget of the Antarctic inland ice', *Antarctic Meteorology*, 361–9

M. F. MEIER (1962), 'Proposed definitions for glacier mass budget terms', *J. Glaciol.* 4, 252–65

M. F. MEIER (1965), 'Glaciers and climate' in *The Quaternary of the United States* (ed. H. E. WRIGHT and D. G. FREY), 795–805

M. F. MEIER and A. JOHNSON (1962), 'The kinematic wave on Nisqually Glacier, Washington', *J. geophys. Res.* 67, 886

M. MELLOR (1959), 'Mass balance studies in Antarctica', *J. Glaciol.* 3, 522–33

M. M. MILLER (1958), 'The role of diastrophism in the regimen of glaciers in the St. Elias district, Alaska', *J. Glaciol.* 3, 292–7

F. MÜLLER (1962), 'Zonation in the accumulation areas of the glaciers of Axel Heiberg Island, N.W.T., Canada', *J. Glaciol.* 4, 302–11

J. F. NYE (1960), 'The response of glaciers and ice-sheets to seasonal and climatic changes', *Proc. R. Soc.* A, 256, 559–84

J. F. NYE (1963), 'The response of a glacier to changes in the rate of nourishment and wastage', *Proc. R. Soc.* A, 275, 87–112

J. F. NYE (1965), 'The frequency response of glaciers', *J. Glaciol.* 5, 567–87; 'A numerical method of inferring the budget history of a glacier from its advance and retreat', ibid., 589–607

G. ØSTREM (1961–62), 'Nigardsbreen hydrologi', *Norsk geogr. Tidsskr.* 18, 156–202

C. M. PLATT (1966), 'Some observations on the climate of Lewis Glacier, Mount Kenya, during the rainy season', *J. Glaciol.* 6, 267–87

A. S. POST (1960), 'The exceptional advances of the Muldrow, Black Rapids and Susitna glaciers', *J. geophys. Res.* 65, 3703–12

A. S. POST (1966), 'The recent surge of Walsh glacier, Yukon and Alaska', *J. Glaciol.* 6, 375–81

G. DE Q. ROBIN (1962), 'The ice of the Antarctic', *Scient. Am.* 861

R. B. SAGAR (1966), 'Glaciological and climatological studies on the Barnes Ice Cap', Geogr. Bull. **8**, 3–47

V. SCHYTT (1962), 'Mass balance studies in Kebnekajse', J. Glaciol. **4**, 281–8

R. P. SHARP (1958), 'The Malaspina glacier, Alaska', Bull. geol. Soc. Am. **69**, 617–47

P. A. SHUMSKIY, quoted by M. MELLOR (1959), 525

P. A. SIPLE (1958), 'Man's first winter at the South Pole', Natn. geogr. Mag. **113**, 439–78

R. P. SUGGATE (1950), 'Franz Joseph and other glaciers of the Southern Alps, New Zealand', J. Glaciol. **1**, 422–9

R. S. TARR and L. MARTIN (1914), Alaskan glacier studies (Washington)

P. C. VISSER (1938), Wissenschaftliche Ergebnisse der Niederländischen Expeditionen in den Karakorum und die angrenzenden Gebiete in den Jahren 1922, 1925, 1929–30 und 1935, Bd. II, Glaziologie (Leiden)

A. T. WILSON (1964), 'Origin of ice ages: an ice shelf theory for Pleistocene glaciation', Nature, Lond. **201**, 147–9

J. WEERTMAN (1961), 'Stability of ice-age ice-sheets', J. geophys. Res. **66**, 3783–92

H. WEXLER (1961), 'Ice budgets for Antarctica and changes in sea-level', J. Glaciol. **3**, 867–72

Chapter 3

The Physical Properties of Ice and Types of Glacier

A glacier is not coherent ice, but is a granular compound of ice and water, possessing, under certain circumstances, especially when much saturated with moisture, a rude flexibility sensible even to the hand. (J. D. FORBES, 1843)

Most of the ice that makes up glaciers and ice sheets originates as newly fallen light snow flakes. The newly fallen snow has a low density as much air is trapped between the hexagonal snow crystals, but as the delicate points of the crystals melt, the snow settles and increases in density. Melting is not the only cause of increase in density, however; other factors include the temperature and the original form of the snow crystals. The crystals themselves are smaller when the temperature is low. As the temperature increases many crystals stick together and fall as large snow flakes. Heavy snowfall only occurs when the temperature is fairly close to freezing-point. When the air is very cold, it can hold little moisture and the snowfalls are lighter.

The newly fallen snow alters in several stages to become glacier ice. The first stage is the change from snow crystals to granular snow. Since the vapour pressure of a snow crystal is greatest at its points, these melt first and the grain becomes more rounded. This process is much more rapid in places where temperatures are near zero centigrade; in such places, the newly fallen snow becomes coarse and granular in a few days. In very cold polar climates such as those of central Greenland or Antarctica, the process may be delayed for years. In intertropical latitudes, on the other hand, the snow often falls as soft hail. It then partially melts under the powerful sun during the day and refreezes at night. By this process, the snow is rapidly transformed into the coarse granular form. Climate, therefore, plays an important part in the change to granular snow.

Another important factor in accounting for the change from fresh powdery snow to ice is the effect of compression. Compression will also tend to be more rapid where the temperature is close to zero, because the individual snowfalls may be heavy. Where the pressure of overlying layers is considerable, the snow has small grains but is compact and very resistant. The density of newly fallen snow is about 0·06 to 0·08, but after two days on a slope it increases to 0·2 in a temperate climate.

The second stage is the conversion of granular snow to firn. After one winter, the snow reaches a density of 0·4 to 0·55. At this stage it may be called 'firn'. There has been some discussion concerning the use of the terms 'firn' and 'névé', which are the German and French respectively for material that could be described in English as 'consolidated, granular snow not yet changed to glacier ice' (M. M. Miller, 1952). Miller suggests that the material be called firn or firn-snow, and the area where it occurs be called the névé. The term 'firn' means literally 'of last year' and this is the simplest way to use the term, to refer to snow that has survived one summer season.

The lower boundary of the firn is not so easily defined. But M. F. Meier (1962) supports the suggestion that it should lie at the point at which the densifying material becomes impermeable to water. This boundary sometimes occurs at a density of 0·55, a density at which Benson and Anderson have found that the rate and mechanism of densification changes. An older suggestion was that the lower limit of firn should be placed at a density of 0·4. This limit does not seem so satisfactory as greater densities than this have been found in temperate glaciers in snow that is less than a year old. For instance, snow on the Cascade glacier in Washington had a density exceeding even 0·55. The definition of firn as material that has survived one ablation season but is not yet impermeable to water seems a satisfactory definition at least for temperate glaciers. A somewhat different definition may be needed for polar ice-sheets. It is important to bear in mind the distinction between temperate and polar glaciers, a distinction discussed later in this Chapter.

The process of change from snow to ice through firn is a continuous one. As the firn increases in age, it also increases in density and crystal size (Table 7)

TABLE 7   *Relationship between age, density and crystal length of firn on the Claridenferner, Switzerland*

Age (years)	1	4	10	12
Density	0·59	0·60	0·68	0·70
Crystal length (mm)	0·7	1·5	2·5	4·5

The physical properties of snow and firn are important from several points of view. Snow is a bad conductor of heat, and it therefore prevents the penetration of cold into the ground when the snow layer is thick. This is important in a consideration of permafrost and periglacial forms (Chapter 21). On the other hand, snow can hold a large quantity of water, up to

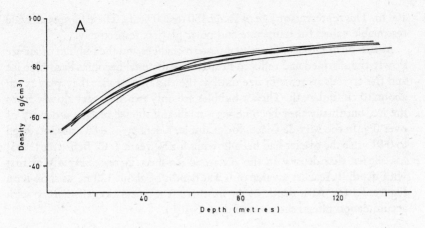

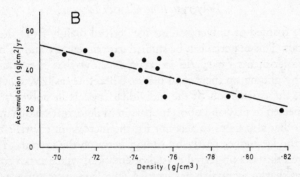

Fig. 3-1 **A** Graph relating density with depth in the Antarctic ice-sheet.
  B Graph relating density with accumulation at 40 m depth in the Antarctic
  ice-sheet
  (John C. Behrendt, *J. Glaciol.*, 1965, by permission of the Glaciological
  Society)

40 per cent by volume or 75 per cent by weight. Snow in this waterlogged
condition may give rise to dangerous and geomorphologically effective
avalanches (Chapter 25). At low temperatures, the snow has the properties
of an elastic body, but at temperatures around zero it shows viscous properties. Thus the snow is able to creep as friction releases heat; the heat melts
the crystal points and allows movement between the grains.

The third stage is the conversion of firn to ice, and the time taken for this
is very variable. In the Claridenferner, the material was still firn after 12
years, and about 25 to 40 years were necessary before it became ice. The rate
is slower in Greenland, where ice of a density of 0·8 was not found above

100 m. This represents an age of about 150 to 200 years. These figures provide reasonable values for temperate and polar glaciers respectively.

As firn turns into ice, the crystal pattern changes and the bubbles of air are slowly transformed and reduced in size. The material becomes hard blue ice and the crystals increase in size until at the end of a glacier they may reach about 10 cm in length. The air bubbles are only expelled very slowly from the ice, but finally they become very small and the ice reaches a density of over 0·9. In the Mer de Glace, for example, a density of 0·88 only increased to 0·91 after the glacier had been flowing for 50 years. J. C. Behrendt (1965) has shown that density in the Antarctic ice-sheet increases rapidly at first with depth. It reaches a value of 0·9 at depths of about 120 m, as shown on Figure 3-1 (a). At a depth of 40 m, there is a correlation of density with accumulation rate as shown on Figure 3-1 (b).

Polycrystalline Glacier Ice

Glaciers are formed of polycrystalline ice derived mainly from the original snow crystals. This material can be studied experimentally and its properties defined. Experiments on single ice crystals show that they can deform plastically by gliding on their basal planes. Therefore, as J. W. Glen (1958a) has shown, the orientation of the individual crystals in polycrystalline ice may be expected to play an important part in its deformation. Ice movement is a process that also helps to account for the increase in crystal size down-glacier, as the ice moves partly by sliding between the crystals. The ice of glaciers often shows a preferred crystal orientation, which is not so marked, however, as that in ice which has formed on a free-water surface. In ice of this type, all the basal planes tend to be parallel. Therefore in using ice for tests in a laboratory, it is important to know the source of the ice. Experiments in which grain structure was studied under polarized light have been made with lake ice, glacier ice and icicles. The results show that all types of ice deform plastically both in tension and compression. It has also been found that the deformation of ice is analogous to that of metals at high temperatures, an analogy which led Glen to carry out experiments similar to those carried out on metals.

Tests have been made at controlled temperatures and pressures to assess the effect of both these variables on the deformation of polycrystalline ice. The curve for ice deformation shows a decrease in deformation rate at first, and then the deformation reaches a steady state for a uniform stress. When the stress is raised above about 4 kg/cm², the creep rate falls initially and then accelerates. This change is caused by some recrystallization. The grain sizes of the samples tested at high pressure are found to be smaller than those that have been deformed by smaller stresses. Recrystallization is probably responsible for the observed creep-rate curve. This process allows the crystals to become aligned in a more favourable pattern for creep, thus accounting for the higher rate of deformation with increasing pressure. The recrystallization produces crystals whose basal planes are preferentially orientated at 45 degrees to the axis of compression, which is parallel to the planes of maximum shear. Plastic deformation of polycrystalline ice, therefore, is partly accomplished by intra-granular slip and partly by recrystallization.

A straight line relationship is found when the logarithmic value of the strain rate, $\dot{\varepsilon}$, is plotted against the logarithm of the stress, τ. This relationship gives the flow law of ice, $\dot{\varepsilon} = k\tau^n$. Glen (1958b) found that the constant n had the value $3 \cdot 17 \pm 0 \cdot 2$ for tests at both $-0 \cdot 02°C$. and at $-1 \cdot 5°C$. This law holds satisfactorily over a stress range of 1 to 10 bars.* S. Steinemann (1954) obtained similar results, but found that at low stresses the steady state took a long time to attain.

Tests have been carried out to study the effect of different temperatures. Glen used temperatures of $-6 \cdot 7°C$. and $-12 \cdot 8°C$., as well as those already mentioned. As the temperature was reduced to $-12 \cdot 8°C$., the strain rate decreased by a factor of 6. Temperature also affects the hardness and strength of polycrystalline ice. The crushing strength of ice is increased as the temperature is lowered. For ice formed from snow at $0°C$., the crushing strength is $18 \cdot 3$ kg/cm^2 but at $-50°C$. the value is $40 \cdot 6$ kg/cm^2, while lake ice has a much higher strength at lower temperatures.

Laboratory and field observations show, therefore, that ice is not a substance of constant viscosity, but a partially plastic substance which can deform by slow creep under stress. The plasticity of ice is seen in the way in which it takes on the form of a rock surface across which it moves. Examples of this are given by J. G. McCall (1960) from observations in the upper tunnel cut through Vesl-Skautbreen, Norway. This small cirque glacier is moving across a head-wall gap, the ice forming the roof of the gap. Grooves impressed upon the ice as it slides across its rock bed above the gap remain imprinted on the ice-roof of the cave for a distance of 50 m. As the glacier only moves at 3 m/year at this point, the grooves are apparently able to survive for about 15 years. The ice, therefore, behaves plastically, moulding itself to the indentations of its floor, but then remaining rigid when the pressure against the floor is relieved. H. Carol (1947) has also studied the plastic deformation of ice in a cave beneath the Ober Grindelwald glacier at a depth of 50 m. The ice was compressed against a rock knob and in close proximity to the rock surface the velocity was increased and its consistency was softened. Carol described it as being like cheese. A plane separated the more rapidly moving ice (72 cm/day) from the slower moving ice above (36·8 cm/day). W. H. Theakstone (1966) has described rather similar plastic contortions in ice caves beneath Østerdalsisen in northern Norway.

The ability of ice to spread under its own weight is a property well exemplified in the character of the ice shelves that float in the Antarctic bays. This spreading illustrates the response of ice by plastic deformation to stresses within it. It is possible to calculate the theoretical profile of an ice-sheet or glacier if the properties and behaviour of ice are taken into account. Nye (1952) has shown that an ideal ice-cap, resting on a horizontal base, would have a parabolic profile, given by $h = \sqrt{2h_0(R-r)}$. h is the height of the surface at a distance x from the edge of the ice-cap of radius R, r is the distance from the point h to the centre, and h_0 is $\tau/\rho g$, τ being the shear stress, ρ the density and g the acceleration of gravity. The results of the calculations agree very well with the surveyed profiles of the Antarctic ice-sheet, suggesting that the assumed properties must closely resemble the real ones.

Classification of Glacier Types

H. W. Ahlmann (1948) has suggested various methods of classifying ice masses. He gives a morphological classification based on the size and form of the ice, a dynamic classification based on the degree of activity of the ice, and finally, a thermal classification. These three types of classification all have a bearing on the geomorphological activity of moving ice on the landscape. The three classifications will be considered in the reverse order, as this is the order of increasing subdivision.

THERMAL CLASSIFICATION

The temperature of an ice mass plays an important part in its morphological activity. This has already been noted in connection with glacier surges and it will be stressed in later chapters. There are two fundamental types of glacier from this point of view: temperate and cold (or polar) glaciers. *Temperate glaciers* are at pressure melting-point throughout their thickness, except in winter when the temperature in the uppermost layers may become temporarily colder. The important point concerning this type of ice mass is that meltwater can be present throughout. Indeed, the temperate character of the ice can often be identified, even if no detailed temperature measurements can be made, by the fact that meltwater issues from beneath the edge of the ice. In this type of glacier, water can circulate freely and ice-dammed lakes and crevasses can remain full of water up to a certain depth (see Chapter 19).

Many of the glaciers of the Alps and southern Scandinavia are temperate in character. This type of glacier, because meltwater can exist at its bed, flows fairly easily, for the ice can readily slip over the wet rock floor (Chapter 4). Glaciers of this type, therefore, move faster than the other type and the erosive action of the ice is probably enhanced as well. Subglacial meltwater can also cause considerable erosion. Another characteristic of the temperate glacier, and one which applies even more strongly to inter-tropical glaciers, has already been mentioned, namely, the rapid rate of firnification. On temperate glaciers, summer melting takes place both in the accumulation and in the ablation zones. The melting in the accumulation zone helps considerably to account for the rapid firnification of the snow.

Cold or polar glaciers differ in very important respects from temperate glaciers. Ahlmann has subdivided this type into two, the sub-polar type and the high polar type. He differentiates the two mainly on their firn characteristics. In the accumulation area of the sub-polar type, the material consists of crystalline firn down to a depth of 10 to 20 m. In summer, the surface can melt and water can be present. In the high polar type, the temperature in the firn remains well below freezing-point and even in summer there is no melting on the surface in the accumulation area. In this type, firnification is a very slow process and does not take place above a depth of 75 m.

From the geomorphological point of view, one of the most important properties of a cold ice-sheet is the absence of meltwater at depth in the ice. Melt-streams, where they occur, for example on the margins of the Greenland ice-sheet, flow on the surface of the ice. They may reach very large dimensions during the short ablation season. A large slush zone also indicates the effectiveness of surface melting in the sub-polar type of ice mass. At the base of the glacier, the ice is well below pressure melting-point and the ice is frozen to the bedrock on which it is resting. This naturally affects the way in which the ice moves over its bed (Chapter 4). It seems probable that ice masses in this condition cannot achieve as much erosion as the temperate type.

The temperature distribution in glaciers and ice-sheets is not simple as it depends on several factors. These include the surface temperature fluctuations, both seasonally and through a longer period, the thickness of the ice, and the geothermal heat flow from below the ice. This problem has been discussed by G. de Q. Robin (1955) for a polar ice-sheet. He shows that a small accumulation on the surface of the ice-sheet may affect its internal temperature. He considers the problem theoretically, assuming a stable ice-sheet of the form described by J. F. Nye (1952), and calculates the temperature at various depths in the ice-sheet. The geothermal heat flow, assumed to be 38 cal/cm^2/year, gives a gradient of 1°C./44 m at the base of the ice-sheet. Another source of heat at the base of the ice-sheet is that resulting from shearing: Nye has shown that most of the movement takes place by shearing in the lowest layers of the ice-sheet. Robin has calculated a shear stress of 0·88 bars at the base of the Greenland ice-sheet, and if the basal movement is 18 m/year, a heat supply could be generated equal to that of the geothermal heat flow. For a movement of 10 m/year, the additional heat supply would be 21 cal/cm²/year. This source of heat must be taken into account away from the centre of the ice-sheet. Ice must move downward in the centre to compensate for the outward movement of ice from the centre, and this will also generate heat. Another consideration is the surface heat supply. The temperatures are lowest at the highest elevations on the surface, but as the ice moves outward and downward, it comes to be overlain by layers of firn that are slightly less cold. Thus there should be an increase in the negative temperature gradient close to the surface outward from the centre of the ice-sheet, provided that no heat reaches the surface from below. The Greenland ice-sheet illustrates these points.

The ice-sheet has a maximum thickness of about 3000 m and an accumulation of about 30 cm/year. The temperature at the surface is about − 29°C., which would give a theoretical basal temperature of − 12°C. This agrees with the views of some glaciologists that the ice-sheet is frozen to its base in its central and southern parts. Seismic observations support the figure and suggest a basal temperature of about − 10°C. The theory suggests that the temperature should increase outward from the centre. Robin argues that if the accumulation on the central part of the Greenland ice-sheet dropped to 10 cm/year, then the temperature at the base of the ice-sheet would rise to zero. Any further decrease in accumulation would allow basal melting. There is, in fact, evidence that the basal ice temperatures are at the pressure melting-point in northern Greenland, where the accumulation is very low. The failure of echoes to return from the base of the ice-sheet in this area supports the theoretical calculations. The idea has been utilized by J. Weertman (1961) in his hypothesis for the formation of cold-ice moraines (Chapter 15).

Robin suggests that an ice-sheet with a basal temperature below the pressure melting-point will require a greater shear stress to induce movement than one that is warm enough for temperatures at the melting-point to exist at the base of the glacier. When the glacier is cold-based, its thickness and surface slope will build up and flow will slowly accelerate at the base. This acceleration of flow may be sufficient to raise the temperature to the pressure melting-point. Then rapid motion could take place to restore the balance of thickness and velocity to the new conditions. These conditions might, however, result in renewed basal freezing. He puts forward this theory as a possible cause for sudden advances of some high polar glaciers, such as Brasvelbreen in Spitsbergen.

Ice-sheets even in very high latitudes can have a temperature distribution such that the ice at their base is at the pressure melting-point. As already noted, this probably applies to the northern part of the Greenland ice-sheet where accumulation and flow are both lower than in the south where the ice is frozen to its bed. It is also probably frozen to its bed near the edges where the thickness diminishes. The same situation is likely also to apply to the Barnes ice-cap in Baffin Island; the significance of this phenomenon will be referred to again when moraine formation is discussed (Chapter 15).

Since Ahlmann proposed the classification of glaciers on the basis of their thermal properties, complications have come to light. Various modifications have, therefore, been subsequently suggested. A. Court (1957) proposed that the terms *permelting*, *refreezing* and *nonmelting* be used instead of temperate, sub-polar and high polar, respectively. Court's terms are meant to signify that glaciers in the first category are permeated by meltwater throughout. The second type has a layer in which water refreezes below a surface layer of melting. The third type never experiences melting on the surface. It is important to note that the whole of one ice mass need not fall solely into one category. Ahlmann used the surface conditions as the criteria for classification, but the temperature gradient is also a valid basis for classification. However, as just discussed, this can lead to complications because high-polar ice-sheets and even the Antarctic ice-sheet can be at the pressure melting-point at their bases. On the whole, however, it seems best to continue to use the widely adopted terms proposed by Ahlmann, as long as it is realized that the whole ice mass need not necessarily fall into a single category.

An example of this fact has been mentioned by F. Loewe (1966) in connection with the Sukkertoppen ice-cap in Greenland. The upper part of this glacier above the firn-line has a temperate character as the temperature at a depth of 4 m is very near freezing-point, because meltwater sinking into the firn releases heat as it refreezes. In the ablation zone, however, where meltwater runs off, the ice attains the mean annual temperature of the air. This is well below the freezing-point and the ice has polar characteristics. Inversions of this type have also been reported from Greenland and Nordaustlandet, Svalbard (V. Schytt, 1964). If these exceptions are allowed for, then the terms proposed can be used conveniently.

DYNAMIC CLASSIFICATION

The activity of glaciers is influenced to a certain extent by their thermal characteristics: a cold-based glacier requires a larger shear stress to induce movement than a temperate glacier. However, the dynamic activity of a glacier is also closely associated with its mass balance, as mentioned in the last chapter. A classification based on the glacier's dynamic activity consists of three main types: 1) active, 2) passive or inactive, 3) dead glaciers. The importance of these three types from the point of view of glacial deposits is considered in more detail in Part III.

Active glaciers are normally fed by a continuous ice stream from an accumulation zone that may lie in a cirque basin or on a plateau. Glaciers fed by accumulation on a plateau may be termed outlet glaciers. Alternatively, some active glaciers may be classed as regenerated in type. These glaciers are fed entirely by ice avalanches falling from an upper accumulation area on to lower ground, where the glacier head consists of large avalanche fans. The Glacier de Nantes in the French Alps is an example of a small regenerated active glacier. It does not move far from its avalanche fan head, but it has formed large terminal moraines that demonstrate its activity. The eastern side of the outlet glacier of Morsárjökull in Iceland is now fed entirely by avalanches from Vatnajökull above (Fig. 3-2). The glacier receives sufficient material to continue down the valley as a normal glacier for several kilometres. It is joined to its western portion which was still connected to the ice-cap above by a continuous ice-fall in 1954. The eastern part of the glacier was severed from direct contact with the ice-cap above by thinning in about 1937 (Plate V).

The dynamic characteristics of glaciers are not directly dependent upon a positive mass balance. Some glaciers can maintain active dynamic movement even with a negative budget, but it is obvious that this state cannot continue indefinitely, as the velocity of a glacier is partly dependent upon its thickness and this in turn on its mass balance. The total budget is more important in determining the relative activity of a glacier. Active glaciers tend to have a large total budget. A glacier in this state can flow actively right down to its snout even if this is retreating up-glacier as a result of a negative budget.

The glaciers on the west side of the Southern Alps of New Zealand provide good examples of active glaciers. In this area, the precipitation is very high and ablation fast; the slope is steep and the glaciers very active. The flow rates measured at the end of the last century are much faster than those measured recently by B. M. Gunn (1964). Suggate (1950) records a maximum rate of 525 cm/day in 1894, 3·2 km from the snout of the Franz Josef glacier. This value compares with 66 cm/day in 1956 measured at about the same distance from the snout. The minimum values given are not so different, but nevertheless indicate a considerable slowing of the glacier over the last 50 years. During the period from 1935 to 1960, the glacier retreated 1260 m, 600 m of this retreat taking place since 1956. Thus although the glacier can still be described as active, and shows characteristics of this state in its lower reaches, its velocity has decreased as it has thinned and retreated over the last 50 years. A dynamically active glacier is one which is flowing fast, whether it is retreating or advancing at its snout. But it will be more active when it is thicker than when it is thinning.

Where the supply of snow to feed the glacier is small, for example on the lee side of a mountain range, the ice may become passive. This may also

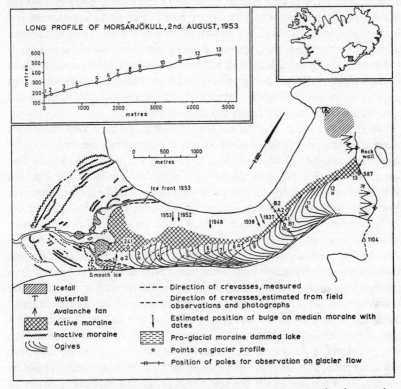

Fig. 3-2 Map of Morsárjökull (J. D. Ives and C. A. M. King, *J. Glaciol.*, 1954, by permission of the Glaciological Society)

occur where the slopes are gentle. The lowlands to the east of the Scandinavian mountains illustrate the type of area where passive ice may be expected to occur. Similarly on flat areas of high ground, where lack of slope inhibits movement, the ice may be passive. The ice under these conditions still receives nourishment in the form of snow accumulation. But the ice is not dynamically active, nor can it be described as dead.

Dead ice is not necessarily immobile according to Ahlmann's definition of the term. He suggests that dead ice is that which no longer receives a supply from an accumulation area. Its movement is restricted to that dependent on the slope. Another definition of dead ice is based on its dynamic property of movement. A static ice mass is described as dead. Ice is most likely to become dead according to both definitions at the maximum distance from the high ground accumulation area. Dead ice is most likely to occur on low ground where flow induced by slope factors will be very small. Ice in this situation is likely to be static as well as dead from the climatological point of view. In this situation, the ice mass may no longer receive a surplus accumulation and lives only on its mass, which is a slowly wasting asset under these conditions. However, an ice mass may become dynamically dead considerably after it is climatologically dead, because it still continues to receive material from its accumulation zone even though this is no longer being actively supplied. In large ice-sheets this difference in time between climatologically dead ice and dynamically dead ice may be several thousand years.

Lobes of dead ice in areas where the ice was at one time very thick may become isolated from their original source. It is sometimes necessary to infer the former presence of dead-ice lobes to account for particular geomorphological features. For example, a dead-ice lobe in the Vale of Belvoir has been postulated to account for the present course of the middle Trent. The river flows in a deep trench, cut anomalously and obliquely to the strike of the rocks between Nottingham and Newark. The ice in the Vale of Belvoir, which lies just east of the Trent trench, had accomplished in its active phase considerable erosion in the soft clays exposed in this area. It would, therefore, have been relatively thick here when the ice-sheet started to decay. It is reasonable to suppose that a dead ice mass would survive in this particular low-lying area. Meltwater flowing along the margin of the dead ice initiated the Trent trench. Dead ice was also very important in the deglaciation of parts of Sweden, such as Norrland, which is an area far from the accumulation zone, particularly during the later stages of deglaciation when the ice centre had again moved to the west of the mountains (C. M. Mannerfelt, 1945). At present, dead ice occurs on the outer margin of the Malaspina glacier in Alaska. The geomorphological significance of dead ice in accounting for a variety of glacial depositional landforms is considered in Chapter 17.

MORPHOLOGICAL CLASSIFICATION OF GLACIERS AND ICE MASSES

The classification of the morphologically distinct ice masses is based essentially on their size and the characteristics of their environment. Altitude in relation to the areal distribution of the ice is an important factor in the classification. Some types have large areas at high altitudes and others large areas at low elevations. The following classification is suggested:

1. Niche, wall-sided or cliff glacier
2. Cirque glacier
3. Valley glacier—Alpine type
4. Valley glacier—outlet type
5. Transection glacier
6. Piedmont glacier
7. Floating glacier tongues and ice-shelves
8. Mountain ice-cap
9. Glacier cap or ice-cap
10. Continental ice-sheet.

1. Niche glaciers in Spitsbergen have been described by G. E. Groom (1959). They consist of a triangular wedge of ice, often with a slightly convex surface, lying in a shallow funnel-shaped hollow in the upper part of the hillside. They develop on steep slopes (up to 42 degrees) and are often associated with rock benches, formed where harder rocks outcrop. A gully runs from the lower end of the glacier to an alluvial fan below. At times, these small ice accumulations are connected to ice on the plateau above, but in other instances they are isolated features. It is thought that the niche glaciers develop independently of the plateau ice-cap above, but this raises the problem of how snow can accumulate on slopes as steep as 42 degrees. The glaciers probably originated as snow patches, which rest between the steep rock slope and the scree beneath. The rock slopes become notched by gullies and it is in these longitudinal gullies that snow drifts can accumulate and develop into niche glaciers. The rock benches give rise to steps on which snow can accumulate in the gullies. The gullies can then be enlarged by nivation processes (Chapter 25) until the snow can accumulate and consolidate to a sufficient thickness to form a small glacier. The niche glacier is thus genetically similar to the rather larger cirque glacier and owes its particular features both to structure and climate. It forms an early stage in cirque glacier development.

2. The detailed work carried out by W. V. Lewis (1960) and those working with him on the cirque glaciers of the Jotunheim in Norway provides a valuable analysis of the characteristics of this type of glacier. Vesl-Skautbreen is a small cirque glacier that has been studied from many points of view and may be taken as an example of this type (Fig. 8-1A). The glacier is less than 1 km in diameter and rests against a steep rock wall. It has a mean surface slope of 26 degrees and lies on a rock floor whose profile is arcuate. The radius of curvature of the bed and back-wall is 240 m and the front end of the glacier rests against a snow-bank, which lies between it and the moraine at its foot. The moraine in turn is separated from the rock bar at the lip of the cirque by a lake. The cirque glacier, therefore, lies in a true rock basin, which is characteristic of many cirques. Further details of the character and movement of this particular glacier are discussed in Chapter 8 in connection with the formation of cirques.

3. When the snow-line is falling in elevation, the ice of the cirque can move out of the cirque basin and down the valley to form a valley glacier. Some-times the ice of several cirques combines to form the valley glacier, as is common in the Alps. This has given rise to the term 'Alpine type', applicable to this form of valley glacier.

Ahlmann has further subdivided the valley type of glacier according to the area-height relationships, giving four subdivisions. His first type of valley glacier is exemplified by the Rhône glacier or Hintereisferner. These glaciers have a considerable proportion of their area a little above their median height. The areas at the highest and lowest levels are reduced to small amounts. The second type is shown by the Grosser Aletschgletscher (Fig. 3-3). This glacier has a large high-level névé basin, so that the greatest area occurs in the upper part of the glacier basin. The third type consists of glaciers that have most of their area at low levels and are largely fed by avalanching from higher snowfields. These include many of the glaciers of Central Asia, originating in high mountains. The Styggedalsbreen in Norway is another example. The fourth group includes glaciers such as those of north-west Spitsbergen which have large lateral tributaries. These have the greatest proportion of their area at levels a little below the mean elevation.

The importance of the relative area of a valley glacier at different levels has already been mentioned. The different behaviour of neighbouring glaciers can be accounted for in terms of these glacier types. Those with a large area in the zone of maximum accumulation will be in a more favourable position than those with only a small area at this level. One essential characteristic of all the glaciers in this category is that they are confined within valley walls throughout their length and terminate in a narrow tongue.

4. The outlet type of valley glacier is similar in its lower reaches to the former type, but it is fed in its upper reaches from an ice-cap and not from an individual cirque or series of cirque basins. The outlet glaciers draining from the ice-caps of Iceland and Norway are good examples of this type. Larger glaciers of essentially the same type drain from the large Greenland and Antarctic ice-sheets. These are wider and may flow with considerable rapidity on steep slopes; they often end in the sea, in which case they must be partly classed in category 7.

The outlet glaciers are very susceptible to the position of the equilibrium

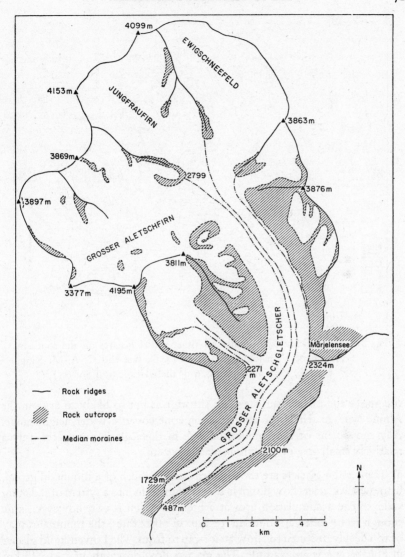

Fig. 3-3 Map of the Grosser Aletschgletscher

line in relation to its source region. This is demonstrated, for example, in Morsárjökull in Iceland and Nigardsbreen and Austerdalsbreen in Norway. At the present time, their equilibrium lines lie near the top of the valley glacier part of the system. But if they were to rise a little higher, then the whole system would become rapidly moribund, as the equilibrium line would

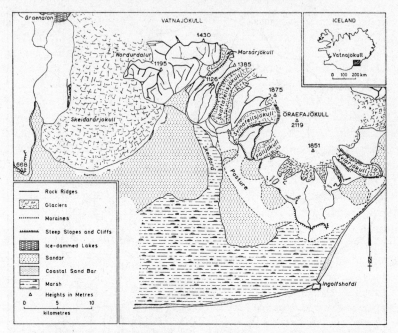

Fig. 3-4 Map of Skeiðarárjökull and other outlet glaciers on the southern
margin of Vatnajökull in Iceland (J. D. Ives and C. A. M. King,
J. Glaciol., 1955, by permission of the Glaciological Society)

rise above the plateau surface. This situation is not so likely to arise in the
Alpine type, in which the accumulation zone covers a wider height range.
Any movement of the equilibrium line in these glaciers involves only a
relatively small area of the whole glacier system.

5. Transection glaciers are those which occupy much of a mountain group,
from which glaciers flow down in several directions into a system of radiating
valleys. The accumulation area at a high elevation is not, however, large
enough to be called a mountain ice-cap. In other cases, the mountains may
be too deeply dissected to allow an ice-cap to form. The Lowenskjold glacier
in Spitsbergen is an example. The glaciers flow down about 10 to 15 km
from the central area of snowfields. Glacier systems of this type are responsible for forming transfluent glacial breaches and cols, which are discussed in Part II. The proportion of the area at the higher elevations is rather
larger in these glacier systems, although the area at the highest levels falls off
more than it does in ice-caps, because of the steeper relief.

6. Piedmont glaciers form when the valley glacier advances out from the containing mountain walls into a lowland beyond. The classic example of a piedmont glacier is the Malaspina glacier in Alaska, which spreads out as a broad lobe on the Pacific coast lowlands. The ice in the lobe is about 600 m thick and it lies in a basin at least 250 m below sea-level. As the glacier spreads out, it maintains a sufficient thickness and surface slope to enable it to flow uphill across the hollow in the coastal foreland (R. P. Sharp, 1958). The dimensions and régime of the glacier have already been commented upon and its stagnant marginal zone noted. This type of glacier is characterized by its relatively large area at the lowest altitude. Nearly one-quarter of the area covers only one-tenth of the height range in another typical example, the Murray glacier in Spitsbergen.

Skeiðarárjökull is another piedmont glacier (Fig. 3-4). It is an outlet glacier flowing south from Vatnajökull in Iceland and spreads out on to the sandur at the foot of the mountains. The glacier is only 8 km wide where it breaks through the mountain wall but it expands into a broad lobe about 25 km along the front. The ice is not dead at its margin as the whole system is fairly active. The accumulation zone of the glacier is on the mountain ice-cap of Vatnajökull but only a relatively small part at a low elevation drains through this outlet glacier. The activity of Skeiðarárjökull is evident in the very large sandur that it has built out in front of its broad snout.

7. Floating ice tongues are at present restricted to high latitudes where glaciers can reach to sea-level. The form that the floating part of the glacier takes depends on the surrounding coastal relief. Where the glacier is confined within a valley, the floating part of the glacier will be no wider than the grounded part. The glacier will lose mass by calving, thus creating icebergs. An example of a glacier ending in a fjord with a floating ice tongue is the Steensby Gletscher in north Greenland. Glaciers of this type are relatively rare in the Arctic, because the glaciers are so heavily crevassed that they usually calve off before their tongues float. The floating part of the Steensby Gletscher tends to disintegrate into separate lobes. The glacier is an outlet glacier draining the inland ice and is 48 km long and 11 km wide near the outlet. It has a low gradient of 1·4 per cent as do the other floating ice tongues in this area. The outermost 8 km of the Steensby Gletscher are floating, and its thickness in this section is only about 75 to 108 m, judging from the 15 to 18 m high ice-cliffs. The disintegration is thought to be caused by a bend in the fjord, which forces part of the glacier to ground while the rest continues to move (Fig. 3-5). The movement causes rotation and splitting up of the floating ice. Two other examples of floating ice tongues in north Greenland are the Petermanns Gletscher and Ryder Gletscher, both of which are flat glaciers with little crevassing.

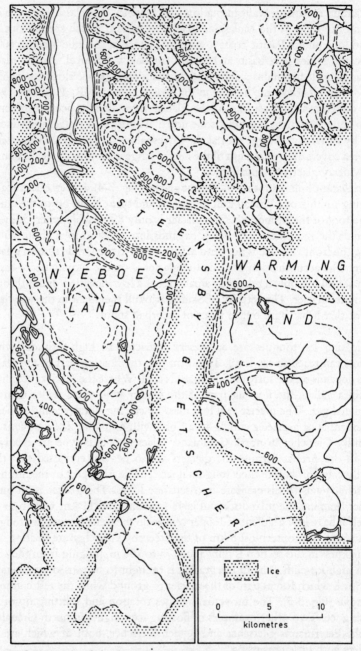

Fig. 3-5 Map of the Steensby Gletscher, north Greenland (F. Ahnert, *J. Glaciol.*, 1963, by permission of the Glaciological Society)

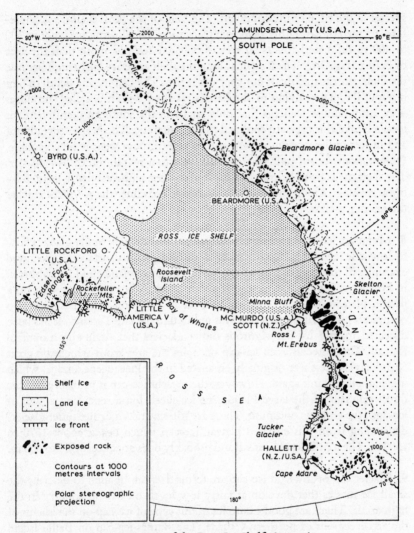

Fig. 3-6 Map of the Ross Ice Shelf, Antarctica

These narrow floating glacier tongues differ greatly from the floating ice-shelves in size. Ice-shelves are more characteristic of the Antarctic: a good example is the Ross shelf. The material of the shelf is partly derived from the outward flow of the inland ice-sheet, but it is mainly supplied by the accumulation of snow on the upper shelf surface. This form of nourishment is particularly important in the Antarctic where the accumulation is much greater at the edge of the continent. The bulk of the Ross shelf ice is, there-fore, formed of firn rather than true glacier ice. The floating ice is about 300 to 400 m thick and covers an area of 550,000 km², or more than the area of France (Fig. 3-6). The ice-cliffs along the shelf edge are fairly high because the shelf consists mainly of not very dense firn. The density at a depth of 6 m is about 0·5. The calving of the barrier, as the ice-shelf is called, produces very large tabular icebergs that only occur in the southern hemisphere, as ice-shelves of this type are confined to the Antarctic. The barrier spreads out and moves under its own weight and its rate of movement is very variable. Observations record rates of from 4 m/year to a maximum of 844 m/year. The shelf melts from below over most of its area, a process that is assisted by brine soaking. Along much of the Antarctic coastline, the ice-sheet reaches the sea in ice-cliffs.

8. Mountain ice-caps are accumulation areas from which outlet glaciers flow to lower levels. They usually rest on upland plateau surfaces. Two good examples of this type of ice-cap are Vatnajökull in Iceland and Jostedalsbreen in Norway. The former ice-cap is about 600 to 760 m thick, resting on a flattish plateau at about 1000 to 1220 m in elevation. Jostedalsbreen also occupies a high plateau area that was uplifted in the late Tertiary. It has been dissected mainly by the vigorous outlet glaciers that drain the mountain ice-cap. The Jostedalsbreen ice-cap occupies the area from which the great Scandinavian ice-sheet originally emanated in the Pleistocene, and to which it eventually shrank again. However, the present ice-cap is probably not the direct remnant of the large Pleistocene ice-sheet, but a renewed growth of ice since the climatic optimum. There are still several vigorous outlet glaciers descending from this elevated plateau ice-cap which lies at about 1370 to 1500 m. The ice thickness is small and probably does not exceed 300 to 600 m.

9. Glacier cap or lowland ice-cap are terms that can be used to describe the small ice masses that develop at fairly low levels on flattish country in the high arctic. There are good examples of this type of ice-cap on the islands of the arctic regions of northern Canada. The Barnes ice-cap on Baffin Island may be taken as an example. Its régime has already been discussed and its dimensions noted on p. 43. It lies on a relatively low plateau surface and has no vigorous outlet glaciers. The ice-cap is not active glaciologically, partly because of its arctic environment and partly because of the gentle relief around it. The Barnes ice-cap is probably frozen to its base at least at its margins, although there is reason to believe that in the central part its base is at pressure melting-point. There are large ice-dammed lakes on the north-eastern edge of the ice-cap.

10. There are only two of the largest type of ice mass at present, namely, the ice-sheets of Greenland and Antarctica. Some details of the dimensions of these ice-sheets have already been given in connection with their régimes. The overwhelming size of the Antarctic ice-sheet compared with any other ice mass is clearly apparent. Its presence alters the whole climate and life of the southern hemisphere. The other two major ice-sheets of the Pleistocene ice-age that rivalled it in size also exerted a similar widespread effect when they were at their maximum extent. At present, there is 85 per cent of the world's ice in the Antarctic and 11 per cent in Greenland (Chapter 1). The Laurentide ice-sheet at its largest covered much the same area as the present Antarctic ice-sheet, but the European one covered only about half this area during the maximum of glaciation.

Conclusions

The character of glacier ice movement depends upon the properties of the ice. The ice in turn depends upon the snow of which it is formed. The external factors that govern its change into ice and the forces exerted on the ice must also be considered. The physical properties of ice have been studied in the laboratory, and a power flow law relating strain rate to stress has been established.

Glaciers can be classified in various ways. The thermal classification is particularly significant in connection with the nature of ice flow. The differentiation between temperate ice, which is everywhere at pressure melting point, and cold or polar ice is very important. The lack of meltwater at the base of the cold type is important and results in considerably slower movement. An ice-sheet need not belong to one type over its whole area, and even large polar ice-sheets may have basal areas at pressure melting-point. Glaciers can also be classified, according to their dynamic character, as active, passive, and dead. Dead ice need not necessarily be immobile, but the term implies that the ice no longer has a source of supply. The ice mass can live on capital for up to thousands of years in large ice-sheets. Morphological characteristics provide a third criterion of classification, ranging from small niche glaciers to continental ice-sheets. Only two of the latter type now exist, but in the Pleistocene period the Laurentide and north European ice-sheets rivalled the largest existing ice-sheet in Antarctica.

REFERENCES

H. W. AHLMANN (1948), 'Glaciological research on the North Atlantic coasts', *R. geogr. Soc. Res. Ser.* **1**, 83 p.

F. AHNERT (1963), 'The terminal disintegration of Steensby Gletscher, North Greenland', *J. Glaciol.* **4**, 537–45

J. C. BEHRENDT (1965), 'Densification of snow on the ice sheet of Ellsworth Land and South Antarctic Peninsula', *J. Glaciol.* **5**, 451–60

C. S. BENSON and ANDERSON, quoted by M. F. MEIER (1962), 260

C. BULL (1957), 'Observations in North Greenland relating to theories of the properties of ice', *J. Glaciol.* **3**, 67–72

H. CAROL (1947), 'The formation of *Roches Moutonnées*', *J. Glaciol.* **1**, 57–59

A. COURT et al. (1957), 'The classification of glaciers', *J.Glaciol.* **3**, 2–7

J. W. GLEN (1955), 'The creep of polycrystalline ice', *Proc. R. Soc.* A, **228**, 519–38

J. W. GLEN (1958a), 'Mechanical properties of ice. I. The plastic properties of ice', *Phil. Mag.* Suppl. 7, 254–65

J. W. GLEN (1958b), 'The flow law of ice', *Un. géod. géophys. int., Symposium at Chamonix 1958, (Publication No. 47, Int. Ass. scient. Hydrol.)*, 171–83

G. E. GROOM (1959), 'Niche glaciers in Bünsow-land, Vestspitsbergen', *J. Glaciol*, **3**, 369–76

B. M. GUNN (1964), 'Flow rates and secondary structures of the Fox and Franz Joseph Glaciers, New Zealand', *J. Glaciol.* **5**, 173–90

R. HAEFELI (1952), 'Observations on the quasi-viscous behaviour of ice in a tunnel in the Z'Mutt Glacier', *J. Glaciol.* **2**, 94–99

J. K. LANDAUER (1959), 'Some preliminary observations on the plasticity of Greenland glaciers', *J. Glaciol.* **3**, 468–74

W. V. LEWIS (ed.) (1960), 'Norwegian cirque glaciers', *R. geogr. Soc. Res. Ser.* **4**, 104 p.

F. LOEWE (1966), 'The temperature of the Sukkertoppen ice cap', *J. Glaciol.* **6**, 179

C. M. MANNERFELT (1945), 'Några glacialmorfologiska formelement', *Geogr. Annlr* **27**, 1–239

J. G. MCCALL (1952), 'The internal structure of a cirque glacier: report on studies of englacial movements and temperatures', *J. Glaciol.* **2**, 122–30

J. G. MCCALL (1960), 'The flow characteristics of a cirque glacier and their effect on glacial structure and cirque formation' in *Norwegian cirque glaciers* (ed. W. V. LEWIS), *R. geogr. Soc. Res. Ser.* **4**, 39–62

M. F. MEIER (1962), 'Proposed definitions for glacier mass budget terms', *J. Glaciol.* **4**, 252–65

M. MELLOR (1959), 'Creep tests on Antarctic glacier ice', *Nature, Lond.* **184**, 717

M. M. MILLER (1952), 'The terms "névé" and "firn" ', *J. Glaciol.* **2**, 150–1

J. F. NYE (1952), 'The mechanics of glacier flow', *J. Glaciol.* **2**, 82–93

G. DE Q. ROBIN (1955), 'Ice movement and temperature distribution in glaciers and ice sheets', *J. Glaciol.* **2**, 523–32

V. SCHYTT (1964), 'Scientific results of the Swedish expedition to Nordaustlandet, Spitsbergen, 1957 and 1958', *Geogr. Annlr* **46**, 243–81

R. P. SHARP (1958), 'Malaspina Glacier, Alaska', *Bull. geol. Soc. Am.* **69**, 617–46

S. STEINEMANN (1954), 'Results of preliminary experiments on the plasticity of ice crystals', *J. Glaciol.* **2**, 404–12

A. W. STUART and C. BULL (1963), 'Glaciological observations on the Ross ice shelf near Scott Base, Antarctica', *J. Glaciol.* **4**, 399–414

R. P. SUGGATE (1950), 'Franz Josef and other glaciers of the Southern Alps, New Zealand', *J. Glaciol.* 1, 422-9

W. H. THEAKSTONE (1966), 'Deformed ice at the bottom of Østerdalsisen, Norway', *J. Glaciol.* 6, 19-21

J. WEERTMAN (1961), 'Mechanism for the formation of inner moraines found near the edge of cold ice caps and ice sheets', *J. Glaciol.* 3, 965-78

J. WEERTMAN (1964), 'The theory of glacier sliding', *J. Glaciol.* 5, 287-303

Chapter 4

Ice Motion

My theory of Glacier Motion then is this: a Glacier is an imperfect fluid, or a viscous body, which is urged down slopes of a certain inclination by the mutual pressure of its parts. (J. D. FORBES, 1843)

The movement of glaciers has been known to the inhabitants of Alpine regions since at least the sixteenth century, from the records of ice invading formerly uncovered ground, and from the changes in position of the most favourable crossings of the glaciers. Actual measurements of the rates of motion were undertaken in the eighteenth century and subsequently: G. J. Hugi recorded that his hut on the Unteraar glacier shifted 1·4 km between 1827 and 1840. An early theory of glacier motion, that of H. B. de Saussure (1760), held that ice moved downhill as a rigid body, sliding on bedrock, but by 1773, an element of plasticity or viscosity in glacier movement had been detected by A. C. Bordier.

Further observations and measurements of ice motion were undertaken by a number of pioneer workers in the mid-nineteenth century—L. Agassiz on the Unteraar glacier, J. D. Forbes on the Mer de Glace, and the Schlagintweit brothers on the Pasterze glacier. Agassiz (1840) established that the centre of a glacier flows faster than its sides, and that the rate of movement is slower in the source regions and in its terminal parts than in between. Forbes (1843) in addition found that glaciers often flow faster where their valleys narrow, that the velocity of flow is greater in warmer weather, and that since the central portions flow faster than the sides (Table 8), it seemed to him likely that the surface layers flow faster than the basal layers. He further argued that glacier ice is plastic under stress, and that whereas some glaciers may be

TABLE 8 *Displacement of Surface Markers on the Mer de Glace near Montanvert measured by J. D. Forbes in 1842*

| Marker | Distance (yards) | | Mean daily motion in summer (inches) |
	From left bank	From right bank	
D$_2$	100	(615)	16·3
D$_4$	230	(485)	20·8
D$_6$	305	(410)	21·2
D$_3$	365	350	23·1

PLATE III   THE TWIN ICE-FALLS AT THE HEAD OF AUSTERDALSBREEN, NORWAY.
The Austerdalsbreen tunnel, excavated in 1959, was located about one-third of the way up the left-hand ice-fall. (C.E.)

PLATE IV   SEDIMENTARY LAYERS (DIRT BANDS) EXPOSED IN AN ICE CLIFF, VESLGJUV-BREEN, NORWAY.
Melted-out debris is accumulating at its foot inside a winter snow bank. (C.E.)

PLATE V MORSÁRJÖKULL, SOUTHERN ICELAND.
Showing well developed ogives below ice-fall and avalanche fan from the ice-cap, and strongly marked median moraine. (C.A.M.K.)

PLATE VI MEDIAN MORAINE AND OGIVES ON AUSTERDALSBREEN, NORWAY.
(C.E.)

frozen to their beds, others may not be, allowing them greater freedom of movement. It is remarkable that these assertions, made over a century ago, have proved to be substantially correct.

The latter half of the nineteenth century witnessed the emergence of various theories of partial or complete plastic deformation of ice under stress (G. Seligman, 1949). F. Pfaff (1874) claimed that plasticity decreased with a lowering of ice temperature, and also, incidentally, was the first to notice, in the firn regions, that vertical as well as horizontal components of surface motion existed. A classic summary of the facts of glacier motion then known, and of the various theories of ice motion propounded, is contained in A. Heim's *Handbuch der Gletscherkunde* (1885), which gives a full exposition of the concept of glacier yielding by intergranular liquefaction and regelation. Theories of plastic flow by the turn of the century were numerous and varied, though none had yet been tested in the laboratory; they included theories of differential movements of ice crystals, deformation of the ice crystals themselves including gliding along crystallographic planes, and rearrangement of molecules within the ice crystals (H. Hess, 1904). On the other hand, many were still convinced that glacier yielding was more a matter of differential flow along internal shear planes, possibly numerous and closely spaced. Such was the opinion, particularly, of H. Philipp (1920). Forbes had long ago demonstrated the existence of such shear planes, and R. T. Chamberlin (1928) successfully measured differential motion along them in the terminal parts of the Brenva glacier.

Studies of glacier flow as a mathematical and hydrodynamic problem began to emerge in the 1920s. Outstanding for the clearness of its exposition was the work of C. Somigliana; it was based, however, on the erroneous assumption that ice is a substance of constant viscosity, as was the work of M. Lagally in the 1930s. Lagally arrived at a figure of 10^{14} poises for the approximate viscosity of glacier ice, and deduced a relation between viscosity (μ), ice thickness (h), ice density (ρ), bedrock slope (α) and velocity (v) of a point on the glacier centre line:

$$v = \frac{h^2 \rho g \sin \alpha}{2\mu} \quad \text{(where } g = \text{gravitational acceleration)}$$

The relationship was checked by observations of v and seismic determinations of h on the Pasterze glacier, and it was concluded that ice behaved as a viscous liquid, a conclusion that has since been firmly refuted.

Methods of measuring glacier flow steadily improved in the twentieth century with the development of modern surveying equipment. In 1928, R. Finsterwalder (1931) was the first to adapt photogrammetric techniques to the precise measurement of ice motion. Equally important to theories of glacier flow is a knowledge of ice thicknesses and bedrock configuration; in this field too, developments in seismic (and later, gravimetric) techniques made possible much more reliable determinations of the form of subglacial rock surfaces. The year 1937 saw the beginning of some very important glaciological investigations of the Jungfraujoch area in Switzerland. M. F. Perutz and others found evidence both of plastic deformation of moving ice, and of laminar flow involving some thrusting along discrete shear planes. Plastic deformation was associated with evidence of strong preferred orientation of the ice crystals, whose basal planes were found to be aligned parallel to the direction of ice flow (Perutz, 1940).

With the Jungfraujoch investigations, we enter the modern period of research into the deformation of ice and the mechanism of glacier flow. Since 1937, field research and laboratory experimentation have led to great advances in our knowledge of the behaviour of ice under stress, and the following sections of this chapter will present the most important results of this work as it bears on the problem of ice motion. There are still, however, considerable gaps in our understanding of this problem, for field measurement has lagged substantially behind the development of theoretical concepts. Moreover, most laboratory and field research on glacier flow has been carried out in connection with temperate glaciers, which yield more readily to stress and flow more quickly than cold glaciers. We need to know much more about the internal movements and temperatures of cold ice-sheets and glaciers. We also need a great deal more information about the mechanism by which ice slides over an uneven surface of bedrock, for this is now known to be the most important contributor, in the case of temperate glaciers, to total glacier movement and ice discharge. The mechanism of basal sliding is much less well understood than the internal deformation of a moving glacier.

Surface Velocity Observations

Maximum rates of ice motion vary enormously from one glacier to another, and over periods of time. The majority of valley glaciers move at speeds of less than one metre a day. For brief periods, however, some glaciers experience forward surges (Chapter 2) in which rates of flow become abnormally high: during part of 1937, for instance, the Black Rapids glacier in Alaska probably attained a speed of 75 m/day, while in the Alps, the Vernagtferner (whose normal maximum velocity is about 5 cm/day) in 1845 reached a speed of 11 m/day. Over longer periods, the most rapidly moving glaciers in the world are those descending steep slopes and fed by large accumulation areas: of the Greenland outflow glaciers, the Jakobshavn Isbræ moves at up to 20 m/day (A. Helland gave such an estimate for this glacier's movement

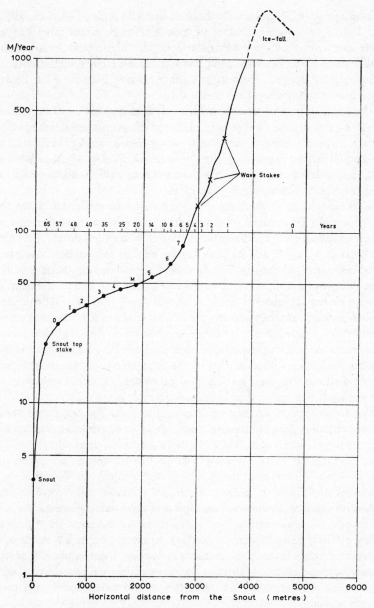

Fig. 4-1 Variation in the rate of flow of Austerdalsbreen, Norway, from the head of the glacier (ice-fall) to its snout (C. A. M. King and W. V. Lewis, *J. Glaciol.*, 1961, by permission of the Glaciological Society)

as long ago as 1877), and the Rinks Isbræ up to 28 m/day (W. Kick, 1957). The fact that the lower portions of these glaciers float and calve into sea inlets also helps them to attain these unusually high rates of movement. Unlike these glaciers of Greenland, most of the Antarctic outlet glaciers are comparatively sluggish. Even in its swiftest flowing part, the giant Beardmore glacier barely moves at a metre a day.

The methods used to measure surface velocities consist of trigonometrical survey by theodolite of stakes bored into the ice to sufficient depth to offset surface ablation, of photogrammetric survey (Finsterwalder, 1931), and of mechanical devices connecting ice to bedrock (R. W. Galloway, 1956). The first, though laborious, is by far the most accurate, and can detect vertical as well as horizontal components of motion.

A recent analysis of the flow of the Saskatchewan glacier (M. F. Meier, 1960) provides an example of the results obtainable by theodolite survey. The probable errors in the determinations of flow were less than 0·06 cm/day for horizontal motion and 0·11 cm/day for vertical movements. Maximum velocities, up to 32 cm/day, occurred at the firn-line, diminishing to less than 1 cm/day near the snout. This reflects the fact that, in a glacier flowing down a channel of uniform cross-section and slope, velocity will be greatest at the firn-line where the volume of ice being discharged reaches a maximum; down-valley, the volume of ice decreases by ablation. Actual glaciers do not flow in channels of uniform cross-section or slope, but nevertheless a longitudinal diminution of velocity below the firn-line is often apparent. Figure 4–1 shows the variation in rate of flow along Austerdalsbreen, Norway, from the point where it leaves the plateau ice-cap to its snout.

Transverse surface velocity profiles for the Saskatchewan glacier show greatest speeds in the central portion and a four- or five-fold decrease within 50 m of the glacier margin. The form of the profiles is parabolic. Figure 4–2 plots the ratio of velocity down-glacier (V_x) to centreline velocity down-glacier (V_c) as a function of distance from centre to margin. The distance S represents side-slip of the glacier; the results represent the average of four transverse velocity profiles, and are typical of most valley glaciers.

Precise surveys also often reveal actual transverse components of flow in valley glaciers. Flow of part of the Saskatchewan glacier, for instance, is directed obliquely towards the southern valley side, a movement that represents ice taking the place of that lost by increased ablation near the rock side. Transverse components of flow also often appear where valleys and their glaciers widen.

Vertical components of flow (those resulting from ice motion, not accumulation or ablation) are more difficult to measure, but must not be neglected. The general tendency is for the ice to have a downward component of

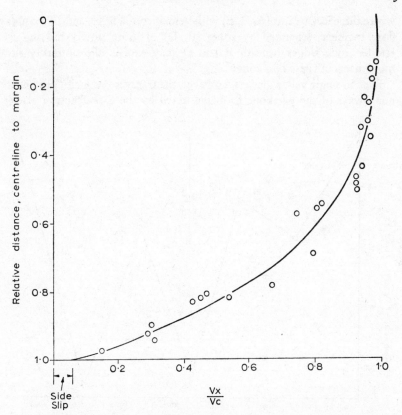

Fig. 4-2 Ratio of down-glacier velocity (V*x*) to centreline down-glacier
velocity (V*c*) as a function of relative distance from the centreline.
Data obtained from four transverse profiles, Saskatchewan glacier
(M. F. Meier, *U. S. geol. Surv. Prof. Pap.* 351, 1960

motion in the firn region, and an upward component near the snout. On
the Saskatchewan glacier, velocity vectors plunge beneath the surface from
above the firn-line to 2 km below it; farther down-glacier, the vectors
gradually begin to rise out of the surface (Fig. 4-7). Near the terminus,
vertical com ponents of motion reach values of 8 mm/day, nearly as much as
the horizontal component.

The Saskatchewan glacier represents a single ice-stream; some large valley
glaciers, however, consist of several ice-streams flowing side-by-side, so that
transverse surface velocity profiles may be more complicated. Figure 4-3
shows the Fröya glacier in north-east Greenland (W. R. B. Battle, 1951),

where three ice-streams (A, C, E) with velocities of 4·5, 10, and 10·5 cm/day flow together, separated by zones (B, D) of more slowly moving ice. Hidden rock ridges beneath B and D may explain the unusually slow movements in these two zones.

Even in single valley glaciers, however, the transverse velocity profiles are not always of the parabolic form illustrated by the Saskatchewan glacier.

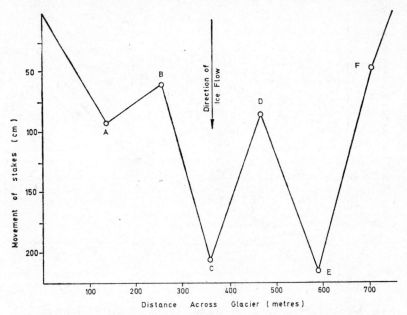

Fig. 4-3 Transverse velocity distribution of the Fröya Gletscher, Greenland, as shown by the movement of six stakes (A to F) over a period of three weeks, August 1949 (W. R. B. Battle, *J. Glaciol.*, 1955, by permission of the Glaciological Society)

Some glaciers exhibit a form of flow termed plug flow or Blockschollen movement (R. Finsterwalder, 1950), in which there is a very rapid increase in velocity close to the valley sides and little differential surface movement in the main central mass of ice. Finsterwalder claimed that Blockschollen flow is mainly associated with fast-moving glaciers in which the ratio of mean velocity to width is greater than about 0·16. Blockschollen movement is typical of many surging glaciers, and often results, if the bed is irregular, in the ice surface breaking up into a series of seracs (as in an ice-fall). The high velocities attained near to the valley-sides in Blockschollen flow have important consequences for valley-side erosion (see p. 111). Blockschollen flow is an extreme form of transverse velocity distribution; there are many glaciers that exhibit a U-shaped transverse velocity profile, in between the plug form of flow (Blockschollen) and the parabolic transverse velocity profile.

TIME-VARIATION OF VELOCITY

Many workers since the time of Forbes have noticed that glaciers do not flow at absolutely constant rates. In the ablation zone, there is a tendency for the ice to move faster in summer than in winter, while in the névé region, the reverse is often the case. Short-period fluctuations also occur, but are much more difficult to measure.

Seasonal variations   More rapid summer movement of ice in the ablation zone is definitely established, for instance, on the Hintereisferner (R. Haefeli, 1948), on the Saskatchewan glacier (Meier, 1960), on the Athabasca glacier (W. S. B. Paterson, 1964), and on the Gorner Gletscher (G. R. Elliston, 1963). Elliston found that velocities were 20-80 per cent greater than the annual average in summer, and 20-50 per cent less in winter. Paterson showed that mean velocities on the Athabasca glacier were 15 per cent greater for the period April-July than for the year. These variations are not confined to temperate glaciers—Paterson (1961) notes that the Sefstrøm Gletscher in north Greenland flowed faster at the beginning of August than at the end of that month when surface melting had ceased—but data from cold ice-sheets and glaciers are still inadequate. Recent work in Antarctica (C. Swithinbank, 1966) has not yet revealed seasonal variations in flow greater than 1 per cent.
   There are several possible causes of more rapid summer flow of temperate and sub-polar glaciers. Higher air temperatures are prevalent in summer, yet of course the ice itself in a temperate glacier cannot fluctuate in temperature (other than in the surface layer affected by winter cold) as it is already at pressure melt-point. But summer is the season of surface melting, and melt-water is thought to play an important part in lubricating the bed of a glacier and thereby enhancing the sliding process. The problem is to what extent surface meltwater may penetrate to the bed of a thick glacier, and whether it could reach the bed of a sub-polar glacier. As truly cold glaciers are frozen to bedrock, the lack of seasonal variation in the flow of Antarctic outlet glaciers is consistent with the meltwater lubrication hypothesis. In the case of a sub-polar glacier, however, the situation is less clear. For the Bersæker-bræ in east Greenland (T. W. Friese-Greene and G. J. Pert, 1965), there is a possible correlation between velocity of ice flow and ablation rates, so that some surface meltwater might be finding its way to the glacier bed in summer at least, possibly down the sides of the glacier.

The faster movements sometimes noted in winter time for the firn regions of temperate glaciers are less clearly understood. In 1942, determination of the velocity of a point on the Jungfraufirn at 3350 m revealed a winter velocity twice as great as that of the summer (Haefeli, 1948). One possible explanation is that the greater thickness of the firn-snow blanket increases the shear stress and therefore raises the flow rate. E. Orowan (1949) suggested that even a 1 per cent weight increase might raise the flow rate substantially, in spite of the lower temperatures in the upper layers. However, surface velocity measurements in firn are reliable only if the marker stakes are drilled to depths greater than the new snow layer (and preferably into ice), for new snow can exhibit considerable movement itself arising from creep and compaction (J. G. McCall, 1952).

Short-interval variations   Variations in rate of flow over periods of hours, days or weeks have been recorded, though extreme precautions must be taken to minimize instrumental error as the movements in question are often minute. J. E. Jackson (1953) has drawn attention to the striking effects of atmospheric refraction changes on theodolite observations, and the theodolite must be protected from the sun. The most precise series of observations concerned with short-interval variations is that given by Meier (1960) for the Saskatchewan glacier. These results were considered to be accurate to within 15 mm/day, and revealed sudden jerky movements of the ice, involving sudden increases in velocity of up to 170 per cent, and sudden decreases up to 230 per cent (giving actual reversals of motion). The fluctuations were irregular and not synchronous from one marker to another. J. W. Glen and W. V. Lewis (1961) also found rapid and irregular changes in the rate of side-slip for Austerdalsbreen (Norway). They suggested that the uneven movement recorded might result from sidewall irregularities or boulders jammed in the ice-rock gap, causing stress to build up until the ice suddenly freed itself with a jerk. But an equally important factor seemed to be weather conditions, for rapid movement often followed on heavy rain and warm air temperatures (one marker moved 5 cm in 8 hours, seven times its normal speed, following a period of rain), suggesting again that meltwater lubrication may be very important in the slip of ice over rock. On Morsárjökull in Iceland, J. D. Ives and C. A. M. King (1955) also noted a possible correlation between rapid ice movement and periods of heavy rainfall (Fig. 4-4). Meier (1960) for the Saskatchewan glacier suggested a relation between velocity and meteorological conditions, but noted that overshadowing such a correlation were other causes of differential motion, namely, localized differential shearing in the ice, and irregular movements of inter-crevasse blocks. Paterson (1964), discussing weekly variations of flow of the

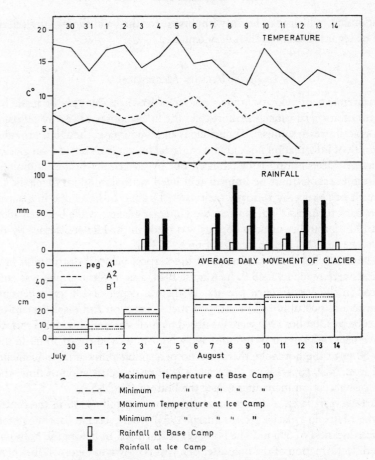

Fig. **4-4** The relationship between precipitation and ice velocity on Morsárjökull, Iceland (J. D. Ives and C. A. M. King, *J. Glaciol.*, 1955, by permission of the Glaciological Society)

Athabasca glacier, found that increased meltwater discharged from the glacier snout tended to occur 3 or 4 days after an increase in velocity had been measured about 2·5 km up-glacier. This suggested that, if an increase in basal meltwater lubrication was responsible for the acceleration of ice flow, the thicker basal water film travelled down-glacier at about 0·8 km/day. J. Weertman (1962) arrived at a similar figure on theoretical grounds, postulating that the thicker zone of basal meltwater is propagated down-valley under the ice as a kinematic wave. The agreement between theory and observation is striking, but many more observations are needed before the relationship, which may be very important for the basal sliding of glaciers, can be regarded as more than very tentative.

Englacial Velocity Measurements

Measurements of velocity inside a glacier or ice-sheet can only be made by means of artificial tunnels or boreholes (or in exceptional cases by utilizing sub-glacial stream tunnels). Tunnelling is an arduous task, though it provides a wealth of information not only on englacial movements but also on glacier structure. Deep boring is best carried out by electrically-heated hotpoints, and the borehole must be immediately lined with aluminium or plastic to prevent its closure by refrozen meltwater. Drilling of boreholes in glaciers dates back to Agassiz' attempts on the Unteraar glacier; while between 1889 and 1909, a number of deep borings was made in the Hintereisferner by A. Blümcke and H. Hess, to depths of up to 224 m.

The first reliable measurements of the vertical velocity distribution in a glacier were made in 1948-9. In August 1948, a near-vertical tube was emplaced in the Jungfraufirn (Perutz, 1950) to a depth of 136 m, where an obstruction, probably solid rock, was met. Distortion and displacement of the tube by October 1949 were measured by inclinometer, showing a surface movement of 38 m (in 14 months) and a bottom movement of 10 m. In the top 50 m of the borehole, there was no perceptible differential movement, and from − 50 to − 136 m there was a continuous retardation of flow, the rate of retardation increasing nearer the bottom (Fig. 4-5a).

A few years later, a deeper borehole was successfully sunk in the centre of the Malaspina glacier (R. P. Sharp, 1953) where seismic data suggested an ice thickness of 595 m. The borehole reached 305 m before the hotpoint failed. Deformation of the hole after one year is shown in Figure 4-5b where net displacement from a vertical straight line is plotted. There was no perceptible differential flow to a depth of 90 m, and only slight variations in velocity down to 285 m, but a marked retardation in flow below that level.

The most complete record of englacial velocity distribution has been provided by boreholes in Salmon glacier, British Columbia (W. H. Mathews, 1959). About one-third of the distance across the glacier surface, a pipe was sunk to − 481·5 m, ending within probably 8·5 m of the rock floor. The velocity at the surface was 22·5 cm/day, 20·7 cm/day at − 240 m, and 12·4 cm/day at the bottom; the vertical velocity profile (Fig. 4-5c) consisted of a smooth curve corresponding closely to J. F. Nye's (1952) theoretical curve for laminar flow of ice resting on a smoothly inclined plane (p.96).

The team investigating the Saskatchewan glacier (Meier, 1960) made many unsuccessful attempts to sink deep boreholes; in the end, one pipe was in-

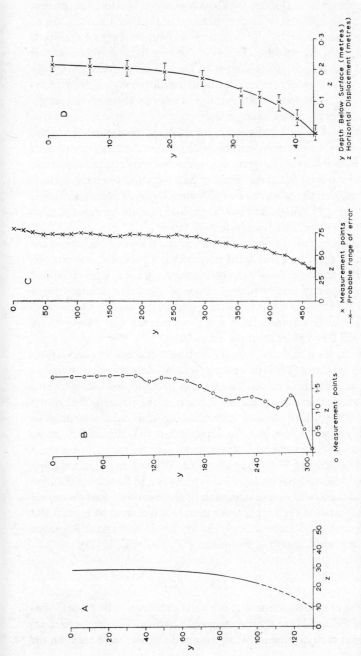

Fig. 4-5 The deformation of vertical boreholes in glaciers:
A Jungfraufirn (M. F. Perutz, 1950): August 1948—October 1949;
B Malaspina glacier (R. P. Sharp, 1953): July 1951—August 1952;
C Salmon glacier (W. H. Mathews, 1959): displacements computed for a one-year period;
D Saskatchewan glacier (M. F. Meier, 1960): August 1952—August 1954.
Note that the vertical and horizontal scales differ in each case. Direction of ice flow left to right. Many of the minor irregularities in the curves for B and C may have no significance, being smaller than the possible errors of measurement

serted to a depth of 50 m (glacier thickness here about 500 m). The interest of this experiment is that very precise measurements of the displacement of the pipe over 2 years revealed differential movement throughout the depth of 50 m and even in the top 10 m (Fig. 4-5d); this conflicts with the previous suggestions of many glaciologists (such as M. Demorest, 1938) that the uppermost 30 m or so of a glacier remain in a rigid 'brittle' condition. The deformation of this pipe showed no sudden change of velocity at any special depth; the velocity simply decreased exponentially.

The data from all these boreholes bear out Forbes' contention that, in a vertical profile, velocities decrease from glacier surface to bedrock. It seems unlikely that further borehole experiments will substantially alter this picture, save possibly in special locations. Further boreholes are urgently needed, however, to test theories of rates of ice deformation in different conditions, and especially in cold glaciers and ice-sheets, for it should be noted that all the boreholes described have been in temperate ice.

Information on englacial velocities from tunnels is limited because of the fact that tunnels must be constructed horizontally or sloping outwards to allow meltwater drainage; consequently, tunnels have been largely confined to ice-falls or relatively steeply sloping cirque glaciers. The velocity distribution given by the Vesl-Skautbreen cirque glacier tunnel is described in Chapter 8. The data obtained from some other tunnels in Norway and Switzerland will be considered on pp. 94 and 102.

A final comment on englacial velocity distributions concerns observations carried out by H. Carol (1947) in a natural subglacial cavity of the Grindelwald glacier at a depth of 50 m. The ice was being forced locally over a bedrock obstruction, causing a sudden increase in the stress applied to the lower layers of ice which resulted in their being squeezed forwards at an abnormally high rate of flow (see also Chapter 3, p. 61). Within 1 m of the bedrock, speeds of flow reached 72 cm/day, whereas higher up in the ice, a more normal velocity of 37 cm/day prevailed. This is one of the very few recorded cases where a basal ice layer has been shown to flow faster than the ice above, and is relevant to the discussion of the extrusion flow theory on p. 98. It is important to note that this example was related to a particular bedrock obstruction and does not imply that the basal ice was travelling more rapidly over the whole width of the glacier (W. V. Lewis, 1947a).

The Internal Flow of Glaciers

The movement of glaciers is made up of two components, internal deformation and basal sliding. These will be considered in turn. The earliest theories of internal flow, such as those of Somigliana and Lagally, were based on the premise that ice behaves as a Newtonian viscous material, in which there is a linear relationship between shear stress and strain rate. This was found to be incorrect, and glaciologists then turned to a plastic flow law (e.g. J. F. Nye, 1951). This assumed that, up to a certain shear stress level (τ_0), no deformation occurred, but at greater shear stresses, an infinite amount of deformation resulted. In other words, the ice was assumed not to be capable of supporting a shear stress $>\tau_0$. The theory of perfect plasticity showed better agreement with observational data than the viscous-type flow law, and it was used, for instance, to calculate the longitudinal profile of the Unteraar glacier (Nye, 1952c) which showed remarkable agreement with its actual measured profile.

The theory of perfect plasticity was a convenient approximation, but only an approximation, and in order to understand glacier flow more exactly, it was obviously necessary to determine how ice deforms under various shear stresses. This can best be done by controlled laboratory experiments. The first such experiments of value were those undertaken by Perutz (1948) in connection with the wartime 'Bergship' project. Perutz showed that ice responds to stress by creep, fairly rapidly at pressure melt-point, and that creep occurred down to temperatures as low as $-20°C$. Further controlled laboratory experiments were carried out by J. W. Glen (1952, 1955) who found that when a load is applied, ice deforms almost instantaneously to a certain strain and then deforms more slowly as time proceeds. This slow deformation is termed 'creep', and it is this that allows glacier ice to flow continuously under stress. The actual mechanisms of creep, mainly intra-granular slip and recrystallization, have been described in Chapter 3. Glen showed that the relation between shear stress (τ) and strain rate $(\dot{\varepsilon})$ is of the form

$$\dot{\varepsilon} = k\tau^n$$

where $n > 1$ and k is a function of temperature. This power law gives a much closer approximation to the way in which glacier ice deforms than the theory of perfect plasticity.

It is useful to recall some of the salient points made in Chapter 3 with regard to the power flow law. A range of values for n between about 2 and 4 has been determined experimentally, depending on shear stress and temperature. Values of n are below 2 at very low shear stresses, and become greater with a lowering of temperature. Cold ice can support greater shear stresses than ice at pressure melt-point and therefore deforms less readily. This is one reason why cold glaciers flow more slowly, and it is the reason why crevasses in cold glaciers tend to be slightly deeper than in the case of temperate glaciers. It was also pointed out in Chapter 3 that, at low shear stresses, experiments have to be continued over long periods before a steady rate of creep is achieved. Laboratory experiments have been continued over periods of months and with shear stresses as low as 0·25 bars, but in actual glaciers, ice is subjected to shear stresses over hundreds of years and the stresses may approach zero in the central surface zones. Such conditions obviously cannot be duplicated in the laboratory.

Glen's power flow law has been tested in studies of ice deformation in actual glaciers, particularly by studying the rates of closure of tunnels. The closure rate depends on the flow law provided that there are no other stresses within the ice. The tunnels on the Z'Mutt glacier (Switzerland) and Vesl-Skautbreen (Norway) gave values of $n = 3$ in close agreement with the laboratory tests. However, two tunnels cut in more actively moving ice, in the Arolla glacier (Switzerland) and Austerdalsbreen (Norway), closed at a greater rate than that given by the flow law because the tunnels were near the bases of ice-falls where the ice was subject to large longitudinal stresses. Meier (1960) has used the data of tunnel closure, the rates of borehole deformation, and various laboratory experiments of Glen and others, to revise Glen's flow law, which he contends does not apply at very low shear stresses. Figure 4-6 plots $\log \dot{\varepsilon}$ as a function of $\log \tau$; the curve which best fits all the data just mentioned is almost a straight line at stresses greater than about 0·7 bars, but curves more strongly at lower stresses. The revised flow law given by Meier is

$$\dot{\varepsilon} = k_1\tau + k_2\tau^n$$

and the best fit is obtained when $n = 4·5$, $k_1 = 0·018$, and $k_2 = 0·13$. The data are for temperate glaciers only.

The range of shear stresses encountered in actual glaciers is between zero and about 1·5 bars; the highest values are located near the bed and sides where retardation of flow is greatest, as already established. It is possible to calculate theoretically the approximate values of maximum shear stress at the glacier bed (τ_b) if the shape and slope of the glacier bed are known. Nye (1952b) took the simple case of laminar flow down a channel of semi-circular section; then

$$\tau_b = \rho g \frac{A}{P} \sin \alpha$$

where A is the cross-sectional area taken perpendicular to the bed, P is the bed perimeter of this cross-section, α is the ice surface slope, ρ is the density of glacier ice (about 0·9) and g is gravitational acceleration. Using estimates of A and P, and applying this simple formula to 16 Alpine glaciers, Nye found values of τ_b ranging from 0·49 bars for the Unteraar glacier at 2100 m in 1945-7, to 1·51 bars for the Grenz glacier at 2700 m in 1948-9. W. H. Ward (1955) found values of 0·42 to 0·92 bars for the Highway glacier

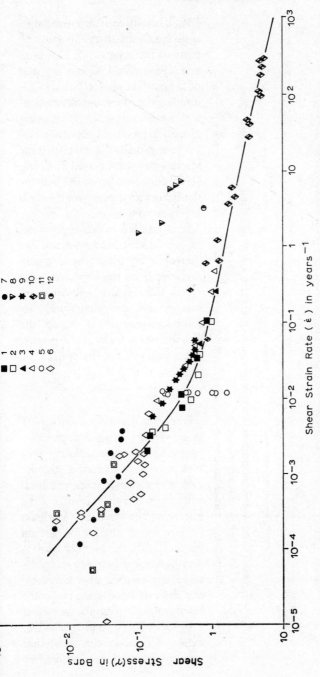

Fig. 4-6 Strain rate as a function of shear stress. Data from various glaciers and laboratory experiments on ice (M. F. Meier, 1960, *op. cit.*)

1. Surface transverse profile, Saskatchewan glacier; laminar flow analysis
2. Surface transverse profile, Saskatchewan glacier; complete analysis by Nye's theory
3. Z'mutt tunnel (J. F. Nye, 1953)
4. Vesl-Skautbreen tunnel (J. F. Nye, 1953)
5. Saskatchewan glacier borehole; complete analysis by Nye's theory
6. Malaspina glacier borehole; laminar flow analysis (R. P. Sharp, 1951-2)
7. Saskatchewan glacier borehole; laminar flow analysis
8. Austerdalsbreen tunnel (J. W. Glen, 1956)
9. Jungfraufirn borehole; laminar flow analysis
10. Laboratory experiments (J. W. Glen, 1955)
11. Malaspina glacier borehole; laminar flow analysis (R. P. Sharp, 1951-4)
12. Arolla tunnel (J. F. Nye, 1953)

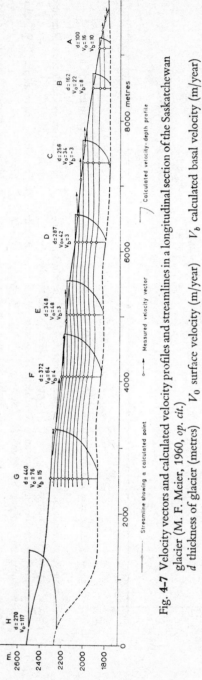

Fig. 4-7 Velocity vectors and calculated velocity profiles and streamlines in a longitudinal section of the Saskatchewan glacier (M. F. Meier, 1960, *op. cit.*)

d thickness of glacier (metres)

V_0 surface velocity (m/year)

V_b calculated basal velocity (m/year)

—— Calculated velocity-depth profile

↑ Measured velocity vector

○ Streamline showing a calculated point

(Baffin Island), with the lower values applying to the sluggishly-moving parts of the tongue. W. H. Mathews (1959) gave 1·42 bars for part of the Salmon glacier, British Columbia. From these and other assessments, it appears that τ_b for moving glaciers is usually in the range 0·5-1·5 bars, in order that motion of the glacier may take place. S. Orvig (1953) found values of 0·4 bars for the Barnes ice-cap, where the ice is barely moving.

It is possible to make estimates of the velocities of movement expected at any given depth within a glacier from the flow law, provided that surface velocities are known and that simplifying assumptions are made about the cross-sectional form of the glacier. For a semi-circular channel,

$$v_0 - v = \frac{x}{n+1} \sin^n \alpha \cdot d^{n+1}$$

where v^0 is the surface velocity, v is the velocity at depth d, and $x = \rho g.k^n$. Thus vertical velocity profiles may be constructed (though clearly one cannot work out basal velocities unless the thickness of the glacier is also known). Mathews (1959) showed that, for the Salmon glacier (p. 90), there is very satisfactory agreement between the measured deformation of a bore-hole and the calculated velocity profile based on the formula above, if $x = 1·25 \times 10^{-4}$ and $n = 2·8$. Figure 4-7 shows the theoretical velocity distributions computed by

Meier (1960) for the Saskatchewan glacier; knowing the approximate thicknesses of the glacier from seismic data, it is also possible to estimate the basal sliding component, which is found to vary between zero and 15 m/year. These values may well be too small, and the fact that in one case the value was negative suggested that the seismic data might be giving too great thicknesses for the ice. The diagram also shows calculated streamlines,

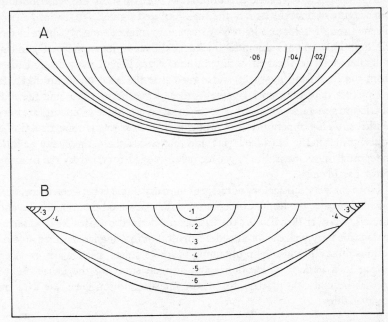

Fig. 4-8 Theoretical velocity distribution (A) and shear stress distribution (B) for a parabolic channel whose width is four times its depth. Velocity and shear stress given in dimensional form (J. F. Nye, *J. Glaciol.*, 1960, by permission of the Glaciological Society)

representing the theoretical paths taken by individual ice particles. These were found to parallel the bedrock profile surprisingly closely.

Theoretical internal velocity distributions for glaciers whose cross-sectional form is not assumed to be semi-circular are more difficult to compute. Nye (1965) has investigated glacier flow in channels of rectangular, elliptic, and parabolic cross-section, assuming that there is no basal slip, and that the ice is isothermal. With these limitations in mind, Figure 4-8a shows the theoretical velocity distribution for a parabolic channel whose width is four times its depth, and Figure 4-8b the calculated shear stress distribution (in dimensionless form). Glen's simple power flow law is adopted, with $n = 3$. An interesting result is that the zone of maximum shear stress at the glacier surface occurs not at the very edge of the glacier but a little way in. If the bedrock sides are steeply sloping, however, it occurs very close to the edge. This theoretically-derived result has been confirmed by observations of crevasse patterns, where the zone of maximum marginal crevassing has often been seen to lie a small distance in from the actual ice edge.

In temperate glaciers, the effect of temperature on shear stress distributions can be ignored since the ice is isothermal. As noted above, Nye assumed an isothermal state for his model used to calculate internal velocity. In cold ice-sheets and glaciers of substantial thickness, the ice is not isothermal, and this will affect internal shear stress distribution. Nye (1959b) analysed some data from the Greenland ice-sheet, and argued that the basal layers are likely to be warmer than the upper layers, because of geothermal heat and the heat of friction generated by shearing in the basal layers. Since strain rates are very sensitive to ice temperature, it is probable that most relative shear motion is concentrated in the basal ice layers, and that, above these, there may be little differential movement, the upper ice being carried forward by the shearing in the basal layers.

From the very satisfactory agreement in many respects between theoretical calculations and the limited observational data obtained from boreholes and tunnels, a power flow law of the basic form derived by Glen in 1952 seems to describe very well the internal deformation of moving ice. Further investigations must concentrate on obtaining more field data, particularly by deep boring in a variety of locations; and also on studying the behaviour of non-isothermal cold glaciers, for whose internal motion very few data are yet available.

OTHER THEORIES OF INTERNAL FLOW

The theory of extrusion flow   Early workers such as Forbes demonstrated that the centre of a glacier flows more quickly than its marginal portions, and deduced therefrom that the surface also moves more quickly than the basal layers. Suggestions that, in certain conditions, the deeper layers of ice might be capable of movement more rapid than the upper layers have often been made, but the idea was not taken up seriously until the years 1937-8, when, independently, two glaciologists developed a more complete theory of extrusion flow and produced apparent evidence in support of it. M. Demorest (1937) suggested that, because of the plasticity of ice, ice was forced out from below the centres of accumulation of west Greenland; at the same time, higher ice moved down to take its place. In 1938, he produced evidence from striated surfaces that basal ice movements were controlled by pressure (see Chapter 6). His theory of extrusion flow was fully expounded in 1942-3; extrusion flow, contrasted with 'gravity flow', resulted from differences in ice pressure, and maximum rates of movement were postulated in the deeper layers, which were said to be more plastic under pressure.

R. Streiff-Becker (1938) put forward evidence from the Claridenfirn. Here the average accumulation over the firn basin was 3·45 million m³/year. Since the surface of the firn was not rising, this amount must be discharged from the basin, and according to Streiff-Becker it passed through a cross-section estimated at 68,000 m². Measurements of surface velocity at this cross-section (average 14 m/year) would allow for the discharge of only about 1 million m³/year; in order to discharge three and a half times this amount, either more ice must be discharged at greater velocity at depth, or the estimate of maximum depth for the cross-section would have to be increased from 110 m to about 300 m. Streiff-Becker chose the former explanation as more likely, and claimed that extrusion flow was taking place.

There is serious doubt that Streiff-Becker's estimates of the cross-section are reliable, and it has also been suggested (W. V. Lewis, in discussion following G. Seligman, 1947) that perhaps not all the ice from the Claridenfirn was escaping solely through the critical cross-section. Extrusion flow has never been encountered in any deep boreholes (a minor instance noted by Carol has already been mentioned) and the concept appears to be physically unsound. W. V. Lewis (1948) asked the very pertinent question 'why the upper layers are not carried forward on the back of the supposedly swifter moving lower layers'. Nye (1952a) pointed out that the upper layers could only be held back by either a longitudinal tensile stress or transverse shear stresses exerted by the drag of the valley sides in the case of a valley glacier. In both cases, the stresses would be many times greater than the ice could support, and the process would also be opposed by the shear stress between ice and bedrock. Geomorphologists such as C. D. Holmes (1937) and S. E. Hollingworth (1931, 1947) have found the theory attractive, in that in certain conditions it was postulated by Streiff-Becker particularly that fast-moving currents of ice would be carried down close to the bedrock floor and theoretically give rise to enhanced rates of erosion at these places. The fact remains that theory is against it, and it has never been shown to occur in practice.

Nye's theory of extending and compressing flow   In considering the internal deformation of glaciers, the basic case considered was one of ice flowing down channels of constant gradient with uniform rate of discharge. In actual glaciers, longitudinal extension and compression frequently occur—the ice tends to thicken or 'compress' in sections where the bedslope is reduced, for instance. The term 'compression' is not meant to indicate actual reduction of volume of the ice but a local thickening of the glacier; conversely with 'extension'. Not only will changes in longitudinal gradient cause compression or extension, but also changes in rates of accumulation or ablation. Nye (1952b) has made a theoretical investigation of the effects of these changes using the simplified flow law of perfect plasticity. If R is the radius of bed curvature, α the surface slope of the ice, and ϕ the rate of ice discharge, then extending flow will occur when the expression $\dfrac{\delta\phi}{\delta x} + \dfrac{\phi}{R} \cot \alpha$ is positive, and compressive flow when it is negative, $\dfrac{\delta\phi}{\delta x}$ being the rate of addition of ice to the upper surface of the glacier over a distance x. If the bed is of even gradient, R would have infinite value, but the left-hand term would still give rise to compressing flow if negative, and extending flow if positive.

Theoretically, under conditions of compressing or extending flow, the slip-lines shown in Figure 7-3 would represent the planes along which the ice would have greatest tendency to shear. The second situation (compressive flow) helps to explain the occurrence of the thrust-planes often seen in glacier snouts. The potential slip surfaces are likely to be brought into use when they happen to coincide with any other zones of weakness in the ice, or when bedrock irregularities or morainic obstacles prompt their formation (see Chapter 7). B. M. Gunn (1964) gives examples of the development of up-sloping thrust-planes under compression. An excellent example of extending flow is given by the upper part of the Austerdalsbreen ice-fall (Plate III) down which the ice is accelerating to speeds of 2000 m a year. Where the ice leaves Jostedalsbreen to plunge down the ice-fall, step-faulting characteristic of extending flow is seen. As it approaches the base of the ice-fall, compression commences and increases in magnitude: the flow has decelerated to about 100 m/year at the base of the ice-fall, and continues to decelerate down-glacier to less than 20 m/year (Fig. 4-1). In the tunnel dug in the lower part of the ice-fall, J. W. Glen (1956) showed that, in the top 20-metre layer of ice, there was a compressive stress of about 3 bars acting parallel to the ice surface, confirmed by the rates of tunnel closure which were abnormally high (see p. 94).

Rotational movements in glacier flow   Rotational slipping in the case of cirque glaciers is now well established by J. G. McCall and others, and will be discussed in Chapter 8. W. V. Lewis (1947b) suggested an extension of the concept to sections of valley glaciers below ice-falls or steeper gradients. Rotation of the Austerdalsbreen ice-fall tunnel excavated in 1955 was occurring so rapidly (14 minutes of arc per day) that its extension to reach bedrock was never accomplished. However, actual rotation of the ice is not the only possible cause of rotation in the tunnel: a uniform shear throughout the thickness of the ice could cause apparent rotation. The tunnel was also being bent convexly upwards. The idea of large-scale rotational movements of the ice in basins below ice-falls, as suggested by Lewis, is undoubtedly too simple, though movements along potential slip-surfaces as postulated by Nye (1952b) for extending and compressing flow may well sufficiently resemble rotational movements. This is one of the many reasons why further detailed studies of surface and englacial ice movements in actual glaciers are required. The implications of these concepts of ice movement, in regard to the development of some erosional features of glaciated valleys, will be considered in Chapter 7.

Basal and Side-Slip

Internal deformation provides one component of glacier flow; sliding over bedrock provides the other. The relative contribution that each makes varies with bed slope, ice thickness, and ice temperature. Direct measurement of side slip is often possible but, in the case of basal sliding, measurements can only be obtained from tunnels or deep boring. The mechanism of sliding, and measured rates of sliding, are of the highest interest to geomorphologists since it is likely that a glacier carrying debris and slipping past bedrock will be able to erode far more effectively than one that is stuck fast to bedrock.

Side-slip may be measured in several ways. If the marginal ice is not too heavily crevassed, a stake may be inserted as deeply and as close to the ice margin as possible, and its motion referred to two fixed points on the adjacent rock wall. It is unusual, owing to marginal crevassing, to be able to find a suitable place for such a stake closer to the ice edge than 1 or 2 m. Alternatively, a stake may be pushed at an angle through the ice edge almost to touch the rock wall, and the vertical and horizontal motion of the end of the stake can be recorded relative to its initial point of contact with the rock wall. The first of these methods gives less erratic movements than the second, since the ice more than 2 or 3 metres from the actual margin flows more uniformly. Since velocities may change rapidly with distance from the ice edge (as in Blockschollen flow, for instance), side-slip must always be measured as close to the ice edge as possible. Side-slip (S_s) is often expressed as a percentage of maximum surface velocity (V_m):

$$S_s = \frac{V_m}{V_s} \times 100 \qquad (V_s = \text{velocity at the ice edge})$$

For the Fox and Franz Joseph glaciers in New Zealand, B. M. Gunn (1964) obtained values of 12 to 21 per cent, with one abnormally high measurement of 68 per cent where the enclosing rock walls were very steep. On Austerdalsbreen (Norway), Glen (1958), and Glen and Lewis (1961) found sideslip of 10 to 20 per cent rising to 65 per cent near the foot of the ice-fall (Chapter 7). They also noted that side-slip was very variable with time, and from place to place, and some of the factors responsible have already been mentioned (p. 88).

The ratio of basal sliding to maximum surface velocity is equally variable. McCall (1952) in the Vesl-Skautbreen lower tunnel found that basal sliding accounted for 90 per cent of movement—and unlike the Austerdalsbreen side-slip measurements, the basal movement was smooth and steady over 12-hour periods, possibly reflecting a relatively even (abraded?) rock floor at the site of measurement. R. Haefeli (1951) in the Mont Collon tunnel obtained a figure of 20 per cent, but this may be too small since it compares movement at the two ends of the sub-horizontal tunnel, and not in a plane perpendicular to the ice surface. In the Jungfraufirn borehole, basal sliding amounted to 26 per cent, assuming that the borehole did in fact reach bedrock, and to 55 per cent for the Salmon glacier. Sliding probably accounts for most of the movement of thin temperate glaciers on steep slopes, and for lesser amounts in the case of thick temperate glaciers on gentle slopes.

There is some evidence (and much speculation) that, in the case of cold glaciers and ice-sheets, basal sliding does not occur at all since the ice is frozen to bedrock. The shear strength of a cold ice-solid interface is of the order of 10 bars if the solid is stronger than the cold ice and if it is wettable, whereas normal basal shear stresses of temperate glaciers are 0·5 to 1·5 bars. Observations by R. P. Goldthwait (1956) in a tunnel in the Thule area of Greenland showed that the ice here was not sliding over its bed, though there was measurable differential flow in the lower ice layers. The bed was strewn with moss-covered boulders which had evidently not been disturbed for some time, for C^{14} dating of the moss gave an age of approximately 200 years. Cold ice tunnels on the Jungfraujoch, studied since 1950, have also shown that cold ice here is not sliding over its frozen bed; furthermore differential motion within the ice is less than would be associated with temperate ice (R. Haefeli, 1963). Altogether, under cold glaciers frozen to their beds, bedrock erosion probably does not occur.

We thus have abundant evidence that an important part of the flow of temperate glaciers consists of basal sliding, and more limited evidence that cold glaciers do not normally slip over their beds. The first analysis of the mechanism of sliding based on direct observation was that of B. Kamb and E. LaChapelle (1964) who studied the process in a tunnel in the Blue glacier,

Washington. The tunnel reached bedrock at 26 m depth, and basal sliding of 90 per cent was measured on a relatively steep slope. If, in the tunnel, the ice was quickly removed from bedrock by excavation, it came away easily, suggesting the presence under natural conditions of a basal water film. The structure and texture of the lowest ice layer (up to about 3 cm thick) was distinctive, and thought to represent a layer in which alternate freezing and melting under the influence of changing pressure conditions—the process termed regelation—had been taking place. This regelation layer also contained abundant rock debris (compare the 'sole' of the Vesl-Skautbreen glacier, p. 201) picked up presumably during the regelation process. In some places, cavities separated the ice from bedrock, the largest being 10 m wide and 20 cm high, and in some cavities, regelation ice spicules were seen weakly attached to the ice under-surface. The observations supported J. Weertman's theory (1957) that *regelation slip* is an important mechanism of basal sliding.

Kamb and LaChapelle demonstrated the process of regelation slip experimentally by hanging a weight attached by wire to a cube frozen into an ice block. Cubes of rock (dunite), plexiglass, and aluminium were used. Under load, the cubes moved slowly through the ice by melting in front (i.e. below them) and refreezing behind (i.e. above them). The shape and size of the ice block were not disturbed, nor were the texture and structure of the ice except along the path of travel of the cubes, where changes in crystal form were apparent, giving features similar to those encountered in the regelation layer at the base of the glacier. Hence, it is thought that, when glacier ice is forced to pass over a bedrock obstacle, melting of the ice occurs around the upstream face where compressive stress is greatest, the meltwater flows around the obstacle and refreezes on the low-pressure downstream side (where there may be a subglacial cavity), and there is a flow of heat from the freezing to the melting places. The energy to keep the cycle going is supplied by the basal shear stress.

The rate of bulk movement of ice made possible by this mechanism was theoretically examined by Weertman in 1957. He proposed a highly simplified model of a glacier bed, in which bedrock protuberances consisted of cubes of dimension L and equal spacing L'. Then, assuming that all heat flow occurs through the obstacle, the speed of regelation sliding (S_a) is given by:

$$S_a = \frac{\tau_b CD}{3H\rho L} \cdot \frac{L'^2}{L^2}$$

where D is the coefficient of conductivity of the rock, H is the heat of fusion (80 cal/gm), ρ is the density of ice, and C is a constant. It will be seen that as the obstacle size becomes larger or the spacing decreases, so S_a becomes smaller. Smaller obstacles allow the establishment of greater temperature gradients, and therefore allow greater regelation slip. Conversely, obstacles larger than 100 cm would reduce S_a to negligible amounts. In 1964, Weertman modified the equation by omitting the factor of 1/3.

Since S_a becomes negligible for larger obstacles, Weertman introduced an additional sliding mechanism, that of creep around obstacles arising from stress concentrations, on the basis of Glen's power flow law. Assuming that the hydrostatic pressure at the base of the ice is sufficiently large to prevent cavity formation on the downstream side of obstacles, Weertman (1957) gave the sliding velocity S_b by this process as:

$$S_b = BL\left(\frac{\tau_b \cdot L'^2}{2 \cdot L^2}\right)^n$$

Increasing the obstacle size on this mechanism actually gives increased rates of sliding. Weertman considered that these two mechanisms together provided a satisfactory explanation of glacier sliding over obstacles of a variety of sizes.

L. Lliboutry (1959) contended that the rates of sliding obtainable by combining the two equations were too small, and that the model of bed form adopted by Weertman was unrealistic. Taking a modified bed form consisting of a series of parallel sine waves of amplitude a and wavelength λ, he showed that the expected velocity of sliding derived from both Weertman's processes is given by:

$$S_a + S_b = \frac{Ba}{2r^2}\left[\frac{2\tau_b}{\sqrt{3}\pi r}\right]^n$$

where r is a roughness factor ($= a/\lambda$). Taking $n = 3$, $B = 0.18$, and $r = 0.1$, Lliboutry found $S_a + S_b = 4$ m/year. This, he considered, was much too small a rate of sliding ('frottement quasi-statique'), which might give rise to glacier polish, but incapable of more significant erosion.

Lliboutry therefore postulated a modification of the sliding theory, rather than changing his model of the glacier bed. He investigated the possibility that the glacier might partially detach itself from its bed: in the sine-wave model, it might only rest on the crests of the waves, and that cavities, normally water-filled in the case of temperate glaciers, might exist in the troughs. The rate of sliding derived was:

$$S_c = \frac{Bz}{2\tau_b^2}\left(\frac{z}{\sqrt{3}\tau_b\lambda}\right)^n \qquad \text{where } z = a\pi^2(\rho gh - p)^2$$

($p =$ prevailing hydrostatic pressure). Taking an arbitrary value of 6·5 bars for the term $(\rho gh - p)$, he obtained a sliding rate of 100 m/year. It is important to note that values of S_c will increase as τ_b, the basal shear stress, becomes smaller. However, as Weertman (1967) has pointed out, there is no means of evaluating the term for basal water pressure $(\rho g h - p)$, nor is it known what thicknesses of water in the troughs might be involved. Moreover, the whole assumption that the glacier bed consists of waves of equal amplitude and all running perpendicular to the direction of ice flow is highly questionable; it is more likely that there will be waves of different amplitudes, and that if the smaller ones are filled with water, the larger ones cannot be, unless all the smaller undulations are submerged, in which case the water-filled cavities would reach very large dimensions. Weertman argues that Lliboutry's model involves indeterminable quantities and unlikely assumptions; and that it only applies when the ice is thin (less than about 10 m) or when it is moving extremely rapidly (as in ice avalanches). It may be valid in the case of a glacier snout resting on a steep slope.

In 1964, Weertman published a revised form of his sliding theory, using the same double mechanism as previously, but allowing for rectangular bedrock obstacles with three different dimensions, and taking into account the effects of obstacles both larger and smaller than the 'controlling obstacle size' (Λ). The latter is the size of obstacle for which $S_a = S_b$. If an obstacle is smaller than Λ, then ice flows around it primarily by regelation; if larger, the ice flows largely by creep. Figure 4-9 plots sliding velocity S and controlling obstacle size Λ against basal shear stress τ_b, using three different values of L'/L ($= r$, the roughness factor). This model gives reasonable agreement with the observations of Kamb and LaChapelle (1964). In their case, with $\tau_b = 0.7$ bars and $r = 9$, S as measured was 5·8 m/year and S from the graph is about 2 m/year. Λ from the graph is about 4 cm, compared with 3 cm as the maximum thickness of the observed regelation layer.

Weertman (1964) also considered the effect of a basal water film on S. Many observations have suggested that basal meltwater acting as a lubricant is an important factor in glacier sliding (p. 87). If the water layer thickness is less than $\Lambda/100$, Weertman showed that its effect on sliding velocity will be negligible; for a water layer of thickness $\Lambda/10$, the increase in sliding velocity might be 20 per cent; while if the water layer were of the thickness (Λ) equal to the size of the controlling obstacles, then truly catastrophic increases in velocity (perhaps ten times) might be expected. Many of the observed variations in rates of glacier flow could thus be explained in terms of basal water layers whose thickness is of the order of one-tenth the height of the controlling obstacle size. For a glacier normally travelling at 100 m/year, with a 1 bar basal shear stress, Figure 4-9 gives $r = 20$ and $\Lambda = 5$ mm. A 20 per cent variation in flow could then be explained in terms of a basal water film only 0·5 mm thick.

Basal meltwater may accumulate from three main sources: 1) meltwater from the glacier surface; 2) ice melted by geothermal heat (about 0·5 cm³/cm²/year); 3) ice melted by basal sliding (about 3 cm³/cm²/year in the case of a sliding velocity of 80 m/year under a basal shear stress of 1 bar). Variations in the amount of meltwater at the glacier bed may explain observed

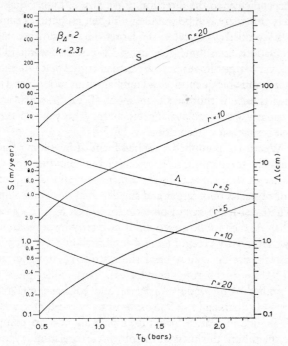

Fig. 4-9 Graph showing sliding velocity (S) and controlling obstacle size (Λ) as a function of the shear stress (τ) for three values of the roughness factor (r) (J. Weertman, *J. Glaciol.*, 1964, by permission of the Glaciological Society)

time-variations in glacier flow (p. 88) and also the catastrophic advances of some glaciers (Weertman, 1962, 1966). Instead of a basal water film of even thickness, it may be more realistic to think of pockets of meltwater at the glacier base which are mobile, and tend to link up as they increase in size. Alternatively, zones of thicker meltwater may be propagated down-valley beneath the glacier as kinematic waves, for which there is some meagre evidence (p. 89). Lliboutry (1964) thought that enlargement of subglacial water pockets beyond a certain critical size might explain many cases of rapid glacier advances. The catastrophic fall of the Allalin glacier on 30 August

1965 (R. Vivian, 1966) is a supreme example. The tongue of this glacier lay on a 30-degree slope and over the last few decades had been gradually thinning. In the week before the disaster, the speed of ice movement had reached 4 m/day, and there had been torrential rain of 56 mm followed by rising temperatures. Crevasses had become water-filled, and sudden surges in the meltwater streams from the snout suggested the existence of numerous subglacial and englacial water bodies. The ice tongue broke away when meltwater had sufficiently reduced ice adhesion on the slope, and precipitated a million cubic metres of ice into the valley below the Mattmark Dam.

Before any theories of glacier sliding can be improved or regarded as acceptable, many more field measurements are required. The greatest weakness in all the models so far formulated concerns the assumed nature of the glacier bed and its roughness. Glacier beds do not consist of rectangular obstacles or sine-wave undulations; hence, the present theories are no more than approximations.

Some Structural Features Arising from Glacier Motion

Crevasses   The theory of the development of crevasses in terms of stress distributions in moving glaciers was first propounded by W. Hopkins in 1862. Crevasse patterns are of several varied types (Fig. 4-10):

A. Chevron crevasses, forming in the marginal zones (old ones may be twisted round as shown);
B. Transverse crevasses, convex up-glacier;
C. Splaying crevasses, beginning longitudinally and approaching the sides at 45 degrees;
D. Radial splaying crevasses, characteristic of the glacier snout.

Often, these types cannot be easily distinguished: for instance, B and C have the same theoretical alignments in the marginal zone.

Crevasses may vary in width from hair-cracks to gaps many metres wide. Of greater significance to the glacial geomorphologist is the matter of crevasse depths, for crevasses are a principal means of penetration of meltwater to the deeper parts of glaciers, and in some cases may enable meltwater to reach the rock floor (as W. V. Lewis (1947) postulated in his theory of rock-step formation beneath thin glaciers). Nye (1955) has shown that, on the basis of Glen's power flow law, the rate of longitudinal extension ($\dot{\varepsilon}$) in a glacier subjected to a longitudinal tensile stress will be given by $\dot{\varepsilon} = B\sigma^n$ where σ is the yield stress in extension. Then the maximum theoretical depth of crevasses (d) will be given by

$$d = \frac{1}{g\rho}\left(\frac{\dot{\varepsilon}}{B}\right)^{1/n}$$

where ρ is the density of ice and g is the acceleration of gravity. $\dot{\varepsilon}$ can be determined by surface velocity measurements from $\dot{\varepsilon} = \dfrac{v_1 - v_2}{x}$ where v_1, v_2 are surface velocities of two points separated by a distance x

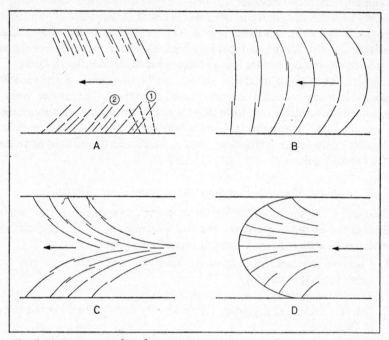

Fig. **4-10** Some examples of common crevasse patterns (R. P. Sharp, *Glaciers*, University of Oregon Press, 1960)
　　A Marginal (1—old rotated crevasses, 2—newly formed crevasses)
　　B Transverse　**C** Splaying　**D** Radial
　　Ice flow from right to left

along a line perpendicular to the crevasse. If $n = 3.3$, $B = 0.1$ years^{-1} bars^{-n}, then $d = 22.8\ \dot{\varepsilon}^{1/3.3}$ where d is in metres and $\dot{\varepsilon}$ is in years^{-1}.

Measured depths of true crevasses rarely exceed 30 m (it is important to distinguish crevasses from moulins (Chapter 6) which may be much deeper). W. Blake (1956) gives 26 m as the maximum depth in the Thule area (northwest Greenland); M. M. Miller (1954) gives 27-30 m for the Taku and Seward glaciers of Alaska, noting that it is not possible to measure by handline the lowest tapering part of the crack. Crevasses exceeding 36 m in depth are known in the Antarctic and Greenland (Miller, 1954) and also in the Alps (F. Loewe, 1955). There are suggestions that, as cold ice can support higher shear stresses, the tensile layer and its crevasses may be deeper in polar than in temperate glaciers, and that crevasses in cold ice may be more sluggish in closing.

Foliation   This is the name usually given to a secondary layering of glaciers produced by deformation; it is to be distinguished from primary layering of sedimentary origin inherited from the firn (see Chapter 8 for a description of sedimentary layering in cirque glaciers). Foliation layers may be up to one or more metres in thickness, and are identified not by dirt inclusions as in sedimentary layering but by differences in type of ice, especially between white bubbly ice and bluish clear ice. Foliation is most intense near to the margins of a glacier, where it often runs roughly parallel to them, near bedrock outcrops, below ice falls, and where different ice streams flow next to each other. The layers may dip at any angle, but often approach the vertical where the foliation is intense. Good examples of foliation are given by N. Untersteiner (1955) for the Pasterze glacier (Austria) and by N. W. Rutter (1965) for the Gulkana glacier (Alaska).

The exact origins of foliation are still in some doubt, but generally it is a structure resulting from deformation of ice during flow. It forms in areas subjected to intense compression or shear, as noted above, and since most of the surface area of a glacier experiences only very low shear stresses, the foliation does not normally originate where it is observed, but in the deepest layers subjected to much greater distortion. The pattern of foliation seen at the surface reflects subsequent differential down-glacier movement and surface ablation. Thus the pattern can be used as a rough measure of the relative flow velocity. A problem not yet solved is the apparent failure of foliation in some instances to form parallel to the planes of greatest shear stress (Meier, 1960).

Ogives   These cross-glacier bands of alternating white and dark ice, curving down-glacier because of differential ice movement, were first described by Forbes (1843) on the Mer de Glace. They are not very common features, but on some glaciers (such as Austerdalsbreen, or the newly-named Vaughan Lewis glacier in the Juneau ice-field of Alaska) are exceptionally striking. They occur only on temperate glaciers below ice-falls, and their spacing (usually 50 to 200 m and decreasing down-glacier owing to compression) is approximately that of one year's flow. The white ice bands consist mainly of white bubbly ice (density c. 0·89), while the dark bands include a high proportion of blue bubble-free ice (density c. 0·91). The formation of these alternating layers is now reasonably well understood (C. A. M. King and W. V. Lewis, 1961; J. E. Fisher, 1962). Each of the dark ice bands represents that portion of the glacier that spent the summer descending the ice-fall. In the process, the ice crystals enlarge by melting and occasional refreezing, much meltwater gathers in crevasses, and dirt accumulates on the stretched ice surface. At the foot of the ice-fall, the much-broken ice reconsolidates by compression into mainly blue bubble-free and dirt-stained temperate ice. In winter, on the other hand, the ice in the ice-fall remains generally frozen, crystals remain small, and the surface will be mostly protected by a mantle of snow. It reconsolidates into mainly white cold bubbly ice.

On the glacier below the ice-fall, differential ablation accentuates the differences between successive bands. The white bubbly ice bands have less thermal conductivity and a greater albedo than the darker predominantly blue-ice bands, and therefore tend to stand up in waves. Moreover, once formed, meltwater tends to collect and aid accumulation of dirt on the dark bands in the troughs, and the coarse-grained dark ice tends to hold dirt better than the smoother fine-grained 'winter' ice.

Conclusion

Glacier motion has been studied in increasing detail over the past century. Modern surveys of surface movements show that rates of flow vary along the length of the glacier and across it. In a single ice-stream, maximum velocities on a transverse profile occur in the central portions. In some glaciers, there is a very rapid change in velocity close to the valley sides, representing a form of flow termed Blockschollen movement. Variations in rates of flow with time have been investigated; in the ablation zone of temperate glaciers, for instance, the ice tends to flow faster in summer than in winter, probably because of the increased lubrication of the bed by meltwater. Some short-period variations may also be related to weather conditions and the availability of basal meltwater. Englacial velocity measurements by means of boreholes or tunnels show that, in a vertical profile, velocities decrease from glacier surface to bedrock, as was anticipated by Forbes in 1843. The theory of extrusion flow must be firmly abandoned.

The movement of glaciers consists of two components—internal deformation or 'creep', and basal sliding. Internal deformation approximates closely to a power flow law in which the exponent has a value usually between 2 and 4. Theoretical models of internal velocity distributions based on this law agree very satisfactorily with observations from boreholes and tunnels. Longitudinal changes in rates of glacier flow are considered by Nye in his theory of extending and compressing flow, a theory that has important implications for glacial valley erosion, as will be shown in Chapter 7. The second component of glacier flow, sliding over bedrock, may be studied in

the slip of a glacier past its sidewall and in the basal sliding directly observed
in tunnels or deep borings. Side-slip has been found to amount to between
10 and 70 per cent of the maximum surface velocity, while basal sliding may
vary from 20 to 90 per cent, in the case of temperate glaciers. Cold glaciers
do not exhibit basal sliding since they are frozen to bedrock, but there may
be relatively rapid shearing in the basal ice layers. Cold glaciers are therefore
unlikely to be effective in eroding the bedrock by abrasion, though it must
be remembered that the distinction between cold and temperate glaciers is
not always a rigid one (see Chapter 3). In temperate glaciers on relatively
steep slopes, basal sliding represents the principal contributor to ice discharge,
and this has important consequences for bedrock erosion. The effect of higher
velocities of basal sliding is to increase the number of boulders being dragged
over an area of bedrock in a given time, and therefore enhance the degree
of abrasion of the surface. Further details of the abrasion process will be given
in Chapter 6. On flat surfaces or up reversed slopes, basal sliding will be
considerably reduced. In such conditions, glacier motion is accomplished not
by longitudinal pressure, which the ice would be incapable of transmitting,
but by a thickening of the ice until the situation is reached at which the centre
of gravity of the ice once again falls downhill sufficiently to permit movement. Movement, however, will be slow since the thicker glacier can discharge more ice for a given velocity than a thinner one.

Basal sliding beneath temperate glaciers occurs by a combination of regelation slip and creep around obstacles arising from stress concentrations.
Weertman and Lliboutry have examined these processes theoretically, using
highly simplified models of bed form. Weertman's revised theory (1964)
shows very satisfactory agreement with the observations of Kamb and
LaChapelle. Both Weertman and Lliboutry consider that basal meltwater is
of great importance in the sliding process and, if present in a film of sufficient
thickness, may cause catastrophic glacier advances. Sudden advances may
also result if warming of the basal ice of cold glaciers brings it into the
temperature range where regelation slip becomes possible.

A study of glacier motion is a highly relevant preliminary to understanding
the origin of many landforms resulting from glacial erosion and deposition,
such as valley steps, rock basins, cirques, and drumlins. The formation of
these and other features will be considered in detail in Parts II and III; and the
implications of both the theory and observation of glacier flow will be
examined in connection with these landforms.

REFERENCES

L. AGASSIZ (1840), Études sur les glaciers (Neuchâtel)

W. R. B. BATTLE (1951), 'Glacier movement in north-east Greenland, 1949', J. Glaciol. 1, 559-63

W. BLAKE (1956), 'The depth of crevasses', J. Glaciol. 2, 644-5

A. C. BORDIER (1773), Voyage pittoresque aux glacières de Savoie Suisse (Geneva)

H. CAROL (1947), 'The formation of Roches Moutonnées', J. Glaciol. 1, 57-59

R. T. CHAMBERLIN (1928), 'Instrumental work on the nature of glacial motion', J. Geol. 36, 1-30

M. DEMOREST (1937), 'Glaciation of the Upper Nugssuak Peninsula, West Greenland', Z. Gletscherk. 25, 36-56

M. DEMOREST (1938), 'Ice flowage as revealed by glacial striae', J. Geol. 46, 700-25

M. DEMOREST (1942), 'Glacier regimens and ice movement within glaciers', Am. J. Sci. 240, 31-66

M. DEMOREST (1943), 'Ice sheets', Bull. geol. Soc. Am. 54, 363-400

G. R. ELLISTON (1963), in discussion following J. WEERTMAN, 'Catastrophic glacier advances', Bull. Int. Ass. scient. Hydrol. 8, 2, 65-66

R. FINSTERWALDER (1931), 'Geschwindigkeitsmessungen an Gletschern mittels Photogrammetrie', Z. Gletscherk. 19, 251-62

R. FINSTERWALDER (1950), 'Some comments on glacier flow', J. Glaciol. 1, 383-8

J. E. FISHER (1962), 'Ogives of the Forbes type on Alpine glaciers and a study of their origins', J. Glaciol. 4, 53-61

J. D. FORBES (1843), Travels through the Alps of Savoy (Edinburgh)

T. W. FRIESE-GREEN and G. J. PERT (1965), 'Velocity fluctuations of Bersækerbræ, East Greenland', J. Glaciol. 5, 739-47

R. W. GALLOWAY (1956), 'Mechanical measurement of glacier motion', J. Glaciol. 2, 642-4

J. W. GLEN (1952), 'Experiments on the deformation of ice', J. Glaciol. 2, 111-14

J. W. GLEN (1955), 'The creep of polycrystalline ice', Proc. R. Soc. A, 228, 519-38

J. W. GLEN (1956), 'Measurement of the deformation of ice in a tunnel at the foot of an ice-fall', J. Glaciol. 2, 735-45

J. W. GLEN and W. V. LEWIS (1961), 'Measurements of side-slip at Austerdalsbreen, 1959', J. Glaciol. 3, 1109-22

R. P. GOLDTHWAIT (1956), 'Formation of ice cliffs', Tech. Rep. Snow Ice Permafrost Res. Establ. 39, 139-50

B. M. GUNN (1964), 'Flow rates and secondary structures of Fox and Franz Joseph glaciers, New Zealand', J. Glaciol. 5, 173-90

R. HAEFELI (1948), 'The development of snow and glacier research in Switzerland', J. Glaciol. 1, 192-201

R. HAEFELI (1951), 'Some observations on glacier flow', J. Glaciol. 1, 496-500

R. HAEFELI (1952), 'Behaviour of ice in the Z'Mutt glacier', J. Glaciol. 2, 94-99

R. HAEFELI (1963), 'Observations and measurements on the cold ice sheet on Jungfraujoch', Bull. Int. Ass. scient. Hydrol. 8, 2, 122

A. HEIM (1885), Handbuch der Gletscherkunde (Stuttgart)

A. HELLAND (1877), 'On the ice-fjords of north Greenland, and on the formation of fjords, lakes, and cirques in Norway and Greenland', Q.J. geol. Soc. Lond. 33, 142-76

H. HESS (1904), Die Gletscher (Braunschweig)

S. E. HOLLINGWORTH (1931), 'The glaciation of western Edenside and adjoining areas and the drumlins of Edenside and the Solway basin', Q.J. geol. Soc. Lond. 87, 281-359

S. E. HOLLINGWORTH (1947), in discussion following G. SELIGMAN (1947), J. Glaciol. 1, 20

C. D. HOLMES (1937), 'Glacial erosion in a dissected plateau', Am. J. Sci. 29, 41-47

W. HOPKINS (1862), 'On the theory of the motion of glaciers', Phil. Trans. R. Soc. 152, 677-745

G. J. HUGI (1842), Über das Wesen der Gletscher (Stuttgart)

J. D. IVES and C. A. M. KING (1955), 'Glaciological observations on Morsárjökull, S.W. Vatnajökull, Iceland', J. Glaciol. 2, 477-82

J. E. JACKSON (1953), 'Surveying on glaciers', J. Glaciol. 2, 235

B. KAMB and E. LACHAPELLE (1964), 'Direct observation of the mechanism of glacier sliding over bedrock', J. Glaciol. 5, 159-72

W. KICK (1957), 'Exceptional glacier advances in the Karakoram', J. Glaciol. 3, 229

C. A. M. KING and W. V. LEWIS (1961), 'A tentative theory of ogive formation', J. Glaciol. 3, 913-39

M. LAGALLY (1934), Mechanik und Thermodynamik des stationären Gletschers (Leipzig)

M. LAGALLY (1939), 'Zur Mechanik eines auf seiner Sohle gleitenden stationären Gletschers', Z. Gletscherk. 26, 193-8

W. V. LEWIS (1947a), 'Some comments on Dr. H. Carol's article', J. Glaciol. 1, 60-63

W. V. LEWIS (1947b), 'Valley steps and glacial valley erosion', Trans. Inst. Br. Geogr. 13, 19-43

W. V. LEWIS (1948), in review of R. F. FLINT, Glacial geology and the Pleistocene epoch, J. Glaciol. 1, 212

L. LLIBOUTRY (1959), 'Une théorie du frottement du glacier sur son lit', Annls Géophys. 15, 250-65

L. LLIBOUTRY (1964), 'Subglacial "super-cavitation" as a cause of the rapid advances of glaciers', Nature, Lond. 202, 77

F. LOEWE (1955), 'The depth of crevasses', J. Glaciol. 2, 511-12

W. H. MATHEWS (1959), 'Vertical distribution of velocity in Salmon glacier, British Columbia', J. Glaciol. 3, 448-54

J. G. MCCALL (1952), 'The internal structure of a cirque glacier', J. Glaciol. 2, 122-30

M. F. MEIER (1960), 'Mode of flow of Saskatchewan glacier, Alberta, Canada', U.S. geol. Surv. Prof. Pap. 351, 70 p.

M. M. MILLER (1954), in discussion on 'The mechanics of glacier flow', J. Glaciol. 2, 339-41

J. F. NYE (1951), 'The flow of glaciers and ice-sheets as a problem in plasticity', Proc. R. Soc. A, 207, 554-72

J. F. NYE (1952a), 'Reply to Mr. Joel E. Fisher's comments', J. Glaciol. 2, 52-53

J. F. NYE (1952b), 'The mechanics of glacier flow', J. Glaciol. 2, 82-93

J. F. NYE (1952c), 'A comparison between the theoretical and the measured long profiles of the Unteraar glacier', *J. Glaciol.* 2, 103-7

J. F. NYE (1955), 'Comments on Dr Loewe's letter and notes on crevasses', *J. Glaciol.* 2, 512-14

J. F. NYE (1957), 'The distribution of stress and velocity in glaciers and ice-sheets', *Proc. R. Soc.* A, 239, 113-33

J. F. NYE (1959a), 'The deformation of a glacier below an ice-fall', *J. Glaciol.* 3, 387-408

J. F. NYE (1959b), 'The motion of ice-sheets and glaciers', *J. Glaciol.* 2, 493-507

J. F. NYE (1965), 'The flow of a glacier in a channel of rectangular, elliptic, or parabolic cross-section', *J. Glaciol.* 5, 661-90

E. OROWAN (1949), in discussion on 'The flow of ice and other solids', *J. Glaciol.* 1, 231-40

S. ORVIG (1953), 'On the variation of the shear stress on the bed of an ice-cap', *J. Glaciol.* 2, 242-7

W. S. B. PATERSON (1961), 'Movements of the Sefstrøms Gletscher, north-east Greenland', *J. Glaciol.* 3, 844-9

W. S. B. PATERSON (1964), 'Variations in velocity of Athabaşca Glacier with time', *J. Glaciol.* 5, 277-85

M. F. PERUTZ (1940), 'Mechanism of glacier flow', *Proc. R. Soc.* A, 52, 132-5

M. F. PERUTZ (1948), 'A description of the iceberg aircraft carrier . . . and some problems of glacier flow', *J. Glaciol.* 1, 95-104

M. F. PERUTZ (1950), 'Direct measurement of the velocity distribution in a vertical profile through a glacier', *J. Glaciol.* 1, 382-3

F. PFAFF (1874), 'Über die Bewegung und Wirkung der Gletscher', *Annls Phys.* 151, 325-36. See also G. SELIGMAN (1948)

H. PHILIPP (1920), 'Geologische Untersuchungen über den Mechanismus der Gletscherbewegung und die Entstehung der Gletschertextur', *Neues Jb. Miner. Geol. Paläont.* 43, 439-556

N. W. RUTTER (1965), 'Foliation pattern of Gulkana glacier, Alaska Range, Alaska', *J. Glaciol.* 5, 711-18

H. B. DE SAUSSURE (1779-96), *Voyages dans les Alpes* (Paris), 4 vols

G. SELIGMAN (1947), 'Extrusion flow in glaciers', *J. Glaciol.* 1, 12-21

G. SELIGMAN (1948), 'The movement of firn and ice in glaciers', *J. Glaciol.* 1, 142-4

G. SELIGMAN (1949), 'Research on glacier flow', *Geogr. Annlr* 31, 228-38

R. P. SHARP (1953), 'Deformation of a vertical bore-hole in a piedmont glacier', *J. Glaciol.* 2, 182-4

R. P. SHARP (1954), 'Glacier flow: a review', *Bull. geol. Soc. Am.* 65, 821-38

R. P. SHARP (1960), *Glaciers* (Condon Lecture Publs, Univ. of Oregon Press)

C. SOMIGLIANA (1927), reported by S. FINSTERWALDER (1926-7), 'Über die innere Reibung des Eises und die Bestimmung der Gletschertiefe', *Z. Gletscherk.* 15, 55-59

R. STREIFF-BECKER (1938), 'Zur Dynamik des Firneises', *Z. Gletscherk.* 26, 1-21

C. SWITHINBANK (1966), 'A year with the Russians in Antarctica', *Geogr. J.* 132, 463-75

N. UNTERSTEINER (1955), 'Some observations on the banding of glacier ice', *J. Glaciol.* 2, 502-6

R. VIVIAN (1966), 'La catastrophe du glacier Allalin', *Revue Géogr. alp.* **54**, 97-112

W. H. WARD (1955), 'The flow of Highway Glacier', *J. Glaciol.* **2**, 592-8

J. WEERTMAN (1957), 'On the sliding of glaciers', *J. Glaciol.* **3**, 33-38

J. WEERTMAN (1962), 'Catastrophic glacier advances', *Un. géod. géophys. int., Symposium at Obergurgl,* 1962 (*Publication No. 58, Int. Ass. scient. Hydrol.*), 31-39

J. WEERTMAN (1964), 'The theory of glacier sliding', *J. Glaciol.* **5**, 287-303

J. WEERTMAN (1966), 'Effect of a basal water layer on the dimensions of ice-sheets', *J. Glaciol.* **6**, 191-207

J. WEERTMAN (1967), 'An examination of the Lliboutry theory of glacier sliding', *J. Glaciol.* **6**, 489-94

Chapter 5

Some Indirect Effects of Pleistocene Glaciation

Many of the Quaternary deposits in all countries . . . indicate a pluvial period just as clearly as the Northern Drift indicates a glacial. (A. TYLOR, 1867)

The ice-cap hypothesis advocated by Mr Croll is a probable cause of a great reduction in the level of the sea through abstraction of water from the sea, and its deposition at the poles in the form of ice. (A. TYLOR, 1868)

The geomorphological consequences of Pleistocene glaciation were by no means confined to the lands covered by ice and the adjacent periglacial zone. The radical changes of climate characteristic of the Pleistocene had far-reaching effects on the earth's water balance, so that the level of the seas rose and fell with the waning and waxing of the great ice-sheets; the discharge of rivers fluctuated according to the varying rates of precipitation and evaporation; and lakes expanded and shrank, or even appeared and disappeared, adjusting themselves like the rivers to the prevailing hydrological conditions. The enormous burden of ice that accumulated in some regions caused actual deformation of the earth's crust, as, on a much smaller scale, did the weight of water represented by the largest of the temporary lakes. Changes in land, sea, and lake levels, as well as changes in fluvial activity, are abundantly documented in the landforms and deposits of both glaciated and extra-glacial regions: this chapter is concerned with these changes, but does not pretend to be more than an outline of a subject whose scope is vast and whose literature is almost as voluminous as that on glacial geomorphology itself.

Pluvial Periods in Extra-glacial Regions

The term 'pluvial' was introduced into geological literature by A. Tylor (1868) to refer to periods of greater rainfall and surface run-off in the Quaternary, which he supposed to have been responsible for deposition of certain sediments in England and northern France. Later, the term was transferred to other regions and drier climates; some years earlier, T. F. Jamieson (1863) had advanced the idea that the lowered temperatures of glacial times had resulted in expansion of lakes. In the hundred years that have elapsed since, the concept that present-day arid and semi-arid regions experienced pluvial and inter-pluvial periods in the Pleistocene has become firmly accepted and efforts are now directed towards establishing the chronology of such events and their correlation with climatic changes in the glaciated areas.

It is now widely agreed that the glacial periods of the Pleistocene were, in many mid-latitude areas, times of increased precipitation, lowered temperatures, and reduced evaporation. The correlation between glacials and pluvials was long established in Western North America with the classic work of I. C. Russell on Lake Lahontan (1885) and the Mono Valley (1889), and of G. K. Gilbert (1890) on Lake Bonneville, but it had been anticipated in the Old World by L. Lartet (1865) who suggested that the glaciation of the Syrian highlands and former high stands of the Dead Sea had been contemporaneous. The correlations between glacials and pluvials are now known to be less simple than these pioneer workers envisaged, for the numbers of established glacial and pluvial events have multiplied, and the climatic conditions prevailing during pluvial and inter-pluvial phases have also been shown to be more complex than was originally thought.

Much of our knowledge of Pleistocene climatic fluctuations in arid and semi-arid areas is based on studies of pluvial lakes. Invaluable records of their fluctuations of level are given by their strandlines and other marginal features such as spits, bars, and deltas, and by their deposits of both littoral and deep-water facies. However, it must be recognized at the outset that lake levels are determined not only by climate, and changes in level may be due to tectonic factors, erosion of the lake outlet (if any), or possible diversion of a river into the lake. In the case of some pluvial lakes, too, entry of meltwater from nearby glacial ice has affected the level, though usually to only a minor extent. Pluvial Lake Lahontan, for example, received meltwater at times, but even if all the ice in its catchment had melted, it could not have supplied more than 20 per cent of the water accumulating there at its maximum extent.

Pluvial lakes were of wide distribution at times in the Pleistocene, and some attained impressive dimensions. The Chad basin once contained a lake ('Megachad') equivalent to four times the size of present-day Lake Victoria; the Dead Sea, whose surface now stands 400 m below sea-level, rose at least once to 30 m above sea-level; the Caspian Sea combined with the Aral Sea, reaching a level 115 m (possibly higher) above its present and backing up the Volga to Kazan, though meltwater from the Central Asian glaciers contributed substantially to this prodigious water body. Since space forbids even a list of the Pleistocene pluvial lakes, three major examples from the Basin-and-Range country of the western U.S.A. are selected for further comment.

Pluvial lakes in the Great Basin   In the Great Basin, the general sequence of Pleistocene climatic change is thought to have been as follows:

```
_____ Interglacial _____ · · · _____ Glacial _____ · · · _____ Interglacial _____
cool ———→ warm ———→ cool ———→ cold ———→ cool ———→ warm
dry ————————→ wet ————————→ moist ————————→ dry ————————→
```

This is necessarily an oversimplification. The tectonic features of the Great Basin formed ideal situations for lake formation when climatic conditions were suitable, and over 120 former lakes have been identified. Gilbert's Lake Bonneville was by far the largest, approaching 52,000 km² or the size of present Lake Michigan, and attaining a maximum depth of 335 m. Its present-day attenuated remnants are Great Salt Lake, Provo and Sevier Lakes. Gilbert concentrated mainly on shore morphology, but though this is impressive, it does not yield as complete or unambiguous a story of the lake fluctuations as do studies of the lacustrine and inter-lacustrine deposits. The record so far deciphered in any detail covers only the 75,000 years or so of the Wisconsinan (last) glaciation; Lake Bonneville undoubtedly existed at times in the pre-Wisconsinan period but the data have yet to be assembled and analysed. The highest stand of the lake at the Bonneville shoreline (1550-1616 m: a precise figure cannot be quoted because of isostatic deformation) was attained on several occasions according to recent work (R. B. Morrison and J. C. Frye, 1965, and references contained therein). Other major shorelines, first demonstrated by Gilbert, occur at about 1460 m (the Provo stand) and 1370 m (the Stansbury level, distinguished by an abundance of tufa). At the highest stands, Lake Bonneville overflowed through Red Rock Pass to the Snake River, and the well developed Provo shoreline represents the level being held by a resistant formation in this outlet. Figure 5-1 shows the fluctuations of level postulated by Morrison and Frye during Wisconsinan time. Their correlations of high-lake phases with glacial episodes should be noted. This is one of the two areas in the Great Basin where a close connection can be established, for glacial ice at times entered Lake Bonneville from the Wasatch mountains. Thus, for example, till and outwash from Little Cottonwood Canyon interfinger with lake sediments laid down at times of high water level, and both the upper and middle Bull Lake (early Wisconsinan or Altonian) tills appear to have been deposited in standing water. The synchroneity of glaciations and lake maxima is difficult to dispute. Radiometric age determinations by C^{14} and Th^{230} have given further confirmation of this correlation (see, for example, R. F. Flint and W. A. Gale, 1958; A. Kaufman and W. S. Broecker, 1965). The lake sediments record alternating phases of deep and shallow water, and occasional intervals of complete desiccation when subaerial weathering and soil formation supervened. The

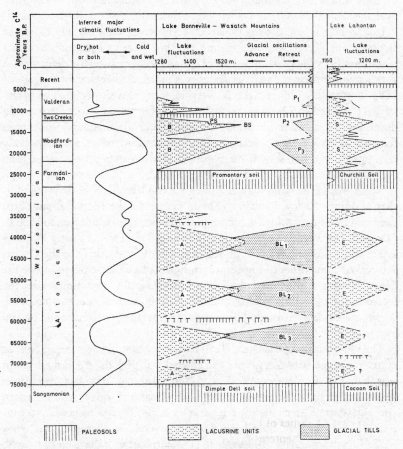

Fig. 5-1 Climatic history, lake-level fluctuations and glacier oscillations in the Bonneville and Lahontan areas, during the Wisconsinan period (R. B. Morrison and J. C. Frye, *Report, Nevada Bureau of Mines*, 1965)

Lacustrine units: **B** Bonneville formation   **A** Alpine formation
            S Sehoo formation   **E** Eetza formation

Glacial units: $P_{1,2,3}$, Upper, middle, lower Pinedale till
            $BL_{1,2,3}$, Upper, middle, lower Bull Lake till

BS Bonneville shoreline last attained (*c.* 13,000 B.P.)
PS Provo shoreline

most important soil recognized in the Wisconsinan sequence (see Fig. 5-1) by Morrison is the Promontory Soil, correlated with the major interstadial, probably 22,000-28,000 B.P., when Lake Bonneville temporarily disappeared.

Lake Lahontan was the second largest Pleistocene water body in the Great Basin (about 22,000 km²; maximum depth 213 m at Pyramid Lake). Its main shorelines today stand at 1332, 1277 and 1213 m (isostatic deformation was negligible since the mass of water was much smaller than in the case of Lake Bonneville). Unlike Lake Bonneville, it never overflowed. The correlations proposed by Morrison are indicated on Figure 5-1. Note the strongly developed Churchill soil, marking a lengthy interval of desiccation between the two main deep-lake periods. Pre-Wisconsinan events are tentatively recognized, but not shown on Figure 5-1; deep-lake periods have been suggested for the Kansan and Illinoian glacials.

South from Lake Lahontan is a third pluvial lake basin of great interest, that of Mono Valley. Today, the basin contains Lake Mono, about 52 m deep and lacking an outlet. At times in the Pleistocene, water rose to 2156 m (Tioga or late Wisconsinan glaciation) and 2190 m (Tahoe or early Wisconsinan, when the lake was 275 m deep and overflowed briefly to the Owens River). At these times of high water, tongues of glaciers emerged from the Sierra Nevada and entered the lake ('Lake Russell'), sometimes with the interposition of outwash deltas, and lake shorelines visibly cut the moraines. Again, the evidence strongly points to contemporaneity of glacial and pluvial events.

The pluvial lakes of the Great Basin have been used to assess the amount of climatic change between glacial and interglacial periods. An early attempt was that of O. E. Meinzer (1922), who concluded that Arizona, for example, was about as moist in pluvial maxima as Oregon is today. More modern computations of precipitation–temperature–run-off relations needed to support a given lake level have been made by several. An example is the work of C. T. Snyder and W. B. Langbein (1962) in Spring Valley, Nevada, who postulate increases of precipitation from the present 300 mm to 500 mm in glacial times, and corresponding reductions in evaporation from 1100 to 800 mm.

Undoubtedly, the potential for further investigations of pluvial lake history both in the Great Basin and elsewhere in the world is very great, and such investigations are destined to shed much light on the events of the Pleistocene, particularly the older events that are so poorly documented in the glacial record. These studies promise eventually to reveal the total number of pluvials and glacials, their chronology and inter-correlation. The principal method of attack will be on the stratigraphy of lake sediments in enclosed basins by deep boring; such sediments must preserve a complete record of changing lacustrine and subaerial conditions since there is little if any possibility of the complete loss of any member except perhaps by wind deflation.

Changing Stream Régimes

The changing climatic conditions throughout the Pleistocene, together with the changes in the relative levels of land and sea to be described in the next sections, had a profound influence on stream activity. With fluctuations in precipitation, and in rates of run-off affected for instance, by the changing vegetation cover, river volumes rose and fell even where they were not directly fed by meltwater. Furthermore, the load of material available for river transport varied, being greatest when vegetal cover was too sparse to restrain downhill creep, when the climate favoured solifluction and other mass movements, or when the rivers were supplied with outwash or were attacking loess-covered areas. Since the hydraulic activity of streams is largely controlled by discharge and load, the streams adjusted their long profiles to the prevailing conditions, and sequences of terraces and valley fills attest to the frequency and magnitude of changes in these conditions. A recent study of stream channel characteristics below an active glacier in Washington (U.S.A.) by R. K. Fahnestock (1963) demonstrates that erosion or deposition of valley outwash is closely related to variations in stream discharge and debris supply. In regions not too far removed from the sea, Pleistocene changes of land and sea-level also affected stream profiles, submergence causing drowning and alluviation of their lower reaches, while emergence resulted in some cases in incision of the lower reaches and the formation of knick points which receded slowly upstream.

The geomorphological evidence of these changes is clear, in terraces, buried channels, and very often also signs that the present-day streams are 'underfit' in that they now occupy valleys too large for the present rates of discharge (see, for instance G. H. Dury, 1960). What is much more difficult is to relate these features to particular climatic events in the Pleistocene, or to use them to establish the nature of climatic changes; for any such attempts to be worth while, reliable evidence on the ages of the features has to be first obtained.

The subject is too vast to elaborate on further, beyond referring briefly to the evidence presented in some of the features of the Thames Valley. The Thames received direct accessions of meltwater and outwash at certain stages (for example, via its tributary the Colne); at other times, variations in load and discharge were determined by climate. The present river is clearly underfit in relation to its floodplain and incapable of now handling the coarse material of which the floodplain is built; its present shrunken size and relatively sluggish activity attest to the changed conditions of the post-glacial period. Higher-level gravel spreads, now forming terraces or plateau capping, are related to previous climatic episodes. The dating of many of these older gravels is fairly well established—the Boyn Hill terrace, it is generally agreed, formed in the Hoxnian (Elster-Saale) interglacial, for it overlies Elster drift at Hornchurch (Essex) and is itself overlain by solifluction deposits of Saale age. The climatic conditions under which the gravels accumulated are less well known, but the impressive extent of aggradation represented by the Boyn Hill terrace, for instance (its valley-floor width approaches 20 km in the Romford area) is consistent with an early interglacial age when a river of sizeable volume was attacking deposits accumulated in its upper and middle reaches and sweeping them downstream; at the same time, base-level was rising in response to melting of the ice-sheets, as will be described in the next section.

Glacio-eustatic Sea-level Changes

The accumulation of snow and ice on land-masses represents storage of moisture and is a vital part of the earth's hydrological cycle. An immediate effect of any increase in world temperatures is the liberation of meltwater from such snow and ice, most of which finds its way to the oceans and causes a necessarily world-wide rise in sea-level. Conversely, any extension of ice-sheets and glaciers results in a drop in sea-level. Such changes of sea-level are described as glacio-eustatic; the term 'eustatic' was introduced by E. Suess in the late nineteenth century to refer to world-wide sea-level changes caused by several possible factors, not only those related to ice-ages. A most difficult problem, in fact, lies in attempting to distinguish glacio-eustatic from other eustatic changes of sea-level, such as might arise, for instance, from changes in the capacity of the ocean basins caused by tectonic activity or deposition of sediments in those basins, from changes in the temperature of the sea-water (a 1°C. change would affect sea-level by about 2 m), or from changes in the level of the sub-oceanic crust as the load of sea-water varies. As sea-level can only be defined in terms of its relation with the land-masses, a further complication arises from the possibility that apparent changes of sea-level may be the result of actual changes in the height of the land arising from crustal deformation. Crustal deformation caused by variations in ice load will be dealt with in the next section. A final consideration to be borne in mind in connection with Pleistocene sea-level changes is that even slight shifts in the earth's axis of rotation would cause changes in the shape and position of the mean sea-level (geoidal) surface relative to the land.

It is evident, then, that to disentangle the effects of purely glacio-eustatic changes of sea-level from the effects of the other factors mentioned is an almost insuperable task. Nevertheless, evidence is available to show that, throughout the Quaternary and in historic times, glacio-eustatic fluctuations of sea-level have taken place.

The first to recognize the possibility of glacial control of sea-level was C. Maclaren (1842), who assessed the change in level resulting from Pleistocene glaciation at 107 to 213 m. A. Tylor's estimate (1872) was 180 m. In 1910, R. A. Daly expounded his hypothesis of glacial control in the evolution of coral reefs, involving the upgrowth of corals from platforms cut during ice-age low sea-levels (see also Daly, 1934). The extent of glacio-eustatic sea-level change during the glacial periods still provides a formidable problem, but several lines of evidence are now converging to suggest a lowering of between 100 and 140 m in each main glacial period. The evidence comes from submerged wave-cut platforms and notches, submerged littoral or terrestrial sediments, and buried channels of coastal rivers. In all cases, however, the possibility must be continually borne in mind that the anomalies of level may be explained by crustal displacements, instead of actual sea-level changes. As examples of the data used in support of low sea stands in the last glaciation, the following may be mentioned. Many oceanic atolls possess basal platforms at about – 100 m, while submarine terraces fringe many coasts at about this level. The buried channels of rivers in western Europe often grade down to levels approaching – 100 m (for instance, the First Buried Channel of the Thames). H. N. Fisk and R. McFarlan (1955) in the Mississippi delta suggest a low sea-level in the last glaciation of – 136 m, though R. W. Fairbridge (1961) considers this figure may be too large owing to insufficient allowance being made for local subsidence and compaction. Fairbridge prefers a figure of – 100 m for the Main Würm and – 70 m for the Late Würm, while J. R. Curray (1965), incorporating some further data, suggests – 145 m for the early Wisconsinan (Main Würm) and – 120 to – 125 m for the late Wisconsinan (Late Würm).

W. L. Donn and others (1962) approached the problem from another standpoint by estimating Pleistocene ice volumes and converting these into amounts of sea-level lowering:

Stage	Estimate A*	Estimate B*
Late Wisconsinan	105 m	123 m
Early Wisconsinan	114 m	134 m
Illinoian	137 m	159 m

* Two estimates of sea-level lowering are given, based on different assumed ice-thicknesses.

They support their reconstruction of Illinoian sea-level stands of – 137 to

– 159 m with new evidence of deeply submerged and buried marine terraces (Newfoundland – 145 m, West Indies – 147 to – 154 m, Argentina up to – 158 m, etc.) on which are found shallow-water shells radiocarbon-dated at more than 30,000 years. However, estimates of ice thickness are so unreliable that the calculated figures of sea-level lowering may be widely in error. Moreover, they make no allowance for the following factors: 1) there

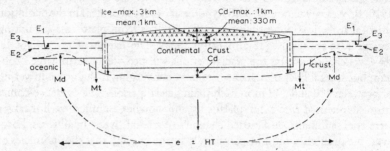

Fig. 5-2 Eustatic-isostatic relationships. The diagram shows a hypothetical ice-loaded continent bound by simatic oceanic crust. Mean ice loading (1 km) depresses crust 330 m (Cd). Crustal displacement (Cd) is effected by elastic compression (e) and/or horizontal transfer (HT) in the deep layers. This may give rise to positive marginal displacement (Md).

Sea-level E_1 is lowered glacio-eustatically to E_2, but may rise after a time-lag (for Md to occur) to E_3. Isostatic unloading of the ocean area by water removal during transfer E_1 to E_2 will also contribute to the height of E_3. The reverse may be expected during deglaciation, when for a deglacial eustatic rise of 100 m, a mean oceanic crustal lowering of 33 m is suggested (R. W. Fairbridge, *Physics and Chemistry of the Earth*, ed. Ahrens, Pergamon Press, 1961)

would be semi-contemporaneous crustal adjustments due to loading and unloading of ice on the land; 2) the ocean floors would themselves respond isostatically to loading and unloading of the sea-water; 3) changes in sea-level automatically cause slight changes in oceanic areas. Figure 5-2 (after R. W. Fairbridge) illustrates some of these eustatic-isostatic relationships. The final sea-level lowering resulting from ice formation would be considerably less, perhaps one-third less, than the simple calculations performed by W. L. Donn and others (1962) suggest. Fairbridge argues, for instance, that melting of all the world's present ice, theoretically equivalent to a column of ocean water 95 m high, would in the end cause a sea-level rise of little more than 50 m. (Even this, however, would flood about one-eighth of the present land area.)

If the precise extent of sea-level lowering in the last (Würm-Wisconsinan) glaciation is still in doubt, the position of sea-level in the preceding glacial and interglacial periods is even more controversial. Donn and others (1962) argue that there has been approximate sea-level stability from one interglacial to another; entirely different is the viewpoint held by F. E. Zeuner

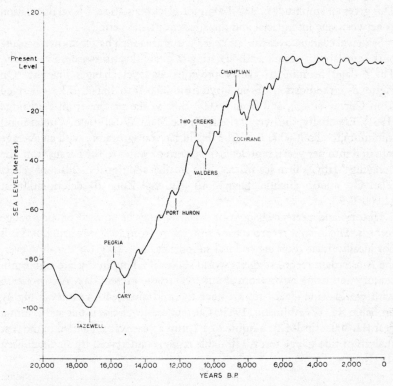

Fig. 5-3 Sea-level change since 20,000 B.P. (R. W. Fairbridge, *op. cit.*, 1961)

(1959) and R. W. Fairbridge (1961) for instance, that there has been a continuous secular lowering of sea-level throughout the Quaternary, on which glacial oscillations have been superimposed. According to this view, late Pliocene–early Pleistocene sea-level máy have been 200 m or more above the present, and in the several interglacials, maximum sea-levels were attained as follows:

Eemian	+ 10 m
Hoxnian	+ 60 m
Cromerian	+100-110 m

Minimum glacial sea-levels according to Fairbridge were:

Würm	– 100 m
Riss	– 55 m
Mindel	– 5 m
Gunz	+ 35 m

This gives an amplitude of 110-120 m for glacio-eustatic sea-level fluctuation as between one interglacial and the adjacent glacial period.

Sea-level changes over the last 20,000 years are much better known because of the availability of radiocarbon dating to establish a time-scale. Eventually, Th^{230} dates promise to clarify previous sea-level changes likewise. On Figure 5-3, the portion of the curve from 20,000 to 10,000 B.P. is derived from Curray (1965), and from 10,000 B.P. to the present from Fairbridge (1961). Essentially, the overall rise from the Main Würm (late Wisconsinan) minimum to about 6000 B.P. (the 'Flandrian transgression', with an average rate of 8 mm per year), represents the return of water to the oceans from the decaying Main Würm ice masses. The smaller additional oscillations reflect relatively minor climatic fluctuations (e.g. the Zone III deterioration at 11,000 B.P.).

Historic and present changes of sea-level may be studied on tide-gauge records. The longest record is for Amsterdam, from 1682 onward, but this is not ideally situated owing to local subsidence. Allowing for this, however, the Amsterdam record suggests world sea-level falling in the late eighteenth century and rising again from about 1880 to the present day. A correlation with world-wide glacial retreat since the mid-nineteenth century is highly probable (H. W. Ahlmann, 1953). Current sea-level rise is put at 1·2 mm a year (H. Valentin 1952); a figure of 1·1 mm a year was suggested in the first analysis of tide-gauge records in stable regions undertaken by B. Gutenberg (1941).

Glacio-isostatic Crustal Deformation

The idea that isostatic deformation of the earth's crust might arise from the weight of thick ice-sheets was advanced by T. F. Jamieson over a century ago. G. K. Gilbert (1890) broadened the concept to include crustal depression by the weight of water in the largest pluvial lakes. Since Jamieson's time, the work of many in Scandinavia, Finland, and North America has proved the correctness of these early ideas. It is known, for example, that glacial loading today keeps the bedrock surfaces of Greenland and Antarctica below sea-level near their centres, while slow adjustment to unloading in Canada, the north-eastern U.S.A., and Fennoscandia is still in progress, even though most of the ice in these areas vanished 7000-8000 years ago. Evidence from tide-gauges, geophysical studies, and tilted Quaternary shorelines is collectively unequivocal. Studies of deformed shorelines in particular have made it possible to calculate rates of uplift consequent upon ice unloading, and to delineate the areas which have been affected by glacio-isostatic deformation, the boundaries of these areas lying roughly parallel to the limits of the last glaciation. The patterns of crustal deformation are revealed in maps of iso-bases, or lines joining points of equal uplift on the affected strandlines. The limits of measurable crustal deformation are represented by so-called 'hinge-lines' (or 'zero isobases') for each shoreline; for instance, in the Great Lakes area, the southern parts of certain Late-glacial shorelines are horizontal, but beyond their respective hinge-lines, they rise northward.

The mechanics of the process of crustal deformation are incompletely understood. When the load of ice exceeds the strength of the crust, subsidence occurs, presumably displacing sub-crustal matter outward, though where this displaced matter is disposed is unknown. There is a time-lag in the pro-cess, but it seems likely that crustal response is complete over a span of 15,000-20,000 years. With deglaciation, the crust is unloaded and gradually recovers its former level (the terms 'crustal rebound' or 'recoil' are often used in this connection). The minimum height and size of ice mass needed to cause crustal sinking is only approximately determinable, for crustal strength varies somewhat. Large modern reservoirs such as Lake Mead and Lake Kariba contain sufficient mass of water to cause crustal subsidence detectable in geodetic surveys. Pluvial Lake Bonneville equivalent in weight to an ice-cap 330 m thick and 250 km in diameter caused 65 m of crustal subsidence, while smaller Lake Lahontan (equivalent to ice 180 m thick and 170 km in diameter) caused only 10 m of subsidence as measured in shoreline deformation. The Bonneville data have been recently re-examined by M. D. Crittenden (1963). The Great Basin, however, may be an area of less than average crustal strength, and the amplitude of warping depends not only on ice thickness but also on the nature of the crustal and sub-crustal material. From evidence available, it seems likely that the maximum subsidence beneath a large ice-sheet should be about 25-30 per cent of the maximum ice thickness, but correspondingly less beneath small ice-sheets, assuming that the density of sub-crustal matter (3·0-3·3) is about 3 or 4 times that of ice.

Fennoscandia   At present, differential uplift is in progress, varying from 9 mm a year in the north of the Gulf of Bothnia to 4 mm a year near Stock-holm and nil around Copenhagen. It should be remembered that these rates are determined from tide-gauge records with respect to mean sea-level, which itself is currently rising at about 1 mm a year. The net emergence results in about 1000 km² of land being added to Finland each century. Studies of the crustal rebound in the Late-glacial and Post-glacial periods are based on warped strandlines, of which an enormous number has now been recognized, and their interpretation in detail is highly complex and controversial. In the centre of the region, a maximum uplift of 500-600 m is deduced, and it is thought that another 200 m or so of recovery will take place before equilibrium is achieved. Thus a total of 800 m of isostatic depression must be invoked, and if we take this as equivalent to 30 per cent of the ice thickness, as suggested above, the maximum ice thickness is likely to have been about 2600 m, which is in accord with evidence from other sources. Maximum rates of rebound probably attained 50 or even 75 mm a year for short periods in the neighbourhood of the Gulf of Bothnia, but they have now slowed down considerably as noted, and complete recovery may take another 10,000 years or so. The lack of equilibrium at present is reflected in the large areas of negative gravity anomalies (up to 30 mgal).

North America   The story of glacio-isostatic deformation is much less well known than in Fennoscandia, for the area is vastly larger, much of it thinly populated, and inadequately mapped. Tide-gauge records, because of the extent of unbroken land compared with Fennoscandia, reveal only a small part of the movements currently taking place. Isostatic recovery at present ranges from nil along a line through Manitowoc–Detroit–Boston, to 4 mm a year along a line Port Arthur–Kingston (Ont.)–Quebec. In the Great Lakes today, water is rising on the southern shores and receding from the northern ones. Just in the area however, where tide-gauge records would be of the highest interest, namely, Hudson Bay, tide-gauges are few and their records unreliable or of short duration. The longest record is for Churchill, since 1928; B. Gutenberg (1941) analysed the first 10 years and deduced a rise of 10-20 mm a year, but since then, doubt has been cast on the validity of the earliest record, and the data since 1940 have failed to show any positive trend. But on archaeological evidence, J. B. Bird (1954) thought a rise of 3 feet a century (= 10 mm a year) likely.

Deformed and uplifted shorelines have been extensively studied, as in Fennoscandia. A line Milwaukee–Cleveland–New York roughly indicates the southernmost limit of measurable up-warping. Maximum uplift is located in Richmond Gulf, Hudson Bay, where it amounts to at least 302 m (W. R. Farrand, 1962). As the Wisconsinan ice here was probably 3000 m thick, about 900 m of recovery ought to be expected, yet already the rate of recovery seems to be slowing down. For example, B. Matthews (1966) in northern Ungava has deduced mean rates of uplift amounting to 60 mm a year for the period 6700-7700 B.P., compared with only 3 mm a year for the period since 4000 B.P. Farrand considers it possible that complete isostatic equilibrium was not attained by the Wisconsinan ice; other points are that the figure of 302 m represents a minimum for post-glacial uplift, since some up-warping must have occurred before shoreline cutting began, and the post-glacial eustatic rise in sea-level must be allowed for.

Gravimetric surveys of the former ice-covered area are far from complete. However, I. Fischer (1959) shows that the pattern of geoidal contours possesses a regular depression of up to 15 m centred over Hudson Bay, and the pattern conforms remarkably to the contour pattern showing the minimum amount of post-glacial uplift deduced from the highest marine limit (see P. B. King, 1965, p. 836).

In the Great Lakes area, the history of crustal deformation and rebound has been most thoroughly investigated, beginning with the classic studies of F. Leverett and F. B. Taylor (1915) and Taylor (1927), followed by others whose work is summarized and evaluated by J. L. Hough (1958). North of the southernmost hinge-line, the former shorelines rise northward (for instance, the Algonquin shoreline from a zero isobase of 180 m in central Michigan to 335 m on the north shore of Lake Huron, rising to greater heights farther north); the oldest shorelines possess the greatest tilts. The zero isobases of successive lake stages are also displaced north through as much as 250 km from the earliest to the latest, showing that the area experiencing isostatic recovery shrank slightly over the 8000-year period since deglaciation. Many of the shorelines are remarkably clear-cut and well developed, suggesting that they were formed in pauses during intermittent isostatic recovery.

British Isles   Recent investigations are demonstrating that the story of glacio-isostatic rebound is here much more complex than earlier workers supposed. Indeed, the older concepts of horizontal raised beaches in Scotland, postulated by the Geological Survey and W. B. Wright (1914), for instance, at 15, 25, 50, and 100 feet (5, 8, 15, and 30 m), are plainly at variance with any idea of isostatic recovery at all, for this ought to lead to deformed rather than horizontal shorelines. Re-mapping of the raised beaches (e.g. J. B. Sissons, 1962) has clearly shown the tilted nature of many, and the fact that many more sets of beaches exist than had been formerly thought. The so-called 25-foot (8 m) beach is now known to reach 14·9 m at Stirling and 12·2 m around Loch Linnhe, but falls to 5·2 m in Lancashire and below sea-level in southern Britain. The highest marine limit yet recorded in Britain is 44 m at Stirling; this is dated at 12,000 B.P. since it merges with outwash of the Perth Readvance. If Fairbridge's estimate of minus 48–50 m for sea-level at this time is accepted, total isostatic uplift of 92–94 m has taken place since that data. Other data in the Stirling area give amounts of uplift totalling 55 m since 10,300 B.P. and 9 m since 5300 B.P. From this it may be inferred that rates of isostatic recovery have slowed from about 24 mm a year (12,000-10,300 B.P.) to 10 mm a year (10,300-5300 B.P.), and finally to 2 mm a year (5300 B.P.-present). Tide gauge data for Britain have been studied by H. Valentin (1953); Dundee shows emergence at the present day in excess of the current eustatic rise of sea-level, and by extrapolation Valentin conjectured that central Scotland might still be rising at 4 mm a year, an estimate not out of keeping with the estimate obtained above from shoreline data. The limit of isostatic recovery probably lies across central Britain in about the latitude of Liverpool.

Investigations of British Quaternary shorelines are still in progress, and are greatly helping to clarify the nature and extent of glacio-isostatic movements. Accounts of some recent work, and a summary of this and the older work by K. Walton, are to be found in *Transactions* No. 39 of the *Institute of British Geographers* (1966).

Conclusion

Pleistocene glaciation was not only the cause of great morphological changes in the glacial and periglacial regions, but affected the whole of the earth in terms of eustatic changes in sea-level, deformation of the earth's crust by the loading and unloading of ice, and variations of temperature, precipitation, and evaporation. The glacial periods were characterized by low sea-levels (probably – 100 to – 140 m in the last glacial period), by increased precipitation and reduced evaporation. The range of oscillation of sea-level between one glacial and the adjacent interglacial was probably about 110-120 m. Pleistocene glacio-eustatic sea-level changes may have been superimposed on a continuous secular lowering of sea-level throughout the Quaternary.

Pluvial lakes in many areas now arid or semi-arid testify to the wetter climatic conditions of the glacial periods in the extra-glacial regions, and the sequences of deposits in these lake basins are most valuable in reconstructing the chronology of the Pleistocene. The changing climatic conditions also had profound effects on stream activity, one of the most obvious morphological effects being the series of terraces and valley fills created in response to changes in river volume and load.

Isostatic deformation of the earth's crust by the weight of thick ice-sheets has been postulated since 1863 and is confirmed by modern studies of rates of land uplift in response to glacial unloading, and by the present depression of bedrock in central Greenland and Antarctica below sea-level by the weight of their ice-masses. Total subsidence beneath large ice-sheets may amount to 25-30 per cent of maximum ice thickness. The morphological effects of recovery following unloading are most clearly seen in the tilting and warping of former marine or lacustrine shorelines.

REFERENCES

H. W. AHLMANN (1953), 'Glacier variations and climatic fluctuations', *Am. geogr. Soc.*, *Bowman Memorial Lectures*, Ser. III, 1-15

J. B. BIRD (1954), 'Post-glacial marine submergence in Central Arctic Canada', *Bull. geol. Soc. Am.* **65**, 457-64

M. D. CRITTENDEN (1963), 'New data on the isostatic deformation of Lake Bonneville', *U.S. geol. Surv. Prof. Pap.* **454** E, 1-31

J. R. CURRAY (1965), 'Late Quaternary history, continental shelves of the United States' in *The Quaternary of the United States* (ed. H. E. WRIGHT and D. G. FREY), 723-35

R. A. DALY (1910), 'Pleistocene glaciation and the coral reef problem', *Am. J. Sci.* **30**, 297-308

R. A. DALY (1934), *The Changing World of the Ice Age* (Yale Univ. Press)

W. L. DONN, W. R. FARRAND, and M. EWING (1962), 'Pleistocene ice volumes and sea-level lowering', *J. Geol.* **70**, 206-14

G. H. DURY (1960), 'Misfit streams: problems in interpretation, discharge and distribution', *Geogr. Rev.* **50**, 219-42

R. K. FAHNESTOCK (1963), 'Morphology and hydrology of a glacial stream—White River, Mount Rainier, Washington', *U.S. geol. Surv. Prof. Pap.* **422** A, 1-70

R. W. FAIRBRIDGE (1961), 'Eustatic changes in sea level' in *Physics and Chemistry of the Earth* (ed. L. H. AHRENS *et al.*) **4**, 99-185

W. R. FARRAND (1962), 'Postglacial uplift in North America', *Am. J. Sci.* **260**, 181-99

I. FISCHER (1959), 'The impact of the Ice Age on the present form of the geoid', *J. geophys. Res.* **64**, 85-87

H. N. FISK and R. MCFARLAN (1955), 'Late Quaternary deltaic deposits of the Mississippi River', *Geol. Soc. Am. Spec. Pap.* **62**, 279-302

R. F. FLINT and W. A. GALE (1958), 'Stratigraphy and radiocarbon dates at Searles Lake, California', *Am. J. Sci.* **256**, 689-714

G. K. GILBERT (1890), 'Lake Bonneville', *U.S. geol. Surv. Monogr.* **1**, 1-438

B. GUTENBERG (1941), 'Changes in sea-level, postglacial uplift, and mobility of the earth's interior', *Bull. geol. Soc. Am.* **52**, 721-72

J. L. HOUGH (1958), *Geology of the Great Lakes* (Univ. of Illinois Press)

T. F. JAMIESON (1863), 'On the parallel roads of Glen Roy, and their place in the history of the glacial period', *Q.J. geol. Soc. Lond.* **19**, 235-59

T. F. JAMIESON (1865), 'On the history of the last geological changes in Scotland', *Q.J. geol. Soc.* **21**, 161-203

A. KAUFMAN and W. S. BROECKER (1965), 'Radiocarbon chronology of Lake Lahontan and Lake Bonneville II, Great Basin', *Bull. geol. Soc. Am.* **76**, 537-66

P. B. KING (1965), 'Tectonics of Quaternary time in middle North America' in *The Quaternary of the United States* (ed. H. E. WRIGHT and D. G. FREY), 831-70

L. LARTET (1865), 'Sur la formation du bassin de la mer morte ou lac asphaltite, et sur les changements survenus dans le niveau de ce lac', *C.r.hebd. Séanc. Acad. Sci.*, Paris 60, 796–800

F. LEVERETT and F. B. TAYLOR (1915), 'The Pleistocene of Indiana and Michigan, and the history of the Great Lakes', *U.S. geol. Surv. Monogr.* 53, 1–529

C. MACLAREN (1842), 'The glacial theory of Professor Agassiz', *Am. J. Sci.* 42, 346–65

B. MATTHEWS (1966), 'Radiocarbon dated post-glacial land uplift in Northern Ungava, Canada', *Nature, Lond.* 211, 1164–5

D. E. MEINZER (1922), 'Map of Pleistocene lakes of the Basin-and-Range Province and its significance', *Bull. geol. Soc. Am.* 33, 541–52

R. B. MORRISON and J. C. FRYE (1965), 'Correlation of the middle and late Quaternary successions of the Lake Lahontan, Lake Bonneville, Rocky Mountain (Wasatch Range), Southern Great Plains, and eastern Mid-West areas', *Nevada Bureau of Mines, Rep.* 9, 1–45

I. C. RUSSELL (1885), 'Geological history of Lake Lahontan, a Quaternary lake of north-western Nevada', *U.S. geol. Surv. Monogr.* 11, 1–288

I. C. RUSSELL (1889), 'Quaternary history of Mono Valley, California', *U.S. geol. Surv. 8th A. Rep.* (1886–7), 261–394

J. B. SISSONS (1962), 'A re-interpretation of the literature on late-glacial shorelines in Scotland with particular reference to the Forth area', *Trans. Edinb. geol. Soc.* 19, 83–99

C. T. SNYDER and W. B. LANGBEIN (1962), 'The Pleistocene lake in Spring Valley, Nevada, and its climatic implications', *J. geophys. Res.* 67, 2385–94

F. B. TAYLOR (1927), 'The present and recent rate of land tilting in the region of the Great Lakes', *Pap. Mich. Acad. Sci.* 7, 145–57

A. TYLOR (1842), 'On the formation of deltas: and on the evidence and cause of great changes in sea-level during the Glacial Period', *Geol. Mag.* 9, 393–9 and 485–501

A. TYLOR (1868), 'On the Amiens gravel', *Q.J. geol. Soc. Lond.* 24, 103–25

H. VALENTIN (1952), 'Die Küsten der Erde', *Petermanns Mitt., Ergänz.* 246, 118 p.

H. VALENTIN (1953), 'Present vertical movements of the British Isles', *Geogr. J.* 119, 299–305

K. WALTON et al. (1966), 'The vertical displacement of shorelines in Highland Britain', *Trans. Inst. Br. Geogr.* 39, 1–145

W. B. WRIGHT (1914), *The Quaternary Ice Age* (Lond., 1st ed.)

F. E. ZEUNER (1959), *The Pleistocene period* (Lond., 2nd ed.)

PART II

Glacial and Fluvioglacial Erosion

We know that the glaciers rub, wear, and polish the rocks with which they are in contact. Struggling to dilate, they follow all the sinuosities, and press and mould themselves into all the hollows and excavations they can reach, polishing even overhanging surfaces. (J. DE CHARPENTIER, 1835)

Chapter 6

Small-scale Features of Glacial and Fluvioglacial Erosion

Few, few believe what I have told,
Men say that I am over bold,
What then? they sneered that Welshmen's tails
Had polished Buckland's rocks in Wales.
(A. C. RAMSAY, 1856, at the Anniversary Dinner of the
Geological Survey)

Rock surfaces in glaciated regions often display clear signs of the former passage of ice over them in the form of markings. These include the familiar striations, channels and grooves of varying dimensions and shapes, and gouges suggesting the extraction of small chips of rock by some agency. Some rock surfaces have been abraded and scoured to a highly polished form, while others, sometimes adjacent to smoothed areas, show signs of plucking. These and other features are now collectively taken as clear evidence of former glaciation, better evidence indeed than the landforms of larger scale such as hanging valleys or apparently U-shaped troughs which in some cases at least admit of other non-glacial explanations. But relatively little is known about the origins of small-scale glacial rock markings: although there is a considerable body of literature describing them, there is considerable disagreement over the exact mechanism of their formation. It is becoming increasingly apparent that moving ice alone may not be the sole agent of formation involved, but that the scouring effects of subglacial water moving under high pressure must be taken into account, and also the abrasive effects of water-saturated ground moraine being squeezed into motion by differential ice pressure. It is clear that in this context it is no longer possible to make a rigid distinction between glacial and fluvioglacial erosion.

Striations, Grooves and Polished Surfaces

L. Agassiz in 1838 was one of the first to ascribe polished and scratched rock surfaces in certain Alpine valleys to abrasion by moving ice whose undersurface was armed with rock fragments of various sizes. The process was vividly described by J. D. Forbes at the side of the Brenva glacier in 1843. At a point where the ice was driven hard against the rock face, he extracted a specimen of basal ice whose face was seen to be 'set all over with sharp angular fragments . . . which were so firmly fixed in the ice as to demonstrate the impossibility of such a surface being forcibly urged forward without sawing and tearing any comparatively soft body which might be below it' (p. 204). In the rock exposed by removal of the ice specimen, grooves, scratches, and a general polishing were visible. Descriptions of striated and smoothed surfaces in various formerly glaciated regions culminated in T. C. Chamberlin's great monograph of 1888, dealing with *The rock scorings of the great ice invasions* in North America, a work which has never been surpassed in the accuracy and detail of its observation.

Striations consist of lines, usually short but very occasionally a metre or more in length, engraved on suitable rock surfaces. They have been recorded in a great variety of situations—on level, sloping, vertical, or curved rock surfaces, for instance—but are most prolific on gently-inclined smooth surfaces which the ice was forced to ascend. They have been recorded not only on open surfaces, but also on the risers of steps and in trenches (M. Demorest, 1938), and in recesses in vertical walls (P. Sheldon, 1926). The clarity of their development depends largely on the character of the rock, for soft, coarse-grained, or brittle rocks do not take the markings well, and on exposed or easily weathered rocks, the markings are rapidly erased. Some striae begin very gradually and slowly deepen, while others begin quite abruptly; again, some fade out slowly and others disappear in a blunt or fractured end. Any individual striation can result from two diametrically opposed directions of ice motion. The larger striations are known as grooves, while at the other end of the scale, the scratches become of minute size. Rock surfaces which appear smoothed and polished to the naked eye are seen to possess innumerable small scratches under the microscope. The degree of polishing depends on the fineness of the abrading material, namely, the rock flour (N. Edelman, 1949), and in some cases a glazed appearance results. Polished rock surfaces do not usually survive long exposure to weathering. F. E. Matthes (1930) attributed the unusual abundance of glacier polish in the region above Yosemite valley to the prevalence of highly siliceous granite which weathers only slowly. Even so, the polish is largely restricted to areas affected by the last glaciation.

On the formation of striations, Chamberlin stated that 'the ice of a glacier has of itself but little abrading and practically no striating power. It can neither groove nor scratch. . . . The little effect that it may produce is of the nature of polishing . . .' (p. 208). The finer scratches are the work of grains of silt or sand, not just embedded in the ice but caught between boulders in the ice and the bedrock, for the physical properties of ice prevent it applying sufficient pressure on the very small surface area of an individual sand grain. Larger scratches and grooves are clearly etched by larger fragments: 'A block of rock that presents a large surface to the embracing ice and but a small area in contact with the rock . . . floor may be held relatively firm and act as a steady graver' (ibid.). Such a situation was actually observed by H. Carol (1947) at 50 m below the Oder Grindewald glacier, where striae 4 mm deep were being carved in the bedrock. If the block is too large and has too great an area of contact with bedrock, the ice will flow over the block, and it will be the underside of the ice which is grooved. Sometimes the cutting tool appears to have rotated, thus presenting a new cutting edge, and giving rise to striations which step sideways or lie *en echelon* (Edelman, 1951).

It is relatively easy to comprehend how rock surfaces are striated and polished, but the larger grooves present quite a different problem and their origin is very obscure. Some of the largest grooves known were described by H. T. U. Smith (1948) in the Mackenzie valley of north-west Canada, where they attain lengths of as much as 12 km, depths of up to 30 m, and widths approaching 100 m (see also Chapter 9). There is no doubt that they are of glacial origin, since they are totally indifferent to structure, they accord in direction with what else is known of ice movement in this region, and they are of similar form to smaller grooves in bedrock and till elsewhere which grade downward into striations. Yet obviously they have not been gouged out by single boulders; possibly groups of boulders packed and frozen together may explain the smaller ones. Another problem is what localized the grooves, for neither structure nor relief appears to have played a part. Although the glacial origin of these particular grooves is undeniable, care must be taken to distinguish grooves of other kinds; thus Chamberlin (1888) thought that some might be merely ice-modified pre-glacial grooves, and others, especially curving and twisting grooves, might be the work of sub-glacial streams.

Striations and grooves have often been used as a means of determining ice motion. As will become apparent, this is by no means a simple matter, nor are the conclusions based solely on them particularly reliable. It is first necessary to appreciate that scratches on rocks or rock surfaces are not always the result of glaciation. Surface markings may result from differential chemical weathering, and a number of other non-glacial processes may produce striations. Agassiz in 1838 considered running water bearing stones along with it, and also avalanches, as possibilities. Chamberlin, at a time when the hypothesis of glaciation by land ice was still in its infancy, thought that some markings might have been made by icebergs or shore-ice during partial submergence. This possibility is in fact a very real one in Arctic regions, where it is still the reason for much uncertainty regarding directions of former ice motion. Many observers have confirmed that mass movements can scratch rock surfaces. C. F. S. Sharpe (1938) gives an example of an alpine mudflow in the San Juan Mountains of Colorado, where blocks exceeding one cubic metre in volume had been moved, and where ledges in the channel containing the mudflow showed excellent striae orientated roughly parallel to the direction of movement. J. L. Dyson (1937) also showed conclusively that snow-slides in Glacier National Park were capable of producing fine striations. Here, fresh scratches up to 10 m long and 6 mm deep were superimposed across older (and probably glacial) striations. Their age, judged by comparing weathering of the scratches and other surface areas, was not more than a few years, and this was confirmed by the fact that in a few cases the stones responsible for the scratching still remained poised at the lower ends of the grooves. A rapidly moving recent snow-slide was suggested as the most probable mechanism. Other examples of the erosive effects of moving snow are given in Chapter 25. In general, any heavy moving mass may cause scratching of the rock surface beneath, and it is important to note that the cutting tools need not be held rigidly in the moving medium (Demorest, 1938, p. 721).

The great bulk of rock scorings in regions known from other evidence to have been ice covered at some former time are, however, undoubtedly caused by glacial action. Scratched surfaces in formerly glaciated regions are often best displayed when the rock can be newly stripped of a till cover or turf and cleaned up with brush and water. Studies of such glacial striations can throw a certain amount of light on directions of ice movement, and they can also provide useful information of the nature and behaviour of the basal ice.

Much detailed work on glacial abrasion marks has been carried out in Finland. As has long been known, a large number of striations must be recorded if there is to be any reliable idea gained of former directions of ice motion. An individual striation will only express ice movement at that point, and local movement will be strongly affected by topographical irregularities. K. Virkkala (1960) quotes striations on a shore cliff at Näsijärvi in southern Finland, where the orientation swings gradually from 315° on the stoss side to 345° on the lee side. Out of 1300 striae measurements around Tampere, he found that 55 per cent bore an azimuth of 300° to 330° (Fig. 6-1), the variation between these values probably reflecting purely local influences on basal ice movement. Bedrock irregularities probably exerted progressively stronger influence on ice movement as the ice thinned during deglaciation. Virkkala (1951) recorded similar local variations in striae trend in another intensive study around Suomussalmi, east Finland (Fig. 6-1), but here there was a much higher concentration of striae directions, 83 per cent trending within 15° of an

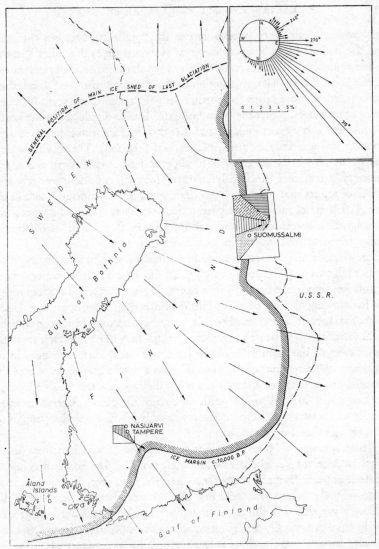

Fig. **6-1** Directions of ice motion over Finland and adjacent areas. The approximate position of the ice margin about 10,000 years ago is indicated (the time of the Baltic Ice Lake and middle Swedish moraines). Glacial striations in the rectangular areas around Tampere and Suomussalmi have been mapped and analysed by K. Virkkala (1960, 1951); in these areas, the vertical line-shading represents the sector in which most striations are orientated; while, in the Suomussalmi area, the horizontal line-shading indicates the sector possibly representing the youngest ice movement, and the stipple indicates a sector possibly representing ice movement locally deflected. Further detail of striae direction in the Tampere region is given in the inset east-west azimuth. Very noticeable was the parallelism between the striae and the chains of lake basins in the area. However, in both studies Virkkala found evidence of distinct sets of striations at variance with the principal sets. In the Suomussalmi area, in addition to the common east-west striae, three other trends appeared directed north, north-west, and south-west. It is clear from other evidence that the main Scandinavian ice-divide lay to the north, north-west, and west of Finland, and that therefore ice movement across the country was generally from these directions (Fig. 6-1). The south-westerly striae may be wholly the result of local deflections. As the north and north-westerly striae are more clearly marked than the common westerly sets, Virkkala argues that they are somewhat younger, and concludes that in the last glaciation, ice moved chiefly from the west; there followed a short phase of stagnation succeeded by renewed ice activity from different ice centres and movement from the north-west and north.

As Virkkala emphasizes, great caution must be used in interpreting striations in this way. To be of any validity, studies must be based on a very large number of striae observations over a reasonably large area, and other evidence on ice movement must be taken into consideration, particularly the dispersion of erratics. It is clear that bedrock irregularities can produce considerable local variations of basal ice movement—which may of course differ from the movement of the main bulk of the ice above—and during changes in ice thickness, the directions of ice movement as a whole may alter. It was also pointed out in Chapter 4 that, even in a valley glacier, the lines of ice flow are not solely directed parallel with the axis of the glacier. Variations in ice supply and wastage will also be accompanied by shifts in the direction of ice motion and therefore striation.

There remains the possibility that crossing sets of striations may be the work of different glacial phases with different ice centres. Sometimes one set of striae appears distinctly fresher than another. However, the question arises as to why the older striae were not erased by the later movement. Such a situation suggests that the later movement was but short-lived and that it could have caused only a very minor amount of bedrock erosion. In other words, crossed striae are unlikely to represent successive glaciations but merely shifts in ice motion during a phase of general deglaciation, as Virkkala concludes

In the examples quoted from Finland, the question did not arise of two possible directions of ice motion differing by 180°. Even for the south-westerly set, a north-easterly source of the ice appears improbable since other evidence adequately locates the Scandinavian ice centres elsewhere as shown on Figure 6-1. But the problem has been encountered, and the form of the striae alone provides no answer to the problem (see p. 136). A particularly

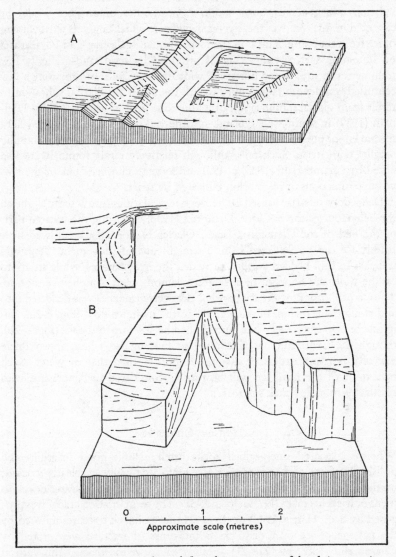

Fig. 6-2 A Striations recording deflected movement of basal ice crossing
irregular bedrock surface
B Eddy-like currents in basal ice as recorded by striations (M.
Demorest, *J. Geol.*, 1938, University of Chicago Press)

doubtful case was described by T. H. Clark (1937) in southern Quebec where striations ran generally north-south. A northern source of ice would be expected in this area, yet there is local evidence which suggests a movement of ice from the south. This is in the form of small-scale crag-and-tail features on limestone surfaces; the 'crag' is provided by hard sand-grains (a few millimetres in size) in the limestone which possess 'tails' of limestone a few centimetres in length fading out towards the north. Similar tails thought to result from glacial modification of a limestone surface were noted by Edelman (1949) in south-west Finland. On the Kakskielaklobb peninsula, hard silicate knobs possess tails up to 2 m long, dying away gradually and running parallel with striae. Such tails, although relatively rarely found, were also noted by Chamberlain (1888, p. 193) and serve to eliminate one of the two possible directions of ice motion indicated by striae.

Deflections of striae caused by bedrock irregularities throw useful light on the behaviour of the basal ice. Demorest (1938) observing the vacated parts of the bed of the Clements Glacier, Glacier National Park, showed how readily ice can be deflected from a straight line of movement. Steps and trenches in thin bedded argillite provided the irregularities, while striations on the walls and floors of these features demonstrated complex ice motion. Thus, a proportion of the ice crossing transverse trenches was deflected into and along the trenches, moving at 90° to the main ice flow (Fig. 6-2a), and in one or two cases, eddy-like currents had been set up in the ice (Fig. 6-2b), though the term 'eddy' is not suited to moving ice whose approximate viscosity is many orders of magnitude greater than that of water. Such behaviour of the basal ice will be further considered in connection with plastically-moulded rock surfaces.

Friction Cracks

Whereas glaciated fine-grained rocks often exhibit good striations and polished surfaces, rocks of medium-size grain are usually less clearly striated; on the other hand, the latter type of rock may have a much clearer development of friction cracks (N. Edelman 1951). The term 'friction crack' was proposed by S. E. Harris (1943) to cover a variety of rock fractures induced by the passage of ice over bedrock. Some types of fracture were noted by Chamberlin (1888, pp. 218-21)—gouges of crescentic shape, semi-vertical fractures with crescentic outcrops whose convexity opposed the ice motion, and chatter-marks, which Chamberlin considered to result from the sort of vibratory motion, well known to machinists, of a cutting tool being forced across a hard surface. Crescentic gouges received detailed analysis from G. K. Gilbert in 1906, who studied numerous examples cut in granite in the

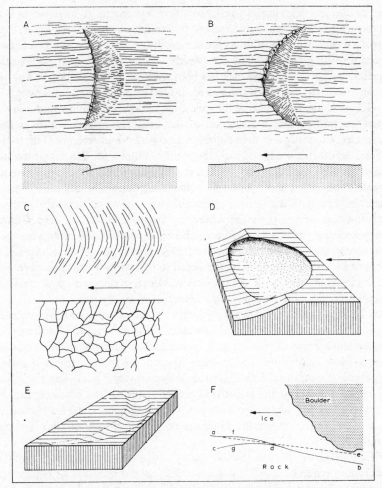

Fig. 6-3 Some small-scale features of glacial and fluvioglacial erosion. The horizontal arrow indicates the direction of ice motion

A Lunate fracture (plan and section) (S. E. Harris, 1943)
B Crescentic gouge (plan and section) (*ibid.*)
C Crescentic fractures (plan and section) (*ibid.*)
D Conchoidal fracture (E. Ljungner, 1930)
E Sichelwannen (F. Hjulström, 1935)
F Formation of a crescentic gouge (G. K. Gilbert, 1906)
   a d e original rock surface
  a f d b rock surface deformed by pressure transmitted from boulder
      in the ice
    c g d conoid fracture
      f g secondary fracture formed when the wedge of rock f g d is
      broken off

Sierra Nevada. Varying in size from a few centimetres to as much as 2 m across, and occurring usually in sets of two to six, one behind the other, their form is shown in Figure 6-3b. As the section shows, they are composed of two fractures, the gently dipping one clearly having been formed before the vertical one, and a chip of rock has been torn out between them. They are most commonly found on the upstream sides of projecting rock bosses.

Gilbert (1906a) argued that they were formed by a sizeable boulder in the basal ice transmitting pressure on to the bedrock surface, which at first deformed elastically and then ruptured along a conoid fracture (Fig. 6-3f). After fracture, the wedge of rock above the fracture tended to straighten and therefore lift upward slightly, when it was cracked off by the moving ice above. A puzzling feature is that the boulder responsible did not usually make direct contact with the bedrock, since the latter shows no sign of any major striation or grooving at the site where the crescentic gouge occurred. Gilbert suggested that perhaps a cushion of sand between the boulder and bedrock was able to transmit the pressure. The localization of gouges on the upstream faces of bedrock projections he attributed to increasing vertical compression of the ice as it passed over the projection, which in turn would increase the pressure on any boulder held in the lower layers of the ice.

F. H. Lahee (1912) was concerned with the second type of fracture noted by Chamberlin, which he termed 'crescentic fractures'. Figure 6-3c shows the characteristic form of these, which he studied on quartzite roches moutonnées in New Hampshire. The fracture outcrops are concave forward and of hyperbolic form, and the axis bisecting the concavity is parallel to the striae. The near-vertical fracture is clean cut near the rock surface, often cutting the actual rock grains, but at depths below a few millimetres, the fractures tend to wander between the grains. Unlike the crescentic gouge, no rock has been removed.

Both crescentic gouges (Sichelbrüche) and crescentic fractures (Parabelrisse) were regarded by E. Ljungner (1930) as the products of conoids of fracture induced by oblique pressure on the bedrock, an adaptation of Gilbert's hypothesis. Ljungner added another form of rock fracture induced indirectly by ice pressure, that of the conchoidal fracture (Muschelbruch) where the fracture plane is concave upward (Fig. 6-3d).

Harris's review of friction cracks added yet another type, the lunate fracture (Fig. 6-3a), very similar to the crescentic gouge except for the fact that the horns of the crescent point forward with the ice motion. It thus became clear that crescent-shaped friction cracks could not be used for distinguishing between two possible directions of ice motion along the axis simply by recording the direction in which the horns pointed, a fact amply confirmed by V. Okko's work in Finland (1950). However, Harris claimed that in all cases of friction cracks, direction of ice motion could be accurately determined by measuring the direction of dip of the 'principal fracture', which is always forward in the direction of ice motion. Not only do friction cracks have this advantage over striations for determining ice motion, but being larger, they are less prone to erasure by weathering, and they are also well developed on many rock types which are not suited to striation. It should be noted that Harris does not include chatter-marks, which have no principal fracture, under the general heading of friction cracks.

In experiments using a file or knife on glass, Harris succeeded in reproducing both crescentic gouges and lunate fractures. Further relevant experiments were described by P. MacClintock (1953): in scoring optical glass with a steel ball bearing, crescentic fractures formed with the horns pointing forward if the ball was forced forward without rolling, but if it was allowed to roll, the horns developed pointed backward. However, the form of the fracture in section did not support Ljungner's hypothesis that such fractures are conoids of percussion.

The idea that the dip of the principal fracture was a reliable indicator of ice movement was not supported by A. Dreimanis's work in Ontario (1953). In the case of crescentic fractures, which accounted for two-thirds of all the friction cracks examined, the majority of the fractures dipped not forward but backward against the ice flow. If Ljungner's theory of their association with conoids of percussion accepted, the backward dip could be produced by a high angle of impact. This would also be expected to produce crescentic gouges with steeply-dipping principal fractures, which in fact Dreimanis finds. The usual angle of dip of the latter he records as 50° to 70°, in contrast to the gentle or low dips said to be typical by Harris and Gilbert. On the other hand, Dreimanis confirmed Harris's claim that the principal fracture of crescentic gouges, regardless of how steeply it dipped, always dipped forward with the ice-flow, and therefore this one type of friction crack may be a good indicator of ice motion at the place where it is found. As in the case of striations, such local indications should not be given a wider regional significance.

Plastically-moulded Surfaces

Glaciated rock surfaces which have suffered smoothing and scouring, as distinct from plucking, often exhibit complexly moulded forms. A recent study by J. Gjessing (1965) of the Portør area of coastal southern Norway provides a good illustration. The area consists of granite intruded by amphibolite dykes; it has obviously suffered considerable abrasion but there are few signs of plucking. It was overridden by south-eastward moving ice in the last (post-Allerød) glaciation. Numerous roches moutonnées occur, heavily striated, and with the striations curving in round the stoss sides. Sinuous grooves furrow the rock, and these also are followed by striae. Striae show deflections around obstructions as Demorest (1938) and Edelman (1949) described elsewhere, and can even be traced into and out of potholes. The problem is to account for the moulding of a hard rock surface into such complex forms, and undoubtedly the processes were connected with the last glaciation.

Plastically-moulded surfaces ('p-forms') have also been examined in detail in northern Norway by R. Dahl (1965). He classifies them as follows:

1. *Cavetto forms*, consisting of channels, oriented with the ice flow, cut on steep rock faces (sometimes as horizontal grooves along vertical walls). They may be up to half a metre deep, have sharp edges, and the upper part may overhang. Striae and crescentic gouges are found within them.

2. *Grooves* occur on open flat surfaces, cut in the hardest of rocks, and have rounded edges to distinguish them from cavettos.

3. *Sichelwannen* are the commonest small trough form (Fig. 6-3e). They may or may not be symmetrical; their axis always coincides with the ice flow (and their horns point forward with the ice); and they are seen on both flat and sloping surfaces. Scratches conformable with the ice flow are found in them, but these need not necessarily be striae.

4. *Curved and winding channels* are often generally reminiscent of fluvial activity. They may arise through the merging of Sichelwannen. Most striking are the deep almost tubular variety whose genesis, according to Dahl, 'seems to require a plastic medium whose movements gave rise to a blasting effect. Ice can hardly be involved here' (p. 127).

5. *Bowls and potholes*. The former are commonest on granite and, curiously, were rarely found on horizontal surfaces. Potholes include both normal fluvial forms found on valley floors and others in anomalous situations, sometimes in rows down valley sides. Bowls and potholes will receive special consideration in a later section (p. 148).

Gjessing considers that four media may be considered as possible sculptors of the p-forms: basal ice containing rock debris, water-soaked ground moraine squeezed between bedrock and the overlying ice, subglacial meltwater under pressure, and ice-water compounds. The first of these has in the past often been accepted as adequate (for example, Demorest, 1938), but there are some p-forms which are hardly compatible with such an origin by ice itself, in the light of recent studies of ice behaviour under pressure, and observations are available to show that basal ice cannot accommodate itself to the more abrupt bedrock irregularities but bridges over them (for example, B. Kamb and E. LaChapelle, 1964). Gjessing himself favours the second medium, water-soaked ground moraine containing ice particles and moving in laminar flow according to differences in ice pressure. The pressure differences would be generated by variations in ice thickness and bedrock irregularities, and the viscous liquid would tend to move in the direction of ice surface slope and ice-flow. There are many observations on the existence and consistency of water-saturated ground moraine; thus Okko (1955) at the southern outlet glaciers of Vatnajökull observed how the basal debris occurred as a 'porridge-like till', and that farther up the glaciers it sometimes appeared to be squeezed upward out of crevasses (pp. 33-34). G. Hoppe (1952) contended that his Veikki-type till ridges represented water-soaked debris squeezed into crevasses. Gjessing argues that the high proportion of clay material in the Ra moraine off southern Norway originated in the subglacial liquid sludge which had scoured the p-forms inland. The questionable point is, however, not the existence of such a substance but its capacity to model hard rock surfaces. We do not know what velocities of movement of the substance might be involved.

There is much evidence to favour the third medium—subglacial meltwater —but before examining this, the fourth medium, that of an ice-water mixture may be briefly mentioned. G. Johnsson (1956) considered the p-forms of southern Sweden to be eroded by rock flour moved by an ice-water mass, and compared the forms produced by normal fluvial or marine water currents. The viscosity of the ice-water paste would vary with pressure and temperature.

Subglacial meltwater as the sculpturing agent of nearly all p-forms is proposed by Ljungner (1930), E. Ebers (1961), and Dahl (1965). Ljungner regarded Sichelwannen, for instance, as purely fluvial forms induced by turbulence (pp. 303-47); potholes and giant kettles (Riesentöpfe) are even more suggestive of fluvial action. A very attractive explanation of Sichelwannen as cavitation forms was put forward by F. Hjulström in 1935. Cavitation erosion has been recognized since 1894, but not until 1930 was it intensively studied, mainly in the field of engineering. It is associated with water moving at very high velocities: at ordinary temperatures it appears above a limiting velocity of 14·3 m/s at sea-level pressure, but if turbulence is introduced, the limiting velocity may be reduced by 10 per cent or more. Metal, glass, and even quartz surfaces are attacked, as hollows within the water mass collapse violently. If the water contains particles, as all subglacial meltwater streams will, these impact on the channel sides at very high speeds, scouring and scratching them to give a sand-blasted effect. Hjulström show, miniature Sichelwannen forms on a metal plate attacked by cavitation, where a protruding metal bolt had set up turbulence. In the case of sub-glacial streams, fast flowing because of hydrostatic pressure, he suggested that a moraine block might act as an obstacle behind which cavitation occurred, locating the Sichelwannen.

Dahl enlarges on Hjulström's suggestions, and considers that p-forms, with the possible exception of the large cavettos which may be ice-cut grooves, are modelled by meltwater moving very rapidly in subglacial tunnels or englacial tubes. The high velocities may be produced by the sudden draining of englacial water-bodies through restricted tunnels. Sichelwannen are the initial cavitation forms which may be gradually modified by corrasion into potholes if the water flow at that point were prolonged. The action of cavitation can explain the curious fact that hard quartz veins in the p-forms are usually worn down as much as the surrounding rock, whereas on other glacially-eroded surfaces they stand up. The patchy occurrence of p-forms Dahl explains by suggesting that only in these places did the subglacial or englacial water contact the bedrock, and the orientation of grooves and the axes of Sichelwannen are said to be controlled by the fact that the meltwater was usually forced in the direction of ice-flow (compare esker tunnels, Chapter 16). Bedrock structure plays no significant part in their orientation, nor in the frequency of their occurrence.

Potholes

The association of potholes with p-forms has already been noted. As is well known, potholes may form in normal stream beds and are not necessarily connected with glaciation at all. However, some potholes occur in such situations that a fluvioglacial origin must be invoked. The literature contains numerous examples. In 1874, W. C. Brögger and H. M. Reusch communicated the results of their survey of 'giants' kettles' in the Oslo region; after removing infill, the largest pothole at Bäkhelagel proved to be 2·4 m in diameter and 10·4 m deep. The walls were smooth, and in some cases were cut in spiral form like 'the impression of a gigantic snail' (p. 756). Smoothly rounded stones were found in the lower layers of infill, one weighing 150 kg. The potholes could not possibly have been related to subaerial drainage be-cause of their positions. Another celebrated instance is the Kailpot beside Ullswater noted by J. E. Marr in 1926. This is nearly one metre deep and of about the same diameter; it lies below steep crags at a height of only 1 m above lake level. In North America, gigantic potholes occur at Taylor's Falls, Minnesota (up to 18 m deep and 3·6 m diameter), while on the steep quartzite bluffs above Devil's Lake in Wisconsin, small potholes are perched in an extraordinary position 120 m above the lake level. Other examples of potholes in anomalous positions are found in the Gletschergarten of Lucerne (one is perched on top of a roche moutonnée), and one noted by K. Faegri (1952) on top of a small island west of Bergen. Many of the larger potholes possess spiral grooves, and contractions or widenings in cross-section.

The hypothesis widely adopted until recently to account for such potholes was the glacier moulin hypothesis, set out by Brögger and Reusch (1874), and applied and amplified by later workers such as W. Upham (1900), G. K. Gilbert (1906b), M. B. Fuller (1925), J. E. Marr (1926), and K. Faegri (1952). In essence, it was held that streams of water pouring down crevasses or other openings in the ice (glacier moulins) would strike the bedrock and create a pothole at that point. If the jet of water struck the hole obliquely, it would spiral up out of it. Although some glacier meltwater streams do behave in this way, there are good reasons for thinking that this is not the correct explanation for most potholes. First, if the glacier was moving to provide the crevasses, how did the waterfall remain in one place long enough to cut a pothole? Secondly, there are clearly problems in explaining situations where potholes occur in great numbers. Thirdly, in all but very thin glaciers, it seems unlikely that the torrent could descend to bedrock in one unbroken fall. Fourthly, the potholes are quite different in detail from the plunge-pool holes normally found beneath waterfalls.

H. S. Alexander (1932) recognized three types of pothole. Plunge-pool holes, the first type, are broad, open, and have flaring walls; they are formed by waterfalls, are not usually narrow or deep, and lack spiral grooves. Secondly, gouge-holes are shallow depressions in stream beds caused by oblique impact of water currents maintaining a planar, not a spiral circulation. Thirdly there are eddy-holes which may be very deep, possess sharp edges, and as confirmed by the spiral grooves on their walls, are produced by a vorticular circulation. They form on stream beds, and it is evident that nearly all potholes described in formerly glaciated regions are of this type —in other words they were the work not of moulin waterfalls but of fast-flowing subglacial or englacial streams.

In experiments with water jets entering a glass cylinder, Alexander showed by study of the movement of loose material that spiral motion does occur, that the energy available for bottom erosion falls off rapidly with increasing depth, and that the angle of the jet can be varied considerably (between 30° and 70°) without much change in the amount of vortex energy generated. However, the vortex energy falls off very rapidly to nil when the jet is vertical: thus spiral grooves in potholes are incompatible with moulin water-falls. The infill of many potholes is probably of no great significance, since it could easily have accumulated after the hole was formed. The so-called grinding stones found by Brögger and Reusch and others may be merely water-worn pebbles subsequently washed into the hole, or even rounded morainic boulders. An important requirement for the formation of deep eddy-holes is that removal of sediment from the hole should at least equal the input, for otherwise the hole will be choked.

As noted in the previous section, potholes are included among p-forms by Gjessing and Dahl. Hjulström and Dahl both consider that cavitation may provide the initial locus for the eddy, and Dahl is convinced that the potholes result from high-energy subglacial or englacial streams striking the bedrock at an angle. C. G. Higgins (1957) and R. Streiff-Becker (1951) put forward similar ideas, and advocated that the moulin hypothesis should now be abandoned.

Potholes of supposed fluvioglacial origin have often been used to determine the limits and extent of former glaciation. Great caution is necessary, though, since potholes are also formed by normal subaerial streams, or even by the overflow waters of ice-dammed lakes, and there have been cases where weathering pits have been taken as potholes. For instance, M. B. Fuller (1925) claimed that there were over 2000 fluvioglacial potholes in part of the Front Range, Colorado, ranging from a few centimetres to 6 m (possibly even 15 m) deep. Although she noted that they occurred almost solely on flat-surfaced granite outcrops and often at joint intersections, and had a weathered appearance, no attempt to explain them as anything other than glacier moulin potholes was made. They were used to delimit the extent of ice in an early Pleistocene glaciation. F. E. Matthes (1930) was careful to distinguish in the Yosemite area between true potholes (as at the lower end of Tuolumne meadows) probably of subglacial origin, and the numerous basin-shaped cavities resulting from strongly localized granite weathering (as at Glacier Point), which he termed 'weather pits'. The latter, in contrast to potholes, had sharply overhanging rims, since enlargement by chemical action is proceeding at or below the level of water standing in them. W. D. Jones and L. O. Quam (1944) strongly refuted Fuller's attempts to use so-called potholes to determine glacial limits in the Rocky Mountains, Colorado; the potholes 'are adequately explained as a pitting in the granite of the sort found in many unglaciated areas' (p. 233). Yet there are cases where true potholes almost certainly do throw light on former ice extent. D. E. Hattin (1958) found numerous eddy-holes in different rock types, possessing smoothed and undercut walls, at over 1200 m on Mount Washington, New Hampshire: these do seem to demand a high-level ice cover in the last main glaciation here.

Plucked Surfaces

Glacial erosion is thought to operate in two distinct ways: by abrasion of rock surfaces (giving them a smoothed or striated appearance) and by pluck-ing or quarrying (also known as joint-block removal). Many consider that the second of these is by far the most important quantitatively; it also provides the tools essential for the first process. Matthes (1930) held strongly to this viewpoint, asserting that abrasive action alone was quite incompetent to account for the excavation of the deep valleys in the Yosemite region and that in tough massive rocks it achieved little but polishing (p. 89). But where the arrangement and spacing of joints was suitable, and where blocks and slabs of rock had already been loosened by chemical and frost-action, block removal by ice on a vast scale became possible, producing a characteristically plucked or quarried surface. Matthes pointed to the abundance and size of angular blocks in moraine, and in the form of erratics, as evidence of the efficacy of plucking. He considered that the process operated best in rocks with moderately coarse joint texture, with perhaps a mean joint spacing of 1 to 7 m, for larger blocks might be beyond the power of the ice to move while smaller material, though quarriable, would not provide such great bulk in total.

A. Cailleux (1952) also contends that glaciers moving over massive un-fractured bedrock may be mainly agents of polishing and minor abrasion. On the other hand, deep ground freezing before the arrival of ice in a region (as M. Boyé, 1950, suggests) may shatter the bedrock; then the advance of a temperate glacier, causing ground temperatures to rise to pressure melting-point or higher, would thaw out the bedrock and transport away the debris.

Evidence that certain rocks or strata are more susceptible than others to block-removal has been produced by many workers. Matthes himself de-monstrated that the narrowing of the lower Yosemite valley into the Merced Gorge marked the point where well-jointed rocks terminate and the un-quarriable El Capitan granite begins. W. O. Crosby (1928) argued that the massiveness of the granite flooring some Alaskan valleys inhibited their deepening by ice. Studying rock surfaces below the Altels and Rosenlaui glaciers, O. Flückiger (1934) found that ice had minimum adhesive and erosive effectiveness when moving down the bedding planes of steeply inclined formations. Another good example of the influence of rock structure on the ability of glaciers to erode effectively was given by I. B. Crosby (1945). Near Bingham in Maine, ice has been able to widen the Kennebec valley where this crosses a zone of hard quartz-mica schist, for the latter possesses good blocky jointing. But it has had little effect in an adjacent part of the valley where this is crossed by a belt of soft, but mainly unjointed, slate. Similar conclusions were reached by J. H. Zumberge (1955) in Minnesota and Michigan. By quarrying, said to be the most important process of glacial erosion, ice was able to excavate rock basins in some highly jointed lava flows and jointed slate beds, but was unable to erode significantly in some massive sills. Till in the area is largely composed of fragments from the quarriable formations only.

In some cases, the sizes of angular boulders in moraine or till can be used to suggest the depths reached by glacial plucking (Chapter 9, p. 225). Demorest (1939) argued that the efficacy of plucking would in fact decrease when the ice had removed rock to depths where joints were fewer; thereafter, the supply of rock fragments to cause abrasion would also diminish and a hypothetical end-stage of erosion would be reached. The argument takes no account of the fact that new dilatation joints would probably develop as erosion proceeded (see p. 160).

The mechanism of glacial plucking is not simply a tug-of-war between moving ice and unbroken rock, for the tensile strength of the latter is in all cases except unconsolidated rocks many times greater than that of ice. Pre-existing fractures in the rocks are essential, and even then, simple ice adhesion may not be very effective. Probably freeze-thaw action plays a vital role in preparing a rock surface for plucking—either as frost action before the advance of the ice (A. Cailleux, 1952) or as subglacial freeze-thaw action. Boulders jammed against the rock bed by moving ice may also be able to gouge fragments from the bed (Chapter 7).

Plucking gives a characteristically shattered and broken appearance to a landscape (unless it is subsequently till-covered), particularly when viewed against the direction of ice flow, and the contrast between surfaces subjected mainly to abrasion and polishing and those affected chiefly by plucking (often adjacent to the former) is a most striking feature of glaciated terrain.

Roches Moutonnées (Stoss-and-Lee Topography)

Roches moutonnées, in which smoothed and plucked surfaces are juxtaposed, are 'the universal earmark of the invasion of an area by glacier ice' (O. Flückiger, 1934). The term was introduced by H. B. de Saussure in 1787 who likened roches moutonnées fields to the rippled appearance of wavy wigs styled moutonnées in his day. As R. F. Flint observes (1957, p. 64), the term has been both misapplied and mistranslated; he suggests it should be avoided, though there is already a mass of literature which uses the term. An alternative is stoss-and-lee topography. The smoothed stoss side is streamlined, with striae curving round it and coming together parallel again on the lee side (for example, Okko, 1955, p. 75), which is usually abruptly broken and shattered. This contrast can be used to indicate approximately former directions of ice motion. Roches moutonnées usually occur in groups dominating considerable areas, as along the Norwegian coast or in the Åland Islands of the Baltic. The individual bosses vary greatly in size, and some writers such as O. D. von Engeln (1938) appear not to limit their size at all, thus extending the category of roches moutonnées to include many glacially streamlined landforms. W. V. Lewis (1947) showed that there is a continuous gradation through major roches moutonnées to valley steps.

Among older theories of their origin, W. M. Davis (1909) regarded them as the ice-moulded residuals of pre-glacial bedrock irregularities, implying that had glaciation lasted long enough, they would have been removed. Matthes (1930) explained them on the basis of variations in joint texture, the ice leaving as protrusions those parts of the bedrock which were more massively jointed. Many roches moutonnées fields do not in fact exhibit any such close relationship to joint patterns. O. Flückiger (1934), supported by von Engeln (1938), attempted to explain the streamlining of the stoss side as the product of moulding by giant stationary waves in the ice (which were not visible at the ice surface owing to ablation). The waves developed on the basis that where two fluids of different specific gravity move over each other, waves form at the contact plane, and in the same way, he supposed that waves would appear when ice flowed over bedrock or when clean upper ice flowed over sluggish dirty basal ice. The ice was said to press harder on the bedrock at the wave troughs and therefore erode more. Such stationary waves have rarely been observed, and as Demorest pointed out in 1939, the vital point ignored by Flückiger is the high viscosity of ice which compels it to move always in laminar flow. Flückiger and von Engeln held that roches moutonnées will be most prominently developed where ice action has been most vigorous and prolonged—they represent the end-stage of the glacial cycle. Demorest adopted a very different viewpoint, regarding them as 'pre-glacial structures which have successfully resisted the last abrasive action of the ice while the areas immediately surrounding them have been cut lower' (p. 600). In 1937, he described roches moutonnées fields in west Greenland where structure (mainly the foliation of the gneiss) had been of paramount importance, giving rise even to 'reversed' forms with steep stoss and gentle lee slopes where the dip was away from the direction of ice advance.

The normal shattered and plucked lee sides of roches moutonnées probably represent subglacial freeze-thaw action. H. Carol (1947) succeeded in 1940-2 in penetrating to a subglacial cavity on the downstream face of a roche moutonnée beneath the Ober Grindelwald glacier. Observing that the presence of the bedrock protrusion was locally increasing the pressure at the base of the glacier (which here was at pressure melting-point) and causing meltwater to form, he argued that in the cavity between ice and bedrock on the downstream face of the protrusion, reduced pressure would encourage the meltwater to refreeze and thus shatter the bedrock. The problems encountered by such freeze-thaw hypotheses are elaborated in Chapter 8. However, a recent study of subglacial cavities under the Blue Glacier, Washington, by Kamb and LaChapelle (1964) indicates that pressure-induced freeze-thaw does seem to occur. Regelation spicules were observed, weakly attached to the under-surface of the ice as it bridged these cavities (which were up to 10 m long and 20 cm high), suggesting alternations of freezing and thawing, probably induced by pressure changes. Whether such small oscillations of a freeze-thaw cycle can be relevant to the large-scale shattering associated with the lee sides of roches moutonnées is a problem which has not yet been adequately answered. Many more studies of conditions in subglacial cavities are clearly needed.

Conclusion

Small-scale rock markings of formerly glaciated regions include such features as striations, polished surfaces, friction cracks, moulded and plucked surfaces, and potholes. Glacial striations and grooves are excavated by angular fragments pressed on to bedrock by larger boulders in the basal ice. In using such features for determining former directions of ice motion, care must be taken to exclude striations caused by non-glacial agents (such as snow slides) and to recognize that striations in one locality only indicate basal ice motion in that locality and need not be in accord with long-term regional ice movement. Crossing sets of striations in the majority of cases do not represent different glacial episodes but merely changes in ice-flow resulting from variations in ice thickness.

Friction cracks provide some of the best evidence of the power of moving ice to erode bedrock. Crescentic gouges and fractures represent elastic deformation and rupture of a bedrock surface by extreme pressure exerted by an ice-impelled boulder. The dip of the principal fracture of a crescentic gouge is in accord with the direction of ice motion.

Bedrock surfaces moulded rather than plucked by flowing ice have only very recently been systematically studied. The forms are complex, and include channels, grooves, bowls, and Sichelwannen (small sickle-shaped troughs). Some may have been eroded by the abrasive effects of water-soaked subglacial till being squeezed into motion by differential ice pressure. The majority are probably cut by subglacial meltwater moving at high velocities under pressure.

Potholes in formerly glaciated regions were often the work of glacier meltwater streams. Some reach gigantic size—depths of 18 m are known—and may possess spiral-shaped grooves in their walls. The glacier moulin hypothesis must be abandoned except for the occasional plunge-pool holes which are different in being shallow and open. The majority of deep pot-holes are eddy-holes cut by fast-flowing subglacial streams or by englacial streams impinging on the subglacial bedrock surface. These forms, if correctly identified, can be used to determine the extents of a former ice cover. Plucked surfaces represent the removal of joint blocks by ice. When the joint blocks are of a certain size (possibly 1-7 m), the ice may be able to remove very great quantities of bedrock. Another important factor may be bedrock shattering by a period of deep freezing prior to ice advance. Well-jointed rocks are much more susceptible to glacial quarrying than massive rocks, and even relatively soft formations if not well jointed may resist ice erosion by plucking.

The formation of roches moutonnées is far from being clearly understood. Probably the smoothed up-glacier slope represents glacial abrasion and polishing, while the shattered down-glacier side represents joint-block removal effected by freeze-thaw action in subglacial cavities. Meltwater passes from the high pressure up-glacier face of the obstacle to the low pressure down-glacier face where it refreezes.

REFERENCES

L. AGASSIZ (1838), 'On the polished and striated surfaces of the rocks which form the beds of glaciers in the Alps', *Proc. geol. Soc. Lond.* 3, 321-2

H. S. ALEXANDER (1932), 'Pothole erosion', *J. Geol.* 40, 305-37

M. BOYÉ (1950), *Glaciaire et périglaciaire de l'Ata Sund nord-oriental (Groenland)* (Paris; Expéditions polaires françaises, 1)

W. C. BRÖGGER and H. M. REUSCH (1874), 'Giant's kettles at Christiania', *Q. J. geol. Soc. Lond.* 30, 750-71

A. CAILLEUX (1952), 'Polissage et surcreusement glaciaires dans l'hypothèse de Boyé', *Revue Géomorph. dyn.* 3, 247-57

H. CAROL (1947), 'Formation of *Roches Moutonnées*', *J. Glaciol.* 1. 57-59. See also W. V. LEWIS, 'Some comments on Dr H. Carol's article', ibid., 60-63

T. C. CHAMBERLIN (1888), 'The rock scorings of the great ice invasions', *U.S. geol. Surv. 7th A. Rep.* (1888), 155-248

T. H. CLARK (1937), 'Northward moving ice in southern Quebec', *Am. J. Sci.* 34, 215-20

I. B. CROSBY (1945), 'Glacial erosion and the buried Wyoming valley of Pennsylvania', *Bull. geol. Soc. Am.* 56, 389-400

W. O. CROSBY (1928), 'Certain aspects of glacial erosion', *Bull. geol. Soc. Am.* 39, 1171-81

R. DAHL (1965), 'Plastically sculptured detail forms on rock surfaces in northern Nordland, Norway', *Geogr. Annlr* 47, 83-140

W. M. DAVIS (1909), 'Glacial erosion in North Wales', *Q.J. geol. Soc. Lond.* 65, 281-350

M. DEMOREST (1937), 'Glaciation of the Upper Nugssuak Peninsula, West Greenland', Z. Gletscherk. 25, 36-56

M. DEMOREST (1938), 'Ice flowage as revealed by glacial striae', J. Geol. 46, 700-25

M. DEMOREST (1939), 'Glacial movement and erosion: a criticism', Am. J. Sci. 237, 594-605

A. DREIMANIS (1953), 'Studies of friction cracks along shores of Cirrus Lake and Kasakokwog Lake, Ontario', Am. J. Sci. 251, 769-83

J. L. DYSON (1937), 'Snowslide striations', J. Geol. 45, 549-57

E. EBERS (1961), 'Die Gletscherschliffe und -rinnen' in Der Gletscherschliffe von Fischbach am Inn (ed. H. FEHN), Landesk. Forsch., München, 40

N. EDELMAN (1949), 'Some morphological details of the roches moutonnées in the archipelago of S.W. Finland', Bull. Commn géol. Finl. 144, 129-37

N. EDELMAN (1951), 'Glacial abrasion and ice movement in the area of Rosala-Nötö, S.W. Finland', Bull. Commn géol. Finl. 154, 157-69

O. D. VON ENGELN (1938), 'Glacial geomorphology and glacier motion', Am. J. Sci. 35, 426-40

K. FAEGRI (1952), 'On the origin of potholes', J. Glaciol. 2, 24-25

O. FLÜCKIGER (1934), 'Glaziale Felsformen', Petermanns Mitt., Ergänz. 218, 55 p.

J. D. FORBES (1843), Travels through the Alps of Savoy (Edinburgh)

M. B. FULLER (1925), 'The bearing of some remarkable potholes on the early Pleistocene glaciation of the Front Range, Colorado', J. Geol. 33, 224-35

G. K. GILBERT (1906a), 'Crescentic gouges on glaciated surfaces', Bull. geol. Soc. Am. 17, 303-16

G. K. GILBERT (1906b), 'Moulin work under glaciers', Bull. geol. Soc. Am. 17, 317-20

J. GJESSING (1965-66), 'On plastic scouring and subglacial erosion', Norsk geogr. Tidsskr. 20, 1-37

J. GJESSING (1967), 'Potholes in connection with plastic scouring forms', Geogr. Annlr 49, 178-87

S. E. HARRIS (1943), 'Friction cracks and the direction of glacial movement', J. Geol. 51, 244-58

D. E. HATTIN (1958), 'New evidence of high-level glacial drainage in the White Mountains, New Hampshire', J. Glaciol. 3, 315-19

C. G. HIGGINS (1957), 'Origin of potholes in glaciated regions', J. Glaciol. 3, 11-12

G. HOPPE (1952), 'Hummocky moraine regions with special reference to the interior of Norrbotten', Geogr. Annlr 34, 1-72

F. HJULSTRÖM (1935), Studies of the morphological activity of rivers as illustrated by the River Fyris (Uppsala)

R. H. JAHNS (1943), 'Sheet structure in granites: its origin and use as a measure of glacial erosion in New England', J. Geol. 51, 71-98

G. JOHNSSON (1956), 'Glacialmorfologiska studier i Södra Sverige', Meddn Lunds geogr. Instn 30, 1-407 (with many illustrations and an English summary)

W. D. JONES and L. O. QUAM (1944), 'Glacial landforms in Rocky Mountain National Park, Colorado', J. Geol. 52, 217-34

B. KAMB and E. LACHAPELLE (1964), 'Direct observation of the mechanism of glacier sliding over bedrock', J. Glaciol. 5, 159-72

F. H. LAHEE (1912), 'Crescentic fractures of glacial origin', *Am. J. Sci.* **33**, 41-44

E. LJUNGNER (1930), 'Spaltentektonik und Morphologie der schwedischen Skaggerrack-Küste', *Bull. geol. Instn Univ. Upsala* **21**, 1-478

P. MACCLINTOCK (1953), 'Crescentic crack, crescentic gouge, friction crack, and glacier movement', *J. Geol.* **61**, 186

J. E. MARR (1926), 'The Kailpot, Ullswater', *Geol. Mag.* **63**, 338-41

F. E. MATTHES (1930), 'Geologic history of the Yosemite Valley', *U.S. geol. Surv. Prof. Pap.* **160**, 137 p.

V. OKKO (1950), 'Friction cracks in Finland', *Bull. Commn géol. Finl.* **150**, 45-50

V. OKKO (1955), 'Glacial drift in Iceland', *Bull. Commn géol. Finl.* **170**, 1-133

C. F. S. SHARPE (1938), *Landslides and related phenomena* (New York)

P. SHELDON (1926), 'Significant characteristics of glacial erosion as illustrated by an erosion channel', *J. Geol.* **34**, 257-65

H. T. U. SMITH (1948), 'Giant glacial grooves in Northwest Canada', *Am. J. Sci.* **246**, 503-14

R. STREIFF-BECKER (1951), 'Pot-holes and glacier mills', *J. Glaciol.* **1**, 488-90

W. UPHAM (1900), 'Giants' kettles eroded by moulin torrents', *Bull. geol. Soc. Am.* **12**, 25-44

K. VIRKKALA (1951), 'Glacial geology of the Suomussalmi area, east Finland', *Bull. Commn géol. Finl.* **155**, 1-66

K. VIRKKALA (1960), 'On the striations and glacier movements in the Tampere region, Southern Finland', *Bull. Commn géol. Finl.* **188**, 159-76

J. H. ZUMBERGE (1955), 'Glacial erosion in tilted rock layers', *J. Geol.* **63**, 149-58

Chapter 7

Glaciated Valleys

The length of time during which these glaciers have existed must have been so great, and there is so much evidence of their dimensions having been formerly greater than now, that a considerable portion of the erosion, or of the widening and deepening of the valley must be attributed to ice-action. But to what extent the valleys in which the Swiss glaciers move were excavated by rivers before the valleys were filled with ice we have no positive data at present for deciding. (Sir CHARLES LYELL, 1866)

The features characteristic of glaciated valleys are among the most easily recognized and the most spectacular results of glaciation. The method of formation of these features is, however, not nearly so obvious, and doubt still remains concerning their formation in many instances. The great glacial troughs, such as those of the Alps and Norwegian mountains, carved out to depths of hundreds or even thousands of metres, demonstrate clearly the immense erosive activity of which glaciers are capable under certain circumstances. Glacial erosion was most intense when thick, well-nourished temperate glaciers were flowing rapidly down steep slopes towards a free outlet. These conditions were realized on the windward slopes of highland regions which received heavy snowfall and which sloped steeply down towards the sea. This situation was particularly characteristic of western Norway, north-west Britain, British Columbia, southern Chile and south-west New Zealand. Under these conditions, fast-flowing glaciers were generated, usually following valleys prepared for them by pre-glacial rivers, and enlarging them into spectacular troughs.

The glaciers that flowed through these valleys originated in various ways. Some of them extended out from cirques as the snow-line fell with increasing cold; others originated on upland plateau areas, forming the outlet glaciers of mountain ice-caps. Vatnajökull in Iceland and Jostedalsbreen in Norway provide examples of this type of ice mass from which many valley glaciers descended and still descend to lower levels. The present outlet glaciers are only very small remnants of their much larger Pleistocene predecessors, whose effects are now clearly revealed in the parts of the valleys from which the glaciers have retreated.

Glaciated valleys develop characteristic transverse and longitudinal profiles. Breached watersheds are another feature of interest associated with the action of large valley glaciers. In many areas, glacial activity has not been sufficiently prolonged for the landforms to develop beyond a youthful or early mature stage; in others, however, there is evidence of the more prolonged effects of valley glaciers on the landscape. Details of the processes operating to produce these forms will be considered.

The Cross-profiles of Glaciated Valleys

One of the most striking characteristics of a well-developed glaciated valley is the U-shape of its cross-profile. The side walls are considerably oversteepened, approaching the vertical in places, as for example in the Austerdalsbreen valley in Norway (Plate IX) and the Lauterbrunnen valley in Switzerland. The valley floor, on the other hand, is flatter than in river valleys, a feature which is partly or wholly the result of deposition of valley train deposits. The extreme flatness of this type of fluvio-glacial deposition is discussed in Chapter 18.

There have been relatively few attempts to measure accurately the geometrical form of a glaciated valley, but one example is that given by H. Svensson (1959), who found that the cross-section of the Lapporten valley in Sweden closely resembles a parabola. The valley cross-profile can be defined by the formula $y = 0 \cdot 000402x^{2 \cdot 046}$, where y is the height and x is the distance from the valley centre. Because the exponent is almost 2, the formula closely approximates to that of a true parabola.

The formation of the glacial parabolic or U-shaped cross-profile could be achieved by the widening of the valley from a preglacial V-shaped form without overdeepening. Alternatively the widening could be accompanied by overdeepening, and when the longitudinal profile is discussed, it will be apparent that the latter can often occur. The deepening of the cross-profile depends on the ability of the glacier to cut into the rock of its bed.

An interesting suggestion has been made by M. Boyé (1950) (see Chapter 6) to account for this aspect of glacial erosion. He considered that the preparation of the valley floor into which the advancing ice is penetrating is important. Permafrost could prepare the ground for effective glacial erosion, by breaking up the rocks in zones where there is plenty of moisture. Slope water would tend to drain to the bottom of the valleys, increasing the water content in this area. It may be suggested, therefore, that the valley bottoms are subjected to particularly intensive frost action and the rocks in this situation are broken up to a greater extent than those on the drier valley sides. The advancing glaciers take advantage of this and remove a greater proportion of material from the valley bottom. The ice thickness and hence the stress of the ice on the bottom is greatest in the valley floor, and this also helps to concentrate glacial erosion in this zone (Fig. 7-1).

Another possible mechanism is the action of pressure release beneath thick ice. W. V. Lewis (1954), D. L. Linton (1963) and others have pointed out that as a glacier bites deeper into its bed, rock of density 2·5 or more is replaced by ice possessing a much lower density (0·9). This reduction of density on the floor of the glacier might lead to the development of horizontal dilatation joints, especially if there are pre-existing stresses in the rock. Breaking up of the rocks in this way would enable the glacier to carry away further material from its bed and increase the depth of the valley. Where glacial erosion has produced a very deep valley, such as that of the Lauterbrunnen or in parts of the Sognefjord system, it seems likely that much of the overdeepening must have been achieved by glacial plucking of previously loosened joint blocks. As was noted in Chapter 6, scouring of the bed would not be able to move such a large volume of material, although this process has no doubt also been active. The preparation of the rock bed to a state which enables the rock to be readily removed by the moving ice is therefore a process worthy of close examination, and one which, if effective, would help to account for the deep U-shape of glacial valleys.

Confirmation of the pressure release process has come from quarrying and mining operations, in which rock is known to expand following the removal of overburden pressure (G. W. Bain, 1931; R. H. Jahns, 1943). The dilatation joints normally develop parallel to the surface from which the load is removed, and thus on valley floors or flat surfaces they tend to be horizontal. There are fine examples of horizontal dilatation jointing in the Rapakivi granite of the Åland Islands in the Baltic, an area which at times was subjected to the pressure of nearly 3000 m of ice. Rapid thinning of the ice allowed sudden reduction of pressure on the rock, which in turn may have caused the development of the dilatation jointing. There are also many examples of dilatation jointing developing parallel with the sidewalls of glaciated valleys. It is possible that the well-known sheet structures in the granite of Yosemite valley in California (F. E. Matthes, 1930) result from rapid excavation of the valley, first by river action and later by ice. As the sheeting develops parallel to the surface being eroded, it helps to perpetuate the same cross-profile during valley enlargement, so that glaciated valleys often show very similar cross-profiles regardless of scale. Many glaciated valleys in Britain could be quoted to show dilatation jointing developed in conformity with a parabolic cross-profile, for instance, Glen Sannox in north Arran.

W. B. Harland (1957) suggested an additional hypothesis, augmenting the possible effect of dilatation in the development of jointing and the breaking up of bedrock. Glacial retreat would not only unload the bedrock, but also ex-

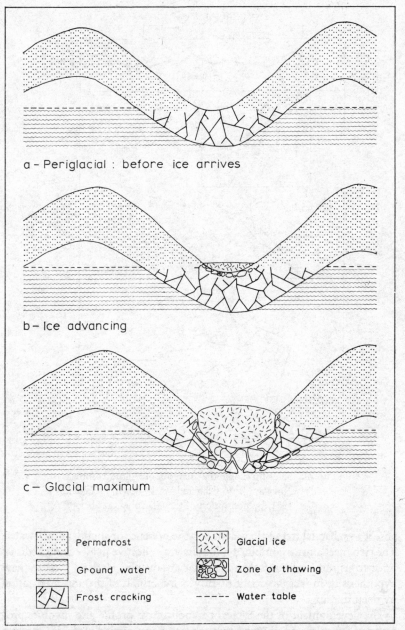

a – Periglacial : before ice arrives

b – Ice advancing

c – Glacial maximum

Permafrost

Ground water

Frost cracking

Glacial ice

Zone of thawing

----- Water table

Fig. 7-1 Diagrams to show the preparation of the valley floor by permafrost and the effects of glacial erosion (J. Tricart and A. Cailleux, *Le Modèle glaciaire et nival*, III, Société d'Edition d'Enseignement Supérieur, Paris, 1962)

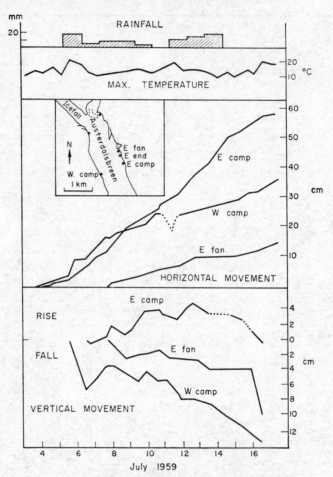

Fig. **7-2** Side-slip along the margins of Austerdalsbreen,
Norway (J. W. Glen and W. V. Lewis, *J. Glaciol.*,
1961, by permission of the Glaciological Society)

pose it to subaerial freeze-thaw action in the presence of abundant meltwater.
The two mechanisms combined might be very effective in loosening bedrock
prior to its removal in a subsequent glacier advance. Harland's freeze-thaw
hypothesis is, in fact, a variant of Boyé's hypothesis of ground preparation
by deep freezing.

The development of the typical parabolic cross-profile also involves processes of abrasion and gouging by rock fragments carried in the ice moving
past the floor and sidewalls. The processes can be examined with reference

to the slip of a glacier past its sidewall. Measurements of side-slip in relation to Austerdalsbreen have been made and analysed by J. W. Glen and W. V. Lewis (1961). At a number of points along the side of this glacier, it was possible to measure side-slip directly, as explained in Chapter 4. The rock is hard gneiss, which is well striated and shows evidence of glacial gouging. Measurements were made at several situations on the glacier (Fig. 7-2). One was near the base of the ice-fall on both sides of the glacier and other measurements were made midway between the ice-fall and the glacier snout. The measurements near the base of the ice-fall showed that the side-slip here amounted to as much as 65 per cent of the maximum centreline velocity (marginal velocity 26 cm/day, compared with centreline velocities of 39·8 cm/day a little farther up-glacier and 19·4 cm/day a little down-glacier) (Glen, 1958). This relatively high side-slip suggests Blockschollen flow (Chapter 4), expected near the foot of a fast-flowing ice-fall. In the lower part of the glacier, the measurement sites were situated at either end of a line of stakes used to measure a transverse velocity profile. At these sites, side-slip amounted to about 20 per cent of the maximum velocity on the profile.

The observations provide important evidence on the way in which a glacier erodes its sidewalls. It was noted that boulders occurred everywhere between the striated and plucked rock walls and the ice moving past them. There were also many boulders on the ice surface, some of which, disturbed by melting, fell into the sidewall gap, adding to the debris there. The movement of these boulders in the gap was clearly responsible for the striations so well displayed on the rockwall there. There were also signs of quarrying activity where blocks had been riven off the down-glacier side of protrusions.

The process of joint-block removal from the valley floor and sides has been examined by J. G. McCall (1960). He notes that the yield stress of ice in shear, assuming perfect plasticity, is only about $\frac{1}{160}$ of that of granite. If a metre cube of granite is considered to be resting on the glacier bed or side, then the theoretical pressure of the ice against the cube consists of a compressive stress on its up-glacier face of a maximum of 20,000 kg. A total maximum shear stress of 30,000 kg would theoretically be exerted on the top and sides of the cube. If the opposing effect of ice on the down-glacier side is ignored, the maximum theoretical stress would therefore be 50,000 kg. If the value actually developed were only half this, and assuming the shear strength of granite to be 160 kg/cm², the force impelling the block would be adequate to shear off a projection in the bedrock having a bearing surface of $\frac{25000}{160} = 160$ cm². Rocks are generally stronger in compression than in tension; the compressive strength of granite, for instance, is about 1400 kg/cm². If the rock cube pressed only on one point on the bedrock, it could crush a protrusion whose area theoretically could be $\frac{25000}{1400} = 18$ cm².

In the case of gouging, McCall shows that up to 16 cm² could be carried away, using the same assumptions. These relative values show that quarrying by the impact of rocks carried in a moving ice stream can be highly effective.

An actively eroding glacier must carry a considerable number of boulders at the ice-rock interface, for it is these boulders that the glacier uses as tools to accomplish its erosion. The tools can be used both for grinding and prising. Their frequent occurrence at the side of glaciers, in the form of lateral moraine deposits, may help to account for the widening that produces the typical U-shaped glacial valley. Frost action on the exposed valley walls will also produce fragments to fall into the glacier sidewall gap. Erosion of this type cannot, however, take place unless the glacier is moving actively along its sidewalls.

The effectiveness of boulders in carrying out erosion increases to some extent with depth beneath the ice. Near the surface, the hydrostatic pressure is small and the ice can flow round the boulder instead of exerting its full pressure against it. At depths of about 22 m, on the other hand, the ice can exert a pressure of about 2 bars, equivalent to its yield stress in compression of about 2 kg/cm² at 0°C., as suggested by McCall (1960) on the basis of the perfect plasticity theory. On this basis, the ice could not exert any greater pressure on the boulder at depths greater than 22 m. The force exerted on the boulder would be limited by the yield strength of ice and would depend solely on the size of the boulder or group of boulders. If a power flow law is adopted, the force that can be exerted on the boulder will increase slightly with depths greater than 22 m, but the question is one that has never been fully investigated.

The actual velocity of the moving ice is a less important factor. This is because the force exerted by ice on the boulder only increases as the $\frac{1}{4}$ power of the ice velocity, according to Glen and Lewis (1961). An increase of ice velocity has relatively little effect on the ice pressure, although it will assist erosion to the extent that the number of boulders carried past a point will increase with the velocity.

It is sometimes possible for valleys in which glaciers have flowed to exhibit V-shaped sections in a generally U-shaped valley. The V-shaped sections are usually associated with reaches of greater gradient than the U-shaped portions, as in the case of the Ötztal, Austria (Plate XI), the valley below Turtagrø, Norway (Lewis, 1947a), and the Aberglaslyn Pass, North Wales (C. Embleton, 1962). The Gorge du Guil in the French Alps, another example, has been examined by J. Tricart (1960). It lies above the Würm snow-line and is 15 km long. It crosses a variety of rock types, but it is particularly narrow where it cuts across massive limestones. Some morainic deposits have been found in the gorge, extending nearly to the bottom of the valley. The gorge predates the Würm glaciation as pockets of moraine occur within small basins along its length. These basins occur where softer rocks have been exploited by the ice. The morainic till has been partially water-sorted, the fine material has been washed out, and it has been consolidated by lime-impregnated waters. Tricart suggests that the gorge was cut by subglacial meltwater flowing under hydrostatic pressure. The glacier in this section of the valley would have been relatively thin, probably less than 300 m, and deep crevasses and moulins would have allowed meltwater to reach the base of the decaying glacier. The meltwater, heavily loaded with abrasive material, would rapidly scour a deep gorge, having a steep V-shaped cross-profile, beneath the ice. The stream that has occupied the valley after the ice retreated has only been able to erode the bottom of the gorge by about 5 m. The evidence does not support a hypothesis that the Gorge du Guil might have been cut by overflow waters from a lake in the main basin upstream of the gorge, for there is no sign of lacustrine deposits in this basin.

These examples illustrate the strong erosive capacity of subglacial streams flowing under thin, heavily crevassed ice on steep slopes (see also Chapter 12). It is their activity that can account for the formation of V-shaped sections within U-shaped glaciated valleys, the V-shaped sections being cut while the whole valley is still occupied by ice. In the situations where glacial meltwater can be particularly active, as in the Guil gorge, the activity of the ice itself is often reduced as a result of thinning and crevassing. Where the slope is gentler and the ice is thicker, meltwater may be unable to reach the bed and is less likely to become concentrated into one stream. Conspicuous fluvio-glacial erosion beneath the glacier is unlikely in this situation. In this way, both U-shaped and V-shaped sections of a glaciated valley can form together, the longitudinal slope being the determining factor in the distribution of the two types of cross-section.

The Longitudinal Profiles of Glaciated Valleys

The essential difference between the long profile of a glaciated valley and that of a river is the irregularity so commonly characteristic of the former. Apart from the U-shaped cross-profile, the most striking features of a glaci-ated valley are the basins and steps that diversify its long profile. The basins frequently contain lakes or have been later infilled with alluvium while the steps often show marked signs of glacial scouring and plucking. The steps are rather similar, but on a much larger scale, to roches moutonnées, as discussed by Lewis (1947b). Other features associated with glaciated valley long pro-files are the hanging tributaries and the truncated spurs that occur along the valley sides. All these features indicate clearly the erosive capacity of the ice, but this action must clearly be selective to produce the observed effects.

In the early years of this century, there was considerable controversy between the protectionist school of thought and those who considered that the features of glaciated valleys were erosional features. The protectionist view was put forward very forcibly by E. J. Garwood in 1910, but it has been generally abandoned in connection with the formation of valley steps and hanging valleys. However, the concept of glacial protection is still valid under some conditions, as will be discussed in Chapter 9.

Before considering the actual characteristics of the landforms, the processes acting subglacially, where the longitudinal profile is fashioned, will be discussed. The side-slip observations can give direct information on the processes operating at the side of the valley, but it is less easy to observe the processes operating at the base of a thick glacier. Information must be obtained indirectly and theoretically, apart from the limited data provided by boreholes and tunnels.

The extent to which a glacier moves over its bed (Chapter 4) is clearly relevant to a study of the formation of rock bars and basins. The conditions under which it can erode its bed require consideration, and in this connection, it is important to differentiate between temperate and cold, polar glaciers. The former are at their pressure melting-point and are able to slide over their beds by regelation and creep. Cold glaciers are frozen to their rock bed and hence differential movement occurs in the lowest layers of the ice rather than at the ice-rock interface. B. Kamb and E. LaChapelle (1964) have studied the basal sliding of the Blue Glacier, Washington. A tunnel was excavated, entering the ice near the top of an ice-fall and reaching bedrock where the ice thickness above the tunnel was 26 m. The movement of the ice relative to the rock was recorded with dial micrometers. A stake, anchored in the ice 10 cm above the bedrock, moved at a mean speed of 1·6 cm/day parallel to the bed. Another stake 150 cm above the bedrock moved 12 per cent faster. Further stakes spaced at 10 cm intervals showed that most of this differential movement took place in the lowest 50 cm; in fact, nearly all the 1·6 cm/day movement took place as slip over the bed. The movement of the surface was almost identical with that at 150 cm above the bed. At the site, the bed had a slope of 22 degrees, and it steepened to 55 degrees down-glacier from the end of the tunnel. The calculated basal shear stress was 0·7 bars and the normal stress, resulting from the weight of the ice, was 1·6 bars. The rate of movement at the base of the glacier was irregular, varying over periods of the order of seconds, and by as much as 10 per cent from day to day. The existence of a regelation layer up to 3 cm in thickness at the base of the ice has already been described in Chapter 4. This layer contained considerably more debris than the overlying ice, but never more than 10 per cent by volume. Similar debris-laden layers have been reported from other glacier tunnels (for example, Vesl-Skautbreen; McCall, 1960), and they indicate that the basal ice can act as an abrasive agent. The debris tends to pile up on the up-glacier side of bedrock protuberances, but where the ice flows over the down-glacier side, cavities may form. The processes whereby the ice flows over and around obstacles have been discussed in Chapter 4.

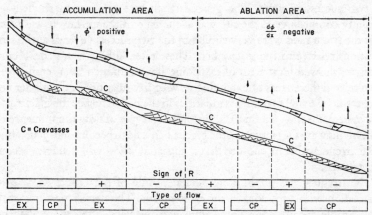

Fig. 7-3 Longitudinal section of an idealized valley glacier. The upper diagram shows the velocity distribution calculated from a simple flow law assuming perfect plasticity. The lower diagram shows the slip-line field and theoretical zones of crevassing. EX—extending flow; CP—compressing flow; R—radius of bed curvature; ϕ—rate of discharge (J. F. Nye, *J. Glaciol.*, 1952, by permission of the Glaciological Society)

In order to account for the irregular erosion at the base of a glacier, by which deep basins and intervening bars may be formed, variations in glacier flow down the valley must be considered. The theory of extending and compressing glacier flow (see pp. 98-100) put forward by J. F. Nye (1952) may throw some light on this matter. Figure 7-3 shows the theoretical pattern of slip planes or potential shear surfaces associated with the two types of flow. Where the surface flow is accelerating or extending, a series of planes of weakness is likely to develop within the ice, curving down tangentially to the glacier bed. In the ablation zone and those parts of the glacier where the flow is decelerating (compressing), the planes of weakness will tend to develop curving upward tangentially from the bed in a down-glacier direction. These latter planes become pronounced near the snout of the glacier and also at the base of steep slopes. It is in these zones, therefore, that the ice may be expected to carry debris up from the glacier floor into the ice and hence it can act more effectively as an eroding agent. In this way, basins at the base of an ice-fall, once initiated, may become self-perpetuating.

The form of the bedrock surface over which the ice is flowing exerts an influence on the ice motion and is in turn itself modified by the ice movement. The former relationship has been shown to be true by Nye (1959) from observations made on Austerdalsbreen. Observations of the longitudinal strain rate were made in identical positions relative to the bed of the glacier in two successive years. The general strain rate in the area studied was strongly compressive as the site was at the base of the ice-fall. It was found that the strain rates were very similar in the two successive years at the same positions relative to the glacier bed. These rates, although in general compressive, showed local areas of extending flow, indicated by a tensile stress. Nye related these areas of extending flow to irregularities in the glacier bed. The results showed that the irregularities in the bed influence the glacier flow in a systematic way, and in turn, the different types of flow can influence the bed. In this way, the features of the glacier bed tend to become self-generating: once irregularities in the bed have appeared, they will tend to become accentuated.

The irregularities of the longitudinal profile of the glacier show some correlation with variations in the cross-profile. The steeper portions of the long profile are often associated with narrower and often more V-shaped sections of the cross-profile, as has already been mentioned. This relationship agrees with Nye's theory of extending and compressing flow. In the steep reaches, where the flow would be extending, bottom movement would not be so conducive to erosion as in those places where the slope was less (or even reversed), giving rise to compressing flow.

The formation of rock steps, bars or riegels, is associated with the basins that usually occur above them. Many possible reasons for the position of rock bars have been suggested: among the most common are changes in lithology, changes in the pre-glacial gradient of the valley, constrictions in the valley walls, and the junction of tributary valleys. J. P. Bakker (1965) has drawn attention to a number of earlier views on the formation of glacial stairways and rock basins. He mentions the confluence of two streams, the variations of rock structure, and the importance of pre-glacial steps in the valleys as possible factors in the formation of glacial stairways. He lays most stress, however, on S. Passarge's views concerning Tertiary differential deep weathering. Evidence of differential deep weathering in immediate pre-glacial times in central Europe is found in the vegetation associated with waterfalls. These falls, of preglacial origin, are associated with rocks of greater resistance to weathering. In the flatter areas between the falls, deep weathering occurred, extending both across and down the valley. Glacial ice advancing over an area of very uneven depth of weathering exploits the weathered zones. By removing the weathered material, the ice can readily form rock basins and stairways. The riegels across the valleys represent the sites of the pre-glacial falls on the unweathered rocks. This process of basin formation does not require the ice to remove fresh and unweathered rock to any great extent. Glacial erosion is particularly effective where deep weathering had taken place in the Tertiary period, and this process may help to account for some of the intrusive troughs, described by D. L. Linton and discussed on page 180.

Good examples of rock bars occur in the valley of Austerdalen, Norway (C. A. M. King, 1959), the uppermost still concealed by glacier ice. There are eleven rock bars in the valley between the point where it joins a tributary of Sognefjord and the snout of the present glacier, which lies just behind a rock bar. A very high valley step underlies the ice-fall, which links Jostedalsbreen ice-cap above to Austerdalsbreen below (Plate III). The whole area is formed of massive gneiss so that the bars cannot be explained in terms of varying lithology. Four rock bars lie between the glacier snout and Veitestrondsvatn and five more are hidden beneath the lake waters. These submerged bars were revealed by echo-sounding traverses across and along the lake. Another rock bar dams up the lake, and the lowest bar dams the lake of Hafslovatn (Figure 7-4).

The rock bars of Austerdalen illustrate several of the situations in which rock bars are frequently found. The uppermost exposed rock bar, near the present glacier snout, occurs where the glacier is forced round a sharp bend. It is noticeable that the bar is highest at the inside of the bend where the flow would be most restricted by a spur extending pre-glacially into the valley at this point. The spur would tend to dam up the ice which would thicken on the upstream side, and the ice would escape over the bar mainly at the outside of the bend. The valley wall above the constriction shows very marked glacial scouring on the outside of the bend, where the flow would be most rapid. The second rock bar, situated near Tungasaeter, forms a steep barrier across the valley. Meltwater, flowing subglacially in the initial stages, has cut a very deep narrow cleft through this bar. The bar probably represents the hang of Austerdalen as a tributary to Langedalen just below the rock bar. Another small rock bar occurs just below the junction of the two valleys. This bar may have been caused by increased erosion under the thicker ice below the confluence of the two glaciers, the classical hypothesis put forward in 1905 by A. Penck. The next major bar is also similarly situated close to the confluence of Austerdalen and Snauedalen, which is an important tributary leading by a col into the adjacent valley of Fjaerland.

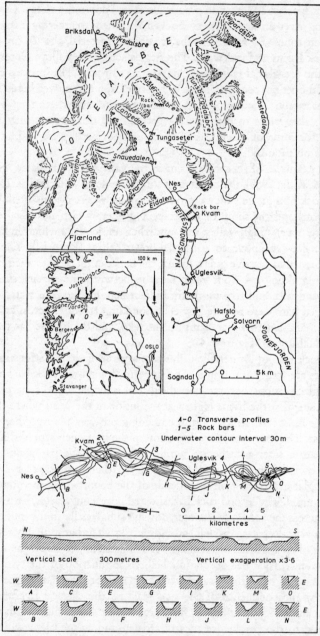

Fig. 7-4 The rock bars in Austerdalen and Veitestrondsvatn, Norway (C. A. M. King, *Geogr. J.*, 1959, Royal Geographical Society)

A further series of five rock bars is hidden beneath Veitestrondsvatn, smaller features than the major rock bar that dams up the whole lake. The basins that separate the bars in the lake have depths of about 60-100 m, and the bars occur where there are bends in the valley. The large rock bar at the end of the lake occurs where the valley is very constricted, before it opens out into the broad basin of Hafslovatn. The latter hangs 168 m above Sognefjord. Ice from the basin reached Sognefjord by two different routes, both of which have large rock bars in their upper reaches. There is a considerable fall of level between Veitestrondsvatn and Hafslovatn, a fall that suggests a thinning of the ice and less vigorous erosion at the point where the bar occurs. Erosion subsequent to ice retreat appears to have lowered the level of Veitestrondsvatn by about 60 m. Supporting evidence for this is found in exposed lake deposits at the head of the lake.

In assessing the factors responsible for the formation of the rock bars, the pre-glacial form of the valley must be taken into account. On the basis of arguments already put forward it seems likely that any pre-glacial irregularity in the long profile will be accentuated rather than removed by glacial erosion. Austerdalen probably possessed well developed knick points, associated with the recent elevation of the pre-glacial erosion surface that forms the conspicuous plateau on which the Jostedalsbreen ice-cap rests. The major knick point of the series probably lay at the site of the present ice-fall. This huge step is about 760 m high and, according to Nye's calculations, the ice cover is probably only about 6 m thick at the top and 46 m thick near the base of the ice-fall.

The situations of the rock bars in this valley may be summarized by stating that they occur both above and below tributary junctions, at bends in the valley and near constrictions in the valley. At all these positions, there would tend to be changes in the velocity and thickness of the ice. These changes would lead to zones of extending and compressing flow. The different types of flow would in turn lead to areas of reduced or enhanced erosion respectively on the bed of the glacier. Once initiated, the steps and bars would tend to be self-perpetuating and increase in magnitude during glaciation. The pre-glacial stream pattern and irregular long profile would determine the position of the bars. In deglacial phases, sub-glacial meltwater streams would begin to cut V-shaped gorges through the bars. These gorges would be further deepened by post-glacial river action, resulting in the eventual draining of any lakes in the intervening rock basins not yet silted up.

Lewis (1947b) has studied the formation of valley steps in a variety of areas, including North Wales and the Cuillin Hills of Skye. He points out that the flat treads of the steps usually show well-marked glacial scouring, while quarrying is more in evidence on the steep risers of the steps. The steps resemble giant roches moutonnées in form. Because of the similarity of the two features, Lewis suggests that they may have a common mode of origin, and contends that different processes operate under thin glaciers from those responsible for the formation of steps under thick ice. When the ice is thin, he suggests that meltwater is able to penetrate to the bed of the glacier (see p. 107 for a discussion of the maximum depth of crevasses in active ice). If it were then refrozen in any cracks in the rocks, shattering of the rock would ensue. This process might operate best on the steep risers of the steps, where plucking is in evidence, and the process would tend to increase the height of the step, but it could not cause the overdeepening of the basin at some considerable distance from the step. When the ice is thick, Lewis suggests that changes in pressure as temperate ice moves over the step could result in alternate melting and refreezing.

Lewis comes to the conclusion that the most effective step formation probably occurs under thin glaciers. This assumes that thaw-freeze is an important process in their formation, a process that is discussed in more detail in Chapter 8. It seems likely that thaw-freeze is not necessarily important under thick glaciers and may not even be significant under thin ice, since temperature changes are very slow beneath ice. The prising action described by Glen and Lewis (1961) is probably more important when the ice is moving at its base, armed with suitable tools in the form of boulders.

The basins separating the bars and steps form the complementary features and are the most important single characteristic of a glaciated valley. The basins indicate the ability of the basal glacier ice to flow uphill, and to erode the rock as it passes over it. If the basins have not been later infilled with alluvium, and if the bar at the lower end has not been cut through by fluvial action, the hollows may contain ribbon lakes, which are characteristic of heavily glaciated mountain areas; the depths of the lakes may be augmented by the deposition of moraine. Lakes of this type can reach considerable depths. Loch Morar, for example, the deepest of British lakes, attains a depth of 315 m. As Linton (1957, 1959) has pointed out, most of the deep glacial lakes of Scotland occur in the west of the country, radiating from the area where precipitation is at a maximum at present. It is likely that these areas would receive the heaviest snowfall under glacial conditions. The glaciers of the western side of the Scottish Highlands would, as a result, have been well nourished, while they also flowed down steep slopes to deep water and were able to achieve the maximum amount of erosion. Where other conditions were also suitable, basin formation could take place effectively. Similar rock basins, whether filled with water to form lakes or with sediment to form valley flats, are characteristic of all mountain groups that have nourished vigorous valley glaciers. There are many good examples in the Lake

District, western Ireland, North Wales and in many other parts of the world.

Not all long valley lakes, however, have necessarily been formed by glacial erosion, despite their appearance on casual inspection. E. Watson (1962) has drawn attention to a ribbon lake that is not of glacial origin in the Tal-y-Llyn valley in North Wales. This straight and deep glaciated valley links a number of pre-glacial drainage lines because the ice has exploited the Tal-y-Llyn fault. Within the trough are several basins, separated by barriers; one of the basins contains the Tal-y-Llyn Lake. The barrier retaining the lake rises about 53 m above the valley floor, and might at first sight appear to be moraine. However, there are outcrops of solid rock visible in the barrier. On closer examination, the rock in these outcrops can be seen to be contorted and shattered, and further, it is not *in situ*. If the valley side above the barrier is examined, it can be shown that the barrier represents a large landslide feature and that the basin is not primarily a glacially eroded landform. It can no longer be regarded as the southernmost rock basin in Wales. Care must, therefore, be taken to confirm that the rock bars do in fact consist of solid rock, as is normally the case.

Some of the largest ice-scoured hollows are those now occupied by the large lakes at the southern boundary of the Alps in northern Italy. The three largest of these lakes are Lake Garda, with a maximum depth of 346 m, Lake Maggiore, 362 m deep, and Lake Como, 410 m deep. These lakes are situated where the glaciers spread out into piedmont lobes and built very large arcuate moraines. The moraines increase the depth of the lakes by adding to the height of the rock bars beneath them. The upper portions of the lakes are narrow, but they spread out into broad expanses near their outlets. The origin of the piedmont lakes may be connected with the strongly compressive flow that must have taken place as the glaciers emerged from the mountains on to the plains. This type of flow has already been associated with possible zones of erosion. By this stage in their courses the glaciers would carry a large amount of debris; they would also be thick and hence the basal pressure would be great. All these factors would, therefore, tend to concentrate erosion in these situations. Similar factors also applied on the northern side of the Alps, where other large glacial lakes occur. A rock basin occurs beneath the piedmont Malaspina glacier (Alaska) at present; conditions in this glacier are similar to those at the Alpine margin in many respects.

Glacial Diffluence and Transfluence: Watershed Breaching by Ice

Where vigorous glaciers are flowing down valleys from which there is no free outlet, the ice level in the valley will rise. This situation may arise either if the outlet is blocked by other ice, or if the level of the ice in the valley is rising more rapidly than the ice can escape down the valley to lower the level. If the ice thickness continues to increase, the glacier surface will eventually rise above the lowest col available, and the surplus ice will then escape across the col into the neighbouring valley. With increasing glaciation, the level may rise to such an extent that all available cols are used by a series of diffluent ice streams. The term 'transfluence' may be used when this stage has been reached. These phenomena were clearly recognized by A. Penck (1905) who drew attention to the examples afforded by the Rhine Valley at Sargans, and also at Zell-am-See.

Glacial transfluence has also long been recognized in Scandinavia, where erratics from Sweden have been carried westward across the main watershed to the Atlantic coastlands of Norway. This resulted from the accumulation of ice over the Baltic and Scandinavia until the ice-shed lay to the east of the main mountain divide. The ice then flowed westward across the latter, carving breaches where the ice flow was concentrated in existing cols. Linton (1949) has demonstrated similar events in the Scottish Highlands where there was also a tendency for the ice thickness to be greatest east of the pre-glacial watershed, and where erratics were also transported from east to west across the backbone of the country.

Many examples of glacial diffluence are known in Britain. In North Wales, pre-glacial watersheds were overridden and breached by ice at the heads of the Nant Ffrancon valley, the Llanberis valley, the Gwynant valley, and the Nantlle valley (C. Embleton, 1962). In the Lake District, an area which was entirely glaciated by local ice moving down the major valleys, R. K. Gresswell (1952) has examined the results of glacial diffluence. In the Langdale valley, for instance, rock steps were initiated by glacial diffluence as well as by varying lithology. Ice moving south down Langdale into the Grasmere basin escaped by two routes to the south. The western stream crossed Red Bank and the eastern one flowed by way of Rydal Water to the Windermere Basin. This latter ice stream was joined by several others. However, it split again into two channels farther downstream. One of these passed south-westward into the Esthwaite Water valley, while the bulk of the ice moved south down the main trough. This ice gouged out the lake to a depth of – 27 m O.D., a depth of 66 m below its present surface. Some ice also escaped eastwards along the Gowan valley, where it was joined by ice flowing down the Kent valley. Thus a complex system of glacial diffluence and confluence was taking place at the stage when the glaciers had shrunk back into their valleys, towards the end of the last glaciation.

The most extensive evidence of glacial diffluence and the most spectacular instances of watershed breaching by ice are to be found, however, in Scotland.

A recent study of Loch Lomond, the largest lake in the kingdom (Fig. 7-5), illustrates a number of aspects of the process (Linton and H. A. Moisley, 1960) The lake attains a length of 36 km and a maximum depth of 190 m. The upper part is long and narrow, but it opens out in the southern portion, which lies in lower country. In pre-glacial times the lake area was drained by streams belonging to at least three different drainage basins according to Linton and Moisley. A watershed extended across the northern end of the lake. From this watershed, a stream drained northward to the Tay by way of Glen Dochart along a zone of weakness in the rocks. To the south of this northern watershed, a stream flowed eastward to the Teith through the valley in which Loch Katrine now lies. A major watershed ran east-west across the present lake from Ben Lomond eastward. South of this watershed, the area was drained by streams flowing south-eastward, possibly two separate streams draining to the Forth and the Vale of Leven. The positions of these pre-glacial watersheds can be identified by the greatly over-steepened valley sides of the breaches through the watersheds, which show considerable evidence of severe ice scouring and moulding.

The ice accumulated in the northern basin to such an extent that it overflowed and breached the surrounding watershed in about six places. The southern breach, just east of Ben Vorlich, was the deepest. This breach diverted much of the drainage of Glen Falloch south into Loch Lomond. The southern basin also became congested with ice, which overflowed across the watershed to the south and south-east. The southern diffluent channel, west of Ben Lomond, was lowered enough to divert the drainage of the upper part of the valley, including Loch Arklet, to the south into the Loch Lomond basin. In this way, the three separate sections of the present lake became linked in one major basin, draining south into the Clyde. The erosion achieved by these diffluent glaciers was such that a valley draining north at levels between 274 m and 458 m was converted into a trough 150 to 180 m deep. This erosion was severely localized, however, and was restricted to the valleys and the cols through which the ice was concentrated by the relief control imposed upon it.

The effects of glacial diffluence depend on the relative power of the original and the diffluent glaciers. The diffluent glaciers that crossed the former cols and formed the continuous trough in which Loch Lomond now lies were able to lower their troughs to such an extent that the drainage was permanently diverted. Not all glacial diffluent channels, however, operate for long enough or carry sufficiently vigorous ice streams to achieve this effect. In these circumstances, the diffluent channel is left on the side of the main valley as an eroded col, which normally has a U-shaped form.

Various stages in the development of a diffluent glacier and its effects on

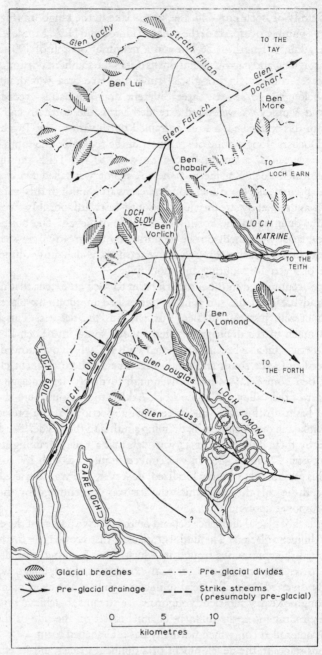

Fig. 7-5 Map of the Loch Lomond area, showing glacial breaches
(D. L. Linton and H. A. Moisley, *Scott. geogr. Mag.*, 1960,
Royal Scottish Geographical Society)

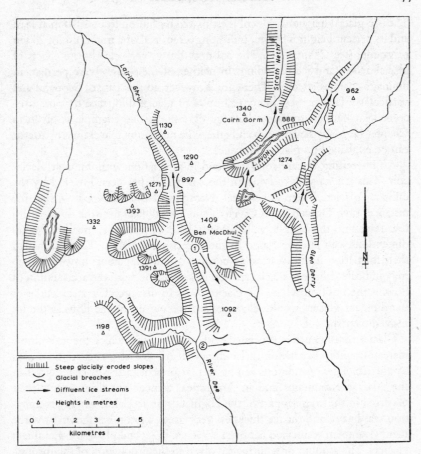

Fig. 7-6 Watershed breaching by ice in the Cairngorms, Scotland (after D. L. Linton, *Trans. Inst. Br. Geogr.*, 1949)

the landscape can be seen in the Cairngorms. A good example is the col or trough, 244 m deep, leading out of the Loch Avon valley (Fig. 7-6). The glacier in the Avon valley became constricted by ice moving across its mouth, causing the level to rise until a col into Strath Nethy was crossed by the ice. The escaping ice then lowered the col and widened Strath Nethy into a glacial trough. The Nethy valley, which was only fed by ice passing through the diffluent col, was not lowered enough to cause any drainage diversion. An equally conspicuous example of glacial diffluence occurred in the neighbouring valley of Glen Dee to the west. The southern outlet of the Glen was also blocked by another powerful ice stream, causing the ice level to rise until ice escaped northward through a striking col into Lairig Ghru.

The pre-glacial height of this col is estimated by Linton to be 1070 m at least and its present height is 840 m, indicating a removal of a minimum of 230 m of granite (see Chapter 11). The col now shows very clearly the effects of glacial action and ice moulding. In neither of these cases were permanent drainage changes effected. There are, however, in this part of Scotland and particularly farther west many examples of glacial diffluence that has proceeded far enough to cause drainage diversions: one example is given in Chapter 12. Many sharp bends and otherwise unaccountable drainage patterns can be explained by glacial diffluence.

An interesting feature is often found in association with diffluent glacier cols. This is a small lake lying in the col, which has been lowered by the diffluent glacier. There are several examples of this in North Wales, for instance, Llyn Ogwen, Llyn Cwellyn and a small lake near the diffluent col that leads into the Nantlle valley. Many lakes of this type, usually of small dimensions, can also be found in the Scottish Highlands. The reasons for their formation in this watershed position are not apparent; however, they may indicate enhanced glacial erosion in association with the acceleration of flow at the position where the glacier starts to diverge. The freer outlet of the diffluent glacier would allow some speeding up of the flow as the ice moved over the col.

Glacier transfluence is associated with conditions in which the ice volume increases rapidly in relation to ice dispersal. All available cols are then used by the ice. These conditions can be seen at present in the glaciated parts of the Alaskan mountains and in Antarctica. Similar conditions must have occurred in the western part of the Highlands of Scotland, where precipitation was heavy and the ice thickness very great. Linton has shown how the main Grampian watershed has been breached in several places by transfluent glaciers. The amount of scouring involved in the production of the through passages reaches 366 m in Glen Tilt, and Loch Treig shows overdeepening to the extent of at least 460 and possibly 550 m (Chapter 11).

The Classification of Glacial Troughs and their Development

Four main categories of glacial trough have been identified by Linton (1963), as follows: Alpine troughs, the Icelandic type, the composite type, and the intrusive type.

1. The main essential of the Alpine type is that the accumulation area of the glacier should be surrounded and overlooked by higher ground. The source of the glacier is often in a cirque or series of cirques, and the glacier occupies a pre-glacial valley. The Haweswater and Ullswater valleys in the Lake

District provide good examples of this type of trough, which is ideally a self-contained unit with no external source of ice and no ice escape by diffluence. These conditions are relatively rarely fulfilled.

2. The ice forming the second type of trough, the Icelandic type, originates on a mountain ice-cap and flows down as a series of outlet glaciers, often through ice-falls. The outlet glaciers from the Jostedalsbreen ice-cap in Norway illustrate the situation. Austerdalsbreen, already referred to in this and earlier chapters, is one of these glaciers that has created a very deep and steep trough of this type. In Britain, there are examples of troughs of the second type in eastern Scotland, where ice accumulated on the high, pre-glacial erosion surfaces, such as the Grampian plateau. Similar flattish upland surfaces provided the gathering ground for the ice that formed the troughs in the Pennines. Bishopdale and upper Wharfedale provide examples of troughs that show marked glacial modification in the Pennines.

3. The third type is the composite type, and describes those systems that did not find the pre-existing valleys sufficient for the discharge of the ice. In the third type, parts of the troughs are eroded entirely by the ice, these parts linking modified portions of pre-glacial valleys. In the composite troughs, glacial diffluence and transfluence have been very active and by this means, new valleys have been created and drainage changes initiated.

Five stages in the evolution of composite troughs are suggested by Linton. Type (a) consists of simple diffluent troughs, for example Strath Nethy, which has already been described. The second type (b) consists of multiple diffluent troughs, such as those leading away from Loch Fyne. The third type (c) consists of troughs formed by simple transfluence. These differ from the first type in that the new trough branches out from the valley head and not from its side, as in the first type. The pass through Beattock summit is a good example of this type. Ice from the north passed south across the former col into Annandale, transferring the watershed 6·5 km to the north in the process. The Nant Ffrancon pass in Snowdonia is another example, where the former watershed has been completely worn away and the watershed displaced eastward. The fourth type of composite trough, (d), is the multiple transfluence system of troughs where the ice spilled over a number of cols at the head of a valley. For example, ice spilled over many cols at the head of Glen Falloch, in the area already described in the discussion of the formation of Loch Lomond. Ice escaped through more than ten cols around the upper part of the Falloch valley, forming watershed breaches and ice-modified cols leading down to glacial troughs. The final group of composite troughs, (e), consists of radiative dispersal systems. This applies to those troughs that display a radial pattern, as in the Lake District, southern Norway, the south-west Highlands of Scotland, and Fjordland in south-west New Zealand. In all these areas, the ice spread out from a centre along pre-existing valleys and also cut across watersheds wherever the ice pressure was sufficient to carry the ice over a col, which was then lowered by erosion.

4. The fourth type of trough is the intrusive or inverse trough. The ice in this type pushes up the valley against the direction of pre-glacial drainage. Such troughs are likely to occur in the lower hill areas where the local ice was not sufficiently powerful to keep out the more powerful ice from more vigorous centres. There are good examples of these troughs in the hill areas near the southern boundary fault of the Highlands of Scotland. Three well marked troughs occur at the south-eastern side of the Kilpatrick, Campsie and Ochil group of hills. Two of these troughs, originally excavated below present sea-level, are now filled with drift, but the third is an obvious feature and forms the section of Clydesdale between Hamilton and Lanark. Upstream of Hamilton, the Clyde is flowing in a deep narrow valley, clearly modified by ice action. Erosion by ice has reduced the valley level to below 60 m at a position far inland. Both the drainage gradients and drainage pattern have also been modified.

The effects of prolonged erosion by valley glaciers in a cold climate have been studied by Linton (1963) in Antarctica where rugged topography has been heavily submerged by ice for a long period. The marginal parts of Antarctica best illustrate these conditions. Under the severe climatic conditions that occur in this area, the snow-line is near sea-level and the volume of ice increases down-valley as tributaries merge, because ablation is largely by calving. The lack of melting prevents loss of ice, as occurs in more temperate climates. For this reason the cross-section of the valley must increase downstream to accommodate the increasing ice volume. Spur truncation will be carried further than in the mountain areas of Britain, for example. Where the ice masses increase greatly downstream, the divides between the glaciers will tend to be cut back and destroyed by this widening process. This tendency increases down-valley in the major troughs to which tributary glaciers drain. There are also good examples of these over-widened glacial valleys in Spitsbergen, where the spurs between two neighbouring valleys have been largely eroded away in their lower portions. The divides between the major valleys in these situations have been shorn of all their smaller spurs and represent long and straight ice-moulded forms. In the Antarctic, these valleys still lie underneath the ice, but it is easy to imagine that this process is taking place actively at present, owing to the volume of ice that fills the valleys. As erosion continues, the divides between two neighbouring valleys become progressively narrower. They are slowly consumed and shortened as erosion continues further. As the divide is traced upward to the nunatak peaks, it becomes sharper, forming an arête where it emerges above the ice. Pressure release, which causes dilatation joints, is a process that helps in the destruction and shortening of arêtes (see Chapter 8). The emergent arête can also be attacked by frost-shattering, which takes advantage of the dilatation joints. Eventually when the divides are destroyed, a pyramidal peak is formed. It is by this process of divide destruction that valley glaciers can most readily widen their valleys. Once the divides have been removed, the ice can combine into one lowland ice-sheet and produce the ice-moulded surface characteristic of such an area (Chapter 9).

Conclusions

It is evident that ice can produce spectacular glacial features by vigorous erosion under suitable conditions. Ice can, however, under other conditions remain almost static and can be more protective than erosional. In order to erode effectively, the ice must exceed a certain minimum thickness and it must flow reasonably fast as a concentrated stream. These conditions are best achieved in valley glaciers that are fed by well-nourished accumulation areas. There must also be a free outlet for the ice to lower ground or out to sea so that its flow can be maintained.

One of the most notable characteristics of glacial erosion is its irregular incidence along the length of the glacial trough. The rock bars and intervening basins, which are so common in glaciated valleys, are evidence of this irregularity of erosion. The explanation for these zones of alternating excessive erosion or reduced erosion must be sought both in the rocks of the valley, in the form of the valley and in the response of the ice to this configuration. Zones of compressing flow, associated with the form of the valley, may well cause an increase in erosion, owing to the nature of the ice movement within the glacier. Additions to the volume of ice by tributary junctions or loss by glacial diffluence must also be considered.

It is likely that the ground is prepared in some way, possibly by deep freezing or deep weathering, before the ice advances over it. Another factor assisting erosion of bedrock by a valley glacier is pressure release. This causes the development of dilatation joints as ice of relatively low density replaces rock of relatively high density. The glacier can both scour and prise material from its bed, although both processes are only achieved effectively with the aid of rock fragments carried near the base or sides of the glacier. Movement past the rock walls at the sides and base of the glacier is essential for active glacial erosion. The presence of meltwater in temperate glaciers aids this basal movement.

REFERENCES

G. W. BAIN (1931), 'Spontaneous rock expansion', *J. Geol.* **39**, 715-35

J. P. BAKKER (1965), 'A forgotten factor in the interpretation of glacial stairways', *Z. Geomorph.* NF **9**, 18-34

J. BLACHE (1960), 'Les résultats de l'érosion glaciaire', *Méditerranée* **1**, 5-31

M. BOYÉ (1950), *Glaciaire et périglaciaire de l'Ata Sund nord-oriental (Groenland)* (Paris; Expéditions polaires françaises, **1**)

H. CAROL (1947), 'The formation of *Roches Moutonnées*', *J. Glaciol.* **1**, 57-59

C. A. COTTON (1941), 'The longitudinal profiles of glaciated valleys', *J. Geol.* **49**, 113-28

C. A. COTTON (1942), *Climatic accidents in landscape-making* (Christchurch, N.Z.)

G. R. ELLISTON (1962), 'Seasonal changes of speed in temperate valley glaciers', J. *Glaciol.* **4**, 289

C. EMBLETON (1962), *Snowdonia* (Geogr. Ass., 'British Landscapes through Maps')

E. J. GARWOOD (1910), 'Features of Alpine scenery due to glacial protection', *Geogr. J.* **36**, 310-39

J. W. GLEN (1958), 'Measurement of the slip of a glacier past its side wall', *J. Glaciol.* **3**, 188-93

J. W. GLEN and W. V. LEWIS (1961), 'Measurements of side-slip at Austerdalsbreen, 1959', *J. Glaciol.* **3**, 1109-22

R. K. GRESSWELL (1952), 'The glacial geomorphology of the south-eastern part of the Lake District', *Lpool Manchr geol. J.* **1**, 57-70

W. B. HARLAND (1957), 'Exfoliation joints and ice action', *J. Glaciol.* **3**, 8-10

R. H. JAHNS (1943), 'Sheet structure in granites: its origin and use as a measure of glacial erosion in New England', *J. Geol.* **51**, 71-98

B. KAMB and E. LACHAPELLE (1964), 'Direct observation of the mechanism of glacier sliding over bedrock', *J. Glaciol.* **5**, 159-72

C. A. M. KING (1959), 'Geomorphology in Austerdalen, Norway', *Geogr. J.* **125**, 357-69

W. V. LEWIS (1947a), 'The cross sections of glaciated valleys', *J. Glaciol.* **1**, 37-38

W. V. LEWIS (1947b), 'Valley steps and glacial valley erosion', *Trans. Inst. Br. Geogr.* **14**, 19-44

W. V. LEWIS (1954), 'Pressure release and glacial erosion', *J. Glaciol.* **2**, 417-22

D. L. LINTON (1949), 'Watershed breaching by ice in Scotland', *Trans. Inst. Br. Geogr.* **17**, 1-16

D. L. LINTON (1957), 'Radiating valleys in glaciated lands', *Tijdschr. K. ned. aardrijksk. Genoot.* **74**, 297-312

D. L. LINTON (1959), 'Morphological contrasts of Eastern and Western Scotland', in *Geographical Essays in Memory of Alan G. Ogilvie* (ed. R. MILLER and J. W. WATSON)

D. L. LINTON (1963), 'The forms of glacial erosion', *Trans. Inst. Br. Geogr.* **33**, 1-28

D. L. LINTON and H. A. MOISLEY (1960), 'The origin of Loch Lomond', *Scott. geogr. Mag.* **76**, 26-37

F. E. MATTHES (1930), 'Geologic history of the Yosemite Valley', *U.S. geol. Surv. Prof. Pap.* **160**, 137 p.

J. G. MCCALL (1960), 'The flow characteristics of a cirque glacier and their effect on glacial structure and cirque formation', in 'Norwegian cirque glaciers' (ed. w. v. LEWIS), R. geogr. Soc. Res. Ser. 4, 39-62

M. M. MILLER (1958), 'Phenomena associated with the deformation of a glacier borehole', Un. géod. géophys. int., Symposium at Toronto, 1957, 4, 437-52

J. F. NYE (1952), 'The mechanics of glacier flow', J. Glaciol. 2, 82-93

J. F. NYE (1959), 'The deformation of a glacier below an ice fall', J. Glaciol. 3, 386-408

A. PENCK (1905), 'Glacial features in the surface of the Alps', J. Geol. 13, 1-19

T. PIPPAN (1957), 'Anteil von Glazialerosion und Tektonik an der Beckenbildung am Beispiel des Salzachtales', Z. Geomorph. NF 1, 71-100

H. SVENSSON (1959), 'Is the cross-section of a glacial valley a parabola?' J. Glaciol. 3, 362-3

J. TRICART (1960), 'A subglacial gorge: La Gorge du Guil (Hautes-Alpes)', J. Glaciol. 3, 646-51

J. TRICART and A. CAILLEUX (1962), Le modelé glaciaire et nival (Paris)

E. WATSON (1962), 'The glacial morphology of the Tal-y-Llyn valley, Merionethshire', Trans. Inst. Br. Geogr. 30, 15-31

Chapter 8

Cirques

I noticed in the valleys of Lanzo, and at heights between 2000 and 3000 metres, hollows in the form of amphitheatres or great cirques, which have generally a shape like the interior of an armchair, or, more elongated, like a sofa. (B. GASTALDI, 1873, in a letter to Sir Charles Lyell)

The cirque has long been recognized as one of the most characteristic forms of glacial erosion. In its fully developed form, it consists of a rounded basin partially enclosed by steep cliffs and sometimes containing a small lake or cirque glacier; the cliffs at the back of the basin may rise to great heights and culminate in jagged peaks or arêtes. Cirques are worldwide in their occurrence, but confined to areas of present or former glacierization. As a consequence of this broad distribution, they have received numerous local names, such as cwm in Wales, corrie or coire in Scotland, kar in Austria and Germany, botn in Norway and nisch in Sweden. The term cirque is the oldest established of these in geomorphology, having been introduced as long ago as 1823 by Jean de Charpentier in a study of the Pyrenees, at a time when the glacial theory was only just being tentatively formulated.

It is today generally accepted that cirques are the result of a number of erosive processes closely associated with the presence of ice in them. Indeed, the view that they are glacially sculptured features is now more than a century old. A. C. Ramsay in 1860 thought it 'probable that such rock basins were ground out by heavy loads of ice' (p. 104), and he was followed in 1873 by Gastaldi who in reference to cirques, asserted that glaciers 'are well able to excavate for themselves deep beds in soft rocks and also in rocks relatively hard' (p. 397). Attempts to explain the development of cirques without appealing to processes of ice erosion encountered grave difficulties. T. G. Bonney (1871,1873, 1877), A. J. Jukes Browne (1877), J. W. Gregory (1915) and others have suggested that they were the products of the work of running water, but failed to explain, for example, the presence of rock basins commonly present in the cirque floors or to consider those cirques which have developed at or near the summits of mountains (the 'crater cirques' of T. C. and R. T. Chamberlin (1911)). Detailed examination of cirques soon showed that their features were not compatible with theories of water erosion. As W. V. Lewis (1938) observes in regard to British cirques, the streams which now drain them are 'trickles by comparison with the size of the cirques which they occupy and as the streams dwindle towards their sources, the great amphitheatres in which they lie have been incised more and more deeply into the ancient upland surface' (p. 250).

Other non-glacial processes have at times been invoked to account for the formation of cirques—solution in limestone areas, wind deflation in arid areas, and, more commonly, mass movements such as landslips—but though these processes may occasionally produce cirque-like features (for example the 'pseudo-cirques' of O. W. Freeman, 1925, or the 'rasskars' described by H. W. Ahlmann, 1919), these are usually not difficult to distinguish from true cirques of glacial origin. Confusion is most apt to arise over cirque-like features resulting from landslipping, and the mass of slipped material may superficially resemble morainic debris, though detailed observation of the size, form, and structure of the slip will indicate its non-glacial origin. R. Common (1965) quotes the case of the Bizle on the north side of Cheviot which he considers has the superficial appearance and correct location of a cirque, but which may well be a rock recess resulting from landslipping.

The close connection between cirques and areas of present or former glacierization, observations of cirques containing cirque glaciers, and the virtual impossibility of devising suitable non-glacial hypotheses, gradually forced universal acceptance of the view that cirques are glacial features, though even now, the processes responsible are not fully understood. W. M. Davis (1909) described the cirques of North Wales as having been 'significantly enlarged and deepened by glacial erosion, with pronounced steepening of the side and head walls, partly by direct glacial action, partly by superglacial weathering', but admitted that 'no one knows just how glaciers can erode retrogressively so as to enlarge as well as to deepen their névé reservoirs'. In the discussion following his paper, Davis referred only briefly to Willard D. Johnson's earlier and very significant contribution to the problem of cirque formation (1904). Johnson's observations of conditions in a bergschrund led him to deduce that alternate freezing and thawing of water in rock joints was a major process in the erosion of cirque headwalls, a process also suggested by A. Helland, one of the pioneers of the glacial theory, as long ago as 1877, but Davis doubted its effectiveness in depth, where temperature variations might be small, and wondered whether pressure changes might aid the freeze-thaw process. As regards cirque floors and their rock basins, ignorance of process was even more striking; vague references to the scooping action of the cirque glacier characterize the literature of the early part of this century.

At the same time, there were those who recognized the connection between cirques and cirque-glaciers, yet allotted to the cirque glaciers a very minor rôle. J. W. Evans (1913) held that cirques were mainly 'the result of frost action facilitated by mountain streams fed by snowfields' and would go no further than admit that 'glaciers have in many cases taken a minor part in [cirque] development', wearing away the floor of the cirque (p. 298). E. J. Garwood (1910) contended that 'ice, on the whole, erodes less rapidly than other denuding agents, and that under certain conditions, it may act relatively as a protective agent' (p. 311). Discussing cirques, he attributed the shattering of the headwalls to frost action above the glacier surface. The glacier itself, occupying the lower parts of the cirque and protecting them from frost action, was held to be responsible only for overdeepening the upper end of the cirque basin.

Clearly, the problem of the origin of cirques could only be settled by intensive studies of those cirques now vacated by ice, by studies of the structure and behaviour of cirque glaciers, and by investigation of the conditions which exist at the contact between ice and bedrock. Another problem which has long attracted the attention of geomorphologists is that of the evolution of cirques and the sequence of landform development which accompanies the enlargement of cirqus, up to the stages where the intersection of one cirque with another begins to result in virtual destruction of the pre-glacial relief.

Cirque Morphology

The term cirque really embraces a whole family of landforms, of very varied appearance and dimensions. At the one extreme, there are small and shallow depressions a few tens of metres across, some containing firn or thin ice, which may represent the beginnings of cirque formation; at the other, there are the great cirques of Antarctica and the Himalayas whose widths may be measured in terms of kilometres and whose backwalls may attain heights of hundreds or even thousands of metres. G. Taylor (1926) quotes a width of 16 km and a backwall 3000 m high for the Walcott cirque on Mount Lister near MacMurdo Sound, while the Western Cwm on Mount Everest has a width approaching 4 km at a maximum, and a headfall which, if the ice were removed, would probably amount to 2800 m in height. However, the largest cirques are not necessarily the most perfectly formed. F. E. Matthes (1900) claimed that several conditions had to be satisfied before maturely-fashioned cirques would result: the cirques must not be so close that in their development they interfere with one another, and the rock type should be reasonably homogeneous in order that the cirque form should not be distorted by the influence of structure. Furthermore, the glacierization of the area must not be so extensive that few if any parts of the landscape remain unsubmerged by ice. The size to which a cirque will grow depends on many factors. Among them, competence of the rock to withstand failure will clearly be important in limiting the heights of the side and head walls; Cwm Cau in the Cader Idris massif (Wales) possesses rock walls up to 430 m high in the volcanic rocks but only 240 m in mudstone. Most of the largest and best developed cirques in Britain, for example, in the Cuillins of Skye or in Snowdonia, are associated with igneous or metamorphic rocks, and this is generally true for large cirques elsewhere in the world. In areas of sedimentary rock, cirques are often only poorly developed, as in southern and central Wales, or in the southern uplands of Scotland. The size of the mountain mass in which cirques are forming will also obviously affect the size to which any one cirque can grow, and the duration of glaciation will be another factor, for as the cirques enlarge, the intervening ridges will be gradually cut away and the cirques themselves will eventually lose their outlines and identities.

There are four elements in a well developed cirque. First, there are the head and side walls, steep and usually shattered; secondly, there is the rock floor, showing evidence of smoothing and polishing, and often of rock basin form; thirdly, W. V. Lewis (1938) has noted that near the junction of head-wall and cirque floor there is sometimes a projecting node of rock; and fourthly, there is the lip of the basin, convexly rounded and often shattered on the down-valley side. Many mature cirques possess roughly the same proportions whatever their size. G. Manley (1959) claims that the ratio between the length of such a cirque and the height of the summit of the backwall above the cirque lip is normally between 2·8 and 3·2 to 1 (thus, for the Western Cwm of Mount Everest it is about 3·2, and for the cirque at the head of Coir' a' Ghrunnda in the Cuillins, a fraction of the size of the Western Cwm, it is about 3·1). In part of Labrador, J. T. Andrews (1965) finds a ratio nearer 2·1 to 1, but this may reflect immaturity of cirque development. The figures cannot be given precisely of course, for the backwall is never of constant height. Fig. 8-1a shows an example of a beautifully proportioned cirque in the Jotunheim of Norway, Vesl-Skautbotn. Sloping at approximately 60 degrees, the headwall has a maximum height of 350 m above the rock floor: the latter is concealed by the small cirque glacier but its elevation has been approximately determined (J. G. McCall, 1952). The ratio of height to length, as defined by Manley, is about 2·8. A small cirque glacier, the subject of detailed investigation, occupies part of the hollow and conceals the lower one-third of the backwall. The almost circular nature of the basin in plan is striking; the spurs enclosing it on each side approach one another and unite in a low moraine-covered rock bar which impounds the small lake. A substantial moraine within the cirque marks the present toe of

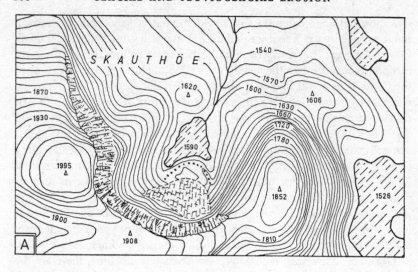

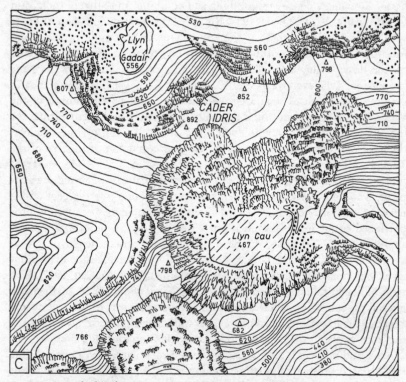

Fig. 8-1 A Vesl-Skautbotn: a cirque in the Jotunheim, Norway
        C Cwm Cau and adjacent cirques, Merionethshire, Wales

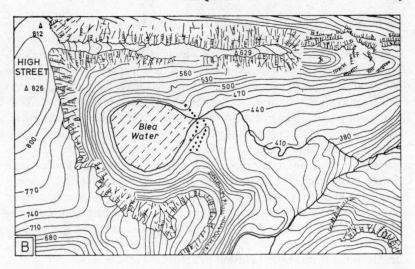

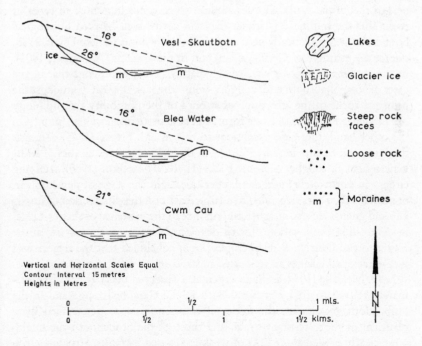

B Cirque containing Blea Water, Lake District, England
The long profiles of each cirque are drawn without vertical exaggeration

the glacier and separates the glacier from the lake. Above the glacier, the rock walls are highly shattered, and the breaking-off of rock fragments by frost action is a process active today and adding material to the inner moraine. The rock floor is not visible, except where a small patch was exposed at the inner end of a tunnel driven through the glacier; here, the rock surface was mostly smoothed, but not highly polished (J. G. McCall, 1960). Vesl-Skautbotn is of the same order of magnitude as the cirque containing Blea Water (Fig. 8-1b) in Westmorland (average length:height ratio = 2·8). Analysis of the morphology is easier in the case of Blea Water where no glacier is present, and comparison with Vesl-Skautbotn suggests that little change in the cirque has occurred since the last ice vacated it. The backwall rises 240-335 m above the lake, which itself is up to 63 m deep; the water surface of the lake in turn lies over 30 m below the central part of the moraine-covered lip of the cirque (W. V. Lewis, 1960). The average angle of slope of the rock wall is between 40 and 50 degrees. Lewis shows that the long profile of the cirque from headwall to lip is markedly arcuate.

The shapes and proportions of many maturely developed cirques seem remarkably little influenced by geological structure, apart from the fact already established that the strength of the rock will limit the size and steepness of the cirque walls. Cirques are known to occur in almost all types of rock. But the outlines of cirques may cut across well-marked lithological boundaries with practically no change, as in Snowdonia (where Lewis, 1938, quotes the example of Cwm-y-Llan) and in the Cuillins (A. Harker, 1901). In the Jotunheim of Norway, M. H. Battey (1960) comments that in the cases of Vesl-Skautbotn and Veslgjuv-botn, 'the lack of variety amongst the principal rocks of the area disposes at once of the possibility that lithology has influenced the positions or forms of these particular cirques' (p. 6). On the other hand, there are exceptions to this. K. M. Strøm (1945) contrasts the Rondane cirques in sparagmite with the shorter steeper cirques of Moskenesøy cut in tougher plutonic rocks. H. R. Thompson (1950), studying cirques in north-west Sutherland, concluded that the stage of development of individual cirques depended to a large extent on the distribution of quartzite and gneiss. He also attached importance to the existence of shatter-planes in controlling the location of well-developed cirques in the gneiss. In the English Lake District, P. H. Temple (1965) concluded that the majority of well-developed cirques were located in structurally weak zones.

L. H. McCabe (1939) in Spitsbergen also observed that cirques were intimately related to rock structure. Both in the Campbell range and in the Stubendorff mountains, the floors of the cirques were determined by a platform of more resistant strata, and those of the Stubendorff mountains were clearly eroded along lines of weakness caused by faults. Structures such as foliation, joints, faults and shatter-planes seem to be relatively more important than lithology in cirque development. Battey (1960) shows how thrust-planes outcropping in the walls of Veslgjuv-botn (Jotunheim) often define the upper and lower edges of tabular masses of rock that have broken away parallel to the free face and form the treads of shallow steps. He also discusses the important part played in cirque morphology by dilatation or pressure-release joints (Chapter 7). These, he suggests, strongly control the headward erosion of the cirque walls, which retreat roughly parallel to themselves by the spalling off of sheets from the free face. In forming thus approximately parallel to the ground surface, they help to maintain the forms of glacial erosion, so that cirques grow in size without any essential change of form.

The influence of pre-glacial relief on the location and morphology of cirques has been much discussed. Since in most cases one cannot establish with any precision the form of the ground surface before cirque erosion commenced, it is impossible to decide between theories which on the one hand attribute relatively little to glacial erosion and much to the influence of pre-glacial relief, and on the other hand, those which postulate radical transformation of the pre-glacial surface. It is generally agreed that though cirques will develop in a great variety of situations, the majority have originated in previously water-eroded features. W. M. Davis (1909) contended that pre-glacial valley heads favoured the formation of a local glacier and the excavation of a cirque. E. de Martonne (1910-11) also thought that pre-glacial valley heads or valley sides would provide suitable locations for the initial collection of snow and its transformation into moving ice; Strøm (1945) indicated that some cirques inherited pre-glacial valleys in the Rondane of Norway; and Harker (1901) made similar suggestions for cirques in Skye.

Clearly, there are many water-eroded forms, both major and minor, which have subsequently been taken over and resculptured by moving ice; the problem is in deciding how much influence the forms of water erosion have had on cirque development. In occasional cases where cirques are aligned along zones of rock weakness, for instance Cwm Cau (Cader Idris), or certain cirques in Sutherland (Thompson, 1950), it may be reasonably argued that pre-glacial river erosion was guided by the structural weaknesses and the heads of valleys thus created later became the sites of cirques. In the Cader Idris massif, one can perhaps obtain some idea of the pre-glacial valley form by examining the head of the small valley which drains west from the summit ridge and contrasting its form with Cwm Cau on the other side of the ridge (Fig. 8-1c).

Although many cirques thus develop out of older water-eroded features, it is now accepted that many also arise through processes of nivation deepening any available hollows, however slight, which are able to collect snow. Lewis (1939) shows that sometimes these nivation hollows themselves are related to previously water-eroded features, but there are obviously numerous other locations in which banks or patches of snow may accumulate and persist. D. L. Linton (1963a) notes that mature cirques may develop in isolation on suitable hill-slopes even though these are almost unindented, and quotes the cirque at about 300 m on the north-east flank of Mount Eagle in County Kerry, the most westerly of British cirques. It is suggested that by various processes (Chapter 25) the snow patch will deepen its hollow, more snow will collect, and eventually ice will form. Once this begins to move, the transition to a small cirque glacier will have begun. In this way, cirques could arise in a great variety of topographical situations; in some cases, pre-glacial forms might strongly affect cirque development, while in others, the cirque might be a virtually independent feature.

A third major factor influencing cirque development, and one at least as important as either rock structure or pre-glacial relief, is the duration of glaciation. Standing on the summit of Galdhöpiggen (2453 m) in Norway (Plate XVI), one can look down on an impressive series of snowfields contained by narrow and jagged arêtes, representing the walls of cirques half submerged by snow and ice. The ridge between Svellnosbreen and Storgjuvbreen is clearly being steadily consumed as the cirques containing these glaciers and standing back to each other, grow in size. Sections of the headwalls surrounding Tverråbreen and Svellnosbreen have already been partly demolished. It is impossible not to admit that the pre-glacial Galdhöpiggen massif is being slowly destroyed by glacial erosion, involving the recession of numerous cirques, and that the degree of destruction is roughly proportional to the duration of glaciation. Willard Johnson admitted as much in 1904 in respect of part of the Sierra Nevada. Viewing the highlands from Mount Lyell, he likened them to 'the irregular remnants of a sheet of dough . . . after the biscuit tin has done its work', and he inferred that 'the suspension of glaciation had suspended as well a process which had threatened truncation of the range' (pp. 571-2). Ahlmann (1919) illustrated the evolution of cirques in the Lofoten Islands. 'On Moskenaesö there occur all stages in the development of a cirque. The first stage is represented by a shallow depression. . . ' The youthful stage is marked 'by a semi-circular cirque with steep, not particularly lofty sides and flattish, scoured, and slightly uneven floor' (pp. 229-30). The excavation of a rock basin indicates maturity, while further lengthening of the cirque owing to headwall recession will cause the formation of a round-headed trough, which according to Ahlmann represents old age.

W. H. Hobbs in 1926 put forward a scheme outlining hypothetical stages in the destruction of uplands by cirque recession (Fig. 8-2a). The 'grooved upland' of youth still preserves much of the pre-glacial surface intact: the classic area is the Bighorn Range of Wyoming, but on a smaller scale, parts of North Wales east of Ffestiniog demonstrate this stage well, as do parts of the island of Arran. Hobb's second stage—the 'early fretted upland'—is well displayed in the Snowdon massif of North Wales, where the smooth-topped ridge is being cut into by well-proportioned cirques. As the cirques enlarge and recede, so a series of arêtes and horns is created, typical of a 'mature fretted upland'. This stage can readily be matched in the Cuillins of Skye (Harker, 1904), in the Galdhöpiggen area of Jotunheim, in the Rondane (Strøm, 1945), in the Lofoten Islands (Ahlmann, 1919), or in the Ötztal Alps of Austria, for instance. Horns may rise from the intersection of three principal arêtes (in the Alps, the Aletschhorn or the Dom, for instance) or of four (Gross Venediger, or the Dent Blanche); examples of more than four principal arêtes converging to a horn are fewer, but the Mönch and the Weissmeis in the Swiss Alps may be cited. It is clear that horns tend to survive long after the arêtes linking them have been destroyed; in some cases, secondary cirques may develop on one or more faces of the horn (for instance, Semeltind and Skardalstind in Jotunheim, Norway). Linton suggests (1963a) that the reason for late survival of horns in landscapes subjected to prolonged glacial erosion may be connected with the formation of dilatation joints. Arêtes might be expected to disintegrate more readily since they will develop two sets of intersecting dilatation joints (Fig. 8-2b) whereas the rock in the horns having been free for some time to expand outwards parallel to three or more free faces, may have reached a state of equilibrium. Thus, 'the assistance given to weathering and removal in the early stages by dilatation joints is considerably diminished when isolation has actually been achieved' (p. 21).

There is less agreement about landform evolution in the later stages of cirque enlargement and recession. Hobbs describes a stage of old age (his 'monumented upland') in which the breaking-down of arêtes creates cols which allow ice to spill over from one cirque into another. Beneath the ice, original cirque floors may coalesce into a surface of glacial denudation, surmounted only by horns. E. Richter (1906) attributed widespread planation (Reusch's 'Paleic Surface') of the Norwegian uplands entirely to this process, but no one would support this extreme view today. W. M. Davis (1900) argued that as the later stages of the glacial cycle were approached, the feeding grounds of the glaciers would become confluent, broken only by nunataks (representing the original horns); he speaks of an ice and snow shield covering a lowland of glacial denudation when old age had been reached. But as C. A. Cotton (1942) has pointed out, 'destruction of moun-

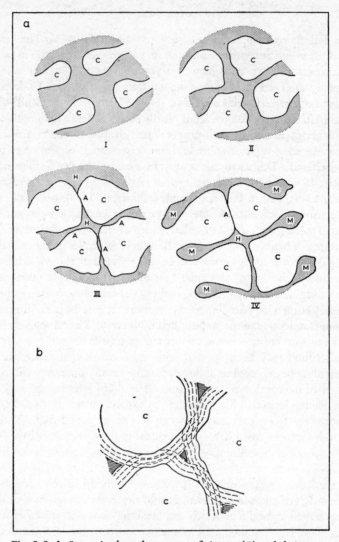

Fig. 8-2 **A** Stages in the enlargement of cirques (C) and their recession into an upland (W. H. Hobbs, *Earth Features and their
Meaning*, Macmillan & Co., New York, 1926)

    I Grooved upland   II Early fretted upland
    III Mature fretted upland   IV Monumented upland
    A Arête   C Cirque   H Horn   M Monument
Preglacial surface shaded

B The development of dilatation joints parallel to cirque
walls aids the destruction of arêtes, where the joints intersect
at shallow angles, and the formation of pyramidal peaks
(horns)

tainous relief by glacial peneplanation is a theoretical abstraction in that no cases of such levelling down of mountains are known with certainty; if a glacial peneplain exists anywhere, it may be expected to exhibit a mammillated cirque-floor-like relief' (p. 195). Strøm (1949) observes that all stages of Hobbs' cycle of glacial erosion can be matched in the Jotunheim of Norway, and that in some areas, cirque glaciers 'have split up or wholly destroyed mountain massifs. . . . As (the cirque) backwalls cut ever farther into the mountain massif where they occur, a horizontal or somewhat rising platform is formed which is frequently difficult to distinguish from an older peneplain' (p. 21). He designates this pediment-like platform built out of cirque floors as a *fly*. E. Dahl (1946) claims that part of the modelling of the Norwegian strandflat took place in this way. In the Antarctic, Linton (1963b) describes the landforms characteristic of prolonged glacial erosion, where the ice, 'in the long time that has been available to it, has been able to gnaw away the greater part of substantial mountains' (p. 281), and where there is represented 'a degree of destruction . . . not even hinted at in such classic areas of glacier study as Alaska or the Alps' (p. 282). The outlines of cirques have often disappeared, their bounding arêtes submerged by ice; ice-sculptured pyramids are the principal surviving masses standing above the ice-fields. One can only conjecture about the form of the land surface beneath the ice in such an area. But quite clearly, prolonged cirque erosion will result in completely radical transformation of the landforms.

Cirque stairways, Kartreppen, or escaliers de cirques, are characteristic of some mountain areas. In Snowdonia, the cirque containing Llyn Glaslyn bears part way up its headwall a less well-formed and smaller cirque directly beneath Snowdon summit, while on the lower side of Glaslyn there is a descent into the much larger cirque holding Llyn Llydaw. The floors of the three basins, measured at their deepest points, lie successively at 808, 562, and 375 m above sea-level. In the Cuillin hills of Skye, another example may be found in the valley descending southward from Sgurr nan Gillean into Harta Corrie. At its head, there is Lota Corrie; then a precipitous step down to a second cirque basin (floor level about 260 m); and a third sharp descent into the head of Harta Corrie. Ahlmann (1919) notes many examples in Norway, such as on the island of Moskenesøy; he suggests that at certain stages of glacierization, ice flowed continuously from the highest cirque in a stairway to the lowest, broken only by ice-falls at the steps between the different cirques. It obviously becomes impossible in some cases to distinguish between cirque stairways and glaciated valleys possessing a series of basins and steps. Indeed, Griffith Taylor (1914) has suggested that typical series of valley steps and basins may originate as isolated cirques in a stairway later modified and united by continuously flowing ice. There are many reasons why, in an area undergoing glaciation, cirques should develop at successively different levels. A rising snow-line might be the cause (see p. 198 for a discussion of the relationships between cirques and snow-lines), or intermittent uplift of the mountain mass (E. Fels, 1929). Cotton (1942) thinks that the cirques of cirque stairways might develop contemporaneously without such changes, lower cirque glaciers forming in lower niches out of avalanche ice descending from higher cirques.

The problem of the origin of cirque stairways is complicated by the multiple nature of Pleistocene glaciation; the cirques that we now see are mostly the products of not one but several glaciations of differing magnitude. However, contemporaneous occupation of cirques in a stairway by separate cirque glaciers has fairly been demonstrated for certain glacial episodes. B. Seddon (1957), for instance, shows that the cirques of Glaslyn and Llydaw on Snowdon, already described as members of a cirque stairway, contained separate cirque glaciers in the Late-glacial period, judging by their individual morainic accumulations on the lips of the cirques at about 610 and 425 m respectively. But it is also clear that these cirques were not formed solely by the Late-glacial ice, but merely reoccupied. Seddon deduces that the size of glacier in each cirque was roughly proportional to the amount of precipitation in their respective accumulation areas. In earlier glaciations of greater magnitude, ice from the higher cirques must have overflowed into the lower ones.

Cirque Orientation and Elevation

In the middle latitudes of the northern hemisphere, it is well known that the majority of well developed cirques face in directions between north and east. Conversely, in the restricted land areas of middle latitudes in the southern hemisphere, a southerly or south-easterly aspect is dominant (see, for example, E. Derbyshire, 1964), although the number of cirques involved is relatively small. The factors determining cirque orientation are several. Structure is held to be locally of importance, and McCabe (1939) considered its influence paramount in controlling cirque orientation during phases of oncoming glaciation in West Spitsbergen, but its influence as a primary factor is denied by others, such as P. H. Temple (1965). Protection from solar radiation is a much more general control, for it determines to a large extent how much snow can accumulate in the cirque and survive through the summer. In the British Isles, most well developed cirques are orientated away from the sun: Lewis (1938) noted that 29 out of 44 British cirques face between north-west and east-north-east. In the Dingle peninsula, a marginal area in respect of Pleistocene glaciation, C. A. M. King and M. Gage (1961) draw attention to an even more pronounced concentration, for out of 34 cirques, all but three face in directions between north-west and east. More detailed quantitative analyses of orientation have been made by V. Schytt (1959) in a comprehensive survey of Swedish glaciers, and for smaller areas, by B. Seddon (1957) and P. H. Temple (1965), for example. Temple has investigated the orientation of 73 cirques in the west-central English Lake District. He distinguishes carefully between the *location* of a cirque on a particular hill mass, and the *aspect* of a cirque. These two parameters are often but not necessarily similar. Seddon's findings for the aspect of cirques in Snowdonia are substantially similar: 75 per cent face between north and east. In part of Labrador, J. T. Andrews (1965) found that 72 per cent of cirques have this orientation.

TABLE 9   *Position and Aspect of Cirques in the West-central English Lake District*

Quadrant/Aspect	Per cent located in each quadrant	Per cent facing each direction
North-east	53·4	52·0
South-east	19·2	19·2
North-west	24·7	23·3
South-west	2·7	5·5

The shade factor may be of most importance in middle latitudes, for in the tropical zone, the angle of the sun's rays is much greater and no clear distinction can be drawn between shady or sunny sides of the mountains. Shade is not the only climatic consideration, however; the strong easterly component in the orientation of British cirques, for instance, requires further explanation. F. Enquist (1917) in the Alps showed that the largest cirques and the largest glaciers lie on the lee side of the mountains with respect to the prevailing snow-bearing winds. In Britain during glacial periods, the main snow-bearing winds were probably from directions between south and west; hollows most favourable for snow accumulation and therefore cirque glacier growth were those facing in the opposite directions, where drifting snow could most effectively collect. In the Front Range of the Rockies in Colorado, cirques are also largest and most numerous on the eastern leeward flanks, and the only surviving cirque glaciers occur on this side too, for during the winter season in which snowfall is now concentrated, winds blow dominantly from the west. Shade and wind direction are therefore the most powerful controls affecting snow accumulation, and, in turn, cirque orientation. There are notable exceptions, however, apart from those cases where structure is a major influence. In the southern half of the Sierra Nevada, as is well shown on the Mount Whitney quadrangle (U.S.G.S. 1:125,000), the greatest number of well-developed cirques lies on the west of the main divide, for the eastern slopes, although representing the lee side with respect to the dominant moisture-bearing winds, are relatively arid because of the rain-shadow effect; snow accumulation on the eastern slopes has therefore never been as great as on the windward side. In detail, however, the position and orientation of cirques on the windward side of these mountains is still closely influenced by the availability of shade.

The elevation of a cirque is closely related to the elevation of the local snow-line, or firn-line. The relationship is a very significant one, since it can be used to estimate positions of the regional or climatic snow-line during the formation of the cirque. The firn-line on a cirque glacier usually occurs about three-fifths of the way between the snout and the upper limit of the ice; for instance, the present small cirque glacier in Vesl-Skautbotn (Jotunheim, Norway) has a firn-line at about 1650 m, a lower ice limit of 1580 m, and an upper limit of 1710 m (Fig. 8-1a). For individual cirques now vacated by ice, rough estimates can therefore be made of the height of the former firn-line (for example, G. Manley, 1959) and from these the climatic snow-line during cirque formation may be obtained. J. T. Andrews (1965) for instance, in the northern Nain-Okak section of coastal Labrador, puts the climatic snow-line at about 240 m above the local cirque firn-line at the time the cirques were forming, and deduces a rise in mean summer temperatures of 6°C. since that time. A crude approximation can even be made using the elevations of the cirque floors, since the climatic snow-line will not usually lie at more than a few hundred metres above these. R. F. Flint (1957, p. 309) shows a map of generalized contours on a theoretical cirque-floor surface for the western United States which reflects in some measure the varying heights of the snow-line over this area, possibly related to the last glacial maximum. In Rocky Mountain National Park (Colorado), the average altitude of the former snow-line implied by the mean level of cirque floors is about 3000 m, compared with 4200 m for the present snow-line. M. M. Miller (1961), studying more than 200 abandoned cirques in the Alaska-Canada Boundary Range, finds that their floor elevations are concentrated at five levels (110, 340, 530, 750, and 960 m), which he relates to former firn-lines in the Wisconsinan glaciation. It is possible that the different levels reflect an intermittently rising snow-line in this period. Miller finds no evidence which might support the view that lithology, bedrock structure, or pre-glacial relief might have been responsible for the successive cirque levels.

Other workers have questioned such views; in some areas the altitudes of cirque floors appear to be too variable and almost random, and it is pointed out that the aspect of the cirque, and therefore the duration and effectiveness of ice erosion, may result even in neighbouring cirque floors having very different altitudes, even though their formation may have been synchronous. In any analysis of cirque-floor elevations as a means of fixing former snow-lines, therefore, it is well to use only well developed cirques and to avoid cirques whose altitude is obviously affected by geology or pre-glacial relief. It should also be borne in mind that a cirque may be composite in the sense that it may have been reoccupied and deepened in successive glacial episodes.

Structure and Behaviour of Cirque Glaciers

An essential preliminary to understanding the origin of cirques is a con-sideration of the nature of cirque glaciers and their movement.

The banded appearance of glaciers was first discussed in some detail by J. D. Forbes (1843). Nearly a hundred years passed before further studies of glacier bands were undertaken. In 1939, G. R. Gibson and J. L. Dyson pub-lished the first detailed map of some of the ice bands of a cirque glacier, the Grinnell glacier of Glacier National Park, Montana, and showed how signi-ficant these bands might be in understanding the mode of accumulation and the nature of ice movement. They regarded the ice bands as annual accumu-lation layers, each representing some 15 to 20 m of winter snow deposited in the firn area, and becoming tilted and slightly deformed as they moved down the glacier. The outcrops of the bands at the glacier surface formed sweeping curves roughly parallel with the back wall; they dipped up-glacier at 12 degrees near the firn-line, and up to 45 degrees nearer the glacier front. They had no evidence as to the disposition of the bands within the glacier. In 1948, J. M. Grove (see J. M. Clark and W. V. Lewis, 1951, and J. M. Grove, 1960) commenced an investigation into the structure of the cirque glacier Vesl-Skautbreen, whose surface banding is unusually simple and striking. It is, moreover, a conveniently small cirque glacier, and one which was simultaneously being studied in its other aspects by a team of research workers. Not only was Mrs Grove able to study the surface banding but to examine the internal behaviour of the banding along two tunnels excavated through the glacier. Her findings on Vesl-Skautbreen, and also on nearby Vesljuvg-breen, are of great significance for all cirque glaciers, and will therefore be presented in some detail.

The Vesl-Skautbreen consists of a series of ice layers, separated by discon-tinuities some of which are emphasized by the occurrence of small quantities of organic or mineral debris. In the past, the discontinuities have been referred to as dirt bands, but this term is undesirable since they are not bands and not always dirty. The ice layers each represent one year's increment of snow in the firn region (a few layers may be composite), and are therefore termed 'accumulation layers'. The discontinuities represent ablation in the summer periods, and are termed 'ablation surfaces'. Where they outcrop at the surface of the glacier, the ablation surfaces yield debris in varying amounts as the surface of the glacier wastes downward. This, together with the fact that the next higher accumulation layer often tends to protrude or hang slightly above the ablation surface, led W. V. Lewis (1947, 1949) to suggest that many ablation surfaces were planes of differential movement, with debris being brought up to the glacier surface by the slow overthrusting of one accumulation layer over the ablation surface beneath. That the ablation surfaces of Vesl-Skautbreen are not thrust planes was conclusively shown, particularly in a simple experiment by J. F. Nye (see J. G. McCall, 1960, p. 49); and it is now agreed that the overthrust appearance is simply the effect of differential ablation. This does not, of course, preclude the existence of thrust-planes on other glaciers, particularly near the termini of valley glaciers, but they are not to be equated with ablation surfaces. As well as the accumulation layers and ablation surfaces, Mrs Grove describes blue layers of ice within the accumulation layers, which she contends result from formation of surface crusts or ice layers by freezing of meltwater in the firn zone; blue veins cutting across the layers probably represent the freezing of meltwater seeping down cracks.

The density of the white ice in the accumulation layers is about 0·9; thus an annual accumulation layer 1 m thick is probably the equivalent of about 15 m of freshly-fallen snow in the firn zone. As this is more than the mean annual snowfall for this part of Jotunheimen, it is clear that drifting, avalanching, and eddy deposition have caused a concentration of snow in this cirque, which is backed by a broad upland plateau to windward. The cirque of Veslgjuv-breen 8 km away shows in some years spectacular effects of eddy deposition of new snow under the backwall. The central tunnel through Vesl-Skautbreen showed that this glacier consisted of a minimum of 88 accumulation layers, as well as an uncertain number of layers compressed in the base of the glacier, and others in the firn zone above the tunnel. The bulk of the ice in this small glacier was therefore probably less than 100 years old. The tunnel showed too that most of the ice was extraordinarily clean, a finding which was borne out also in the case of Veslgjuv-breen. What debris there was in the glacier, apart from that in the basal layer next to the rock floor, was clearly not originating from the rock floor but from debris fallen or collected on the firn zone, especially as the result of rock falls from the backwall, and thereafter passing down into the glacier. One ablation surface in particular, outcropping a little above the central tunnel entrance, was, on the left hand half of the glacier, a prolific yielder of debris which must have represented a series of large rock falls on the firn zone about eighty years ago.

The greatest amount of rock debris in the glacier was found in the basal layer or 'sole' next to bedrock. The ice here seen at the inner end of the central tunnel, had a blackish appearance, containing material ranging from rock flour to large boulders; it was also sheared and slickensided indicating a tearing movement on the rock bed. Since the ice here was also bubble-free, it probably originated not from compaction of firn but from freezing of meltwater (compare the regelation layer of B. Kamb and E. LaChapelle, 1964, discussed in Chapters 4 and 7), the meltwater seeping down the back of the glacier. Rock debris fallen into the gap between the upper part of the glacier and the headwall was almost certainly the source of most of the debris in the glacier sole.

The accumulation layers have been deformed by ice movement. In the firn zone, the latest accumulation layers dip down the glacier at about 30 degrees, and as shown in Fig. 8-3a, the dips of the layers exposed below the firn-line are nearly horizontal at their outcrops, but steepen so as to dip up-glacier farther down the ice. At the entrance to the central tunnel, they dip up-glacier at 26 to 28 degrees, but below this, the dip once again flattens out, becoming virtually nil near the glacier snout. The section provided by the central tunnel showed a synclinal arrangement of the ice layers; at its inner end, they curve so as to meet the rock bed tangentially. Thus, the deformation of the layers by glacier movement suggested in itself how that movement was taking place. Clearly, the layers were gradually being rotated as they passed down the glacier; at the base of the glacier, they were also being stretched and bent upward by basal drag.

More precision was given to such theories of ice motion in this glacier by J. G. McCall's careful theodolite surveys of the changing position of pegs both on the ice surface and in the tunnels. From these records, McCall was able to interpolate the velocity distribution and to plot velocity profiles (Fig. 8-3b). The five velocity profiles here shown indicate the shapes and positions of originally straight section lines (a-a to e-e) after undergoing seven years (= c. 2500 days) movement. The flow lines were based on observed velocities plus an assumed regular curve (approximating to the bedrock surface) for the lower limit of ice movement. The flow lines show downward movements in the firn zone, movement parallel with the surface at the firn-line (which lies between stakes V and VI), and pronounced upward movements in the lower tongue. The velocity vectors reveal the more rapid movement of the surface layers of the firn (profiles a-a and b-b), the retardation of the basal ice (profiles b-b through e-e), and a very slightly lower rate of movement for the surface ice than for the deeper ice in profiles c-c and d-d. The hypothesis which best explains the observed surface and internal movements of the ice, and the deformation of the accumulation layers, is that of rotational sliding. Suggested by W. V. Lewis in 1947 and elaborated in

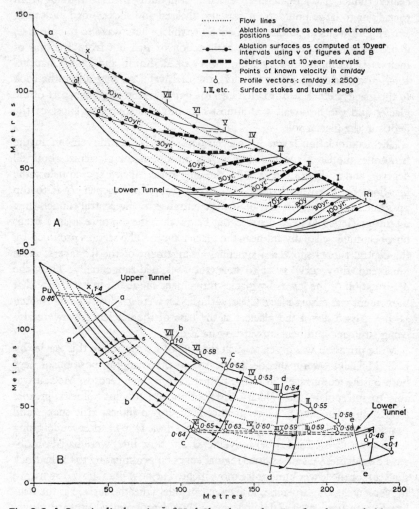

Fig. 8-3 A Longitudinal section of Vesl-Skautbreen showing flow lines and ablation surfaces (J. G. McCall, "Norwegian Cirque Glaciers', Royal Geographical Society, *Research Memoir*, 4, 1960)
B Velocity distribution in a longitudinal section of Vesl-Skautbreen (J. G. McCall, *op. cit.*)

1949, the hypothesis implies sliding of the glacier along a roughly arcuate bedrock floor and, simultaneously, rotation around an approximately horizontal axis. These points are consistent with the observed rotation of the accumulation layers, with the forward and downward motions recorded in the firn zone, and with slight upward movement near the toe of the glacier. Furthermore, the central tunnel, excavated originally with a slight outward slope to facilitate meltwater drainage, over a period of time gradually tilted to become horizontal, and later developed an inward slope; lowering of the tunnel floor near the entrance became necessary to prevent water accumulating in the inner parts of the tunnel. The motivating force behind such rotational movements appears to be quite simple; the main weight of winter snow accumulates on the upper part of the glacier beneath the headwall, whereas the lower part of the glacier in summer experiences most losses by ablation. Thus the overall surface of the glacier is effectively steepened, and rotation occurs to restore equilibrium when the forces are sufficiently large to overcome friction between the ice and bedrock.

Although rotational sliding can explain much of the observed behaviour of Vesl-Skautbreen, it does not explain everything. That the accumulation layers are deformed indicates some degree of creep in the ice; the observed angles of dip of the accumulation layers in the central parts of the glacier surface are greater than would be expected if it were a question solely of the rotation of a rigid mass; and Fig. 8-3b shows that forward motion of the ice ceases almost entirely at point F against the moraine and that in sections c-c and d-d, the locus of maximum velocity occurs neither at the surface nor near the bed, but within the lower half of the glacier. It is apparent that some force system within the glacier is slightly retarding the surface ice movement; McCall argues that a very weak form of extrusion flow is superimposed on the rotational sliding because the surface zone is behaving as a sloping dam beneath and against which the interior ice is pushing. The surface zone is slightly more rigid since in the lower part of the glacier it is braced against the terminal moraine at its foot, and since it is colder in the winter and early summer than the deeper ice. The result is that the lower part of the glacier tends to bulge slightly; and this in turn is sufficient to allow ice movement along the bed virtually to cease at point F by the moraine.

These theories of ice movement have been developed in respect of Vesl-Skautbreen, but it is highly probable that they apply directly to most other simple cirque glaciers, and less directly to compound types of cirque glacier. Thus, observations on the Arapaho cirque glacier in the Front Range of the Rocky Mountains, the largest in Colorado, show that it too is moving primarily by rotational sliding (H. A. Waldrop, 1964); studies of the double cirque glacier of Veslgjuv-breen also reveal rotational movements, and some slight bulging of the ice surface in the zone of maximum dip of the accumulation layers (J. M. Grove, 1960). The significance of these observations and theories with respect to cirque morphology will be considered in the next section.

The Problem of Cirque Erosion

Cirques possess a number of features which can only be explained in terms of glacial erosion or by processes closely associated with ice. The principal features are the rock basins in cirque floors, the rock lip or threshold at the cirque exit, the steep head and side walls, and the presence of smoothed and abraded rock surfaces mainly but not exclusively on the cirque floors, and of broken and shattered surfaces especially on the head and side walls. The processes responsible for these features are still matters of controversy, and their investigation is rendered difficult by the fact that direct observation of processes operating beneath a glacier is usually impossible. The main processes now generally considered responsible for cirque formation are those of abrasion, joint-block removal, and freeze-thaw action or frost riving. Nivation processes play an important part in the initial stages of cirque formation, but their consideration will be deferred to Chapter 25.

ABRASION

The nature of this process has already been considered in relation to glaciated rock surfaces and glaciated valleys in Chapters 6 and 7. Here, we are specifically concerned with the abraded surfaces of cirque floors and sidewalls, and in this connection, the subglacial rock surface exposed at the inner end of the central tunnel through Vesl-Skautbreen is very instructive. Unlike the rock surfaces which may be readily examined in ice-free cirques, its features obviously owed nothing to subaerial weathering. The surface remained damp and unfrozen throughout the year, suggesting that frost riving was not a relevant process in its erosion, at least at the present. Abrasion and joint-block removal were more plausible possibilities as sculpturing methods. With respect to the first of these, projecting parts of the rock surface were seen to be smoothed (but not polished) and striated. A plumb-bob suspended from the ice above was observed to move exactly in the direction of striation. Short-period shear-meter records indicated an even rate of sliding of the ice over the rock bed. McCall and others who examined the surface considered that a convincing case could be made for regarding abrasion as the chief erosional process operating here today, and noted that as much as 90 per cent of the motion of this glacier was contributed by basal sliding.

Surfaces showing abrasion and striation are most commonly associated with cirque floors and thresholds, but are also to be found, as Lewis noted in 1938, in places on cirque headwalls. It is here that they are much more difficult to explain. W. Dort (1957) noted striated surfaces high on the head-walls of cirques in northern Idaho, and attempted to explain them as being carved in a period when the cirques were over-filled with ice—when even the divides above the cirque headwalls were submerged by moving ice.

Abrasion by moving ice is thought to be responsible for the basin form of cirque floors. The hypothesis of rotational slip is very relevant to this problem. W. V. Lewis in 1949 suggested that 'when a section of a glacier is moving in a rotational slip, the bedrock beneath must be moulded into a basin fitted to this rotational movement' (p. 156). Gibson and Dyson in 1939 thought that the rotational movements were the effect and not the cause of the rock basin, but Clark and Lewis (1951) were convinced that the tendency to rotate was inherent in certain glaciers and that this might, under favour-able circumstances, lead to the scouring-out of basins in the rock floor beneath. The fact that the long profile of the rock floor in the Vesl-Skautbreen cirque is nearly arcuate has already been illustrated (Fig. 8-1a), though the information on the ice-covered segment is meagre. Lewis (1960, pp. 97–98) shows how the long profile of the cirque holding Blea Water (Westmorland) also approximates closely to an arc of a circle (Fig. 8-1b); on physiographical evidence, he restores the probable form of the ice surface of a small glacier in this cirque. With a surface slope of 13 degrees or more, such a glacier would probably not be in equilibrium, but would tend to rotate about an approximately horizontal axis 600 m above the cirque floor, until its surface slope were reduced to 7 or 8 degrees; 'there would be no difficulty in pro-ducing the required force not only to move the bottom debris uphill out of the corrie basin, but also to scour this basin actively' (p. 99). Such concepts of rotating cirque glaciers are undoubtedly oversimplified—more complex forms of ice motion are most certainly involved in larger cirque glaciers—but such simple concepts form a useful working hypothesis which must be further tested.

Lewis considered that the relatively high velocities of the basal ice in a simple rotational slip were a great advantage in excavating the rock basin. In the profile c–c measured in a direction perpendicular to the rock floor at c (Fig. 8-3b), for instance, the velocity of the ice is slightly greater next to bedrock than it is at the glacier surface. In addition, the very rapid diminution of the rate of basal ice movement in Vesl-Skautbreen (resulting from the 'damming' effect described on p. 203), from 6·4 mm/day at the inner end of the central tunnel to less than 1 mm/day at the toe of the glacier (stake F), can be related to the reduced amount of corrasion on the cirque threshold (at the toe of the glacier) as compared with the Skautbreen basin floor which lies at least 12 m below its threshold. The rotational slip hypothesis provides a simple mechanism for the uplift of ice over a cirque threshold and for the evacuation of debris uphill from the basin floor.

JOINT-BLOCK REMOVAL

The details of this process and its effects are also presented in a wider context in Chapters 6 and 7. It should be recalled that it is not a question of ice simply 'plucking' rock from the bedrock surface, since the yield stress in shear for ice at 0°C. is only 1 kg/cm². By attachment to the bedrock surface, ice could only remove material already loosened by frost-riving, for example, or well-jointed rock. At the inner end of the Vesl-Skautbreen central tunnel, the bedrock surface showed no signs of frost-riving or weathering. It was smoothed and striated in places as already described, but it also possessed surface angularities which led Lewis (1954, p. 419) to describe its appearance as 'freshly fractured when the all-pervading coating of rock flour was removed'. The basal ice of the glacier did not adhere to the rock but could be pulled off easily and cleanly, so that McCall found it difficult to imagine how any further plucking of rock could take place, even though it might have occurred in the past.

Lewis in 1954 was the first to consider the possible role of dilatation joints in joint-block removal by ice (Chapter 7); with J. W. Glen, he argued that 'the effects of glacial quarrying on such [jointed] rocks might be far greater than a simple calculation based on the relative strengths of ice and rock would suggest' (1961, p. 1118). As already mentioned on p. 191, Battey showed the probable existence of dilatation joints roughly parallel to the cliff faces of the Vesl-Skautbreen and Veslgjuv-breen cirques of Jotunheim, Norway, and considered their rôle a most important one in cirque development. Linton (see p. 193) contends that they play a vital part in the des⎽⎽uction of arêtes and divides, while also assisting in the survival of pyramidal peaks. It is conceivable, therefore, that dilatation joints developing in cirques greatly assist joint-block removal by ice from the headwalls and floor, and that the surface exposed in the Vesl-Skautbreen central tunnel was one formed in the recent past by massive joint-block separation along dilatation joints parallel with that surface, and later abraded and striated by continuing ice-flow over it.

Another process that may play a part in preparing the bedrock surface for erosion, particularly block removal, by cirque glaciers, is that of pre-glacial frost action. L. Lliboutry (1953), for instance, has argued that the rock in the cirques is first broken up under periglacial cold conditions. Later, the ice cleans out the debris and smooths the rock floor.

Repeated freezing and thawing of water in rock crevices has often been suggested as the mechanism behind the shattering of cirque walls since the observations of Lorange on cirques in Norway quoted by A. Helland in 1877 (p. 164). In 1904, W. D. Johnson gave evidence that the most important concentration of freeze-thaw action at the back of a cirque glacier was associated with the bergschrund. He descended an unusually deep bergschrund in a glacier on the north of Mount Lyell in the Sierra Nevada, reaching the base of the cirque headwall at a depth of 45 m. For the last 6-10 m of his descent, he observed that the rock wall of the cirque was exposed, the ice forming the other wall of the bergschrund 2 or 3 m away. 'The rock face, though hard and undecayed, was much riven, its fracture planes outlining sharply angular masses in all stages of displacement and dislodgement. Several blocks were tipped forward and rested against the opposite wall of ice; others, quite removed across the gap, were incorporated in the glacier mass at its base. Icicles of great size, and stalagmitic masses were abundant; the fallen blocks in large part were ice-sheeted; and open seams in the cliff face held films of this clear ice' (p. 574). Meltwater was evidently penetrating any available crevices in the rock wall, and Johnson concluded that the change of air temperatures in the bergschrund across freezing-point, perhaps occurring diurnally in the summer, was the effective agent of rock disintegration.

Although workers such as G. K. Gilbert (1904), A. Penck (1905), and W. H. Hobbs (1910) welcomed this hypothesis of cirque wall shattering and erosion, it was not long before several difficulties became evident. I. Bowman (1920) and R. E. Priestley (1923) contended that bergschrunds were not always present in cirques still occupied by glaciers today and that many bergschrunds failed to reach bedrock. R. von Klebelsberg (1919) observed that meltwater was absent from certain bergschrunds penetrated by military tunnelling operations in the eastern Alps. Priestley and others found it difficult to apply the hypothesis, with its emphasis on activity within a bergschrund no more than 30-60 m deep, to cirques with headwalls many hundreds of metres in height. T. C. and R. T. Chamberlin (1911) thought that diurnal temperature changes would be small and ineffective, and wondered whether seasonal changes might be more significant. But once an enclosed bergschrund is occupied by relatively dense cold air, how can warmer air ever displace it?

In 1938 and 1940, Lewis put forward certain modifications of W. D. Johnson's original bergschrund hypothesis, notably in suggesting that much of the meltwater appearing in the bergschrund was of external origin, from either summer rain or the melting of winter snows high above the bergschrund, and that such meltwater could seep down narrow cracks far below the limits of the deepest bergschrunds. Moreover, he rightly observed that the position of the bergschrund on a cirque headwall would vary slightly each season and to an even greater extent during the waxing and waning of a glacial period when the ice surface would rise and fall with changes in the glacier's budget. The observations of Lewis and others on cirque glaciers in humid maritime climates such as those of Iceland or Norway have shown that in normal years meltwater is plentiful. Indeed, the supply of meltwater is no problem on any temperate glacier in the summer season. In the case of polar glaciers, meltwater is more difficult to provide; Priestley (1923) and others have recorded melting taking place on rock outcrops in Antarctica when air temperatures were well below freezing-point, but the amount of melting can only be very slight in such conditions (see also Chapter 3).

A most serious problem encountered by the bergschrund hypothesis in its application to temperate glaciers is that of the freezing of the meltwater, and the removal of latent heat (80 cal/g at 0°C.) in the process. Prior to 1950, it had too readily been assumed that temperatures in bergschrunds would fluctuate considerably above and especially below freezing-point, aiding freeze-thaw action. McCabe (1939) was one of the few who had made actual measurements of temperatures in bergschrunds; he recorded (in August) 0°C. at a depth of 7·6 m in one case, and 0·5°C. at 1·8 m in another, both in Spitsbergen. These figures were not unexpected, for H. V. Sverdrup (1935) had shown that the temperature at depths of more than 10 m in the firn of Isachsen's Plateau, Spitsbergen, was 0°C. even in winter. As noted in Chapter 3, this is typical of temperate glaciers which consist of ice whose temperatures at any depth are at pressure melting-point for that depth, apart from a surface layer of cold ice, up to 20 metres or so in thickness, which is cooled down by the cold atmosphere well below freezing-point in winter. There is no way by which meltwater can be re-frozen simply by contact with ice at pressure melting-point; and the deeper one goes in a bergschrund, the less possibility will there be of any air of sub-zero temperatures penetrating from the outside atmosphere.

An ingenious attempt to overcome these problems was made by C. D. Holmes in 1944. His hypothesis did not depend on either meltwater or cold air penetrating deep behind a cirque glacier, but on local thawing and freezing at the base of isothermal ice induced by variations in ice pressure, for the pressure exerted by 30 m of ice, for instance, will lower the pressure melting-point fractionally (by 0·0192°C.). Cirque glaciers vary seasonally in thickness and thus small changes of temperature will occur at the contact between isothermal ice and rock. A thin film of meltwater produced during winter when the snow-ice mass is thickest would penetrate any available rock

PLATE VII (*above*) SEDIMENTS (seen as a white band across the photograph)
DEPOSITED IN PLUVIAL LAKE LAHONTAN, NEVADA.
(C.E.)

PLATE VIII (*right*) TIOGA MORAINE (last glaciation), WALKER CREEK VALLEY, SIERRA
NEVADA, CALIFORNIA.

The recent nature of this moraine is reflected in the coarseness and lack of
weathering of the surface material. (C.E.)

PLATE IX   AUSTERDALSBREEN, NORWAY.
Showing truncated spurs and hanging valleys. Thin morainic debris rests on glacially-scoured rock in the left foreground. (C.E.)

PLATE X   VALLEY AT TURTAGRØ, NORWAY.
Showing glacial trough carved below a higher level bench, itself glacially scoured. The difference in level between the river and the highest peaks (Skagastölstind) is about 1500 m. (C.E.)

crevices and refreeze when pressure was later reduced. The process, Holmes argues, need only be very slow—0·5 m of rock removed from the cirque floor every 200 years would lead to a cirque 1000 m deep during the probable period of Pleistocene ice activity. But are such slow and minute temperature changes at all effective in rock shattering? Evidence brought forward by W. R. B. Battle (1951, 1960) suggests not (see also H. R. Thompson and B. H. Bonnlander, 1956).

In an attempt to provide more data on conditions in bergschrunds, W. R. B. Battle carried out a number of temperature measurements in various berg-schrunds in Greenland, Norway, and Switzerland between 1948 and 1953 (Fig. 8-4, a and b). Using thermographs and freeze-recording units, he studied particularly the bergschrunds at the head of Tverråbreen (Jotunheim), at the Jungfraujoch, and at the head of Glacier 32, Pangnirtung Pass, Baffin Island. In the Tverråbreen bergschrund, air temperatures fluctuated only from 0°C. to −1°C. in August 1950, and from 0°C. to −2°C. in August 1951, even though this was a fairly open bergschrund. There was no sign of any daily temperature cycle. At the Jungfraujoch, the fluctuations were −2°C. to −5°C. in March 1951, and 0 to −1°C. in August and September 1951. The rate of temperature change was in all cases very slow, apart from one drop of 1·2°C. in 6 hours (Jungfraujoch, March 1951). Nor did the small fluctuations of air temperature inside the bergschrunds always accord with the much greater changes of external air temperatures: on 19 March 1951, at the Jungfraujoch, external air temperatures fell by 5°C. at a time when berg-schrund air temperatures rose by nearly 2°C. In the bergschrund of Glacier 32, air temperatures at a depth of 27 m ranged from +0·5°C. to −3·7°C. The positive temperature recorded here was unusual but of short duration.

The freeze-recorders confirmed the thermograph evidence though they did not operate very satisfactorily; their failure to respond to changes of temperature which frequently only amounted to 1°C. suggested that such small changes might not be very effective in rupturing rocks. Moreover, it became clear that under certain conditions water might be supercooled to −1°C. without freezing. Battle conducted a number of laboratory experiments on the freezing of water-saturated porous stones, which showed that only when temperatures fell below −5°C. did marked deterioration set in. On the other hand, smaller falls of temperature sufficed to affect cracked non-porous rocks, provided the drop in temperature occurred at a rate of at least 0·1°C. per minute. This allowed freezing of water to occur first at the top of any crack, producing a closed system in which pressure could build up (see also Chapter 20). But changes of 0·1°C. per minute had never been recorded in bergschrunds—the maximum was probably not much more than 0·004°C. per minute. Such experiments also serve to dispose of C. D.

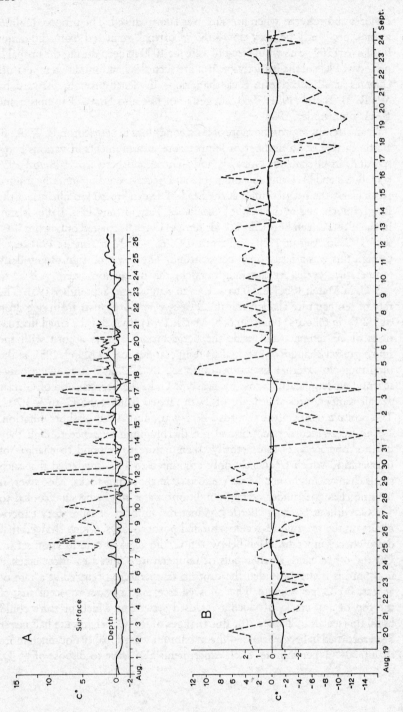

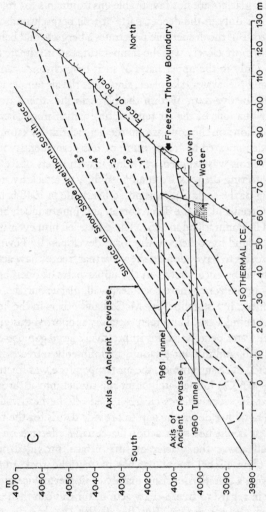

Fig. 8-4 A Temperature record from the Tverråbreen bergschrund, August 1951 (W. R. B. Battle, 'Norwegian Cirque Glaciers', Royal Geographical Society, *Research Memoir*, 4, 1960)
B Temperature record from the Jungfraujoch bergschrund, August–September 1951 (W. R. B. Battle, *op. cit.*)
C North–south cross-section of the glacier on the south face of the Breithorn, showing isotherms (°C), tunnels and water-filled cavern (J. E. Fisher, *J. Glaciol.*, 1963, by permission of the Glaciological Society)

Holmes' hypothesis of subglacial erosion in which the postulated temperature fluctuations are fractional and probably very slow.

Reviewing Battle's work, it is clear that deep or closed bergschrunds belonging to temperate glaciers are not favourable environments for freeze-thaw action, and yet it is only in the deeper parts of true bergschrunds that the rock wall is exposed. But the deeper one penetrates a bergschrund behind a temperate glacier, the more closely will the temperatures approximate that of the isothermal ice. Only in the upper parts of such bergschrunds can re-freezing of any incoming meltwater take place, since here the air temperature may be several degrees below zero. Within the bergschrund, there are no large-scale or rapid fluctuations of the air temperature which might favour alternate freeze-thaw action; on the contrary, in the bergschrunds examined, meltwater from outside appeared to freeze more or less permanently on to the rock wall wherever this was exposed, covering it with layers of ice and thereby effectively protecting the rock from all possibility of attack by frost-riving. This ice coating of the rock wall was noted by Battle in 1948 behind the Grif Gletscher in Greenland, in the Tverråbreen and Jungfraujoch berg-schrunds 1950-51, and in Baffin Island in 1953. These facts led him eventually to abandon the bergschrund hypothesis in the form developed by Lewis or Johnson; by 1953, he seems to have adopted the view that freeze-thaw action was much more likely to be encountered in the upper parts of open berg-schrunds, or generally in the zone where rock headwall, glacier surface, and freely-circulating air meet. Investigations by McCall and others in the head-wall gap or randkluft behind Vesl-Skautbreen certainly supported this view. Plenty of rain and meltwater penetrated this impressively large gap between the firn and the rockwall, and in winter would undoubtedly refreeze since air circulation was rapid. Even in summer, the air temperature was fraction-ally below freezing ($-0.6°C.$) immediately after the tunnel into it through the firn was excavated. Rock fragments clearly prised off the headwall by freeze-thaw action were falling down the gap to provide debris for the basal sole of the glacier. But could not such action be equally effective on the exposed headwall well above the glacier or firn surface, provided melt-water was available from higher snow patches to wet the rock surface occasionally? In this way, a glacier consisting mainly of isothermal ice would play a protective rôle in regard to freeze-thaw action, for this can then only take place at or above the ice surface. But the glacier also has the very important function of a transporting agent, removing frost-shattered material and preventing the accumulation of protective talus against the rock wall. And as Battle observed (Thomson and Bonnlander, 1956, p. 768), as the glacier has thinned or thickened, so more or less of the headwall would be exposed to destructive action.

Battle's deductions were based on his studies of temperate glaciers. A new approach to the problem was made by J. E. Fisher (1955, 1963) who examined the conditions pertaining to cold glaciers or firn, where atmospheric cold is penetrating through the ice or snow mass to its contact with the rock. Even in the Alps, the ice or snow above the firn-line may be at temperatures much below zero; 'it is to these areas of cold ice, bergschrunds or no bergschrunds, that one must look ... to understand processes of cirque erosion' (1963, p. 514). In 1954, Fisher organized the excavation of a tunnel through cold ice on the north slope of Monte Rosa, at 4240 m. When the rock face was reached after 92 metres, the temperature was − 13°C.; sub-zero temperatures probably characterize the ice-rock interface here down to an elevation of 3800 m, below which isothermal ice may be expected. The junction between cold ice and isothermal ice at the rock face, marked by the 0°C. isotherm, was almost encountered by a later tunnel (completed 1961) on the Breithorn, at a little over 4000 m (Fig. 8-4 c). In neither case was there any gap (bergschrund or randkluft) between ice and rock. The only meltwater encountered was in a tunnel driven in 1960 on the Breithorn (a little below the level of the 1961 tunnel) which, because it struck a large water-filled cavern in the ice, never reached the rock face. The origin of the very considerable quantity of water here was a mystery. It was inconceivable that it could represent meltwater seeping down from above, for it was present in too large a quantity, would have had to penetrate cold firn (temperatures down to − 6°C.) in which no cracks were visible, and moreover the quantity of water in the cavern actually increased in winter. Fisher suggested that it represented water forced up by hydrostatic pressure from lower levels where it was derived from melting of isothermal ice. It may be suggested that if, in other situations, meltwater were similarly forced up so as to reach the rock wall behind cold ice, refreezing could occur and rock disintegration would thus be induced at that level, but this remains a speculation, and the circumstances on the Breithorn may have been exceptional. Fisher preferred to associate rock wall shattering with the 'boundary between cold ice, frozen to the bedrock, and wet isothermal ice. This boundary is invited by external conditions to oscillate slowly back and forth, with powerful shattering and removal of rock as a result' (1963, p. 518). Such oscillations, he thought, would be measured in terms of perhaps hundreds of years, the result of minor climatic fluctuations, a very different time-scale from the daily fluctuations postulated by W. D. Johnson and sought for by Battle. Fisher further suggests that this mechanism, the concentration of freeze-thaw action at the 0°C. boundary in certain cold glaciers may provide an explanation for the approximate uniformity in level of cirque floors in any given area, for the location of the 0°C. horizon will be related to the prevailing mean atmospheric temperatures.

Thus cirque wall shattering, according to Fisher, will be associated with those types of cirque glaciers possessing cold ice overlying isothermal ice. Those built entirely of cold ice or entirely of isothermal ice will not normally provide favourable environments for freeze-thaw action at depth. In the varied climatic conditions of the Pleistocene, most cirque glaciers probably passed through phases when cold ice overlay isothermal ice.

Conclusion

The glacial origin of cirques is firmly accepted today. The principal morphological features of a well-developed cirque include steep and shattered headwalls, a rock threshold sometimes capped with moraine, a rock basin of circular outline enclosed by the headwalls and the threshold, and a ratio of height to length of about 1 to 3. Many cirques show a relationship with rock structure, but some do not. The majority have developed by glacial action out of previously water-eroded features or from nivation hollows. Enlargement of cirques leads to the development of arêtes and horns, and in the later stages of this process, the greater part of original mountain masses may be destroyed to leave only residual pyramidal peaks rising above an irregular cirque-floor surface. Cirque elevation and orientation are closely controlled by meteorological factors.

Headwall shattering by freeze-thaw action is one of the chief factors in cirque development, perhaps even the most important one. The association of this process with the bergschrund, however, is not entirely justified; it is apparent now that the most favourable environment for freeze-thaw action may well be the exposed rock wall above the glacier or firn limits, or in the upper parts of any headwall gap or randkluft between the glacier and the rock wall. Alternatively, the process may be associated, as Fisher believes, with particular climatic conditions when the boundary between cold ice and isothermal ice in a glacier migrates up or down part of the headwall, sapping it intensively within this zone. This might tend to undercut the headwall and thereby cause its retreat, a possibility noted by McCall at Vesl-Skautbreen. Headwall sapping by freeze-thaw action will serve to enlarge the cirque overall, whereas abrasion by rotationally-moving ice will tend mainly to deepen it. The relative potency of the two is difficult to judge; Flint (1957) thinks it generally probable that the walls retreat far more rapidly than the floors are deepened and McCall (1960) concluded that the Vesl-Skautbreen cirque had been enlarged primarily in a horizontal or headward manner by sapping. Hence, cirque-floor elevations may not change radically during glacierization. Nevertheless, the rock basin form of cirque floors does indicate at least a minimum of deepening by abrasion, and the rotational slip hypothesis seems best adapted to explain this feature. The third process, joint-block removal, is one which may operate over the whole surface area of a cirque, but it will be particularly relevant to the erosion of the headwall where the existence of well-developed joints will be of the greatest help to freeze-thaw action.

REFERENCES

H. W. AHLMANN (1919), 'Geomorphological studies in Norway', Geogr. Annlr 1, 1-148 and 193-252

J. T. ANDREWS (1965), 'The corries of the northern Nain-Okak section of Labrador', Geogr. Bull. 7, 129-36

M. H. BATTEY (1960), 'Geological factors in the development of Veslgjuv-botn and Vesl-Skautbotn', in 'Norwegian cirque glaciers' (ed. w. v. LEWIS), R. geogr. Soc. Res. Ser. 4, 5-10

W. R. B. BATTLE (1960), 'Temperature observations in bergschrunds and their relationship to frost shattering', in 'Norwegian cirque glaciers' (ed. w. v. LEWIS), R. geogr. Soc. Res. Ser. 4, 83-95

W. R. B. BATTLE and W. V. LEWIS (1951), 'Temperature observations in bergschrunds and their relationship to cirque erosion', J. Geol. 59, 537-45

T. G. BONNEY (1871), 'On a cirque in the syenite hills of Skye', Geol. Mag. 8, 535-40

T. G. BONNEY (1873), 'Lakes of the north-eastern Alps and their bearing on the glacier erosion theory', Q.J. geol. Soc. Lond. 29, 382-95

T. G. BONNEY (1877), 'On Mr Helland's theory of the formation of cirques', Geol. Mag. 14, 273-7

I. BOWMAN (1920), The Andes of Southern Peru (Am. Geogr. Soc. Spec. Publ. 1) 295

T. C. and R. T. CHAMBERLIN (1911), 'Certain phases of glacial erosion', J. Geol. 19, 193-216

J. DE CHARPENTIER (1823), Essai sur la constitution géognostique des Pyrénées (Paris)

J. M. CLARK and W. V. LEWIS (1951), 'Rotational movement in cirque and valley glaciers', J. Geol. 59, 546-66

R. COMMON (1965), 'Slope failure as a factor in rates of erosion', Rep. Br. geomorph. Res. Grp, Symposium at Bristol, 1965, 3-5

C. A. COTTON (1942), Climatic accidents in landscape-making (Christchurch, N.Z.)

E. DAHL (1946), 'On the origin of the Strand Flat', Norsk geogr. Tidsskr. 11, 159-71

W. M. DAVIS (1900), 'Glacial erosion in France, Switzerland, and Norway', Proc. Boston Soc. nat. Hist. 29, 273-322

W. M. DAVIS (1909), 'Glacial erosion in North Wales', Q.J. geol. Soc. Lond. 65, 281-350

E. DERBYSHIRE (1964), 'Cirques, Australian landform example no. 2', Aust. Geogr. 9, 178-9

W. DORT (1957), 'Striated surfaces on the upper parts of cirque headwalls', J. Geol. 65, 536-42

F. ENQUIST (1917), 'Der Einfluss des Windes auf die Verteilung der Gletscher', *Bull. geol. Instn Univ. Upsala* **14**, 1–108

J. W. EVANS (1913), 'The wearing down of rocks', *Proc. Geol. Ass.* **24**, 241–300

E. FELS (1929), 'Das Problem der Karbildung', *Petermanns Mitt., Ergänz.* **202**, 85 p.

J. E. FISHER (1953), 'Two tunnels in cold ice at 4000 m on the Breithorn', *J. Glaciol.* **2**, 513–20

J. E. FISHER (1955), 'Internal temperatures of a cold glacier and conclusions therefrom', *J. Glaciol.* **2**, 583–91

R. F. FLINT (1957), *Glacial and Pleistocene geology* (New York)

J. D. FORBES (1843), *Travels through the Alps of Savoy* (Edinburgh)

O. W. FREEMAN (1925), 'The origin of Swimming Woman Canyon, Big Snowy Mountains, Montana, an example of a pseudo-cirque formed by landslide slipping', *J. Geol.* **33**, 75–79

B. GASTALDI (1873), 'On the effects of glacier erosion in Alpine valleys', *Q.L. geol. Soc. Lond.* **29**, 396–401

G. R. GIBSON and J. L. DYSON (1939), 'Grinnell Glacier, Glacier National Park, Montana', *Bull. geol. Soc. Am.* **50**, 681–96

G. K. GILBERT (1904), 'Systematic asymmetry of crest lines in the High Sierra of California', *J. Geol.* **12**, 579–88

J. W. GLEN and W. V. LEWIS (1961), 'Measurements of side-slip at Austerdalsbreen, 1959', *J. Glaciol.* **3**, 1121

J. W. GREGORY (1915), 'The geology of the Glasgow district', *Proc. Geol. Ass.* **26**, 162

J. M. GROVE (1960), 'The bands and layers of Vesl-Skautbreen', in 'Norwegian cirque glaciers' (ed. W. V. LEWIS), *R. geogr. Soc. Res. Ser.* **4**, 11–23; 'A study of Veslgjuvbreen', ibid., 69–82

A. HARKER (1901), 'Ice erosion in the Cuillin Hills, Skye', *Trans. R. Soc. Edinb.* **40**, 221–52

A. HARKER (1904), 'Tertiary igneous rocks of Skye', *Mem. geol. Surv. Gt Br.*

A. HELLAND (1877), 'On the ice-fjords of north Greenland, and on the formation of fjords, lakes, and cirques in Norway and Greenland', *Q.J. geol. Soc. Lond.* **33**, 142–76

W. H. HOBBS (1910), 'The cycle of mountain glaciation', *Geogr. J.* **35**, 146–63 and 268–84

W. H. HOBBS (1911), *Characteristics of existing glaciers* (New York)

W. H. HOBBS (1926), *Earth features and their meaning* (New York)

C. D. HOLMES (1944), 'Hypothesis of subglacial erosion', *J. Geol.* **52**, 184–90

W. D. JOHNSON (1904), 'The profile of maturity in Alpine glacial erosion', *J. Geol.* **12**, 569–78

A. J. JUKES BROWN (1877), 'The origin of cirques', *Geol. Mag.* **4**, 477–9

B. KAMB and E. LACHAPELLE (1964), 'Direct observation of the mechanism of glacier sliding over bedrock', *J. Glaciol.* **5**, 159–72

C. A. M. KING and M. GAGE (1961), 'Note on the extent of glaciation in part of west Kerry', *Ir. Geogr.* **4**, 202–8

R. VON KLEBELSBERG (1919), 'Glacialgeologische Erfahrungen aus Gletscherstollen', *Z. Gletscherk.* **11**, 156–84

W. V. LEWIS (1938), 'A melt-water hypothesis of cirque formation', *Geol. Mag.* **75**, 249–65

w. v. LEWIS (1939), 'Snow patch erosion in Iceland', *Geogr. J.* **94**, 153-61

w. v. LEWIS (1940), 'The function of meltwater in cirque formation', *Geogr. Rev.* **30**, 64-83

w. v. LEWIS (1947), 'Valley steps and glacial valley erosion', *Trans. Inst. Br. Geogr.* **14**, 19-44

w. v. LEWIS (1949), 'Glacial movement by rotational slipping', *Geogr. Annlr* **31**, 146-58

w. v. LEWIS (1954), 'Pressure release and glacial erosion', *J. Glaciol.* **2**, 417-22

w. v. LEWIS (1960), 'Norwegian cirque glaciers', *R. geogr. Soc. Res. Ser.* **4**, 104 p.

d. l. LINTON (1963a), 'The forms of glacial erosion', *Trans. Inst. Br. Geogr.* **33**, 1-28

d. l. LINTON (1963b), 'Some contrasts in landscapes in British Antarctic Territory', *Geogr. J.* **129**, 274-82

l. LLIBOUTRY (1953), 'Internal moraines and rock glaciers', *J. Glaciol.* **2**, 296

g. MANLEY (1959), 'The late-glacial climate of North-west England', *Lpool Manchr geol. J.* **2**, 188-215

e. DE MARTONNE (1910-11), 'L'érosion glaciaire et la formation des vallées Alpines', *Annls Géogr.* **19**, 289-317; **20**, 1-29

f. e. MATTHES (1900), 'Glacial sculpture of the Bighorn Mountains, Wyoming', *U.S. geol. Surv. 21st A. Rep.* (*1899-1900*), 167-90

f. e. MATTHES (1930), 'Geologic history of the Yosemite Valley', *U.S. geol. Surv. Prof. Pap.* **160**, 137 p.

l. h. MCCABE (1939), 'Nivation and corrie erosion in West Spitsbergen', *Geogr. J.* **94**, 447-65

j. g. MCCALL (1960), 'The flow characteristics of a cirque glacier and their effect on glacial structure and cirque formation', in 'Norwegian cirque glaciers' (ed. w. v. LEWIS), *R. geogr. Soc. Res. Ser.* **4**, 39-62

m. m. MILLER (1961), 'A distribution study of abandoned cirques in the Alaska-Canada Boundary Range', *Proc. 1st int. Symp. Arctic Geol.*, *1960*, **2**, 833-47

a. PENCK (1905), 'Glacial features in the surface of the Alps', *J. Geol.* **13**, 1-19

r. e. PRIESTLEY (1923), *Physiography* (*Robertson Bay and Terra Nova Regions*), *Rep. Br. Antarct. Exped.* (1910-13)

a. c. RAMSAY (1860), *The old glaciers of Switzerland and North Wales* (London)

e. RICHTER (1906), 'Geomorphologische Beobachtungen aus Norwegen', *Sber. wien. Akad. Math. Naturw.* **105**, 1

v. SCHYTT (1959), 'The glaciers of the Kebnekajse massif', *Geogr. Annlr* **41**, 213-27

b. SEDDON (1957), 'Late-glacial cwm glaciers in Wales', *J. Glaciol.* **3**, 94-99

k. m. STRØM (1945), 'Geomorphology of the Rondane area', *Norsk geol. Tidsskr.* **25**, 360-78

k. m. STRØM (1949), 'The geomorphology of Norway', *Geogr. J.* **112**, 19-27

h. v. SVERDRUP (1935), 'Scientific results of the Norwegian-Swedish Spitsbergen Expedition in 1934; part III', *Geogr. Annlr* **17**, 53-88

g. TAYLOR (1914), 'Physiography and glacial geology of east Antarctica', *Geogr. J.* **44**, 365-82, 452-67, and 553-71

g. TAYLOR (1926), 'Glaciation in the south-west Pacific', *Proc. 3rd Pan-Pacific Congr.*, *Tokyo*, 1924

P. H. TEMPLE (1965), 'Some aspects of cirque distribution in the west-central Lake District, northern England', *Geogr. Annlr* **47**, 185-93

H. R. THOMPSON (1950), 'Some corries of north-west Sutherland', *Proc. Geol. Ass.* **61**, 145-55

H. R. THOMPSON and B. H. BONNLANDER (1956), 'Temperature measurements at a cirque bergschrund in Baffin Island: some results of W. R. B. BATTLE's work in 1953', *J. Glaciol.* **2**, 762-9

H. A. WALDROP (1964), 'Arapaho glacier, a sixty-year record', *Univ. Colo. Stud. Ser. Geol.* **3**, 1-37

Chapter 9

Glacial Erosion in Areas of Low Relief

A very remarkable action of existing glaciers is to chafe and polish the rocks over which they are pushed or dragged.... The fact is certain that ... there is continued contact between the supporting rock ... and the glacier itself. Its stupendous unwieldy mass is dragged over the rocky surface, it first denudes it of every blade of grass and every fragment of soil, and then proceeds to wear down the solid granite, or slate, or limestone, and to leave most undeniable proofs of its action upon these rocks. (J. D. FORBES, 1843)

The features most frequently associated with glacial erosion are those that have been described in the last two chapters. These are features associated with areas of considerable relative relief, where the ice is generally confined to the valleys; hence its erosive power is concentrated and can be exerted effectively. The forms already considered will be contrasted with those produced by ice action in areas of low relief. The essential difference between these two environments is the relative relief and not the absolute relief. Glacial erosion on an elevated plateau will not differ so significantly from that on a lowland plain as it will from glacial erosion in an area of steep relief. The main difference between glacial erosion in these two contrasting environments is, therefore, the difference between the action of glaciers in areas of steep relief and the action of ice-sheets in the areas of low relief.

In the upland areas of low relief, the ice will be situated mainly in the ice accumulation zone. This may produce rather different results from the effects of ice erosion in a lowland of gentle relief, where the ice will usually be in the ablation zone, except in very cold climates. The lowland area is more likely to become an area of glacial deposition when the ice has retreated. Therefore the effects of glacial erosion in lowlands can often only be studied by examining the subdrift forms.

The early controversy between the glacial erosion school of thought and the glacial protectionists was partly the result of a failure to appreciate that ice can behave both as an agent of erosion and an agent of protection according to the characteristics of the area concerned. The most important variables in this connection are the relief of the area and the nature of the glacier régime. Where the régime is active, in areas with heavy accumulation and rapid ablation, the flow will be correspondingly fast and the ice will be able to effect considerable erosion where the relief features are suitable. On the other hand, where both accumulation and ablation are small and temperatures low, the ice will mostly move slowly and erosion, especially in areas of low relief, may well be negligible.

The temperature of the ice plays an important part and will determine whether the ice-sheet belongs to the cold or temperate category. Cold ice masses have temperatures below the pressure melting-point and therefore are normally frozen to their beds. Meltwater is confined to the surface or surface layers and cannot penetrate to the bed of the ice-sheet. Ice masses of this type will be considerably less effective as agents of erosion than those that can move over their beds on a thin film of water. Temperate ice masses have this characteristic, as they are at the pressure melting-point throughout their thickness. It should be remembered, however, that different parts of an ice-sheet may have different characteristics. It has been suggested that parts of the Barnes ice-cap and the Greenland ice-sheet may be at the pressure melting point in their central parts. This is because of the greater pressure of the thicker ice and the accumulation of geothermal heat. Near their margins where the ice thickness decreases, the ice becomes colder as atmospheric cold can penetrate farther and the pressure is less. Thus in these parts, the ice is frozen on to its bed and basal sliding ceases.

Basal ice velocity cannot be measured directly under an ice-sheet so that indirect observations and theoretical relationships must be used to obtain information concerning ice movement and ice behaviour. C. Bull (1957) in his study of the Greenland ice-sheet measured the surface slope of the ice and related this to the ice thickness and the shear stress at the ice-sheet bed. These types of measurements will eventually provide the quantitative information necessary to assess the way in which an ice-sheet can erode its bed. There are, however, complications, for example those apparent in Greenland where the shear stresses were found to differ in the southern and northern parts of the ice-sheet. The differences were thought to be caused by the base of the ice being at the pressure melting-point in the north, but frozen to the bed in the south.

S. Orvig (1953) has studied the variation of shear stress on the bed of the Barnes ice-cap in Baffin Island. The thickness of the ice was measured by gravimetric survey on the southern lobe of the ice-cap, where the ice reached a maximum thickness of 460 m. The slope of the ice surface was also measured and from these observations, the average shear stress was found to be about 0·4 bars, the value increasing in a north-easterly direction. This increase was supported by indications of more active flow in this direction. Towards the southern margins of the lobe, the ice was retreating. The low value of the shear stress of 0·4 bars in the southern part is less than that suggested by Nye for valley glaciers (Chapter 4). It would be associated with small strain rates and hence low ice velocities, especially as the ice in this area is thought to be colder than its pressure melting-point and hence frozen to its bed. Other investigations confirmed that the ice was nearly stationary. In this instance, therefore, there is good evidence that the ice-cap was acting in a protective capacity. It is, however, possible that the shear stress values were somewhat too low as they were not measured along lines of maximum slope.

The higher shear stress values in the north-east were associated with faster flow, which may have been caused by the recent lowering of ice-dammed lakes, impounded along this margin of the ice-cap. The lowering would remove a considerable amount of support from this part of the ice-cap and induce more rapid flow in this direction. There may be a significant link between the presence and height of ice-dammed or other marginal lakes and the movement of the ice-cap.

J. F. Nye (1959) has calculated the velocity distribution along a vertical line in the Greenland ice-sheet; the results probably apply to much of the Antarctic ice-sheet also and possibly to the large Pleistocene ice-sheets of Northern Europe and America. In this study, he made the very important distinction, as far as geomorphological effect of the ice is concerned, between the velocity of movement on the bed and the differential movement within the ice. He comes to the conclusion that in the ice-sheets of Greenland and Antarctica, the essential movement may be considered as almost entirely made up of rapid shearing in the lower layers. Some simplifying assumptions were, however, made in arriving at this conclusion. The reason for this result is partly the relatively higher temperature in the lowest layer and the dependence of the strain rate to a considerable degree on the temperature. The higher temperature causing this effect results from the accumulation of geothermal heat and heat generated by the shearing in the basal layers. Although the temperatures are higher near the base of the ice, they may well still be considerably below freezing-point. The lower layers will respond more effectively to the stresses within the ice-sheet for these reasons. The bottom layer in which shearing is concentrated could be as much as 100 m thick or it could be much thinner than this value, and the actual base is frozen to bedrock in the marginal areas as noted above.

Lower shear stresses are not necessarily associated with lower velocities. If the temperature increases, the shear stress may be reduced while the velocity remains the same. A temperate ice-sheet probably increases in basal velocity of flow towards its margin. The rates of surface flow also differ greatly from place to place around the margin. This variation has been clearly demonstrated around the Antarctic continent. Observations of ice-flow here have shown that some parts of the marginal regions have flow rates that are at least an order of magnitude greater than the flow rates along wide adjacent stretches. The higher rates of flow are characteristic of the outlet glaciers and vary from 30 to 300 cm/day with an average of 100 cm/day. In the wider, slower section the ice-sheet flows at about 4 cm/day. Under these circumstances, those parts of the underlying area in which the ice-flow is concentrated must be undergoing considerably more rapid erosion than the neighbouring stretches where the ice is moving much more slowly.

In summarizing the factors that affect the way in which ice erodes the ground over which it is flowing, the importance of ice thickness and velocity should be stressed. Observations of ice velocities confirm that ice-sheets flow considerably more slowly than active glaciers. Ice-sheets on the whole will exert less erosive effect on their bed and this effect will be more uniformly applied over a considerable area. Differential erosion will be mainly confined to exploiting irregularities within the bedrock and will not depend so much on the varying rates and character of glacial flow. Under favourable conditions, however, even ice-sheets may be effective agents of erosion. Such conditions are fulfilled when the ice-sheet has a ready outlet so that its velocity can be maintained to the edge. These conditions apply particularly in cases where ice-sheets descend over areas of fairly low relief into the sea. The well-nourished ice-sheets, flowing westward over the north-west highlands of Scotland and calving into deep water in the Atlantic provide an example. These conditions may be contrasted with those that occurred on the east of Scotland, in the Irish Sea and on the eastern side of Scandinavia.

In eastern Scotland the ice moving down into the low ground came into contact with ice from Scandinavia, and this impeded and reduced its flow. Anglesey also provides an example, where ice erosion does not appear to have significantly altered Tertiary and early Pleistocene planation surfaces (C. Embleton, 1964). This lack of effective erosion was related to the great distance of the Irish Sea from the ice sources and the blocking effects of the Welsh mountains and glaciers. However, where the Irish Sea ice was funnelled through gaps in the hills in the Dee and Mersey estuaries it may have been able to gouge out hollows in the bedrock (R. K. Gresswell, 1964). In eastern Scandinavia and the lowlands to the east, the ice moved down on to low ground and could not escape by calving, so it built up into a very slowly moving, extensive mass. In areas of this type, where the ice was not able to move fast, whether because of blockage or distance from its source, the ice-sheet would not be expected to be an effective agent of erosion.

Landforms of Glacial Erosion in Areas of Low Relief

One of the most characteristic forms of glacial erosion in areas of low relief is the 'knock and lochan' formation discussed by D. L. Linton (1963). It occurs both at low and high elevation and is particularly well displayed in the north-west highlands of Scotland. The ice in this area was moving fairly actively over an outcrop of Lewisian gneiss westward to the Atlantic. As a result of this glacial activity, the landscape now consists of many ice-moulded knobs, with intervening lochans which have been eroded along lines of structural weakness in the rocks. The features often show strong structural control: faults, weak dykes and other zones of weakness were exploited by the ice. Where the structural control is strong, as in the Lewisian gneiss, the direction of ice movement is not immediately obvious from the landscape lineations.

In many other areas, however, the ice imposes its direction of flow on the landforms it creates. Such features have been called glacial fluting, especially where the features are of small dimensions and considerably elongated. Good examples occur in the lowlands of central Canada, such as north-eastern Alberta where the features have formed on a large scale. The glacial flutings and lineations are eroded in the Pre-Cambrian granite-gneiss of the Canadian Shield. They occur in the area north of Lake Athabasca and are parallel to the glacial striations, but at an angle to the bedrock trends. They trend at about 60 degrees east of north. In this area differential erosion is subdued compared with the fluting pattern. Dykes cutting across the flutings have been eroded to the same extent as the softer metasediments into which they were intruded.

In the Mackenzie District the glacial flutings resemble giant grooves (see also Chapter 6). The ridges are 3 to 8 m above the troughs, about 91 m apart, and up to 1·6 km long. It seems likely that they are elongated parallel to the ice movement. The fluting wave length has a preferred value of 90 to 120 m in a variety of rock types and a second smaller preferred spacing value of 180 to 214 m. It is, therefore, thought that the movement of the ice was in some way responsible for the fluting pattern (C. P. Gravenor and W. A. Meneley, 1958). A possible mechanism could be found if there were parallel bands of varying pressure zones in the ice. The material would be eroded from the high pressure zones to a greater extent than from the low pressure zones. The solid rock flutings merge into forms composed of drift, and both types may be erosional in character. The forms in both instances show that the ice was moving actively in the low relief areas in which they occur. This situation supports the view that ice-sheets probably erode more effectively on ground of low elevation than on high ground in the accumulation area when the relief at both elevations is of low amplitude.

Larger streamlined forms also occur. These forms resemble drumlins in shape, but they differ from true drumlins in that they are composed of rock. The features have been termed 'rock drumlins'. Linton (1959) has described features of this type on the plateaux of well-jointed Lower Old Red Sand-stone lavas of Lorne. Similar features are also formed a little farther south in the harder Dalradian schists and quartzites. The rounded streamlined hills are formed in the more resistant rocks within the series. Similar though larger features occur in south-west Iceland, where they are up to 8 km long. There are also examples in northern Iceland, where the alignment clearly reflects the direction of ice movement. The ice must have been about 500 m or more thick to have moved sufficiently rapidly to produce these ice-moulded forms in this area of relatively low relief.

The hummocky glaciated surface characteristic of ice-sheet erosion is especially well developed in areas of old rocks on the ancient shields, for example around the Baltic and on the Canadian shield. These are areas where relative relief rarely exceeds several tens of metres. They occur in Karelia in Finland, in the Norrland interfluve area of Sweden, central Finland, central Labrador, parts of Quebec, and north of the Great Lakes of America. Most of these areas are plateaux of about 100 to 600 m in elevation and relatively undissected by through valleys. The relief is rather haphazard and the drain-age confused. Any structural weaknesses have been clearly exploited and revealed by the ice action. The hills often have fairly steep slopes of about 10 to 15 degrees and the intervening hollows show feeble drainage. The hollows are often peaty or contain standing water in innumerable small lakes. The exposed rocks often show very clear evidence of glacial modification in the form of polished and striated surfaces, some in the form of roches moutonnées. There are also in places small patches of blocky moraine, usually in the sheltered hollows in the lee of the bosses. These moraine patches are usually thin and discontinuous and are formed of ground moraine.

The features eroded by actively moving ice sheets are generally asymmetri-cal, having steeper proximal slopes than distal slopes. The amount of erosion has been shown by R. H. Jahns (1943) to be considerably greater on the lee side. He based his views on a study of the sheet structure of the granites in New England. He contended that these structures were pre-glacial and were such that the sheets become thicker and flatter with increasing depth. It is thus possible to use the structure to arrive at an estimate of the depth of glacial erosion (Chapter 11). The increase of thickness of the sheets with depth is consistent. The size of the boulders derived by ice from the different sheets therefore gives a clue to the depth from which the boulders were derived. The results of the study suggest that the bosses have been lowered by about 3 to 4·5 m on the north or proximal sides and that the depth of erosion was rather greater on the east and west slopes and on the summits (Fig. 11-4). The lee or south slopes seem to have suffered the maximum erosion, which was achieved largely by plucking, rather than by abrasion as on the other faces.

The plucking and quarrying that were active on the distal slopes appear to have removed a maximum of 30 m of rock. These figures indicate the minimum amount of erosion accomplished by the ice-sheet in this area.

The largest boulders denote plucking to the greatest depth, and seem to have come from a depth of about 29 m where the sheets were about 4·5 to 6 m thick. The largest boulder was from a sheet 8·2 m thick, and has remained fresh and angular. Such large blocks must have been derived by quarrying from considerable depths on the lee of the bosses. These quarried faces are fairly common features of the area. The methods of estimating glacial erosion gave consistent results in this area when the observations were confined to one particular granite outcrop that lay athwart the ice-flow. The results suggest that erosion by a continental ice-sheet is not so effective as that produced by powerful valley glaciers under favourable conditions, for such glaciers can erode hundreds or even thousands of metres vertically.

The relative inability of ice-sheets to erode the ground over which they are passing has been demonstrated by observations made by J. T. Andrews (1965) in north-west Baffin Island. A magnetite dyke outcrops in the Striding Valley and sampling lines were set up at right angles to the dyke and parallel to the valley axis. All the occurrences of boulders of magnetite within a 5 m by 5 m square were counted at geometrically increasing distances of 10, 20, 40, 80, 160 m, and so on, from the dyke. The results for observations in the direction of ice movement were plotted on double logarithmic paper. The relationship between the number of magnetite boulders and distance from the outcrop is a straight line on the graph. The rate at which the number drops off is extremely rapid, suggesting negligible transport after 4000 m; even by 400 m, the number of boulders has fallen to 4 per cent of the number in the first sample area. In the direction opposite to the ice-flow, there were no boulders after a distance of 100 m.

The results of this experiment suggest that, on this upland plateau, there was only a minor amount of glacial movement and the ice was not capable of seriously modifying the bedrock over which it was moving. Nor was it able to transport material eroded from the bedrock for any great distance. The conditions of the ice in this area were such that it must have been frozen to its bed, at least in the marginal zone. The mean annual temperature is now about $-10°C$. at an elevation of 580 m in the vicinity of the magnetite dyke. The internal conditions of the ice when the boulders were deposited would suggest an internal movement of only 0·3 m/year. This value agrees with the observations made on the southern Barnes ice-cap, where the basal shear stress is as low as 0·4 bars.

From a study of orientation patterns of surface boulders in such areas, Andrews has come to the conclusion that there may have been slight differential movement between the ice and the bedrock or till, even when the ice was frozen to its bed, although this movement was probably not at all vigorous. As the ice increased in thickness, however, the basal ice may have reached pressure melting-point and this would have facilitated movement between it and the bedrock or till, and also the development of preferred orientation in the underlying bouldery till deposits.

In some situations, ice-sheets are capable of differential erosion. J. Tricart and A. Cailleux (1962) have drawn attention to some of these situations. The margin of the Canadian shield provides good examples of forms which have been termed 'roxen lakes' by W. M. Davis (1920) after Lake Roxen in Sweden. The Great Bear Lake and the Great Slave Lake as well as many other smaller lakes occupy analogous situations to Lake Roxen at the margin of the Canadian shield. In Europe, the Gulf of Finland is elongated along the junction of the Baltic shield and the Russian platform with its cover of Palaeozoic rocks. The ice-sheets have eroded these weaker rocks, which consist of schists, clays and limestones, very vigorously as is indicated by the large quantity of this material in the moraines of the ice-sheets. In some instances, pre-glacial erosion had prepared the ground in advance, but in many instances the state of the rocks was such that this explanation is improbable. The lake basins are eroded in areas where the ice is advancing over an area of low relief on which rocks of different resistance are exposed. The softer rocks are eroded to form the basins.

In other areas, the pre-glacial relief consisted of upstanding scarps formed of more resistant rocks. In this instance, the advancing ice-sheet might cause erosion which would reduce rather than enhance the pre-glacial relief. An example of the reduction of relief is to be seen in the Finger Lakes region of western New York (K. M. Clayton, 1962). The hard rocks of this area form the Allegheny scarp, which is normally above 460 m high. It faces north towards the oncoming ice in the area south of the Lake Ontario basin, and it has been very severely reduced in elevation by glacial erosion. In a large embayment, eroded by the ice spreading out from the Lake Ontario basin, the height of the scarp has been reduced in some areas to less than 300 m. The ice-sheet has also destroyed the pre-glacial watershed.

Another example of ice-sheet erosion reducing the relief is found in south-east England (Fig. 11-3). To the south-west of Hitchin, where active ice erosion did not occur, the Chalk forms the very well marked scarp of the Chiltern Hills. This scarp has probably never been over-run by ice, at least in the Saale and Weichsel glaciations. Where the scarp is well developed, it has subsidiary benches formed of the Melbourne Rock in front of the main scarp and the Chalk Rock behind it. As the feature is traced north-eastward from the Hitchin gap, the scarp form becomes completely modified. This modification could be in part the result of the early Pleistocene marine transgression, but the disappearance of the Melbourne Rock bench from the scarp slope until north Norfolk is reached about 80 km to the north-east shows that this is not the only cause. The form of the main scarp is also considerably modified. It loses the steep scarp slope and becomes set back by many kilometres (see Chapter 11). Across much of Cambridgeshire and Norfolk, it forms a gentle slope, which is drift-covered in places and becomes a plateau-like surface, covering an extensive area. Farther north-east, the scarp disappears altogether and the streams draining north-west to the Fens rise 40 km east of the Gault, which outcrops west of the Chalk. The scarp does not reappear until Swaffham and then continues from this point to the Norfolk coast. Further evidence that this modification of the scarp was caused by erosive activity of the ice-sheet is provided by the glacial deposits. The fabrics of the East Anglian tills indicate that ice moved from the overdeepened Fenland basin to the north-west of the Chalk and spread out over the ground to the south-east, while the large quantity of chalky material in the tills of East Anglia bears witness to the effectiveness of ice-sheet erosion of the upstanding chalk scarp.

Erosion may have been assisted in this area, and also in the Finger Lakes region, by an increase of ice thickness in the basins from which the ice was flowing over the surrounding areas of upstanding resistant rocks. Thus the smoothing and erosion of the upstanding parts of the pre-glacial landscape have been accompanied by extra deep scouring of the softer rocks that lay in the path of the ice before it reached the scarp features. Subsequent glacial deposition and the post-glacial infilling of the basins have partially masked the increase of relief related to the differential erosion of the softer rocks.

Another good example of the differential effects of glacial erosion, now hidden by subsequent deposition, has been described by A. Straw (1958, 1963). The subdrift contours of the Vale of Belvoir in the East Midlands show clearly the effect of scouring by the ice-sheet moving south across the area from the lowlands of the Vale of York and Humber area. The ice was able to exploit the soft clay outcrops that run north-south to the west and east of Lincoln. In doing this, the ice-sheet was able to destroy almost all trace of the presumed north bank of the proto-Trent, which in pre-glacial times is thought to have flowed across the Vale of Belvoir through the Ancaster Gap. The material eroded from the soft clay areas has been dumped on the rising ground to the south, where rather harder rocks outcrop. The drift forms the south Nottinghamshire Wolds, where the drift thickness exceeds 30 m.

Straw (1958) has also drawn attention to the sub-drift form of the Ancholme Valley, which occupies a clay strip in Lincolnshire between the outcrop of the Lincolnshire Limestone on the west and the Chalk of the

Lincolnshire Wolds on the east. The sub-drift surface has been eroded to a little below sea-level to the south of Lincoln (Fig. 9-1). To the north of the Witham, the sub-drift valley is very flat over a considerable distance at about present sea-level. Had it been formed by fluvial action working to a lower base level, the valley would probably have been asymmetrical, reflecting the dip of the rocks. The valley is, in fact, symmetrical in form, suggesting that the southerly moving ice-sheet was the formative agent. The ice removed the soft clays and carried them south to mingle with the drift of the area to the south, in which much Jurassic clay is incorporated with the chalk.

These examples from the south of England show that ice-sheets, moving far from their mountain sources, can still exploit the soft rocks that lie across their paths. Under favourable conditions, they can also cause serious modification to harder rocks that are standing above the general elevation. It cannot be generally assumed that ice-sheets, at the limit of their extent, must necessarily be incompetent as agents of erosion. They will be selective in their erosion, at times increasing the relief and at times decreasing it.

In their source regions, on the other hand, ice-sheets may act in a protective rôle at times. The Lake District is normally thought of as an area deeply modified by glacial erosion. It is certainly true that the valleys of the area show very clearly the influence typical of valley glacier erosion. In the inter-fluve areas and on the flattish hill tops, however, the evidence of glacial erosion is much less strong, and in fact there is evidence that in these areas the ice has been largely protective. The recognition by several geomorphologists of a series of pre-glacial planation surfaces in this area supports this view (R. B. McConnell, 1938). The same applies to neighbouring areas, including the Howgill Fells and the Southern Uplands of Scotland.

The inability of the ice to effect much erosion or to remove the previously formed planation surfaces of the flatter parts of these areas can probably be accounted for partly by their relatively low relief. The fact that the ice-flow was concentrated in the valleys in such a way that the ice on the hill tops could not move effectively is also important. In the Howgill Fells, the inability of the ice to erode was partly caused by the blocking of the escape routes for local ice by other ice from neighbouring more active centres. This resulted in a reduction of velocity and the ice could no longer erode effectively. The same argument can be applied to the flatter summit areas of the Pennine Hills, on which there is little evidence of glacial erosion. Linton (1959) has drawn attention to the inability of ice to erode effectively on the high plateaux of the Cairngorms region of Scotland. He cites the presence of tors as evidence of the weakness of glacial erosion in these areas. He argues that the tors were formed by deep weathering in a warmer pre-glacial climate and were subsequently exhumed before the ice formed (see

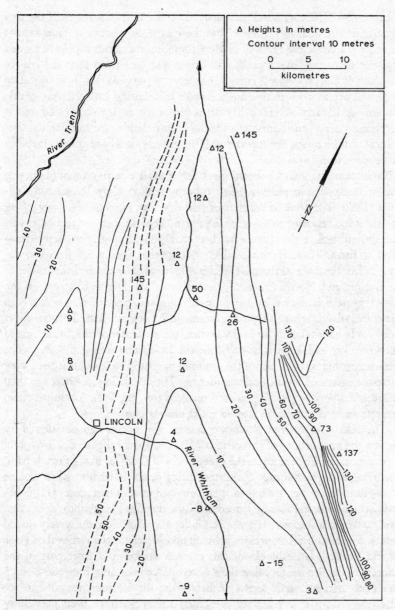

Fig. **9-1** The sub-drift contours of the Vale of Ancholme, Lincolnshire (A. Straw, *E. Midld Geogr.*, 1958)

Chapter 24). The tors stand undestroyed on the upland surfaces. If the tors are in fact pre-glacial, it is clear that the ice on the plateau surfaces cannot have been an effective agent of erosion, though some authorities have argued that the tors may have a periglacial origin and date from after the last ice advance. If Linton's view is correct, however, it suggests that ice in its source area will act as a protective agent where it is resting on relatively gentle slopes, particularly where it has no ready access to lower ground and is receiving only a small amount of snow accumulation. Ice in these circumstances will be relatively thin and will move only slowly so that its inability to erode is to be expected.

One reason why an ice-sheet is probably a more effective agent of erosion on lowlands than in plateau areas has been mentioned by Tricart and Cailleux (1962). The zone of permafrost that develops in front of an advancing ice-sheet has less time to form in the upland regions, because the ice covers the ground before the climate has been cold long enough for deep permafrost to form. There is considerable evidence that deep permafrost had developed in the path of the ice-sheets by the time they reached the lowlands in Europe and North America. The existence of deep permafrost on the lower ground is shown by the long-distance transport of fragile fossils in clays in Poland, where they have been carried for 300 km. This transport could only have taken place in areas where the ground moraine, incorporated into the ice, had been frozen before the ice-sheet reached the area. The permafrost helps to break up the ground, as frost and ground-ice are very effective agents of mechanical weathering. The advancing ice-sheet can then more effectively erode the prepared material (see Fig. 7-1). These processes operate most effectively in jointed rocks such as granite.

The relative effectiveness with which erosion occurs can be assessed by noting the proportions of the various strata crossed by the ice that are found in the moraine. For example, the green rocks of Queyras in the French Alps, intrusive into schists lustrés, form upstanding peaks under normal erosion of non-glacial character, owing to their low rate of disintegration. This is because their joints are widely spaced and they therefore break into blocks that are too large for streams to transport. Under glacial erosion, the widely spaced joints, prepared by frost action, provide no obstacle to the removal of large blocks. Indeed, large blocks of this material make up a large part of the moraines, despite the fact that their outcrop occupies only between 2 and 5 per cent of the area. It seems probable that the ice itself cannot operate effectively on a low relief surface unless the rocks have been previously prepared by frost or other action. The character of the joints is of particular importance in this respect.

Another suggestion of Tricart and Cailleux concerns the effect of a general lowering of temperature of about 20°C. in the rocks near the surface with the oncoming of glacial conditions. They suggest that such a general lowering of temperature will cause contraction of the rocks, and that this may initiate joints in previously sound rock. This would prepare the way for frost action and subsequent removal by the ice-sheet when it reached the area.

Ice would first accumulate on the upland areas of low relief, when the level of the snow-line fell below the plateau surface. The snow would prevent the formation of permafrost and help to account for the relatively limited amount of glacial erosion where the relative relief was low. The uplands of Norway provide a very good example of this (K. Strøm, 1948). The most conspicuous of the upland low relief surfaces of Norway is the broad plateau on part of which the Jostedalsbreen ice-cap lies. The elevation of this surface during the Plio-Pleistocene period, combined with the falling temperature, may have initiated the formation of the great Scandinavian ice-sheet. This upland plain, lying at an elevation of about 1800 m, has suffered relatively little glacial modification. It remains as a fairly level and undissected surface, showing only minor glacial moulding in areas where it is exposed away from the ice-cap remnants. It is, however, very deeply bitten into around its margins by the outlet glaciers that pour with very high velocity through the ice-falls around the edge of Jostedalsbreen.

The contrast between the high plateau surface (the 'fjell') and the very serrated Jotunheim Mountains shows how the slow moving ice-cap has protected the flat surfaces. The mountains of Jotunheim, on the other hand, stood out pre-glacially as mature hills above the plateau surface, while during the glacial period, their summits stood up above the ice as nunataks. They have been deeply sculptured by cirque and valley glaciers, and their summits have been severely riven by frost action to form jagged peaks. The upper hill slopes now consist of piles of huge boulders leading up to frost-shattered ridges.

In this part of Norway a great variety of glacial landscapes lies close together. There is the broad fjell, showing slight evidence of glacial action in the form of roches moutonnées and hummocky bosses, but in general little modified by deep erosion. Above the plateuax, the isolated mountain groups show evidence of strong cirque formation and the peaks are characteristically frost-shattered nunataks. Cutting into the fjell are the deep fjords (Chapter 10), which demonstrate clearly the immensely powerful action of glaciers when they are steep, fast-flowing and thick. The difficulty of generalizing about ice erosion is apparent when all these forms are found in close proximity as in Norway and other mountain areas of this type. Greenland, Labrador, Scotland and parts of the south-west of the South Island of New Zealand provide other examples of similar assemblages of glacial forms.

Conclusions

In assessing the geomorphological rôle of glacial erosion in areas of low relief, whether they be at high or low elevation, each area must be considered individually. It is probably generally true, however, that ice-sheets are more effective agents of erosion in lowland areas than in upland areas. The reasons for this may differ from one area to another. One reason is that the high-level areas of low relief are normally those that are relatively resistant to erosion, and for this reason are upstanding. Other significant reasons include the effect of permafrost on the low ground in front of an advancing ice-sheet and the susceptibility of certain rocks to glacial erosion. Ice thicknesses also tend to be greater in the lowland areas than on the elevated plateaux. The examples that have been quoted from the Midlands of England and elsewhere illustrate these points. That ice-sheets can erode effectively in some areas of low ground has been demonstrated by the differential erosion suffered by scarps, because of their upstanding position relative to the approaching ice.

One of the most characteristic landscapes of ice-sheet erosion in areas of relatively hard, homogeneous rocks with minor lines of weakness, is the hummocky knock and lochan relief, characteristic, for example, of parts of the north-west Highlands of Scotland and parts of western Connemara in Ireland as well as widespread areas of the Canadian and Baltic shields. In these regions, deposition is limited to thin spreads of basal moraine, deposited in the lee of the bosses.

A general relationship exists between the forms produced by the ice-sheets and the general structural zones. The zone of fjell and fjord in western Scandinavia is succeeded eastward by a wide belt of hummocky boss country. This in turn gives way to the fringing roxen lakes and other water bodies along the margin of the shield. Farther from the ice centres, the glacial land-forms become predominantly depositional. It is not clear how far this rela-tionship depends on the pre-existing relief and structure and how far it depends on the zones of different action within the ice-sheet. It is likely that both factors have combined to produce the observed effects. J. B. Bird (1959) has shown that large ice-sheets can be protective over large areas in Canada. V. Tanner (1938) has given similar evidence for Finland. Extensive surfaces of planation have been preserved undissected and little modified in Swedish Lapland, where many pre-glacial features have survived. The pre-glacial relief and structure play a large part in the resulting ice-sheet erosional effects in areas where the landscape is fairly accidented.

REFERENCES

J. T. ANDREWS (1965), unpub. Ph.D. Thesis, Univ. of Nottingham, Chapter 4

J. B. BIRD (1959), 'Recent contributions to the physiography of Northern Canada', Z. Geomorph. NF 3, 151-74

C. BULL (1957), 'Observations in north Greenland relating to theories of the properties of ice', J. Glaciol. 3, 67-71

K. M. CLAYTON (1965), 'Glacial erosion in the Finger Lakes region (New York State, U.S.A.),' Z. Geomorph. NF 9, 50-62

W. M. DAVIS (1920), 'A Roxen lake in Canada', Scott. geogr. Mag. 41, 65-74

C. EMBLETON (1964), 'The deglaciation of Arfon and southern Anglesey, and the origin of the Menai Straits', Proc. Geol. Ass. 75, 407-30 (see pp. 428-9)

C. P. GRAVENOR and W. A. MENELEY (1958), 'Glacial flutings in central and northern Alberta', Am. J. Sci. 256, 715-28

R. K. GRESSWELL (1964), 'The origin of the Mersey and Dee estuaries', Geol. J. 4, 77-86

R. H. JAHNS (1943), 'Sheet structure in granites: its origin and use as a measure of glacial erosion in New England', J. Geol. 60, 71-98

D. L. LINTON (1959), 'Morphological contrasts of Eastern and Western Scotland', in Geographical Essays in Memory of Alan G. Ogilvie (ed. R. MILLER and J. W. WATSON)

D. L. LINTON (1963), 'The forms of glacial erosion', Trans. Inst. Br. Geogr. 33, 1-28

R. B. MCCONNELL (1938), 'Residual erosion surfaces in mountain ranges', Proc. Yorks. geol. Soc. 24, 76-98

M. M. MELLOR (1959), 'Ice flow in Antarctica', J. Glaciol. 3, 377-84

J. F. NYE (1959), 'The motion of ice sheets and glaciers', J. Glaciol. 3, 493-507

S. ORVIG (1953), 'On the variation of the shear stress on the bed of an ice-cap', J. Glaciol. 2, 242-7

R. P. SHARP (1954), 'Glacier flow: a review', Bull. geol. Soc. Am. 65, 821-38

H. T. U. SMITH (1948), 'Giant glacial grooves in Northwest Canada', Am. J. Sci. 246, 503-14

A. STRAW (1958), 'The glacial sequence in Lincolnshire', E. Midld Geogr. 2, 29-40

A. STRAW (1963), 'The Quaternary evolution of the Lower and Middle Trent', E. Midld Geogr. 3, 171-89

K. M. STRØM (1948), 'The geomorphology of Norway', Geogr. J. 112, 19-27

V. TANNER (1938), 'Die Oberflächengestaltung Finnlands', Bidr. Känn. Finl. Nat. Folk (Helsingfors), 86

J. TRICART and A. CAILLEUX (1962), Le modelé glaciaire et nival (Paris)

Chapter 10

Glaciation in Coastal Environments

The mountains in Kangerdlugssuak Fjord, though a little lower than those in the sound by Upernivik Island, descend precipitously to the water at angles of eighty degrees from heights of full 1000 metres. The glacier at the head terminates in a steep wall advancing into the fjord. . . . Greenland is intersected by many large fjords which . . . pierce deeply into the country. In front of the fjords near the open sea there is a 'Skärgård' of larger and smaller islands. (AMUND HELLAND, 1877)

There are certain coastlines in the world where the effects of glaciation are dominant, and where a distinctive environment results from the juxtaposition of glacially sculptured landforms and the sea. Among the features encountered in such coastal regions are fjords, representing glacial troughs now partly occupied by the sea, the fjärds and föhrdes of glaciated lowland coasts, and the partially submerged platforms known as strandflats.

Fjords

Fjords are best developed on the coasts of British Columbia, southern Alaska, southern Chile, eastern Canada (especially east Baffin Island), Greenland, Norway,* Iceland and Spitsbergen, the south-west of South Island (New Zealand), and finally Antarctica where to a considerable extent they are still occupied or buried by ice. They are more common on west-facing than east-facing coasts, they occur mostly in higher latitudes (more than 45 degrees from the Equator), and they are usually backed by rugged and dissected highlands. Fjords have the features associated with glacial troughs, possessing a cross-profile of catenary form. As the lower part of the trough is submerged, only echo-sounding can provide adequate detail of the trough floor, which normally appears broad and relatively flat, though this may be the result of sediment infilling. The steepness of the trough sides, suggesting intensity of glacial down-cutting, is one of the most impressive features of many fjords, for example the Naeröyfjord in Norway or Milford Sound in New Zealand. Again in common with many glacial troughs, a cross-profile of composite form is evident, showing successive stages in the deepening of the trough by

* It should be noted that the term 'fjord' in the Norwegian language is used in a much broader sense, including lakes and coastal inlets of other types.

both ice and water erosion. In the case of the Sognefjord, the largest of the Norwegian fjord systems, remnants of former high-level valley floors can be clearly seen above the steep sides of the main central trough—the bench at 400-500 m on the east of the entrance to Aurlandsfjord is a good example —providing evidence for the reconstuction of former fluvial valley systems draining west (see pp. 242-3).

Sounding has shown that fjords nearly always possess submerged rock basins, separated by rock bars which are termed thresholds when they lie at or near the sea entrance. Some claim that the threshold, with the associated rock basin behind, is the principal diagnostic feature of a fjord, most clearly distinguishing it from the ria of a submerged non-glaciated coast. The Sognefjord possesses one of the deepest known rock basins in the world. The main fjord head begins at Årdal, and the floor sinks evenly and gently to a depth of 640 m at the junction of Lustrafjord, 940 m at the entry of Aurlandsfjord, over 1000 m off Balestrand, and finally touches 1308 m between Höyangsfjord and Vadheimsfjord. Even this is not rock bottom: the flatness of the fjord floor in a transverse sense here (Fig. 10-1c) almost certainly indicates some sedimentary infilling. Depths in excess of 1200 m continue to be recorded as far west as the entry of the small Böfjord, beyond which the Sogne swings to the south-west and shallows rapidly to less than 200 m. The great Sognefjord has an entrance less than 200 m deep and only 3 km wide. The threshold is marked by the occurrence of numerous rocky islets or skerries; the threshold is clearly a rock bar and not morainic accumulation. Thus behind it lies an enclosed rock basin, 1100 m in depth. It is not clear why the Sognefjord should attain so much greater depths than those of neighbouring fjords.

The deepest fjord in northern Norway is Tysfjord, 727 m, enclosed by a bar little more than 200 m deep which in turn is followed seawards by a further rock basin descending to – 660 m. Other deep fjords in the world are

TABLE 10   *Maximum Known Depths, in metres, below present sea-level, of some Fjords*

Vanderford, Vincennes Bay	Antarctica	2287
Skelton Inlet	Antarctica	1933
Sognefjord	Norway	1308
Hardangerfjord	Norway	870
Upernavik Icefjord	Greenland	1055
North-West Fjord, Scoresby Sound	Greenland	1450
Chatham Strait	Alaska	883
Messier Channel	Chile	1288
Baker Fjord	Chile	1244

listed in Table 10. In the case of the fjord system of north-east Greenland, one of the most remarkable and spectacular of any, the fjords are all enclosed by rocky thresholds rising to within 400 m of the sea surface and often much less —Vega Sound is practically blocked by rocky islands with only shallow straits between. The deepest fjords are usually also the longest (Sognefjord threshold to Årdal, 160 km), probably reflecting the fact that the largest glaciers with many tributaries were capable of the greatest amount of erosion.

Fjords display diverse relationships with rock structure, and occur in a great many different types of rock, from Pre-Cambrian to Tertiary in age and including both sedimentary and igneous formations, but it is the older crystalline rocks that are the most competent to form the deeply cut and steep-sided troughs. The rectilinear pattern of fjord coastlines has for long been held as evidence that the fjords developed along two or more intersecting lines of fracture—R. I. Murchison urged this connection in 1870— and J. W. Gregory (1913) adopted the extreme and now generally abandoned viewpoint that fjord systems were largely of tectonic origin, the result of fracturing under tension; subsequent fluvial and glacial action was held responsible for only slight modification and evacuation of fault-shattered debris. In 1927, he applied this view in detail to the fjords of the Hebrides. Inadequate geological mapping still makes it difficult to disprove the tectonic theory in many areas. H. W. Ahlmann (1941) studying north-east Greenland found that the influence of faulting was difficult to assess, but pointed out that geological mapping to date had failed to reveal any major longitudinal faults, and that in any case 'tectonic phenomena . . . as orientating factors of the erosion agencies, [must] be carefully differentiated from the results of such agencies' (p. 171).

B. A. O. Randall (1961), examining the Lyngenfjord area of northern Norway, concluded that there was a very close relationship between fracture pattern and valley direction. A vector diagram of valleys and fjords in South Lyngen totalling 386 km in length revealed primary concentrations on bearings of 10-20° and 290°, with several secondary maxima. The most important directions of fracture (based on analysis of 95 fractures) were 10-20° and 40-50°. The latter corresponds to a secondary concentration of valley and fjord directions. Randall concluded that valley direction intimately depends on the local fracture pattern, and suggested that the actual sites of valleys (and fjords) are lines of fracture, thus supporting Gregory's contention. This view has been strongly challenged by R. Nicholson (1963). Nicholson's mapping in the Glomfjord district of northern Nordland shows that normal faults or veins are few in number; thrusts are present, but appear to be no more significant in terms of ease of erosion than ordinary bedding contacts. Fjords such as Glomfjord or Melfjord cut across contacts between metamorphics and the granite-gneiss, and no fractures or lines of fracture were visible on fjord walls, though these were exposed up to heights of 1000 m. Both Randall and Nicholson exclude joints from consideration, for, as indicated in Chapter 7, there is good evidence that joints of dilatation type may result from the excavation of deep glacial valleys and may therefore be of secondary origin.

Most workers would now agree that fractures are not necessary for fjord formation. All elements of rock structure, including strike direction and foliation, may affect fjord direction, and erosion merely selects the most important weaknesses in any region.

EROSIONAL DEVELOPMENT OF FJORDS

Since most fjords pass inland into normal glacial troughs, and both possess common characteristics, there is now no dispute that glaciation is responsible for the main features of fjord physiography. The extent of penetration of the sea today is simply determined by the depth of glacial excavation of the trough floor, for glacial erosion, unlike fluvial erosion, is not controlled by a base-level, though sea-level does control the point at which the ice in a fjord begins to float. The magnitude of the rock basins now inundated by the sea (1100 m in the Sogne) attests the power of glacial erosion in these localities, for no other mechanism is available for the extraction and uphill removal of rock debris from these basins assuming that the basins are of erosional origin and that they are totally enclosed by solid rock. The fjord coasts all lie in areas conducive to the development in the Pleistocene or at present of thick and rapidly moving glaciers, descending steeply from highlands well nourished by snow. As well as gouging out deep basins, the trunk glaciers also cut down more rapidly than tributary glaciers, thus producing many examples of hanging valleys, and frequently the size of the glaciers was such as to enable them to spill over previous divides and create new routes for the discharge of ice. Thus the Nordenskjold glacier in north-east Greenland once moved through Mystery Valley towards Is Fjord, and in the same area, there is a good example of a partially demolished watershed between Franz Joseph Fjord and the head of Dusén Fjord.

Attempts have been made to disprove the existence of deep enclosed basins in fjords by claiming that the thresholds are merely moraines or the results of landslips, for example. Some thresholds may indeed not be solid rock, as at the mouth of Yakutat Bay, Alaska, but the dredging up of occasional erratics or loose boulders from a threshold by no means disproves that the feature consists of solid rock beneath superficial debris. Many thresholds are undoubtedly solid—witness the hard Devonian conglomerate of the Sognefjord threshold, or the rocky islets which often top the threshold. The thresholds of some fjords are much too enormous to be regarded as moraines and their low angle of back slope into the fjord basin also militates against this view. It is unfortunate, however, that echo-sounding does not distinguish between solid rock and morainic thresholds.

Nor are the basins simply the result of regional back-tilting. The large amount of tilting necessary would be clearly revealed in the attitudes of other land forms. On the other hand, a degree of back-tilting is involved in many fjords as a result of incomplete isostatic recovery since the last glaciation, or in the case of fjords bordering Greenland and Antarctica which are still depressed by the weight of present ice. Some fjord basins will in time become somewhat shallower, but the amount involved is nowhere near that needed to account for the basins.

In short, the basins must be regarded basically as the result of glacial erosion. The question then arises why the maximum deepening by ice should normally occur in the central parts of the fjords and why a threshold should exist at the fjord mouth. It is not often a question of differential rock resistance; the deepest parts of the north-east Greenland fjords described by Ahlmann (1941) lie in crystalline and sedimentary rocks of pre-Devonian age, whereas the shallower outer parts of these fjords lie in a belt of younger and generally less resistant rocks. R. Koechlin (1947) contended that the threshold marked the approximate termination of the fjord glacier rapidly melting in contact with salt water. W. A. Don Munday (1947) thought that thresholds were the result of decreased erosion at valley mouths where the ice was able to spread out and its velocity thus became much reduced. Clearly also glacial erosion will cease at the point (the grounding line) where the lower glacier, reduced in thickness, begins to float and loses contact with the ground. Another explanation of the threshold was given by A. Cailleux (1952) who with M. Boyé emphasized the part played by deep freezing of the ground prior to the advance of ice in a region, as discussed in previous Chapters. It can be argued that the depth of freezing and rock shattering would decrease rapidly to nil near the sea coast and that therefore in approaching the coast, a glacier would be able to erode less deeply, giving rise to a threshold. But the depth of rock basins behind thresholds seems much too great for this to be the sole explanation; the glaciers gouging out rock basins must have cut deeply below even the maximum depth of previous frost shatter. The same disadvantage attaches to J. P. Bakker's hypothesis (1965) of deep pre-glacial weathering to prepare the basins for glacial erosion (see Chapter 7): weathering to depths of many hundreds of metres is unlikely.

A recent study by A. P. Crary (1966) of an Antarctic ice-covered fjord, Skelton Inlet, has thrown some light on the development of fjord basins.

Seismic and gravimetric observations enabled the probable profiles of the bedrock and ice under-surface, and the thicknesses of sediment infilling, to be deduced (Fig. 10-1a). At the southern end of the section illustrated, the fjord glacier merges with the Ross Ice Shelf. Point A represents the grounding line of the glacier, marked by a slight but sudden fall in ice surface elevation; from A to B the ice seems to be just touching sediments (morainic and marine, probably), as though a recent increase in ice thickness has occurred. The ice in the section A-B, alternatively, may be separated from the sediment by a water layer too thin to be detected by geophysical means. From B seawards, the floating ice gradually thins. The amount of seaward thinning can be deduced from the formula

$$11 \cdot 17 \, \frac{\delta t}{\delta x} = \frac{\tau_s}{\gamma}$$

where δt is the change in ice thickness along the fjord over a distance x, τ_s is the shear stress (in bars) on the sidewalls, and γ is the half-width of the fjord (in km). Uncertainty of the values of τ_s reduces the usefulness of this relationship, but some estimates, ranging from 0·8 to 1·2 bars, are made by Crary from studies of crevasses and ice movement, using Glen's power flow law. The floating ice thins down-valley as the sidewall shear stress decreases or as the valley widens. If, for any reason, sea-level relative to the land falls slightly, or if the ice thickens owing to increased accumulation, then part of the floating ice will become grounded and bedrock erosion will be possible. Since the inner thicker part of the floating ice will ground first, it will tend to erode a basin that shallows seaward.

It should be stressed that fjords are not the result of glacial troughs later being subjected to submergence. It is quite clear that the floors of fjords were eroded to their present levels when the land stood at approximately its present height with respect to sea-level. The post-glacial rise of sea-level, of the order of 100 m, is minute in comparison with most fjord depths, and is largely irrelevant. G. K. Gilbert in 1910 correctly maintained that glaciers entering the sea would continue to displace sea water and exert normal pressure on their beds until flotation actually occurred. For ice to erode the floor of the deep Sognefjord basin, it would have to be about 1600 m thick, assuming sea-level to have been 100 m lower than at present, though evidence from erratics and striated surfaces at over 500 m above sea-level on the islands in the mouth of the fjord indicate that at times, the ice in the deep basin behind must have been at least 1800 m thick.

There is abundant evidence, some of which has already been indicated, that the fjord glaciers were largely guided in their paths by well-defined valley systems. In north-east Greenland, J. H. Bretz (1935) thought that the

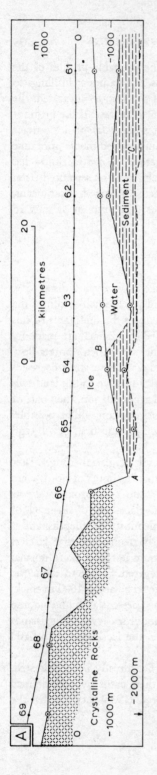

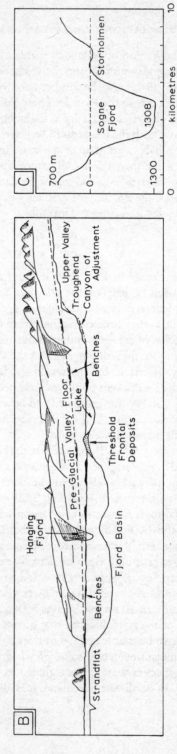

Fig. 10-1 A Section along Skelton Inlet, Antarctica, based on gravimetric and seismic observations (A. P. Crary, *Bull. geol. Soc. Am.*, 1966)

B Diagrammatic section along a Norwegian fjord, showing assemblage of characteristic features (J. Gjessing, *Norsk geogr. Tidsskr.*, 1966)

C Cross-profile of the deepest fjord basin in Norway, middle Sognefjord (J. Gjessing, *op. cit.*)

PLATE XI (*left*) PART OF THE ÖTZTAL, AUSTRIA.
Showing V-shape and interlocking spurs in the lower part of a major glacial trough. (C.E.)

PLATE XII (*above*) YOSEMITE VALLEY, CALIFORNIA.
El Capitan on the left rises 1200 m above the valley floor which is itself underlain by up to 300 m of sediment. (C.E.)

PLATE XIII   THE CIRQUE CONTAINING VESLGJUV-BREEN, NORWAY.
The difference in altitude between the lake and the summit of Galdehöe is 400 m. Note the winter snow banked against the foot of the rock walls, and the pattern of dirt in the glacier. (C.E.)

PLATE XIV   CIRQUE ON HAFRAFELL, ICELAND.
Little ice or firn now remains beneath the present-day winter snow. Note the moraine ridges, and the frost-shattered debris. (C.E.)

ice had taken advantage of stream-eroded valleys cut into the plateau margins. Ahlmann (1941) pointed in this area also to the pattern of fjords, which strongly suggests original fluvial action. In some places elbows of capture and reversed drainage can still be picked out. Late-Tertiary and early Quaternary uplift caused rapid incision of the streams, producing well-developed valleys for the glaciers subsequently to occupy. In western Scotland, S. Ting (1937) described the existence of a pre-glacial valley system draining south and west, south of Skye and Lewis, and generally north farther north. As J. A. Steers describes (1952), large parts of these valleys were glaciated and subsequently became fjords or sea-lochs. It is in western Scotland that the deepest British rock basins are found, such as Loch Morar (depth 308 m), and in the Inner Sound, east of Skye (324 m) (A. H. W. Robinson, 1949).

Although most workers attribute the greater part of fjord erosion to ice, some maintain that the ice merely modified valleys already of substantial depth and carved by other agents. A novel suggestion, not yet fully evaluated, is that of J. H. Winslow (1966), who discusses the undoubted problems encountered by the simple glacial hypothesis. Among these problems are the need to explain longitudinal fjords running across the general direction of ice movement from land to sea, the explanation of the deep submarine extensions of many fjords on the continental shelf (see p. 245), and the apparent existence of well-developed fjords in locations where ice accumulation must have been limited. A good example of this last problem lies in the Faröe Islands, where many of the fjords are paired head-to-head with only low ground separating them (for instance, at Klaksvík). To account for these and other anomalies, Winslow suggests that fjords represent glacially-modified raised submarine canyons. The fjords are said to have acquired many of their distinctive characteristics as submarine canyons before glaciation, and were eroded initially by submarine processes, possibly turbidity currents, before uplift and glaciation.

In Norway, J. Gjessing (1956, 1966) has attempted to separate features of fluvial and glacial erosion in connection with fjord and fjord valley development (Fig. 10-1b). He accepts the glacial origin of the fjord basins and intervening rock bars sometimes capped with morainic deposits. Another feature of the fjord valleys, and often a very impressive one, is the abrupt trough end. Many fjord valley heads in Norway share this characteristic: examples are the head of Isterdalen at the Trollstigveien, the southern end of Flåmsdal terminating in a 400 m cliff leading up to Myrdal and the Hardangervidda plateau, and the Måbodal gorge which in one direction passes down to Eidfjord and in the other heads in the giant Vöringfoss where the river leaps 150 m from the Hardangervidda at Fossli. The steep ends of the valleys heading in the Jostedalsbreen plateau are genetically similar,

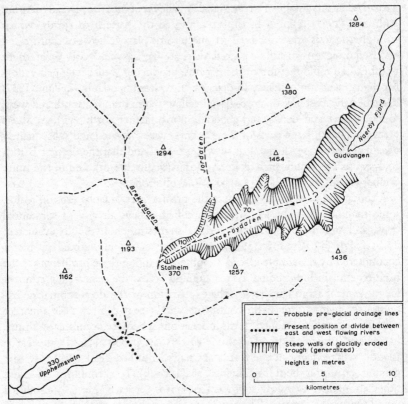

Fig. 10-2 The region around Naeröydalen, Vestland, Norway. A glacially eroded
trough, partly submerged to form Naeröyfjord, ends abruptly at Stalheim;
its excavation has disrupted the pre-glacial drainage and displaced the watershed between east and west flowing rivers by at least 20 km

though in many cases they are still submerged by glacier ice, as in the ice-falls
of Austerdalsbreen or of Brixdalsbreen. Many previous workers regarded
these trough ends as pre-glacial (or possibly interglacial) knick points of
water erosion, retreating up-valley after a period of Tertiary and Pleistocene
uplift. Gjessing argues, however, that the trough ends are of Pleistocene
glacial origin, that the pre-glacial valley profile had no trough end (Fig.
10-1b), and that this feature has developed successively in glacial and interglacial episodes. The trough ends often possess a striated and ice-worn
appearance, and are clearly at variance with the work of the present streams
which are cutting 'canyons of adjustment' into them. The cutting of these
notches into the trough ends may have been partly the work of subglacial

streams (as in the case of gorges through valley steps discussed in Chapter 7) and partly the result of waterfall cutting when the melting ice-margin lay near the brink of the falls during deglaciation. H. Holtedahl (1967) considers that subglacial stream erosion was of major importance in the formation of the canyons at the heads of the Måbo, Flåm and other fjord valleys.

A small area around Stalheim above the head of Naeröyfjord illustrates some of these relationships (Fig. 10-2). Pre-glacial drainage was to the west through the valley of Uppheimsvatn; Brekkedalen and Jordalen at one time joined near Stalheim and escaped by this route, whose valley floor is still visible at 370 m around Stalheim. The older view involved diversion of this drainage by river capture into Naeröydal, and the subsequent modification of the latter by glacial action. The alternative view, urged by O. Holtedahl (1960, p. 520) and others, is that the Naeröydal trough and fjord beyond are basically the result of glacial erosion. The 200 m step or trough end at Stalheim was eroded by rapidly flowing ice escaping to the north-west and the Sognefjord, and the step was notched to a small extent by waterfall erosion. The pre-glacial watershed was displaced by 10-16 km. Basins of glacial erosion lie in Naeröyfjord, for instance in the section (up to 60 m deep) between Gudvangen and the islands by Naeröy village, and in the fjord valley above Gudvangen where they are infilled by alluvium.

Skjaergård and Strandflat

Extending seawards from many fjord coasts there is often to be found a zone of low rock islands (or skjaergård) and partially submerged platforms of erosion (strandflats). The features are undoubtedly best developed off the Norwegian coast, where the strandflat and skjaergård appear around Stavanger and continue (with breaks) to the North Cape region, reaching a width of 50 km off Ranfjord (latitude 66°). Other coasts possessing similar off-shore features are those of western and southern Iceland (there are typical skjaergård coasts off Breiðafjörður and Faxaflói, but in some other parts the strandflat has been covered by the sandar), Spitsbergen (strandflat up to 10 km wide), and western Greenland (up to 30 km wide at Gothaab).

The Norwegian strandflat is essentially a horizontal cut into the land mass (K. Strøm, 1948), though there are probably several different closely-spaced levels involved. Platforms at 8 to 18 m and 30 to 40 m have been claimed for instance, as have submarine platforms at shallow depths, and it also seems likely that isostatic tilting has occurred. Smöla and Fröya are good examples of large strandflat islands each measuring 200-300 km², with most of their surfaces below 20 m in height, and surrounded by countless small islets and reefs. Some platforms seem to be virtually horizontal, or possess seaward slopes as small as 1 in 1000; cases of slope towards the mainland probably represent incomplete isostatic recovery. The strandflat often ends abruptly at its inner margin, though it can be traced a short way up some fjords as narrow benches (for example, Sognefjord). It cuts indiscriminately across geological structure, and also across some fjord rock basins where these extend seaward of the mainland. The skjaergård islets merely represent the post-glacial accident of partial submergence of certain strandflat levels. These islets should be distinguished genetically from the skerries of south-eastern Norway and Sweden, for instance, where the irregular glaciated rock surface *slopes* gently outward beneath the sea.

Both strandflat and skjaergård are intimately connected with glaciation. The rounded knobs of hard crystalline rock forming the islets bear witness to severe glacial erosion, and the strandflat levels themselves may be largely of Pleistocene age. Marine action has played a part which is not easy to assess. Although the bevelling of the rock structures and the abrupt inner margin strongly suggest wave action, the platforms are very broad, cut in very resistant rocks, and possess little or no seaward slope. Simple wave erosion over such a width and gradient is very unlikely, if not impossible. Speculation about the evolution of the Norwegian strandflat is of long standing and it is impossible here to present all the different suggestions. Nansen's work published in 1922 provides a survey of previous views, and presents his own hypothesis, which includes a pre-glacial period of denudation to a peneplain, followed by wave-cutting at various sea-levels in the Pleistocene. In cold periods, wave-cutting was facilitated by sea-spray freezing in rock crevices and thus loosening material for waves to transport. A novel suggestion was that of E. Dahl (1947) who claimed that much of the strandflat surface was one produced by the coalescing cirque and valley floors of coastal glaciers, the walls of the cirques and valleys having been demolished to subdued forms and the whole then somewhat smoothed by ice-sheet erosion. Outlines of former cirques can perhaps be detected on parts of the strandflat. H. Holtedahl (1959), working on the strandflat in the Møre-Romsdal area, also concluded that erosion by coastal glaciers played an important part in the modelling of the strandflat. Strøm (1948) emphasized marine abrasion, showing how the strandflat was broad and well-developed only on wave-exposed coasts (contrast the east- and west-facing coasts of Moskenesøy, for example). He regarded the several wave-cut levels as having been modified by ice erosion in successive glaciations, the last glaciation having reached only as far as the inner parts of the strandflat. The width of the strandflat in many places, however, demands some earlier period of preparation and lowering of the land surface near to present sea-level before Pleistocene wave-cutting commenced, and in this respect pre-glacial peneplanation (as Ahlmann strongly invoked in 1919) has much to commend itself, unless one accepts the view that much of the strandflat is a cirque-floor surface.

An interesting feature of the strandflat and the continental shelf beyond is the series of deep submarine troughs and basins that often continue the lines of fjords far out from the mainland. The features are beautifully displayed on the maps prepared by O. Holtedahl (1940): Storfjord, for instance (Holtedahl, Sheet 5), has extensions reaching out 80 km from the large coastal islands. These troughs may be of glacial origin, but it is far from clear how or why ice extending this distance from the mainland on an almost flat surface, presumably as a piedmont ice-sheet, should be able to erode selectively, or even to erode at all. J. H. Winslow (1966) has suggested that the submarine troughs are not glacial features but fall in the category of submarine canyons. Alternatively, they may be tectonic features.

Fjärds

Fjärds are coastal inlets associated with the glaciation of a lowland coast, and therefore lack the steep walls characteristic of glacial troughs. They may possess rock basins, though much shallower than in the case of fjords, and can in this way be separated from non-glacial forms such as rias. A good example of a fjärd coast is that of Maine, U.S.A., north and north-east of Portland. Numerous narrow inlets and islands produce a highly irregular coastline, yet neither the peninsulas in their distal portions nor the islands rise much more than 30-50 m above the sea. Glacial erosion has played an important part in determining the present form of these fjärds. Most of them are developed on weaker belts of Palaeozoic sedimentary and metasedimentary rocks trending approximately north-south and coincident with the direction of ice outflow from the interior of New England, suggesting differential glacial erosion. The submerged floors of the fjärds are typically irregular, as can be readily appreciated from the submarine contours of the U.S.G.S. topographical maps (see, for instance, the Boothbay Quadrangle, 15 minute series), and shallow rock basins and rock bars as well as morainic accumulations, are present. Where the ice-flow was impeded, as at Mount Desert Island (a resistant granite mass), notches and troughs indicating more intensive glacial erosion appear, and Somes Sound, unlike other coastal inlets, has a close resemblance to a true fjord, with one wall rising 300 m above the bottom of the sound.

Fjärds are also typical of the southern Swedish coast. E. Werth (1909) differentiated them from fjords and linked them with the system of ribbon lakes and valleys inland whose pattern is one radiating from the interior to the coasts and accordant with ice-flow. He also suggested that subglacial streams followed these routeways, further eroding them, in their endeavour to reach the edge of the ice and regions of lesser ice pressure. The ice, the main agent which sculptured the fjärds, was clearly guided in most cases by pre-glacial valley systems, as in the case of fjords; and again, it was able to erode without respect to base-level. The present state of submergence of the fjärds is therefore an accident, contingent on the amount of glacial erosion, the amount of sea-level change since glaciation, and also the degree of post-glacial isostatic recovery.

Föhrden (Transgressed Tunnel Valleys)

The term 'föhrde' has never been adequately defined, but is often used to refer to the sea inlets of the coasts of eastern Denmark, and part of the Baltic coast of Germany. It does not help the cause of clarity that many of these inlets have the local name of fjord. Like fjärds, föhrdes are associated with submerged lowland coasts which have also been overrun by ice, but the essential distinction between fjärds and föhrdes (though Werth, 1909, disputed this) lies in the fact that the former are principally the result of ice erosion whereas föhrdes were eroded mainly by meltwater streams. In eastern Jylland, good examples of föhrdes are given by Vejle Fjord, Flensburg Fjord, and Limfjord. Their shapes are very variable, but usually they narrow inland from wide mouths. Their banks are low, composed of glacial deposits, and their depths are usually less than 20 m. They represent the drowned terminations of the tunnel valleys (Rinnentäler or Tunneldale), cut by subglacial streams escaping from under the Baltic ice (Chapter 12). Hence their long profiles may be of irregular form and distinct from those of simple rias.

Conclusion

Although there is still some disagreement on the amount of erosion to be attributed to the ice itself, fjords represent partially submerged glacial troughs. The degree of submergence is determined by the amount of glacial deepening, and only to a minor extent by the post-glacial rise in sea-level. Fjords contain impressive examples of rock basins enclosed by thresholds at their entrances. The thresholds can in some cases be proved to consist of solid rock but further investigation is necessary. The basins, occasionally exceeding 1000 m in depth, are explained by various hypotheses, involving either direct glacial erosion or glacial erosion following a period of preparation of the ground by deep freezing or weathering. Fjord glaciers will erode bedrock as far down-valley as the point of flotation. Beyond this, they thin gradually as sidewall shear stresses decrease or as the valley widens. Crary (1966) associates the reversed slope from the basin to the threshold with this thinning of the floating ice down-valley. His hypothesis can also explain multiple basin formation. The inner ends of fjord valleys, sometimes at the termination of the sea inlet but more often farther inland, are frequently represented by huge steps or trough ends. These have been interpreted by Gjessing (1966) as glacial rock steps, into which meltwater and subsequently normal rivers have cut waterfall notches.

Certain features of fjords are not glacial but inherited from a previous valley system. Many fjords bear a close relationship with lines of structural weakness, though Gregory's extreme view (1913) of the tectonic origin of fjords cannot now be entertained. Surviving features of pre-glacial river-cut valleys have been described, and one recent suggestion (Winslow, 1966) regards fjords as glacially-modified raised submarine canyons.

The strandflat with its accompanying skjaergård is closely linked with fjords and the glaciation of coastal areas. Its origin is uncertain; it may represent a pre-glacial peneplain subsequently trimmed by wave-action at various closely-spaced levels, and modified to a greater extent by ice erosion. Fjärds are inlets on a lowland coast modified by glaciation; föhrdes represent the drowned terminations of valleys cut by subglacial meltwater.

REFERENCES

H. W. AHLMANN (1919), 'Geomorphological studies in Norway', *Geogr. Annlr* 1, 3-148 and 193-252

H. W. AHLMANN (1941), 'The main morphological features of north-east Greenland', *Geogr. Annlr* 23, 148-82

J. P. BAKKER (1965), 'A forgotten factor in the interpretation of glacial stairways', *Z. Geomorph.* NF 9, 18-34

J. H. BRETZ (1935), 'Physiographic studies in East Greenland', *Am. geogr. Soc. Spec. Publ.* 18

A. CAILLEUX (1952), 'Polissage et surcreusement glaciaires dans l'hypothèse de Boyé', *Revue Géomorph. dyn.* 3, 247-57

R. L. CAMERON (1965), 'The Vanderford submarine valley, Vincennes Bay, Antarctica' in *Geology and Palaeontology of the Antarctic* (ed. J. B. HADLEY), *Am. geophys. Un.*, Washington, 211-16

A. P. CRARY (1966), 'Mechanism for fjord formation indicated by studies of an ice-covered inlet', *Bull. geol. Soc. Am.* 77, 911-29

E. DAHL (1947), 'On the origin of the strand flat', *Norsk geogr. Tidsskr.* 11, 159-71

G. K. GILBERT (1910), *Harriman Alaska Expedition*, 3 ('Glaciers')

J. GJESSING (1956), 'Om iserosjon, fjorddal- og dalendedannelse', *Norsk geogr. Tidsskr.* 15, 243-69

J. GJESSING (1966), 'Some effects of ice erosion on the development of Norwegian valleys and fjords', *Norsk geogr. Tidsskr.* **20**, 273-99

J. W. GREGORY (1913), *The nature and origin of fjords* (London)

J. W. GREGORY (1927), 'The fjords of the Hebrides', *Geogr. J.* **69**, 193-216

H. HOLTEDAHL (1959), 'Den norske strandflate, med saerlig henblikk på dens utvikling i kystområdene på Möre', *Norsk geogr. Tidsskr.* **16**, 285-303

H. HOLTEDAHL (1967), 'Notes on the formation of fjords and fjord-valleys', *Geogr. Annlr* **49**, 188-203

O. HOLTEDAHL (1940), *Dybdekart over de Norske kystfarvann med tilgrensende havstrøk* (Norske Videnskaps-Akad., Oslo)

O. HOLTEDAHL, ed. (1960), *Geology of Norway* (Norges geol. Unders. **208**)

R. KOECHLIN (1947), 'The formation of fjords', *J. Glaciol.* **1**, 66-68

W. A. DON MUNDAY (1948), 'Formation of Norwegian fjords', *J. Glaciol.* **1**, 202

F. NANSEN (1922), *The strandflat and isostasy* (Oslo)

R. NICHOLSON (1963), 'A note of the relation of rock fracture and fjord direction', *Geogr. Annlr* **45**, 303-4

B. A. O. RANDALL (1961), 'On the relationship of valley and fjord directions to the fracture pattern of Lyngen, Troms, N. Norway', *Geogr. Annlr* **43**, 336-8

A. H. W. ROBINSON (1949), 'Deep clefts in the Inner Sound of Raasay', *Scott. geogr. Mag.* **65**, 20-25

J. A. STEERS (1952), 'The coastline of Scotland', *Geogr. J.* **118**, 180-90

K. M. STRØM (1948), 'The geomorphology of Norway', *Geogr. J.* **112**, 19-27

H. R. THOMPSON (1953), 'Geology and geomorphology in southern Nordaustlandet, Spitsbergen', *Proc. Geol. Ass.* **64**, 293-312 (strandflat, p. 298)

S. TING (1937), 'The coastal configuration of western Scotland', *Geogr. Annlr* **19**, 62-83

E. WERTH (1909), 'Fjorde, Fjärde und Föhrden', *Z. Gletscherk.* **3**, 346-58

J. H. WINSLOW (1966), 'Raised submarine canyons: an exploratory hypothesis', *Ann. Ass. Am. Geogr.* **56**, 634-72

Chapter 11

Quantitative Aspects of Glacial Erosion

*Then why is it that the rivers that flow from glaciers are so muddy? I venture to say ...
that glaciers, by erosion, seriously affect their beds. . . . . The mud of rivers is chiefly
derived from this incessant ice waste, and that is why it is so unearthy, so clean, fresh,
and impalpable.* (A. C. RAMSAY, 1864)

There has long been controversy over the ability of glaciers and ice-sheets
to erode their rock beds. Although the hypothesis of glacial protection is
today only retained in certain circumstances (see Chapter 9), there are still
considerable differences of opinion as to how much glacial erosion has
occurred in particular localities or regions. Various methods have been
employed in attempts to quantify glacial erosion; none has yet provided
either accurate or wholly reliable figures, but some have given useful
approximations.

Erosion of Artificial Markers by Re-advancing Ice

The classic experiment of this type dates back to 1846, when F. Simony cut
a·cross 3 mm deep on a rock in front of the Dachstein glacier. The glacier
advanced in 1856, scraping the rock and, after retreat, left it bare and polished
with no sign of the original mark. Others have occasionally succeeded in
obtaining similar short-period measures of glacial erosion: A. de Quervain
in 1920 reported bedrock erosion by ice of 0·5 to 1·5 mm in six months;
O. Lütschg gave evidence of larger amounts in 1926. Instead of carving
marks of limited depth, de Quervain used small boreholes drilled out and
filled with a measured depth of wax. But as a method of measuring glacial
erosion, it is unreliable and limited in its application. Much depends on being
able to predict a glacier advance over a particular spot, so that although many
markers have been set up, few results have been obtained, especially as the
last hundred years have witnessed a general recession of glaciers. The few
results obtained include an unknown quantity of rock weathering before the
ice advanced, as R. von Klebelsberg (1943) stressed; they are not representative
since the terminal ice is not the main agent of glacial erosion, and extrapolation from necessarily short-period measurements is dangerous.

Measurements of Silt Discharged by Glacier Meltwater Streams

Where meltwater streams emerge, usually sub-glacially, at the glacier terminus, they appear milky white in colour owing to the high quantity of fine sediment (rock flour) in suspension, the colour sometimes giving the name to the river (for example, Hvitá = White River, in Iceland). R. S. Tarr (1908) described how the Kwik river in southern Alaska emerged from a subglacial tunnel: 'So heavily burdened is it that a pail of water dipped from the surface has an inch of mud and sand on its bottom in a minute or two' (p. 90). The streams are obviously the main agents evacuating solid debris from the glacier basins behind, and the silt content of the streams in general reflects the amount of erosion in progress in these basins. However, not all the debris produced by glacial erosion is carried away by the meltwater streams; in particular, the larger boulders will be piled up in moraines, but estimates of the quantity of rock involved in morainic accumulation can be made from sample studies, and the quantity of debris accumulated here in any given period of time is usually surprisingly small compared with the loads of the meltwater streams. For the latter, we have to assess not only the suspended load but also material in solution, and the bed load. Problems, however, arise. Measurements of stream discharge and load should ideally be kept regularly and frequently for a prolonged period to even out short-term variations (see G. Østrem et al., 1967, pp. 277-9), and all outlets of the glacier basin must be covered. Even if this be done, it must be recognized that not all the material in the stream need be the product of glacial erosion—some can be produced by subaerial erosion above the ice surface (a consideration which also applies to the morainic accumulations), and some may be the result of subglacial stream erosion.

One of the early applications of the method was by H. F. Reid (1892) to the Muir glacier, Alaska, where the silt removed by the stream was equivalent to an overall annual lowering of the subglacial rock surface by 19 mm. The figures employed, however, were very uncertain. S. Thorarinsson in 1939 gave more reliable estimates of subglacial erosion for Hoffellsjökull. Eighteen samples were taken from the main Hoffellsjökull river under normal conditions, and also two from the Skeidará jökulhlaup of 1938. The samples were taken in various locations—large turbulent meltwater streams provide extremely difficult conditions for representative sampling—at distances up to 4000 m from the ice margin. Silt content varied closely with discharge. The average suspended load of the main Hoffellsjökull river was 1337 mg/l, and six samples gave 271 mg/l for the average quantity of material in solution. Comparable figures for a nearby non-glacier fed river were 766 and 52 mg/l respectively. Using a figure of 3000 m for the total annual precipitation less evaporation, Thorarinsson showed that 2800 m³ of sediment were being removed annually from each square kilometre of Hoffells-jökull, which, assuming a specific gravity for the rock debris of 1·5, is equivalent to 4200 tons or a layer 2·8 mm thick. But this makes no allowance for the bed load of the meltwater rivers, which could not be measured. Possibly the last figure should be doubled (= 5·6 mm) to allow for this. This represents a rate of denudation five times greater than in an adjacent non-glacierized area where denudation was calculated on the same basis. At the other extreme, the denudation represented by the Skeidará jökulhlaup of 1938 (see Chapter 19), though occurring over the short period of the glacier burst only, amounted to 15 or 20 times that of the normal rate for Hoffells-jökull.

In the French Alps, the rate of denudation achieved by the small St. Sorlin glacier has been investigated over several years (J. Corbel, 1962). The glacier covers 3·5 km² and its meltwater stream carries an average of 1500 mg/l. Employing a useful simple formula to give the amount of erosion in cubic metres per square kilometre per year ($\frac{4E.T}{100}$, where $E =$ depth of meltwater stream in decimetres and $T =$ weight of solids in mg per litre), Corbel arrives at a figure of 1200. To this he adds 1000 m³/km² annually for the quantity of material being piled up by the ice as moraine, giving a total of 2200 m³/km². The corresponding figure for Hoffellsjökull according to Thorarinsson may be in excess of 5600 m³: it should be remembered that the St. Sorlin glacier is a small one with little activity at the present day. Large active glaciers may have rates many times greater than these, but reliable estimates are relatively few. Corbel (1959) gives some examples: Muir glacier, Alaska—5000 m³; Heilstugubreen, Norway—1400 m³ (a figure which is almost certainly too small); Hidden glacier, Alaska—30,000 m³ (which is almost certainly too large). The mean rate for erosion by active glaciers probably lies in the range 1000 to 5000 m³, which should be compared with what we know of rates of fluvial erosion. For the Mississippi, the figure is 50 m³, for the Colorado above Grand Canyon, 230 m³, for the Hwang Ho 1000 m³. Again, these figures must not be thought accurate, but they do suggest that glacial erosion is several times more potent than fluvial erosion.

Reconstructions of the Pre-glacial Surface

If it were possible accurately to restore the form of the land surface before glacial erosion commenced, we would have measures of the amount of erosion during the period of glaciation at any one locality, and of the variations in depth of erosion from place to place. But in nearly every case, no two workers are agreed on the form of the pre-glacial surface; moreover, the possible effects of glacial erosion are inextricably entangled with those of fluvioglacial erosion, fluvial erosion in interglacial and post-glacial periods, and the work of other agents of denudation. Since the evidence on which pre-glacial surfaces are reconstructed and the circumstances of glaciation are so varied, it is only possible to discuss the method by examples.

Reconstructions of pre-glacial surfaces have often been attempted in connection with the excavation of glaciated valleys. W. M. Davis in 1909 wrote that 'the depth of glacial erosion in a main valley is roughly indicated by the discordant altitude of the hanging lateral valleys . . . but . . . allowance must be made for the glacial erosion of the lateral valleys' (pp. 340-1). He might also have added that glacial erosion is not the only possible process in over-deepening valleys and creating hanging tributaries. But he assumed that ice erosion was the cause of this overdeepening, and quoted figures of 60-180 m as a minimum measure of the deepening of the main Snowdonian valleys in Wales by ice.

In Yosemite valley, F. E. Matthes (1930) attributed a greater proportion of main valley deepening to pre-glacial rejuvenation, recognizing several stages of river down-cutting before ice erosion finally modified the valley to its present spectacular form (Fig. 11-1). Extrapolating sections of tributary long profiles unmodified by glacial erosion, he deduced that Yosemite valley had been deepened in three stages by pre-glacial river down-cutting, the last being termed the Canyon stage and represented, for instance, by the floor of Bridalveil creek above Bridalveil Falls. At this stage, Yosemite canyon had a floor level 180-370 m above its present alluvial floor. The latter is now known to be underlain by up to 600 m of sedimentary fill (B. Gutenberg et al., 1956), so that glacial erosion of the bedrock floor of Yosemite valley may amount to as much as 800-1000 m in places. Part of this figure, however, may represent deepening by interglacial river action, to estimate which is a constantly recurring problem.

A third example of pre-glacial valley-floor reconstruction and its use in estimating the depth of glacial erosion is taken from part of the Mur valley in Austria. H. Spreitzer (1963) contrasts the north-bank and south-bank tributary valleys of the Mur east of Tamsweg, establishing that ice flowed down the former to join the trunk glacier of the Mur, but flowed up the southern tributaries to escape in part into the Drau basin beyond. In these southern tributaries, glacial erosion is thought to have been relatively weak since the ice was moving slowly uphill, and the pre-glacial valley-floors are therefore well preserved, only notched by river-cut gorges where the main Mur valley is approached. But in the case of the northern tributaries, ice erosion was much more intense, and comparing these valleys with their

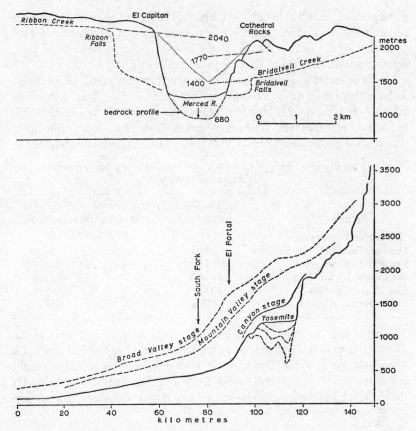

Fig. 11-1 Transverse and longitudinal profiles of Yosemite valley, California (C. Wahrhaftig, Guidebook for Field Conference I, *Congr. int. Ass. Quatern. Res.*, 1965)

southern counterparts, Spreitzer claims that the former have been deepened by ice in the Pleistocene by about 150 m. Once again, however, it is very difficult to assess how much deepening should be attributed to interglacial periods.

Minimum estimates of glacial valley erosion are, of course, given by rock basins, which must be wholly the work of ice; on this basis, F. A. Kerr (1936), for example, argued that the fjord valleys of northern British Columbia had been ice-deepened by at least 600 m. Similarly, Sognefjord (Chapter 10) bears witness to ice erosion of at least 1100 m in depth. These are minimum figures for ice erosion, but they cannot be contested. Likewise, the giant grooves described by H. T. U. Smith (1948) in north-west Canada are certainly the work of ice, and it is here calculated that a minimum of 1 km³ of rock has been removed to form the grooves over an area of 130 km², equal to a layer nearly 8 m thick if spread out evenly.

Just as attempts have been made to reconstruct pre-glacial valley-floors, so have there been attempts to restore the form of pre-glacial watersheds which were subsequently breached by ice. The work of D. L. Linton (1949) in Scotland provides some interesting figures. Table 11 shows the amount of glacial erosion involved in creating breaches through certain watersheds.

TABLE 11   *The Depths of Ice Erosion represented by some glacially breached Watersheds in Scotland*

Location	a) Pre-glacial col level (m)	b) Present floor level (m)	c) Depth of ice erosion implied (m)
The Saddle (Cairngorms)	900	815	85
Head of Glen Derry (Cairngorms)	900	750	150
Lairig Ghru (Cairngorms)	1070	840	230
Glen Eagles (Ochil Hills)	340-370	270	70-100
Loch Treig (Grampians)	670-700	150	520-550
Geldie-Feshie col (Grampians)	670	385	285

The pre-glacial col levels are obtained by extrapolating supposed pre-glacial slope remnants on shoulders above the ice-carved troughs. The method is essentially that of W. M. Davis who in 1909 described the destruction of the col at the head of the Nantlle valley in North Wales and claimed that the divide here had been lowered by ice erosion. Figures corresponding to those in the table above would be:

a) 460-530   b) 230   c) 230-300

It is not known whether the cols were significantly deepened by non-glacial processes in interglacial (or even pre-glacial) periods, but this appears unlikely, and there is no doubt from other evidence that ice did move across these watersheds and through these breaches. The depths of glacial erosion are impressive when the nature of the rock involved is considered—the granite of Lairig Ghru, the metamorphic rocks around Loch Treig, or the slates and igneous rocks at the head of Nantlle.

So far, the reconstructions of the pre-glacial landscape and amounts of erosion have been in terms of vertical differences in height. But it is also possible, though less frequently, to attempt to restore the pre-glacial positions of certain landscape features in a horizontal sense. Examples will make the reasoning clear. The Driftless Area of Wisconsin is bordered roughly on the south-west in Iowa by an escarpment of the. Niagara Formation whose

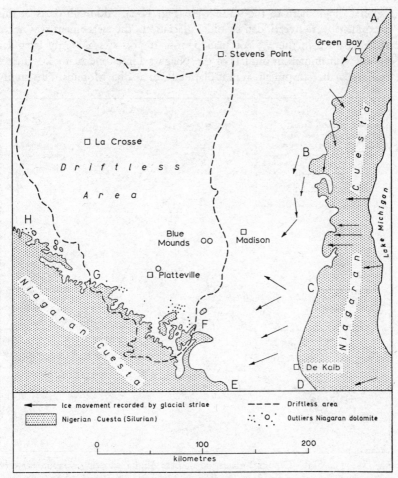

Fig. 11-2 The Driftless Area of Wisconsin, Minnesota, Iowa and Illinois, and the
Niagaran cuesta (L. Martin, 1916)
   A-B simple escarpment form, no outliers, ice moving parallel with
      escarpment
   C-D, E-F escarpment deeply buried in drift, ice moving across the Niagaran
      outcrop
   F-G highly irregular unglaciated escarpment, with numerous outliers
   G-H irregular escarpment affected by pre-Wisconsinan ice only.
      Irregular form may have developed since the Illinoian glaciation

regional dip is here to the south-west (Fig. 11-2). Although parts of this escarpment were overridden by older glaciations, the escarpment was never crossed by ice in the Wisconsinan period. It possesses a highly irregular outline with numerous outliers of the Niagara Limestone many kilometres from the main escarpment, as near Platteville or at Blue Mounds. This highly

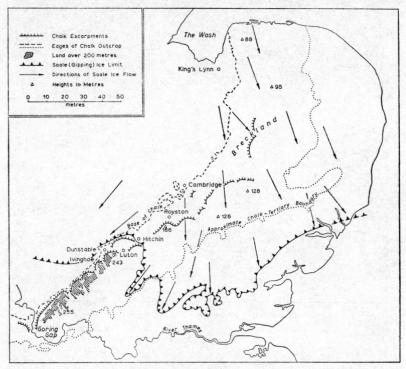

Fig. 11-3 The Saale (Gipping) glaciation of East Anglia, showing relation between directions of ice flow (as determined by R. G. West and J. J. Donner, Q. J. geol. Soc. Lond., 1956) and the Chalk escarpment

crenulated escarpment may be contrasted with the smoothly outlined escarpment, also of the Niagara Formation, in eastern Wisconsin and Illinois which was overridden by ice in the last glaciation, the ice moving generally southward and parallel to it. L. Martin (1916) argued convincingly that the smoothness and simple outlines of the Niagaran escarpment in the east were the results of glacial erosion removing any existing irregularities; to do this would, by comparison with the form of the escarpment in the Driftless Area, involve the trimming back of parts of the eastern escarpment by as much as 8 or 16 km.

In south-eastern England, there are also conspicuous differences between the glaciated and unglaciated portions of the Chalk escarpment (Fig. 11-3). In the Chilterns, from Goring Gap to Ivinghoe Beacon, the main scarp stands within 1·5 km of the basal outcrop of the Chalk, except in the vicinity of the principal wind-gaps. But north-east of Dunstable and Hitchin, the main

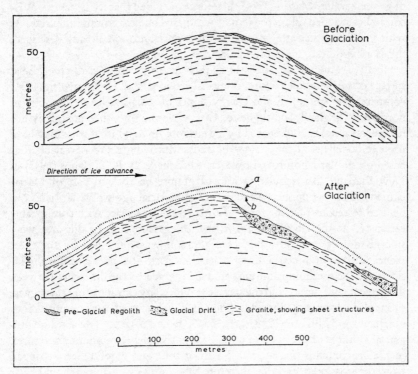

Fig. 11-4 Sheet structures (dilatation joints) in a granite hill, Massachusetts, and their use in reconstructing the pre-glacial form of the hill (R. H. Jahns, *J. Geol.*, 1943, University of Chicago Press)

escarpment is found progressively farther from the basal outcrop—at Royston, it is more than 8 km distant, and in the Breckland area up to 12-14 km. The escarpment also changes markedly in form, from a prominent and sharply defined edge south-west of Dunstable, to a very gentle and subdued form, much obscured by drift, in Cambridgeshire and East Anglia, as explained in Chapter 9. The difference is probably the result of glacial erosion of the Chalk escarpment in these latter areas—in other words, ice has displaced the line of the escarpment by up to 14 km.

In limestone areas, the relative abundance of caves as between glaciated and non-glaciated sections has been used to determine depths of ice erosion. The limestones of eastern (glaciated) Wisconsin are conspicuously lacking in caves or sink-holes, even allowing for some obliteration of these features by surface drift. On the other hand, the same formations in the driftless portions of Iowa and Minnesota possess abundant cavern systems down to the level of ground water. Martin (1916) therefore concluded that ice in eastern Wisconsin has stripped off up to 60 m of weathered and cavernous limestone. As with so many other attempts to measure glacial erosion already described, this is a hypothesis, not a proven fact.

Finally, we may look briefly at endeavours to reconstruct the pre-glacial surface of smaller-scale features of glacial erosion. S. E. Hollingworth (1931) compared the smoothed and craggy slopes of stoss-and-lee topography in north-west England, and suggested that while the smoothed areas may not have been appreciably lowered by ice erosion, up to 30 m of rock may have been removed from the craggy uplands. Much more precise calculations based on similar landforms, however, were made by R. H. Jahns (1943) in New England. An undulating area of granite bedrock in part of Massachusetts displays fine sheet structures of pre-glacial age whose attitudes can be used to reconstruct the form of the pre-glacial surface with considerable certainty. At the close of the Tertiary, the granite hills of this area were smooth domes, whose surface configuration was controlled by the sheet planes (dilatation joints). During glaciation, rock was plucked from the lee sides, and the minimum amount removed is given by the difference between the present land surface and the surface produced by extrapolating the highest sheet plane from the stoss side (Fig. 11-4). Figures of rock removal thus obtained for 8 hills are: 12, 8, 9, 12, 33, 9, 15 and 8-12 m. The absolute pre-glacial altitudes of the hills are unknown, so that these are minimum figures, but Jahns contends that at least 3-5 m of rock and regolith were stripped from the stoss sides, and that therefore the figures given above should be at least increased by this amount. Confirmation of the depths of erosion is provided by the fact that one huge boulder must have been plucked from a granite sheet 8 m thick, and that sheets of such thickness are not encountered until a depth of nearly 30 m below the surface is reached (Chapter 9).

Computations of the Volume of Glacial Drift

The value of this method lies essentially in its application to regions rather than individual localities. If, for a given region, we can determine the volume of glacial deposits and also determine with some precision the area from which they were eroded originally, we have theoretically a measure of glacial erosion in that source area. But there are numerous difficulties. Calculating volumes of drift in any region is by no means easy unless very detailed information on the form and position of the sub-drift surface is available. Some of the drift may have been removed by subsequent denudation and leaching, and some may have been carried out of the region beyond the coast and will now be invisible (Davis, 1909, p. 305). It is nearly always impossible accurately to identify the limits of the source region of the drift, and we cannot by any means be sure that the bulk of the drift has in fact been quarried by ice and not by other agents of erosion or a combination of these. Yet, regarding the last point, there is some evidence that most drift deposits are not simply re-worked alluvium or pre-glacial soils; many workers have found that drift consists predominantly of freshly quarried material. P. G. H. Boswell (1916), for instance, analysing the North Sea Drift in East Anglia showed that 'the large size and freshness of the mineral grains, together with the abundance of more easily decomposable ones, point to the fact, borne out by the evidence of the contained boulders, that the ice did not merely scour off the weathered material from the earth's surface, but that fresh rocks were also broken down' (p. 94).

As a method of estimating glacial erosion, it has long attracted attention. A. M. Hansen in 1894 thought that the rock basins in Scandinavia could be refilled and replenished many times over with the Scandinavian detritus now lying in such immense quantities on the North European plain, and later workers produced figures to indicate that the quantity of drift on this plain was equal to an overall lowering of 150-300 m for Scandinavia. In a similar way, R. F. Flint (1957) calculated that overall glacial erosion of the Laurentian Shield, deduced from the amount of drift lying on and around it, was of the order of 10 m. Previously in 1930, he used the thinness of the drift and its strikingly local composition over the state of Connecticut as evidence that glacial erosion was here of very limited amount. This view was supported in New Hampshire by J. W. Goldthwait and F. C. Kruger (1938). Analysing the till stones, they showed that most of the coarse material in the New Hampshire drift lies within a few kilometres of its original source. The average thickness of the drift is 3-5 m, and there is evidence that part of this (possibly 1 m) consists of pre-glacial weathered material. In the north and east, some drift may have been carried out to sea, but there is no such way of disposing of drift in the south and west, where the ice was moving towards Connecticut and Rhode Island. The conclusion is that total glacial erosion in New Hampshire cannot have averaged more than about 3 m.

In eastern England K. M. Clayton (1965) has measured the volume of the Chalky Boulder Clay deposited during the Riss (Saale) glaciation, as 300 km³, or possibly 400 km³ if allowance is made for later leaching and denudation of the original mass. The deposit is essentially a local one in eastern England, containing little far-travelled material, and its source must have been the Chalk outcrop of Cambridgeshire and East Anglia since the directions of ice-flow are known. The quantity is, as far as it is possible to judge, commensurate with the amount of erosion of the Chalk cuesta and the displacement of its scarp several kilometres from the basal Chalk outcrop, as mentioned earlier.

These examples concern ice-sheet erosion in regions of pre-glacial low relief. Volumes of drift and corresponding depths of erosion may be much greater in other environments: the enormously deep (over 300 m) drift-filled hollows at the southern ends of some of the Finger Lakes in New York State (H. L. Fairchild, 1934) may well be related to the intensive glacial erosion which the Finger Lake and adjacent valleys have suffered. In the majority of highly dissected regions, however, it is neither easy to determine the sub-drift surface, to ascertain precisely the source region of the drift, nor to ensure that all the drift produced by glacial erosion has been included in the balance sheet.

Conclusion

All methods of estimating glacial erosion are therefore characterized by doubts and uncertainties. But the evidence accumulated so far certainly suggests that active glaciers may be extremely potent agents of erosion, working at many times the speed of rivers of equivalent dimensions in terms of water discharge, while ice-sheets moving sluggishly on surfaces with little regional slope may be comparatively powerless even given the whole span of the Pleistocene.

Attempts to measure amounts of glacial erosion in a given locality or region have included 1) the use of artificial marks on rock surfaces later scraped by advancing ice; 2) measurements of the silt content of glacier meltwater streams and of the area of the respective glacier basins; 3) reconstructions of pre-glacial or interglacial land surfaces; and 4) estimates of the volume of glacial drift in a given region and its comparison with the area of the source region of that drift. The first two methods apply to present-day glacierized regions, the last two to regions of Pleistocene glaciation. The least suspect and most meaningful results are probably those given by the second method, but further studies involving collection of data over longer periods are urgently required.

REFERENCES

P. G. H. BOSWELL (1916), 'The petrology of the North Sea drift and upper glacial brick-earths in East Anglia', *Proc. Geol. Ass.* **27**, 79-98

K. M. CLAYTON (1965), 'Glacial erosion and deposition in eastern England', *Rep. 7th Conf. int. Ass. quatern. Res. (Colorado, 1965), Abstracts*, 69

J. CORBEL (1959), 'Vitesse de l'érosion', *Z. Geomorph.* NF **3**, 1-28

J. CORBEL (1962), *Neiges et glaciers* (Paris)

W. M. DAVIS (1909), 'Glacial erosion in North Wales', *Q.J. geol. Soc. Lond.* **65**, 281-350

H. L. FAIRCHILD (1934), 'Seneca valley physiographic and glacial history', *Bull. geol. Soc. Am.* **45**, 1073-110

R. F. FLINT (1930), 'The glacial geology of Connecticut', *Bull. Conn. State geol. nat. Hist. Surv.* **47**, 73

R. F. FLINT (1957), *Glacial and Pleistocene geology* (New York)

J. W. GOLDTHWAIT and F. C. KRUGER (1938), 'Weathered rock in and under the drift in New Hampshire', *Bull. geol. Soc. Am.* **49**, 1183-98

B. GUTENBERG, J. P. BUWALDA, and R. P. SHARP (1956), 'Seismic explorations on the floor of Yosemite Valley, California', *Bull. geol. Soc. Am.* **67**, 1051-78

A. M. HANSEN (1894), 'The glacial succession in Norway', *J. Geol.* **2**, 123-44

S. E. HOLLINGWORTH (1931), 'The glaciation of western Edenside and adjoining areas, and the drumlins of Edenside and the Solway basin', *Q.J. geol. Soc. Lond.* **87**, 281-359

R. H. JAHNS (1943), 'Sheet structure in granites: its origin and use as a measure of glacial erosion in New England', *J. Geol.* **51**, 71-98

F. A. KERR (1936), 'Quaternary glaciation in the Coast Range, northern British Columbia and Alaska', *J. Geol.* **44**, 681-700

R. VON KLEBELSBERG (1943), 'Die Alpengletscher in den letzten Dreißig Jahren (1911-41)', *Petermanns geogr. Mitt.* **89**, 23-32

D. L. LINTON (1949), 'Watershed breaching by ice in Scotland', *Trans. Inst. Br. Geogr.* **17**, 1-16

O. LÜTSCHG (1926), 'Beobachtungen über das Verhalten des vorstossenden Allalin-gletschers im Wallis', *Z. Gletscherk.* **14**, 257-65

L. MARTIN (1916), 'The physical geography of Wisconsin', *Bull. Wisc. geol. nat. Hist. Surv.* **36**, 1-549

F. E. MATTHES (1930), 'Geologic history of the Yosemite Valley', *U.S. geol. Surv. Prof. Pap.* **160**, 137 p.

G. ØSTREM, C. W. BRIDGE, and W. F. RANNIE (1967), 'Glacio-hydrology, discharge and sediment transport in the Decade Glacier area, Baffin Island, N.W.T.', *Geogr. Annlr* **49**, 268-82

A. DE QUERVAIN (1919), 'Über Wirkungen eines vorstossenden Gletschers', *Vjschr. naurf. Ges. Zurich* **64**, 336-49

H. F. REID (1892), 'Studies of Muir Glacier, Alaska', *Natn geogr. Mag.* **4**, 19-84

F. SIMONY (1871), 'Die Gletscher des Dachsteingebirges', *Sber. Akad. wien. math. naturw. kl.* **63**, 501-36

H. T. U. SMITH (1948), 'Giant glacial grooves in Northwest Canada', *Am. J. Sci.* **246**, 503-14

H. SPREITZER (1963), 'Grössenwerte des Ausmasses der glazialen Tiefenerosion (vornehmlich am Beispiel des oberen steirischen Murgebietes)', *Mitt. naturw. Ver. Steierm.* 93, 112–19

R. S. TARR (1908), 'Some phenomena of the glacier margins in the Yakutat Bay region, Alaska', *Z. Gletscherk.* 3, 81–110

S. THORARINSSON (1939), 'Hoffellsjökull, its movements and drainage', *Geogr. Annlr* 21, 189–215

Chapter 12

Erosion by Glacial Meltwater, and Glacial Diversion of Drainage

Where engaged in gorge cutting, these meltwater streams work with great rapidity, for the volume is great, the sediment load heavy, and therefore, with sufficient grade to prevent deposit, they are active agents of erosion. (R. S. TARR, on the ice-marginal streams in the Yakutat Bay region, Alaska, 1908)

Meltwater may be found on all parts of glacier surfaces below the firn limit, but above this limit it is relatively rarely encountered because of the great permeability of the firn and snow. It is also not usually present on cold polar or sub-polar glaciers, except very temporarily and locally, and in these cases it soon refreezes without travelling any significant distance. On temperate glaciers, however, surface meltwater streams whose direction of flow is determined by the ice surface slope may attain considerable dimensions, melting out gullies 10 m deep or more which possess all the characteristic features of normal stream systems such as meanders. On flat or gently sloping glacier surfaces, the streams may be broad and sluggish, but on more steeply sloping glaciers, meltwater flow may attain high velocities, for the ice channel offers little frictional resistance and the supraglacial stream usually carries little or no load. The volume of water will vary rapidly according to the diurnal air temperature cycle and the weather conditions (warm rain greatly accelerates melting), and will obviously differ between summer and winter.

These supraglacial streams sooner or later lose themselves down crevasses in the ice, or down cylindrical melt-holes known as glacier mills, moulins, or Mühle. They then become englacial streams flowing in tubes or down shafts, and whenever ice pressures permit, will eventually reach the base of the ice mass to become subglacial streams. Well-developed systems of sub-glacial drainage are more typical of stagnant than of active ice, though even on steeply sloping active glaciers, it is rare to find continuous supraglacial drainage reaching to the snout. Most meltwater generated on glaciers reaches the snout finally by a subglacial route and emerges from a cave, sometimes under considerable pressure. On the glacier surface, abandoned moulin holes and channels mark former supraglacial stream courses, for as melting proceeds, streams will be swallowed by the ice at progressively higher positions in their courses.

The great bulk of water emerging subglacially at glacier margins is therefore of surface origin, and after leaving the surface, follows devious routes within or beneath the glacier before escaping. Basal melting does occur beneath most glaciers and ice-sheets, excepting only cold glaciers frozen to bedrock. The chief sources of heat at the ice base are the slow escape of earth heat, and small amounts generated by friction resulting from ice movement (Chapters 3 and 4). But these sources are small, and contrary to the views of some workers such as R. G. Carruthers (1939), produce insignificant quantities of meltwater compared with surface melting. H. Hess (1935) showed that the average basal melting for Alpine glaciers was less than 5·3 cm per year, whereas loss of surface ice by melting often amounts to several metres a year.

Of the routes followed by meltwater, those termed supraglacial and englacial will leave relatively few traces once the ice has vanished. Subglacial streams however, may leave very definite evidence of their existence either in the form of stream-bed deposits (see Chapter 16) or in the form of channels eroded in bedrock or drift. Meltwater may also drain along the borders of ice-sheets or glaciers for short distances, and these marginal streams will also usually leave similar traces of their existence. As meltwater streams may attain high velocities and great size, they may be very active erosive agents when armed with sand and silt particles washed out from other glacial deposits. O. D. von Engeln (1911) described how marginal drainage of the Hidden Glacier in Alaska cut in shales and sandstones a gorge 1·6 km long and up to 3 m deep in less than 3 years. The part played by subglacial streams in producing plastically moulded bedrock surfaces and potholes has been described in Chapter 6. The following sections will describe the various forms of channels cut by meltwater and the significance and uses of these channels in interpreting the history of glaciation of an area.

Glacial Lake Spillways

Meltwater lakes may be impounded by ice blocking existing drainage outlets or by drift barriers (see Chapter 19). Relatively few meltwater lakes exist today, since those held up by drift barriers have in many cases been drained by rapid downcutting of the outlet stream where this flows across the barrier; and ice-dammed lakes are not very common since most glaciers on which meltwater is plentiful lie on high ground where drainage from the ice is not impeded. Two sizeable examples of present-day ice-dammed lakes are the

Conn and Bieler Lakes on the north-east side of the Barnes ice-cap, Baffin Island. There is, however, much evidence that Pleistocene ice-dammed lakes were in some areas both numerous and extensive. Such lakes drained by a variety of routes—over the ice surface, into fissures in the ice, subglacially (a very common means of escape), or over bedrock or drift cols to escape into an ice-free area. In the latter case, the overspilling lake waters cut channels in the cols and carved out deeper valleys beyond.

Until recently, most channels recognized as being the result of meltwater erosion in the British Isles were interpreted as spillways of ice-dammed or drift-dammed lakes. This hypothesis has its origins more than a century ago; in 1863, for instance, T. F. Jamieson described the evidence for ice-dammed lakes in Glen Roy which drained through and scoured out cols at several different heights. The most detailed discussion of glacial lake spillways in Britain, however, dates from 1902 when P. F. Kendall claimed that most of the meltwater channels in the Cleveland Hills were cut by overflowing waters from a series of glacially impounded lakes around the margins of the Hills. Indeed, assuming that the channels were lake spillways, he used them as a chief line of evidence for the former existence of ice-dammed lakes in the area, and regarded them as 'perhaps the most impressive memorials of the Ice-Age that our country contains' (p. 481). Four types of channel were recognized: 'direct overflows' such as Newton Dale in the Cleveland Hills, where the water flowed away from the ice across a former watershed; 'severed spurs' where a minor watershed is breached; 'marginal overflows' formed along the ice edge; and 'in-and-out channels' cut in a hillside by meltwater flowing round a projecting lobe of ice. In most cases, Kendall assumed that lakes of various sizes provided the principal sources of meltwater to cut the channels.

A number of features was held to be characteristic of these 'overflow valleys'. They were said to be entirely independent of the natural drainage and very seldom received tributaries with accordant floor level. Their transverse profiles exhibit steep sides and broad flat floors, while their long profiles usually steepen rapidly downstream. Sometimes meanders of considerable amplitude are present, and the lower end of the valley may 'open out inconsequently upon a steep hillside' (Kendall, 1902, p. 485) below which a mass of gravel may have accumulated.

Two channels which display many of these features and which were claimed to be typical of Kendall's direct overflow type by A. R. Dwerryhouse (1902), were later surveyed in detail by R. F. Peel (1949) (Fig. 12-1a and b). All the typical features listed by Kendall may be clearly seen; there is, however, an additional feature of great significance, namely the anomalous long profiles of the bedrock floor which in one case rises to a point 12 m higher than the supposed intake of the channel, a feature which led J. B. Sissons (1958a) to the view that these and other channels in southern Northumberland were not in fact glacial lake spillways (see p. 280).

Kendall's views have been presented in some detail, because they have had a profound and to some extent undesirable influence on many subsequent British studies of meltwater channels. Many workers assumed that all meltwater channels were spillways, and postulated ice-dammed or drift-dammed lakes simply from the existence of the channels. Dwerryhouse's work already mentioned presented such an interpretation; so did F. W. Harmer (1907), who for instance regarded Goring Gap of the Thames crossing the Chilterns as cut by overflow from a supposed Lake Oxford. Ironbridge Gorge, intaking at 94 m and now carrying the diverted River Severn, was regarded by L. J. Wills (1924) as the overflow route of a supposed Lake Lapworth. The Ferryhill Gorge across the Magnesian Limestone cuesta in County Durham was said by A. Raistrick and K. B. Blackburn (1932) to have drained a lake standing at 120 m in the Wear valley to the north. And in eastern England, several workers including Kendall have postulated a Lake Humber and a Lake Ouse, barricaded by ice in the Vale of York and by Scandinavian ice along the east coast, and overflowing in the later stages by the remarkable through-valley transecting Norfolk and linking the Little Ouse and Waveney drainage near Diss at a level of about 24 m O.D.

There is no doubt that some meltwater channels in Britain are true overflows and that some glacial lakes did exist, though, as discussed in Chapter 19, the evidence for such lakes must be independent and not consist solely, as so often in the past, of the supposed overflow channels themselves. In the basin of the Warwickshire Avon, F. W. Shotton (1953) has given detailed evidence (Chapter 13) for the existence of a 'Lake Harrison' during the advance of the Riss (Saale) ice, impounded between the Jurassic escarpment and three ice lobes (western Severn valley ice, northern ice, and north-eastern Chalky Boulder-Clay ice), and stretching from Leicester and Birmingham to Moreton-in-the-Marsh. Lake clays attaining a maximum altitude of 123 m near Moreton, and an erosion terrace at about 120 m mapped by G. H. Dury (1951) from Moreton towards Daventry, suggest a maximum lake level at about these heights. Three gaps in the Jurassic escarpment are considered by Shotton as possible overflow routes: the Watford Gap near Daventry at 137-140 m, the Fenny Compton gap (123 m), and the Dassett gap (125 m), the latter two converging to join at Banbury in the Cherwell valley. Clearly all three cannot have functioned simultaneously, and it seems unlikely that the higher level Watford Gap was ever used since there is no means of blocking off the other two lower level gaps. However, the levels of the Fenny Compton and Dassett gaps are so close that simultaneous use of these by overflowing lake

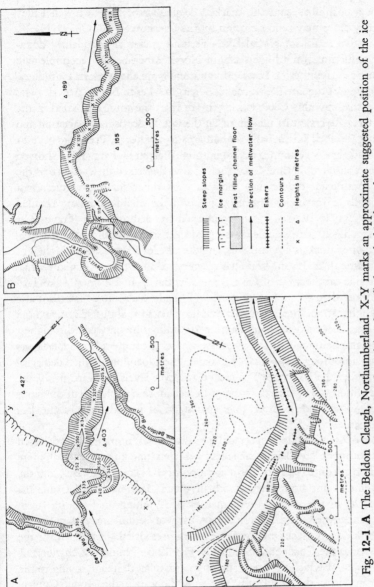

Fig. 12-1 **A** The Beldon Cleugh, Northumberland. X–Y marks an approximate suggested position of the ice margin when meltwater last flowed (southward) through this channel **B** The East Dipton Channel, Northumberland (A & B after R. F. Peel, 1949; J. B. Sissons, 1961) **C** The Deuchrie channel, southern Scotland (J. B. Sissons, *Scott. geogr. Mag.*, 1961, Royal Scottish Geographical Society). Spot heights and contour heights are in metres for all three diagrams are in metres

waters is a possibility, and the marked deepening of the Cherwell valley which they enter may reflect erosion by the overflow.

The erosive action of glacial lake overflow waters is abundantly documented, sometimes on a huge scale, in North America. One example will be taken from the Interior Lowlands, and one from the western Cordillera.

The largest of the late Pleistocene ice-margin lakes in North America was Lake Agassiz, covering 280,000 km^2 (more than the combined areas of the present Great Lakes) at its maximum in the late Wisconsinan glaciation and occupying the Red River basin and an area to the north. Present-day remnants include Lakes Winnipeg and Manitoba. Warren Upham (1896) showed that for a considerable time Lake Agassiz overflowed southward across the present continental divide into the valley of the Minnesota River, past Granite Falls to reach the Mississippi. The intake of this overflow channel (Upham's 'River Warren') is now marked by shallow Lake Traverse at 299 m (Fig. 12-2). The channel was not deepened below this owing to resistant granite outcrops downstream, and the lake level was held at this height for some time. J. A. Elson (1965) has recently mapped the associated Campbell shoreline on the west of Lake Agassiz, rising from 299 m at Lake Traverse to more than 430 m farther north in Canada as a result of later isostatic tilting. Higher strandlines show that overflow began at about 322 m, and the channel was deepened in stages (sixteen strandlines lie between the highest and the Campbell stage) as the overflow waters encountered resistant layers in Palaeozoic sedimentary rocks. The channel was abandoned when deglaciation permitted the lake waters to escape at lower levels eastward into Lake Superior. Elson shows that the story is in fact more complex than this since Lake Agassiz existed at more than one stage of ice advance in the late Wisconsinan.

In the western Cordillera, spectacular erosive effects of meltwaters released by the bursting of late Pleistocene lakes have been long known, though there has been great controversy over the location of the lakes, their sizes, and the magnitude and duration of the floods involved. The plateau of Columbia River basalts in the south-eastern quarter of Washington State has been scoured, channelled, and largely swept bare of sedimentary deposits by immense floods of glacial meltwater over an area of 5000 km^2. This is the area known as the Channelled Scablands (Fig. 12-3). Numerous channels or 'coulees' none of which now contains any permanent drainage owing to the dry climate, cross the Scablands. The most famous is the Grand Coulee, 80 km long, up to 8 km wide, and 300 m deep at its head, but there are others almost equally immense. Some of the coulees are braided and they possess normal features of fluvial erosion such as dry waterfalls (up to 120 m high and 5 km wide), abandoned high-level meanders, and interchannel islands

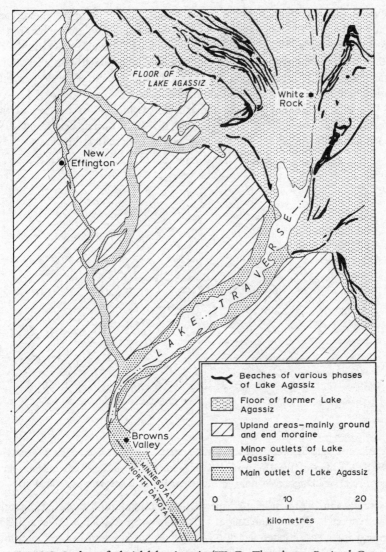

Fig. 12-2 Outlets of glacial lake Agassiz (W. D. Thornbury, *Regional Geomorphology of the United States*, John Wiley & Sons, New York, 1965)

with cappings of river-deposited gravel. Additionally, there are unexpected and highly unusual features such as giant current ripples and gravel bars noted by J. T. Pardee in 1942, which attain heights of 10 m and wavelengths of 80 m, and, most disturbing of all, rock basins (with or without lakes). One such basin is that containing Rock Lake, 11 km long and 80 m deep, while other basins up to 30 m deep lie on divide crests.

Theories to account for this remarkable assemblage of features in the Scablands have been many. I. C. Russell (1898) and Morris Leighton (1919) both generally suggested the work of glacial meltwaters; but it was J. H.

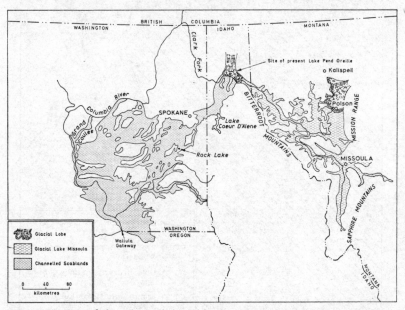

Fig. **12-3** Map of the Channelled Scablands and Pleistocene lake Missoula in Washington, Idaho, and Montana (W. D. Thornbury, *op. cit.*, 1965)

Bretz who in 1923 first excited speculation by invoking a brief but gigantic flood caused by sudden bursting of a glacial Lake Spokane. I. S. Allison (1933) argued for a more extended flood caused by an ice-dam in the Columbia Gorge, while R. F. Flint (1938) favoured normal erosion by meltwaters and a period of valley filling followed by partial re-excavation to account for the anomalous gravel deposits. W. H. Hobbs' theory (1943) of glacial erosion is not now supported.

The balance of opinion currently regards Bretz's theory of a sudden flood as most probable, but places the source of water in a glacial Lake Missoula (Fig. 12-3). The latter was first documented by J. T. Pardee (1910). It resulted from a Cordilleran ice-lobe blocking the Clark Fork of the Columbia River at Lake Pend Oreille (Idaho) and forming a lake over 7500 km², which rose to 1340 m in the Bull Lake glaciation and 1280 m in the Pinedale. The lake stood 600 m deep against the ice which was itself 1100 m thick. According to work by J. H. Bretz, H. T. U. Smith, and G. E. Neff (1956), the ice-dam burst seven times in the Wisconsinan period alone, releasing catastrophic floods across the Scablands. The coulees were used and re-used by these floods which may well have occurred also in pre-Wisconsinan glaciations. When Lake Missoula burst at its maximum level, Pardee showed that up to 40 km³ of water were discharged per hour, and probably 1500 km³ in a few days. The Missoula flood also entered glacial Lake Columbia farther downstream, raising its level by 150 m and emptying through the Grand Coulee.

Repeated floods of such magnitude are capable of explaining all the features of the Channelled Scablands. Probably cavitation erosion associated with enormous velocities of escaping water was responsible for excavating the rock basins. Parallel phenomena at the present-day, of smaller but still catastrophic proportions, are represented by the Icelandic jökulhlaups (see Chapter 19).

Proglacial Drainage

The term connotes drainage away from the ice, and therefore includes those examples of spillways in the last section. However, proglacial drainage need not depend on ice-border lakes, and the majority of proglacial drainage systems simply act as sluiceways for meltwater directly from the wasting ice. Volumes of meltwater often reach very high levels, and fluctuate considerably. Heavy loads of debris give rise to extensive outwash deposits farther downstream (see Chapter 18). The courses of proglacial rivers are usually determined by pre-glacial stream channels.

In Britain, the middle Thames acted as a proglacial drainage system in the Saale glaciation, carrying meltwater from Oxfordshire through Goring Gap into the London Basin. The valley is now characterized by its thick infill and the series of gravel terraces produced during successive incisions of the river at different phases of the Pleistocene. The present river is a much shrunken remnant, occupying only a small part of its former floodplain, and its sluggish flow contrasts forcibly with the coarse gravels underlying the floodplain and in the terraces.

Central North America possessed the most gigantic system of proglacial drainage in the world, focussing on the Mississippi-Missouri-Ohio Rivers, and collecting meltwater at times from nearly 3200 km of ice margin. Space does not permit a complete description of the system. One of the smaller members of the proglacial drainage net was the Illinois River. With its tributaries the Fox, Des Plaines, and Chicago Rivers, it collected water from successive ice borders in Illinois. During the Valders phase, when an ice lobe occupied the northern two-thirds of Lake Michigan, the Illinois River was fed by huge discharge from the Calumet stand of glacial Lake Chicago (Bretz, 1959). As a proglacial drainage route, the Illinois River probably functioned longer and more continuously than any other outlet for glacial drainage in the Great Lakes area. In its lower section below Hennepin, its valley is remarkably wide and straight–projecting spurs are rare; it has a gentle gradient and very thick gravel infill. It is in this section following an earlier course of the Mississippi (Dubuque-Clinton-Hennepin-Illinois River), but the present characteristics of its valley result in no small measure from its function as a major proglacial sluiceway.

Marginal and Sub-marginal Meltwater Channels

The surface of glacier ice, where it abuts against a hillslope, is often gently convex, owing to increased melting at the ice margin. As noted by I. C. Russell in 1893, this is a result of radiation of heat from the adjacent exposed ground, which itself is a better absorber of insolation than the ice. Thus a longitudinal hollow, one side of which is formed by the ice, may mark the ice margin and meltwater may collect in it (Plate XIX). The quantity of melt-water is not usually great since its source is purely local, though there are cases where drainage from other more distant sources is diverted along the ice margin (as along the north side of the Tasman glacier, New Zealand). The direction of movement of the meltwater will be controlled by the longitudinal gradient of the ice margin. If there is no gradient, the meltwater will accumulate as a marginal lake which may drain by way of a spillway or over, into, or beneath the ice. If the ice margin slopes longitudinally, meltwater will drain along the margin so long as there are no fissures in the ice, and will carve a marginal channel, partly in ice, partly in bedrock. In the early stages, only a narrow shelf may be cut in the hillside, but if the ice margin remains relatively stable in position, a true rock-cut channel may be eroded which deepens downstream. As the ice will normally possess many fissures, and will rarely make a watertight seal with the hill slope, the meltwater will not adopt a marginal course for any great distance, rarely more than 1 or 2 kilometres, beyond which the water will disappear into or beneath the ice. Marginal channels are therefore characteristically short, comparatively straight, and run along a hillside making only a small angle with the contours. They are commonest on long gently-sloping hillsides, especially where the slope is veneered in drift, for then the meltwater can quickly erode a channel.

Marginal drainage of present-day glaciers in Alaska and Greenland has been vividly described in the classic reports of R. S. Tarr (1897, 1908, 1909) and O. D. Von Engeln (1911). They make it clear that simple marginal drainage is not as common as has sometimes been supposed. There is first of all the case where a marginal stream is diverted away from the ice margin by rock spurs. Tarr (1909) illustrates this on the eastern margin of Hayden glacier (Alaska): 'Below Floral Pass, a marginal valley is developed . . . but at one point the ice presses against a low spur of the mountain, forcing the marginal drainage into a narrow rock gorge for an eighth of a mile. The stream then emerges into a broader valley, with one ice wall, and within a quarter of a mile disappears in an ice tunnel. . . . Below the ice tunnel for more than a mile there is no visible stream in the marginal valley whose bottom is hummocky, moraine-covered ice into which drainage disappears' (p. 83). Later, he relates how drainage from the Orange and Variegated glaciers 'flows in a marginal channel with ice for one wall and a gravel cliff for the other; then, where the ice crowds up against a granite spur, it enters a rock gorge more than half a mile long and 200 feet deep [in which] the glacial torrent rushes with great velocity and in a succession of falls' (p. 142).

Both Tarr and Von Engeln emphasize that under certain conditions there is no true marginal drainage, but instead a system of sub-marginal drainage. This is described as flowing 'under the ice but near its lateral edge' (Von Engeln 1911, p. 110), and is claimed to be more normal than marginal drainage when the ice is stagnant, or when the ice mass is active but stable in position. Sub-marginal streams may flow parallel with (but underneath) the ice margin, or obliquely beneath the ice; their exact courses are possibly influenced by structures (such as crevasse patterns) in the ice. There may be many sub-marginal streams beneath each margin of a glacier, often forming an anastomosing system leading down to truly subglacial or englacial drainage systems: their channels are of course only revealed after ice wasting uncovers them. Renewed advance of a glacier possessing mainly sub-marginal and subglacial drainage has been seen to result in the blocking or constriction of this drainage, and the reappearance of truly marginal streams: this was noted by Von Engeln (1911) following sudden advances of the Atrevida, Hidden, Variegated, and Galiano glaciers in Alaska. Large marginal streams are nearly always associated with rapidly advancing valley glaciers.

In interpreting meltwater channels after deglaciation, great care must be taken to distinguish sub-marginal from truly marginal channels. In cases of doubt, a sub-marginal origin must be considered to be at least as likely as a marginal origin. In studies of meltwater channels in Sweden, C. M. Mannerfelt (1945) finds the sub-marginal type the more common, and J. Gjessing (1960) in central Norway found it was 'not possible to separate with certainty

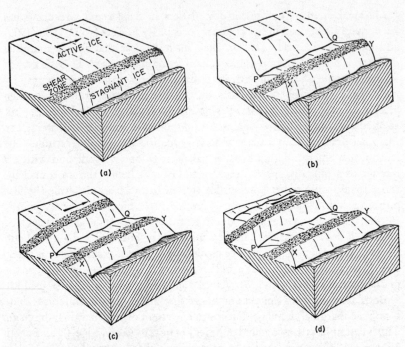

Fig. 12-4 Stages in the development of marginal drainage channels (V. Schytt,
Geogr. Annlr, 1956)

(a) Stagnation of the glacier margin owing to thinning; active ice
begins to shear past the dead ice

(b) Debris melted out from shear zone protects ice from ablation and
causes ridge X-Y to form. A new depression P-Q is melted out on
the glacier side of this ridge

(c) Further thinning of the glacier causes development of new shear
zone. Stream P-Q continues to deepen its ice channel

(d) A second ridge and another melt depression begin to form by
differential ablation. Stream P-Q will shortly melt down to bedrock
and become superimposed on it

the strictly marginal parts of the channel systems from the sub-marginal and
subglacial parts' (p. 462).

Another complication in the lateral drainage of ice masses has been noted
by V. Schytt (1956) along the northern border of the Moltke glacier in
Greenland. Here there was not just one line of drainage marking the ice
margin but other parallel drainage channels within the ice. The latter appeared to be forming by differential ablation along the outcrops of dirt-laden
shear planes (which thus controlled their position and spacing) formed be-

tween the active ice of the glacier and its stagnating margins (Fig. 12-4). The innermost channel, just forming, was relatively shallow and cut entirely in ice; the next nearer the ice margin was deeper and possibly had melted down through the ice to bedrock; while the ice edge was marked by a truly marginal channel with one ice wall. Away from the ice margin, a series of older and now largely abandoned drainage channels, parallel to the ice margin, could be seen. Thus simultaneous formation of several lateral channels is possible.

In spite of the difficulty of distinguishing truly marginal channels, many workers have found them of great help in reconstructing conditions of deglaciation. The main uses of marginal channels are threefold: they can provide estimates of ice margin gradient (which obviously cannot be steeper than the mean longitudinal slope of the channel floor); they may throw light on the annual melting rhythm of the ice; and, as noted already, large marginal channels suggest active advancing ice during their formation.

Estimates of ice margin gradients provide means of reconstructing the form of the ice cover and also of correlating one marginal channel with another. As far back as 1908, J. L. Rich estimated the ice margin gradient in part of the Finger Lakes area, from the slopes of the Johnson Hollow meltwater channels, as 1 in 80. In Scandinavia, Mannerfelt (1945) has calculated ice surface slopes in many areas. In Arådalen (eastern Norway, latit. 63°N.) he computes ice slopes between 1 in 33 and 1 in 100 depending on the part of the ice tongue selected; on Högfjället (Sweden, latit. 62°N.) the mean slope is 1 in 30. Mannerfelt (1949) stresses that care should be exercised in this procedure, for marginal drainage naturally tends to undermine the ice border, and thus erode channels which slope more steeply than the ice surface; even more important is not to confuse marginal with sub-marginal channels which may have slopes very much steeper than the ice margin. To avoid this pitfall, J. B. Sissons (1958b) in the Eddleston valley of southern Scotland used only the longest and most gently graded channels of the area to calculate the former ice slope (1 in 50, north-east) and thereby to establish ice margin positions when certain channels were in use. In this area, marginal and sub-marginal channels are numerous but mainly small, less than 5 m deep and 30 m wide; the smallest ones mapped were narrow benches about 10 m wide, cut by water flowing in channels one side of which was ice and the other bedrock or drift.

In north-west Wales, marginal channels were formed along the border of Irish Sea ice (C. Embleton, 1964b), and are distinguished from sub-marginal channels by their very gentle slopes, some having gradients of only 1 in 300, such as that near Tregarth, south of Bangor. This example is about 2 km long, flanked on its northern side by drift mounds, and from an intake at 114 m slopes gently south-west to Rhyd-y-Groes where there is a small outwash flat at 110 m. In the same region, a good example of probable sub-marginal drainage occurs between Llanrug and Caernarvon, where anastomosing channels, each from 3 to 15 m deep, are cut in drift and possess gradients (1 in 25 to 1 in 45) that cannot possibly reflect a marginal origin. A series of channels near Conway with gradients between 1 in 20 and 1 in 50 was interpreted as marginal by Embleton in 1961, but the steeper members of the group are now thought to be sub-marginal and subglacial.

Mannerfelt (1945, 1949) has claimed that the spacing of marginal channels in some sequences may have been controlled by the amount of annual ablation which the ice surface suffered, each shallow channel being used in one spring and summer only and being abandoned in late summer when subglacial drainage became dominant. In northern Sweden, he found the most complete series of lateral drainage channels and terraces on the west slope of Gråsidan, where no less than 78 consecutive marginal channels were measured, with an average vertical spacing of 3·5 to 4 m. Figures for other areas ranged from 3 to 5 m. Sissons (1958b) in southern Scotland deduced comparable figures of 3 to 3·6 m which he also suggested represented annual thinning of the ice. However, caution is again necessary. It has already been explained how, according to Schytt, several marginal channels can actively co-exist along or just within an ice border. Moreover, some workers have encountered vertical spacings much too large to be the result of annual ablation (for example, the figure of over 20 m arrived at by C. Holdar, 1957), and both Mannerfelt (1960) and Sissons (1961a) have recently expressed doubts about the validity of earlier claims.

Subglacial Channels

Although active subglacial streams cannot be observed directly, there is no doubt as to their existence beneath many glaciers and the importance of their functions. I. C. Russell (1893) deduced that 'the drainage of the Malaspina glacier is essentially englacial and subglacial' (p. 238) from observations of meltwater descending moulins and crevasses, and later escaping from the ice margin: Fountain stream 'comes to the surface through a rudely circular opening, nearly 100 feet in diameter, surrounded in part by ice. Owing to the pressure to which the waters are subjected, they boil up violently, and are thrown into the air to the height of 12 to 15 feet, and send jets of spray several feet higher. The waters are brown with sediment, and rush seaward with great rapidity, forming a roaring stream fully 200 feet broad' (ibid.). When the subglacial tunnels are abandoned by their streams, they can sometimes be located and explored, as for instance in the case of a small network of tunnels beneath the Martin River glacier in Alaska investigated by L. Clayton (1964). There is also clear evidence of subglacial (or sometimes englacial) drainage provided by the manner in which many glacial lakes periodically drain: the well-known Märjelen See drains regularly into or beneath the Aletsch Gletscher (see Chapter 19), and G. de Boer in 1949 described the subglacial draining of a small lake beside Leirbreen in Norway. On a vastly larger scale, the Icelandic jökulhlaups represent great volumes of meltwater forcing a way beneath the ice. The work of subglacial streams is equally evident. On the one hand, their depositional activity is reflected in esker systems (Chapter 16), while their erosional activity has given rise to channels of unusual form, cut often in anomalous positions.

The maximum thickness of ice beneath which subglacial streams can exist will depend both on the degree of activity of the ice and on its temperature conditions in depth. In stagnant highly-fissured ice at pressure melting point throughout, meltwater streams can flow beneath 100 m of ice or more. Sissons (1963) states that 'many channels and some eskers . . . prove that meltwaters penetrated 300 feet beneath glacier ice' (p. 110); Embleton (1964a) found evidence in the Wheeler valley (North Wales) for the existence of subglacial streams beneath about 120 m of ice. W. H. Mathews (1964) showed that meltwater in the South Leduc glacier (British Columbia) penetrated at least 110 m below the glacier surface, the location being about 2 km from the glacier snout and at the glacier's lateral margin. Sub-zero temperatures in the basal parts of the glacier would preclude much deeper penetration, as would increasing ice pressure in the case of an actively flowing glacier. Glacier moulins have been explored to depths of 60 m (for example, G. Dewart, 1966) and occasionally a little deeper. It seems therefore that Von Engeln was essentially correct when in 1911 he postulated that meltwater drainage in thick active glaciers would be sub-marginal and englacial rather than truly subglacial.

The simplest form of subglacial channel is that termed the subglacial chute by Mannerfelt (1945, 1949). These were cut by meltwaters plunging directly down a hillslope beneath the ice. Often, they were fed by marginal or sub-marginal channels, whose meltwaters after flowing along a hillside suddenly abandoned that course for one more directly related to the bedrock slope. Sometimes there is evidence that the meltwaters once continued farther along the slope before the chute was opened, and it seems probable that the opening and location of many chutes is dependent on the development of crevasses in the ice. In other cases, pre-glacial stream valleys have been used and enlarged as chutes, as shown by the way in which some marginal or sub-marginal channels terminate at such a stream valley (Sissons, 1961a, p. 20). The chutes can normally be distinguished from post-glacial gullies by their relations to other meltwater channels, their abrupt beginnings and endings (sometimes part-way down a slope), the absence of alluvial fans from their lower ends, and the fact that they are often now dry. Mannerfelt considered that chutes might open and operate mainly in the later part of each melting season, but it is clear that in most cases the chutes were developed over much longer periods. Subglacial chutes have only been recognized as such in Britain since about 1957, but since then many workers (for example, Sissons 1958b, R. J. Price 1960, Embleton 1964, D. Q. Bowen and K. J. Gregory 1965) have mapped and described them. Their height range is usually no more than 100 m, supporting the theory of crevasse control (E. Derbyshire, 1961).

More complex forms resulting from subglacial drainage have long been known to exist in parts of northern and central Europe covered by the last ice-sheet. In north Germany and Denmark, there are the valley systems known as the Rinnentäler (channel valleys) and Tunneldale (tunnel valleys) respectively. They represent subglacial stream courses carrying meltwater westward and southward to the ice margin during the last glaciation (Fig. 12-5). The tunnel valleys of Denmark are up to 75 km long (A. Schou, 1949), relatively narrow, and up to 100 m deep, with flat floors and sharply defined sides in transverse profile. Some can be traced from the föhrden of the Baltic coast (p. 246) continuously to the sandy outwash plains of western Jylland. In long profile, their floors are characteristically irregular, for the subglacial streams, being under pressure, were able to force their way uphill in places and excavate depressions. As a result, ribbon lakes are frequently found occupying parts of the valley floors, such as Hald Sö and Lake Tjele near Viborg. Streams may follow parts of the tunnel valleys, but are usually diminutive in size compared with the valleys which the swollen glacial torrents carved. The system of tunnel valleys continues southward into Schleswig and north Germany, where their subglacial origin has been particularly studied over many years by P. Woldstedt (1926 et seq.). The valleys (Rinnentäler) tend to be orientated in the direction of regional ice slope, and therefore generally perpendicular to the ice margins, for the subglacial waters were forced out under the ice towards regions of lesser ice pressure. At the ice margins where hydrostatic pressure and therefore the load-bearing capacity of the streams were suddenly reduced, low outwash cones were built up at the points where the torrents emerged from the ice. As the outwash cones accumulated, so the terminal base-level of the streams was slowly raised, and as a result, the streams began to flow englacially and finally even supraglacially over collapsed ice filling the Rinnentäler (E. M. Todtmann, 1936). As in Denmark, the floors of the Rinnentäler are uneven in long profile, and Woldstedt considers that the majority of the north German lakes lie in such depressions (Rinnenseen).

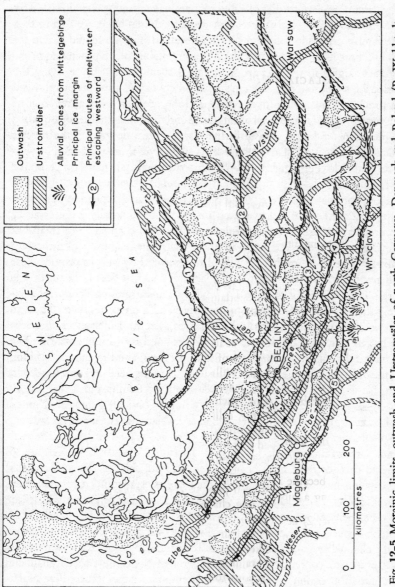

Fig. **12-5** Morainic limits, outwash and Urstromtäler of north Germany, Denmark and Poland (P. Woldstedt, *Norddeutschland und angrezende Gebiete im Eiszeitalter*, K. F. Koehler Verlag, Stuttgart, 1955)

In Britain, modern studies of subglacial drainage systems were initiated by J. B. Sissons (1958a), whose attention was first drawn by R. F. Peel to two meltwater channels in southern Northumberland (Fig. 12-1a and b, and see p. 265) whose floors possess an up-and-down long profile. By mapping probable ice margins in the area from marginal channels, and by studying the relationships of these two channels with others in the area, Sissons showed that previous explanations of the humped long profiles—such as reversals of meltwater flow, post-glacial erosion of one end of the channel, or erosion of a channel across a col previously formed by a stream flowing in the reverse direction—were unsatisfactory, and that only a hypothesis of erosion by subglacial water flowing under hydrostatic pressure could adequately account for the features. Fig. 12-1c shows another example of a channel in East Lothian, the Deuchrie channel, where the subglacial hypothesis is applied by Sissons. It possesses a clear up-and-down long profile, is of unusually large capacity compared with nearby marginal and sub-marginal channels, and small but definitely subglacial eskers lie on its floor. Two other examples of meltwater drainage systems in southern Scotland investigated by Sissons provide clear evidence that they were initiated subglacially. The first lies on the north side of the Tinto Hills (Sissons, 1961b) where an anastomosing system of relatively shallow channels receives at its upper end a series of eskers. These in turn emerge from the lower end of another system of meltwater channels, the largest of which crosses the Hills in a col and is of humped form. The intimate relationship between channels and eskers, and the anomalous long profile of the col channel point conclusively to a subglacial origin. The second example is in the Carlops area, a little south of Edinburgh. Here, Sissons (1963) has mapped and re-interpreted a complex braided system of channels which, he argues, developed first subglacially; later, as the ice cover wasted, the meltwater became concentrated in open ice-walled channels. The evidence supporting a subglacial origin is fourfold: 1) several channels are so situated that ice must have been present on both sides of them when they were being cut; 2) at least one channel possesses a humped long profile; 3) one side of a large channel is cut into by a number of gullies which can only be meltwater features and themselves formed under ice as subglacial chutes; 4) the presence of eskers, some with a thin cover of ablation moraine, in the channel system. The braiding of the channels around rocky knolls and hills points to contemporaneous formation of these sub-parallel channels. The initiation of the whole system is thought to have occurred by superimposition of englacial streams on an area of varied relief during the last period of ice downwasting. R. J. Price (1963) developed this hypothesis as an alternative to that which explains up-and-down channels as simply the product of meltwaters flowing under hydrostatic pressure. He suggests that if an englacial stream by melting down gradually encountered parts of the underlying bedrock relief, it would not only become superimposed on these but would also probably develop an irregular long profile, for its rate of channel down-cutting would be much faster in ice than in bedrock. Two advantages of this hypothesis are that it accounts for the non-existence of channels in some areas (where the meltwater was flowing englacially), and that it avoids the necessity for postulating meltwater flowing subglacially beneath great thicknesses of ice.

Following Sissons' lead, many other workers in Britain have found strong evidence of the work of subglacial meltwaters in erosion and deposition. Subglacial drainage systems have been described from the Tweed basin (Price, 1960 and 1963), the Cheviots (Derbyshire, 1961), Eskdale in Yorkshire (Gregory, 1965), North Wales (Embleton 1961, 1964a and 1964b), South Wales (Bowen and Gregory, 1965), and elsewhere. At the same time, similar studies have been made in various parts of Europe, such as that of Gjessing (1960) in eastern central Norway, and in North America (for example, Sissons, 1960b).

Col Channels

Meltwater channels are sometimes cut into existing cols; they may be of great size and raise special problems. As a particular type of meltwater channel, they were recognized by Mannerfelt (1945, 1960) who named them Sadelskäror or col gullies. A classic example occurs south-west of Drommen peak in northern Sweden (latitude 63°). Into the col separating Drommen from another peak to the south-west has been carved the spectacular Drom-skäran, a canyon which begins abruptly with a 40-metre high cliff. The floor of the canyon at the foot of the cliff is eroded into a shallow rock basin, beyond which the canyon floor drops rapidly through a vertical interval of 110 m, with sidewalls about 50 m high. Mannerfelt visualizes the formation of the col channel as initiated by a subglacial stream forced through the col by hydrostatic pressure. Later, during downwasting of the ice, the hills and the col between them were gradually uncovered, and from the ice on the south-east of the col, meltwater began to pour across the col to regain the ice on the far side which, owing to regional ice slope, was at a slightly lower level. As downwasting proceeded, the col was more deeply eroded; a water-fall with plunge-hole at its foot developed and receded headward towards the crest of the col. A shallow ice margin lake fed the waterfall since strand-lines appear leading into the col, and a delta accumulated at the lower end of the canyon where another ice lake existed. Dromskäran is thus interpreted as a subglacial channel later functioning as a normal glacial lake spillway, though the dimensions of the lake were small and the water was essentially provided by the melting ice around.

A subglacial origin for col channels has been suggested by many other writers, though examples of such channels linked with lake strandlines as in the case of Dromskäran appear to be rare. The great size of some appears to demand more meltwater than might be locally available, some of them possess humped or irregular long profiles, eskers sometimes lead into the heads of the col channels suggesting that ice and not a lake lay behind them, and some col channels possess either double intakes or double outlets which are not easy to explain on a subaerial hypothesis. Derbyshire, studying col channels in the Cheviot Hills and in Labrador, concluded that the majority were cut by subglacial waters flowing under pressure, of such volume that they were not derived from their own immediate vicinity (Sissons, 1961a), and flowing in the direction of lesser ice thickness and therefore less pressure. Gjessing (1960) comments in regard to northern Atnedalen and Gudbrandsdalen (central Norway) that in accordance with the regional ice slope to the north and north-west, 'all subglacial water in the area between the culmination zone (i.e., the ice divide) and the present watershed (north and north-west of the culmination zone) was forced to ascend to a pass at the present watershed on its way to the valleys on the other side' (p. 443). Thus subglacial col channels, as in the case of the larger subglacial drainage systems generally, can be used to indicate approximately the slope of the ice surface and variations in regional ice thickness. The Sarn Galed and Sarn Adda col channels in Flintshire (Embleton, 1964a) provide good examples; both possess humped profiles, and were eroded by subglacial meltwaters being forced generally southward across cols in the direction in which the Irish Sea ice of this area was thinning. Similarly, the two Northumbrian spillways described by Peel (1949) and reinterpreted by Sissons (1958a) are essentially col channels carrying meltwater generally eastward across cols towards areas of thinner ce cover.

Open Ice-walled Channels

Present-day meltwater streams can sometimes be observed flowing in open ice-walled channels. Russell (1893) described the courses of the Osar, Kame and Kwik rivers on the eastern margin of the Malaspina glacier: 'Each of these issues from a tunnel and flows for some distance between walls of ice. The Kame . . . issues as a swift brown flood partially choked with broken ice, from the mouth of a tunnel and flows for half a mile in an open cut between precipitous walls of dirty ice 80 to 100 feet high' (p. 239). B. Roberts (1933) noted similar behaviour of the Kreppa emerging on the north of Vatnajökull in Iceland, while R. P. Sharp (1947) described how the Wolf

Creek (Yukon) emerged from an englacial tunnel to flow for 3 km between walls of ice. These ice-walled channels clearly originated by collapse of ice-tunnel roofs, and this is probably the commonest mode of origin, but there is also the possibility to be borne in mind that downcutting by supraglacial streams to bedrock can produce ice-walled trenches. In either case, the ice walls will guide the course of the stream and give rise to meltwater channels, after the ice has vanished, in anomalous positions. C. P. Gravenor and W. O. Kupsch (1959) have identified channels in east-central Alberta which they claim are of this type. Some have been filled with till, others are broad open troughs with sand and gravel on their floors. Some feed into esker systems, which might suggest a subglacial origin, but while the smaller ice-walled channels may have been true tunnels, the width of the larger trenches (more than 1 km) and the paucity of till in some of them, suggest that they were open to the sky. They developed in a pattern of parallel and intersecting elements during disintegration of stagnant ice over the area. In southern Illinois, Morris Leighton (1959) contends that the tendency for streams to be arranged in trellis pattern over Illinoian drift reflects the pattern of open crevasse meltwater channels developed in the stagnating Illinoian ice. Sissons (1960b) has put forward the view that some of the large channels in the Syracuse area of New York State may have operated as ice-walled channels after thinning of the ice and tunnel roof collapse. Limited ice thickness and the huge volumes of Great Lakes waters using them make it difficult to regard them simply as subglacial, while their locations and directions are not explicable in terms of marginal drainage. Some, moreover, such as the lower Butternut valley, contain outwash terraces which are unlikely to have formed in tunnels. In the lower Oneida valley, dead ice deposits show that a tongue of ice persisted here and was cut across by meltwater from the west giving rise to a transverse ice-walled channel, later infilled by sand and gravel. Ice-walled channels are also envisaged in east central Norway by Gjessing (1960) who describes how the subglacial rivers gradually melted 'the ceilings of the chambers and tunnels, so that drainage ascended up through the ice' (p. 471).

Superimposed Channels

The possibility that meltwater channels with humped long profiles might be explained by superimposition of englacial streams has already been noted. But this is only one particular case of a process whereby any stream flowing on or in ice will lower its bed rapidly by solution and will eventually encounter bedrock. If the rate of ice channel lowering is not too rapid, the rock not too resistant, and the ice in firm contact with bedrock, then the stream will be superimposed on the rock. The process has been hinted at by many workers in the past. W. O. Field (unpublished work quoted by Flint, 1957, p. 168) observed how a meltwater stream from the Casement Glacier, Alaska, had been superimposed across resistant bedrock in 1935, and when next seen in 1941, had succeeded in cutting a gorge 8 m deep and 15 m wide. R. S. Tarr (1908) in the Yakutat Bay region referred to streams leaving the ice margin and cutting a gorge across a rock spur (p. 100); Von Engeln (1911) in the same area also commented on this phenomenon and stated that the stream would be permanently entrenched in such a course provided its rate of downcutting were equal to or greater than the rate of ice wasting (p. 128). The application of the process to formerly glacierized areas was not common until recently; J. W. Goldthwaite's reference (1938) to superimposition of meltwater streams in New Hampshire was an exception. In Britain, R. Common (1957), Embleton (1961), Price (1960, 1963), and Sissons (1961a, 1963) have utilized the hypothesis to explain certain channels. Price has made the most complete analysis of the hypothesis in respect of meltwater channels in the Tweed basin. Channels resulting from superimposition may have long profiles of three forms: 1) continuously and gently sloping; 2) gently sloping at first, becoming markedly steeper down-channel; 3) up-and-down character. The last has been discussed already and demands superimposition of englacial rather than supraglacial streams. The second is characteristic of a stream superimposed on a rock outcrop and plunging beneath the ice once it has crossed the obstruction, while the first is typical of a channel cut by a supraglacial stream, whose downstream gradient accorded with the ice surface slope and which by downmelting has happened to encounter bedrock in part of its course. Supraglacial streams are usually of clear sparkling water and their channels in the ice seldom contain much debris, as noted by Russell (1893) on the Malaspina glacier, but when they reach the basal ice layers, a great deal more debris will become available for bedrock erosion. The meandering habit of supraglacial streams is well known, and the hypothesis of their superimposition may explain the occurrence of meandering channels on slopes where the form of the ground would not permit meandering streams to develop marginally or extra-marginally (Sissons, 1961a, p. 29).

Summary: The General Significance of Meltwater Channels

Studies of meltwater channels throw much light on the conditions pertaining to episodes of deglaciation, provided they are correctly interpreted. In the past the majority of channels were regarded as glacial lake spillways or as marginal channels, but it is now clear that other types of channel, especially sub-marginal and subglacial types are equally if not more important, and in those areas where the meltwater was not able to escape proglacially, were responsible for carrying most of the drainage during ice wasting. It is particularly important not to interpret sub-marginal channels as marginal, for gross errors in reconstructions of ice margin positions and gradients can thereby arise. If truly marginal channels can be identified, however, they can be of the greatest use in such reconstructions, bearing in mind that under certain conditions the marginal zone of an ice mass may possess more than one line of lateral drainage; and in a few instances, closely spaced sequences of minor channels may even reveal the annual wasting rhythm of the ice. But marginal channels are not the only indicators of the form of the ice. Subglacial river systems are often directed by variations in ice pressure and therefore tend to seek out regions of thinner ice; from this, it may be possible to deduce the regional ice surface slopes. Finally, the highest meltwater channels in a region may be used to indicate tentatively the minimum heights of the firn line during their formation.

Meltwater channels can also suggest the condition of the ice during deglaciation. Large marginal rivers are most typical of active and advancing glaciers; sub-marginal drainage is often associated with active but stable glaciers, while subglacial drainage is more characteristic of stagnant ice or the thinner zones of active glaciers. Open ice-walled channels develop in the final stages of ice disintegration. The depths to which meltwater can penetrate ice depend on the degree of activity of the ice, the increasing ice pressure with depth, and the ice temperatures at different levels. Stagnant ice at pressure melting point permits meltwater to penetrate for depths of 120 m and possibly much more.

Subglacial drainage systems have been extensively studied in recent years, and their behaviour is now better known. They are responsible for impressive and unusual features of both erosion and deposition. They consist of both local drainage chutes, and the more extensive and complex systems fed by a wider melting area and controlled by ice-pressure variations. Subglacial channels may be distinguished by humped long profiles, by their anomalous positions in the landscape (which may rule out most other possibilities), and by their relation to esker systems. Humped long profiles are thought to have been cut by subglacial water being forced up or down hill by hydrostatic pressure, or to have resulted from superimposition of englacial streams. As ice melt proceeds, tunnel roofs collapse and open ice-walled channels replace the former subglacial drainage. Finally, it should be emphasized that many meltwater channels are of composite origin, not only having been re-used and re-excavated in successive glaciations but also having developed in different positions in relation to the ice. Thus channel systems may be initiated by subglacial drainage, pass into a phase of open ice-walled channels, and finally be occupied by proglacial streams.

Glacial Diversion of Drainage

Glaciation may interfere in several ways with existing river systems, both directly and indirectly. The simplest case is that of ice damming a former valley, but caution should be exercised in applying this hypothesis too readily since glacier ice, as shown above, is not often the impenetrable barrier to water that it might seem. Obstruction by moraine is common, as is complete plugging by drift and obliteration of former valleys. Glacial erosion can also radically derange the pre-glacial drainage by lowering certain valley floors or cols, thereby permitting new routes to be adopted by rivers post-glacially. Finally, river systems may be affected by glacio-isostatic movements consequent on the loading or unloading of the earth's crust during glaciation and deglaciation. Circumstances of glacial diversion vary greatly from one area to another, so that few generalizations are profitable; but generally drainage diversion by ice or drift obstruction produces the most spectacular rearrangements of river systems in regions of low relief. The processes will be discussed by reference to contrasting examples, both large and small in scale.

1) THE BARABOO-DEVIL'S LAKE AREA, WISCONSIN (Fig. 12-6)

This small area on the margin of the so-called Driftless zone in southern Wisconsin provides a classic model of drainage diversion. In the Cary phase of the Wisconsinan glaciation, the Wisconsin River was ponded by ice at Wisconsin Dells, the resultant lake overflowing north and west by the Black to the Mississippi. Below the Dells, the pre-glacial course of the Wisconsin River, now in part plugged by drift and reconstructed from well records in the plains sections (W. E. Powers, 1946) took it through the north rim of the Baraboo syncline (in a gap now utilized in the reverse direction by the Baraboo River), and through the South Range in a gap not now followed by any river but occupied by Devil's Lake. The latter is impounded both on the north and on the south-east by moraines of the Cary ice. With the melting of the Cary ice, the Wisconsin River was unable to resume its former course because of these morainic barriers and drift plugging, and now makes a detour over drift-covered country round the east of the Baraboo Ranges. At Devil's Lake, the former valley of the Wisconsin River is filled with at least 86 m of morainic and fluvioglacial material.

2) THE NORTH EUROPEAN PLAIN (Fig. 12-5)

Rising in the central European uplands are the great rivers Weser, Elbe, Oder, and Vistula which drain to the Baltic or North Seas. Their pre-glacial courses across the lowland of North Germany and Poland are not known, having been completely obliterated and altered by successive ice sheets and their deposits. The outstanding valley systems of the present-day lowland are the Urstromtäler (Germany) or Pradoliny (Poland), great meltwater

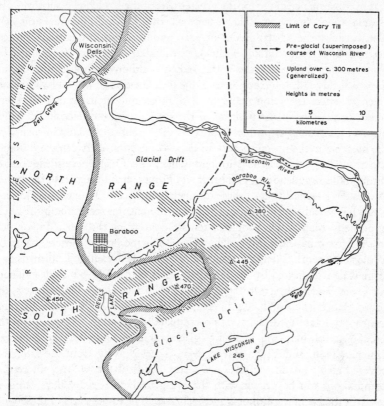

Fig. 12-6 Map of the area around Baraboo and Devil's Lake, Wisconsin, showing present and pre-glacial courses of the Baraboo river

channels formed peripherally to the Baltic ice at various stages, and now carrying lengthy segments of the four great rivers just mentioned, together with many lesser streams. The Urstomtäler, cut largely into the sandy Geest outwash, are margined by moraines on their northern sides and are extremely capacious—that south of the Brandenburg moraine is thought to have once carried a river forty times the volume of the present Elbe, and possibly larger, the water coming both from melting ice to the north and normal land drainage to the south. Of the five main Urstromtäler (numbered on Fig. 12-5), the southernmost passing Wroclaw and Magdeburg and possibly continued

in the lower Weser, is generally considered to date from the Warthe glaciation, while the remainder operated at various stages of Baltic ice retreat in
the Weichsel. All the principal Weichselian Urstromtäler, with the exception
of that lying north of the Baltic Ridge, converge on the lower Elbe which
was in consequence in use for the greater part of the Weichsel glaciation. The
Vistula, Oder, Spree, and Havel were all forced to follow courses directed
westward by the Urstromtäler in part, and now present curious zig-zag
patterns as a result. None of the Urstromtäler, however, is occupied by
continuous drainage, for the floor gradients, although very gentle, do not
show a steady fall. Interruptions are due to formation of low outwash cones
(such as where the Rinnentäler (p. 278) break into them on their northern
sides), oscillations in the position of the ice margin, and deposits of wind-
blown loess. A major problem concerns the origin of the Durchbruchtäler,
the breaches in the moraines and the cross-connections between the different
Urstromtäler, such as the segments where the Oder successively breaks
through the Brandenburg, Frankfurt and Pomeranian moraines to the north.
Woldstedt's explanation of these channels as subglacial Rinnentäler flowing
south has already been mentioned, and at present this seems the most likely
hypothesis. Many other problems remain concerning the detailed evolution
of the Urstromtäler; although the general scheme presented here, based on
Woldstedt's work, is largely accepted, the complexity of the landforms on a
large scale is indicated by the fact that W. Behrmann (1950) finds evidence
for seven Weichsel retreat stages in the Berlin area alone. Nevertheless, the
broad principles of drainage diversion by barriers of both ice and moraine
are clearly established. There is no comparable situation farther east in the
U.S.S.R., for here the main pre-glacial drainage lines of the Dniester,
Dnieper, Don, and Volga Rivers were directed away from the later ice-
covered areas, but in the northern Great Plains of the U.S.A., interesting
comparisons can be drawn with the pre-glacial west-east valleys draining
towards the ice in the Dakotas and Nebraska. The lower parts of these were
blocked and largely obliterated by successive glaciations from the Kansan
onwards, and the present Missouri River to which they are now tributary
in these states is essentially marginal to the limits of glaciation, as in the north
German Urstromtäler, though far less is known about the details of its glacial
history.

3) THE TEAYS (Fig. 12-7)

This now-vanished river system of Ohio, Indiana, and Illinois provides
an unrivalled illustration of drainage diversion resulting from drift filling
of former valleys (L. Horberg, 1945, 1956). Prior to the Kansan glaci-

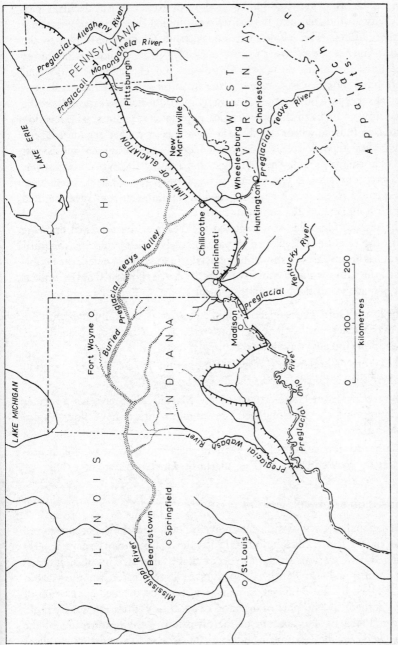

Fig. 12-7 Map of the pre-glacial Teays river system. Parts of this system were later integrated to form the present Ohio river system. The Mississippi above St. Louis is shown in its pre-glacial course

ation, as far as is known, the Teays flowed from the Appalachians past
Fort Wayne (Indiana) into Illinois where it joined the ancestral Mississippi
(Horberg, 1950). The whole of its course from Chillicothe (Ohio) to the
Illinois River at Beardstown is now buried by drift, a hundred metres or
more thick in Indiana as W. D. Thornbury and H. L. Deane (1955) and
others have shown. Oil, gas and water well records, together with geophysical surveys, have located it precisely. No important streams follow the
course now and it has little topographical expression: the one small exception
to this is in Indiana where the Wabash River has by chance encountered the
Teays valley in two places, and in each widens markedly as it crosses the soft
infill (Thornbury, 1958). The infilling of the Teays valley was not, however,
a one-stage process; it was begun in the Kansan (or earlier?), parts were
subsequently re-excavated and deepened in the Illinoian, and later reburied
beneath Illinoian and Wisconsinan drifts.

The present Ohio has now replaced the Teays as the principal drainage
line of the region; it developed in part to receive meltwater as a marginal
stream to the ice-sheets terminating on its northern side. In detail, the evolution of the Ohio is much more complex, and it is thought that the present
river is made up of the following segments:

a) the lower Ohio, of pre-glacial origin and of small size compared with the
Teays farther north; there is controversy as to whether it headed near
Madison (Indiana) or south-east of Cincinnati.

b) the short Wheelersburg (Ohio)–Huntington (West Virginia) section
where the Ohio still follows the Teays closely.

c) the section from Huntington to New Martinsville represents a former
tributary of the Monongahela, reversed in direction of flow by ice
blocking.

d) the headwaters consist of the Monongahela and Allegheny, which pre-
glacially were tributary to the Ontario-St. Lawrence system.

4) DIVERSION BY GLACIAL EROSION

The erosional activity of diffluent and transfluent glaciers has been referred
to in Chapter 7. After dissolution of the ice, the drainage may be permanently
deranged if valley floor levels have been suitably altered. This possibility has
long been noted—A. Penck (1905) thought it the reason for the curious
course taken by the Rhine at Sargans—but the first detailed application of
the hypothesis to a number of instances of anomalous drainage was by D. L.
Linton (1949a, b). His account of the diversion of the headstreams of the
river Geldie in the Grampian Highlands of Scotland provides an excellent

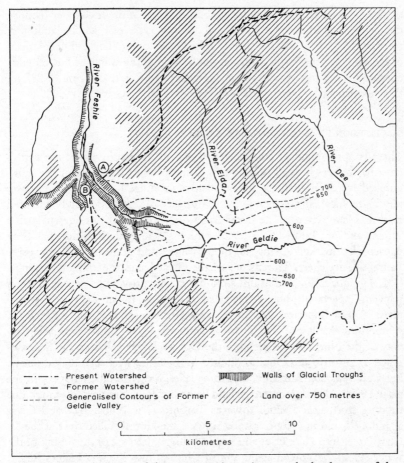

Fig. 12-8 The diversion of the upper Geldie to become the headstream of the Feshie, resulting from glacial breaching of the former watershed at A-B (D. L. Linton, *Scott. geogr. Mag.*, 1949, Royal Scottish Geographical Society)

illustration (Fig. 12-8). Mapping of former valley floor remnants shows that the Geldie once possessed headstreams rising close to the former watershed marked A-B, and converging where the present divide between the Geldie and the Feshie is situated. Linton convincingly refutes the possibility that the Feshie has captured the Geldie, particularly on the grounds that the geological structure does not favour the breaching of the former watershed A-B by headward erosion of the Feshie. On the other hand, the features and position of the breach in the watershed are entirely in accord with a hypothesis of glacial erosion in which ice accumulated in the headwater basin of the Geldie, was impeded in its normal outflow to the east, and began in part to cross the former watershed A-B at its lowest point. This diffluent stream of ice, entering the Feshie basin on the far side of the watershed, eventually effected the cutting of a breach in the watershed to a total depth of 285 m, thus permanently diverting the upper reaches of the Geldie.

5) DIVERSION BY GLACIO-ISOSTATIC CHANGES OF LEVEL

There is now abundant evidence testifying to the sensitivity of the earth's crust to changes in load (Chapter 5). The tilting of segments of the crust, induced by differential ice loading and unloading, has various important physiographical effects, such as the disturbance of certain delicately balanced river and lake systems. The Great Lakes of North America represent such a system and are located in an area where glacio-isostatic tilting has been particularly marked owing to the advances and retreats of the massive Laurentide ice-sheet. The history of the Great Lakes is extremely complex (J. L. Hough, 1958) and in this section it is only proposed to outline certain relevant aspects of their post-glacial evolution. In an early post-glacial phase, Lakes Superior, Michigan, and Huron stood at the same level, submerging Sault Ste. Marie to a depth of 15 m, and overflowing by way of the North Bay outlet and the Ottawa River valley to the St. Lawrence. Lakes Erie and Ontario drained to the St. Lawrence as they do now, but were separate from the other three lakes. Slow isostatic recovery in the period 9500-4000 B.P. raised the land in the north with respect to a hinge-line running from near Duluth towards Toronto. The North Bay outlet was abandoned, and in early Nipissing times, Lake Huron spilled over at the St. Clair River into Lake Erie; and Lake Michigan, because of the raising of the water level, overflowed at the site of Chicago into the Des Plaines River. Downcutting of the Chicago River outlet was small owing to a bedrock bar, but the St. Clair outlet in till was more easily lowered in stages until eventually, about 3000 years ago, the Chicago outlet was abandoned. The fall in water level exposed the Sault Ste. Marie bar between Superior and Huron, so that these two lakes now differ in level by 7 m. Thus isostatic tilting resulted in the drying-up of the North Bay-Ottawa outlet and the transference of drainage to the St. Clair and Chicago outlets; differential erosion rates later caused the abandonment of the latter. But the situation remains critical, for isostatic uplift is still in progress; moreover, the Chicago outlet has been artificially resurrected, though carefully controlled.

Alteration of drainage by glacio-isostatic tilting is not of course confined to the Great Lakes area, though the effects are here on a grand scale and well documented. Similar changes in lake outlets and drainage directions are known to have occurred in the area of the Baltic ice and especially in Finland, where isostatic recovery is also still in progress, raising the area near the head of the Gulf of Bothnia 60 to 70 cm a century more than in southern Finland.

REFERENCES

I. S. ALLISON (1933), 'New version of the Spokane flood', *Bull. geol. Soc. Am.* **44**, 675-722

W. BEHRMANN (1950), 'Die Umgebung Berlins nach morphologischen Formengruppen betrachtet', *Erde* **1**, 93-122

G. DE BOER (1949), 'Ice-margin features, Leirbreen, Norway', *J. Glaciol.* **1**, 332-6

D. Q. BOWEN and K. J. GREGORY (1965), 'A glacial drainage system near Fishguard, Pembrokeshire', *Proc. Geol. Ass.* **76**, 275-81

J. H. BRETZ (1923), 'Glacial drainage on the Columbia Plateau', *Bull. geol. Soc. Am.* **34**, 573-608

J. H. BRETZ (1932), 'The Grand Coulee', *Am. geogr. Soc. Spec. Publ.* **15**

J. H. BRETZ (1959), 'The double Calumet stage of Lake Chicago', *J. Geol.* **67**, 675-84

J. H. BRETZ, H. T. U. SMITH, and G. E. NEFF (1956), 'Channelled scabland of Washington: new data and interpretations', *Bull. geol. Soc. Am.* **67**, 957-1049

R. G. CARRUTHERS (1939), 'On northern glacial drifts', *Q.J. geol. Soc. Lond,* **95**, 299-333

L. CLAYTON (1964), 'Karst topography on stagnant glaciers', *J. Glaciol.* **5**, 107-12

R. COMMON (1957), 'Variations in the Cheviot meltwater channels', *Geogr. Stud.* **4**, 90-103

E. DERBYSHIRE (1961), 'Subglacial col gullies and the deglaciation of the north-east Cheviots', *Trans. Inst. Br. Geogr.* **29**, 31-46

G. DEWART (1966), 'Moulins on Kaskawulsh glacier, Yukon Territory', *J. Glaciol,* **6**, 320-1

G. H. DURY (1951), 'A 400-foot bench in south-eastern Warwickshire', *Proc. Geol. Ass.* **62**, 167-73

A. R. DWERRYHOUSE (1902), 'The glaciation of Teesdale, Weardale, and the Tyne valley, and their tributary valleys', *Q.J. geol. Soc. Lond.* **58**, 572-608

J. A. ELSON (1965), 'Western strandlines of glacial Lake Agassiz', *Rep. 7th Conf. int. Ass. quatern. Res. (Colorado, 1965), Abstracts,* 126

C. EMBLETON (1961), 'The geomorphology of the Vale of Conway, with particular reference to its deglaciation', *Trans. Inst. Br. Geogr.* **29**, 47-70

C. EMBLETON (1964a), 'Subglacial drainage and supposed ice-dammed lakes in North-East Wales', *Proc. Geol. Ass.* **75**, 31-38

C. EMBLETON (1964b), 'The deglaciation of Arfon and southern Anglesey, and the origin of the Menai Straits', *Proc. Geol. Ass.* **75**, 407-30

R. F. FLINT (1938), 'Origin of the Cheney-Palouse scabland tract, Washington', *Bull. geol. Soc. Am.* **49**, 461-523

294   GLACIAL AND FLUVIOGLACIAL EROSION

R. F. FLINT (1957), *Glacial and Pleistocene geology* (New York)

J. GJESSING (1960), 'Isavsmeltningstidens drenering, dens forløp og Formdannende virkning i Nordre Atnedalen', *Ad Novas* 3, 492 p.

J. W. GOLDTHWAIT (1938), 'The uncovering of New Hampshire by the last ice-sheet', *Am. J. Sci.* 36, 345-72

C. P. GRAVENOR and W. O. KUPSCH (1959), 'Ice disintegration features in western Canada', *J. Geol.* 67, 48-64

K. J. GREGORY (1965), 'Proglacial Lake Eskdale after sixty years', *Trans. Inst. Br. Geogr.* 36, 140-62

F. W. HARMER (1907), 'On the origin of certain cañon-like valleys associated with lake-like areas of depression', *Q.J. geol. Soc. Lond.* 63, 470-514

H. HESS (1935), 'Die Bewegung im Innern des Gletschers', *Z. Gletscherk.* 23, 1-35

W. H. HOBBS (1943), 'Discovery in eastern Washington of a new lobe of the continental Pleistocene glacier', *Science* 98, 227-30

C. HOLDAR (1957), 'Deglaciations förloppet i Torne Träsk-området efter senaste nedisningsperioden med vissa tillbakablickar och regionala jämförelser', *Geol. För. Stockh. Förh.* 79, 291-528 (with English summary)

L. HORBERG (1945), 'A major buried valley in east-central Illinois and its regional relationships', *J. Geol.* 53, 349-59

L. HORBERG (1950), 'Bedrock topography of Illinois', *Bull. Ill. St. geol. Surv.* 73

L. HORBERG (1956), 'Pleistocene deposits along the Mississippi valley in central-western Illinois', *Rep. Invest. Ill. St. geol. Surv.* 192

J. L. HOUGH (1958), *Geology of the Great Lakes* (Univ. Illinois Press)

T. F. JAMIESON (1863), 'On the parallel roads of Glen Roy, and their place in the history of the glacial period', *Q.J. geol. Soc. Lond.* 19, 235-59

P. F. KENDALL (1902), 'A system of glacier lakes in the Cleveland Hills', *Q.J. geol. Soc. Lond.* 58, 471-571

M. M. LEIGHTON (1919), 'The road-building sand and gravel of Washington', *Bull. Wash. geol. Surv.* 22, 307 p.

M. M. LEIGHTON (1959), 'Stagnancy of the Illinoian glacial lobe east of the Illinois and Mississippi rivers', *J. Geol.* 67, 337-44

D. L. LINTON (1949a), 'Watershed breaching by ice in Scotland', *Trans. Inst. Br. Geogr.* 15, 1-16

D. L. LINTON (1949b), 'Some Scottish river captures re-examined', *Scott. geogr. Mag.* 65, 123-31

D. L. LINTON (1951), ibid. 67, 31-44

C. M. MANNERFELT (1945), 'Några glacialmorfologiska formelement', *Geogr. Annlr* 27, 1-239 (with English summary)

C. M. MANNERFELT (1949), 'Marginal drainage channels as indicators of the gradients of Quaternary ice-caps', *Geogr. Annlr* 31, 194-9

C. M. MANNERFELT (1960), 'Oviksfjällen: a key glaciomorphological region', *Ymer* 80, 102-13. See also *Guidebook to Excursion E.SW.7, 19th int. geogr. Congr.*, Stockholm 1960, 29-40

W. H. MATHEWS (1964), 'Water pressure under a glacier', *J. Glaciol.* 5, 235-40

J. T. PARDEE (1910), 'The glacial Lake Missoula', *J. Geol.* 8, 376-86

J. T. PARDEE (1942), 'Unusual currents in glacial Lake Missoula, Montana', *Bull. geol. Soc. Am.* **53**, 1569-99

R. F. PEEL (1949), 'A study of two Northumbrian spillways', *Trans. Inst. Br. Geogr.* **15**, 75-89

A. PENCK (1905), 'Glacial features in the surface of the Alps', *J. Geol.* **13**, 1-19

W. E. POWERS (1946 [1947]), 'The Dells and Devil's Lake region, Wisconsin', *Chicago Nat.* **9**, 74-86

R. J. PRICE (1960), 'Glacial meltwater channels in the upper Tweed drainage basin', *Geogr. J.* **126**, 483-9

R. J. PRICE (1963), 'A glacial meltwater drainage system in Peebleshire, Scotland', *Scott. geogr. Mag.* **79**, 133-41

A. RAISTRICK and K. B. BLACKBURN (1932), 'The Late-glacial and Post-glacial periods in the North Pennines', *Trans. nth. Nat. Un.* **1**, 16-36 and 79-103

J. L. RICH (1908), 'Marginal drainage features in the Finger Lakes region', *J. Geol.* **16**, 527-48

B. ROBERTS (1933), 'The Cambridge expedition to Vatnajökull, 1932', *Geogr. J.* **81**, 289-313

I. C. RUSSELL (1893), 'Malaspina Glacier', *J. Geol.* **1**, 217-45

I. C. RUSSELL (1898), 'The great terrace of the Columbia and other topographic features in the neighbourhood of Lake Chelan, Washington', *Am. Geol.* **22**, 362-9

A. SCHOU (1949), 'The landscapes' in *Atlas of Denmark* (ed. N. NIELSEN) (Copenhagen)

V. SCHYTT (1956), 'Lateral drainage channels along the northern side of the Moltke Glacier, North-west Greenland', *Geogr. Annlr* **38**, 64-77

R. P. SHARP (1947), 'The Wolf Creek glaciers, St Elias Range, Yukon Territory', *Geogr. Rev.* **37**, 26-52

F. W. SHOTTON (1953), 'The Pleistocene deposits of the area between Coventry, Rugby and Leamington, and their bearing on the topographic development of the Midlands', *Phil. Trans. R. Soc.* B, **237**, 209-60

J. B. SISSONS (1958a), 'Sub-glacial stream erosion in southern Northumberland', *Scott. geogr. Mag.* **74**, 163-74

J. B. SISSONS (1958b), 'Supposed ice-dammed lakes in Britain, with particular reference to the Eddleston valley, southern Scotland', *Geogr. Annlr* **40**, 159-87

J. B. SISSONS (1960a), 'Some aspects of glacial drainage channels in Britain. Part I', *Scott. geogr. Mag.* **76**, 131-46

J. B. SISSONS (1960b), 'Subglacial, marginal and other glacial drainage in the Syracuse-Oneida area, New York', *Bull. geol. Soc. Am.* **71**, 1575-88

J. B. SISSONS (1961a), 'Some aspects of glacial drainage channels in Britain. Part II', *Scott. geogr. Mag.* **77**, 15-36

J. B. SISSONS (1961b), 'A subglacial drainage system by the Tinto Hills, Lanarkshire', *Trans. Edinb. geol. Soc.* **18**, 175-93

J. B. SISSONS (1963), 'The glacial drainage system around Carlops, Peebleshire', *Trans. Inst. Br. Geogr.* **32**, 95-111

R. S. TARR (1897), 'The margin of the Cornell Glacier', *Am. Geol.* **20**, 139-56

R. S. TARR (1908 [1909]), 'Some phenomena of the glacier margins in the Yakutat Bay region, Alaska', *Z. Gletscherk.* **3**, 81-110

R. S. TARR (1909), 'The Yakutat Bay region, Alaska: physiography and glacial geology', *U.S. geol. Surv. Prof. Pap.* **64**

W. D. THORNBURY (1958), 'The geomorphic history of the upper Wabash valley', *Am. J. Sci.* **256**, 449-69

W. D. THORNBURY and H. L. DEANE (1955), 'The geology of Miami county, Indiana', *J. Geol.* **48**, 449-75

E. M. TODTMANN (1936), 'Einige Ergebnisse von glazialgeologischen Untersuchungen am Südrand des Vatna-Yökull auf Island (1931-34)', *Z. dt. geol. Ges.* **88**, 77-87

W. UPHAM (1896), 'The glacial Lake Agassiz', *U.S. geol. Surv. Monogr.* **25**

O. D. VON ENGELN (1911 [1912]), 'Phenomena associated with glacier drainage and wastage, with especial reference to observations in the Yakutat Bay region, Alaska', *Z. Gletscherk.* **6**, 104-50

L. J. WILLS (1924), 'The development of the Severn valley in the neighbourhood of Ironbridge and Bridgnorth', *Q.J. geol. Soc. Lond.* **80**, 274-314

P. WOLDSTEDT (1920), 'Die Durchbrüche von Schtschara und Bug durch den westrussischen Landrücken', *Z. Ges. Erdk. Berl.* (1920), 215-25

P. WOLDSTEDT (1926), 'Probleme der Seenbildung in Norddeutschland', *Z. Ges. Erdk. Berl.* (1926), 103-24

P. WOLDSTEDT (1935), 'Bemerkungen zu meiner geologisch-morphologischen Übersichtskarte des norddeutschen Vereisungsgebietes', *Z. Ges. Erdk. Berl.* (1935), 282-95

P. WOLDSTEDT (1950), *Norddeutschland und angrenzende Gebiete im Eiszeitalter* (Stuttgart)

PART III

Glacial and Fluvioglacial Deposition

Notwithstanding their seeming disarray, there is more order about these drifts than might be thought. (R. G. CARRUTHERS, 1947)

Chapter 13

General Considerations of Glacial and Fluvioglacial Deposition

Say when, and whence, and how, huge Mister Boulder,
And by what wond'rous force hast thou been rolled here?
Has some strong torrent driven thee from afar,
Or hast thou ridden on an icy car?

(P. DUNCAN, c. 1840)

Glacier ice and its associated meltwater deposit a wide range of materials in many different forms. Some general points concerning glacial deposits are considered in this chapter and the following chapters in Part III deal with specific forms.

In any study of glacial drift, it is essential to examine the processes at work on the edge of and beneath present-day ice masses, but it should be borne in mind that modern conditions do not necessarily closely resemble those that operated around the Pleistocene ice-sheets of Europe and North America. The present-day Greenland ice-sheet, for example, is not forming moraine on its eastern margin, and whereas most of the Antarctic ice-sheet today ends in the sea, most of the borders of the northern hemisphere Pleistocene ice-sheets lay on land. The present-day ice masses are on the whole situated on old hard rocks from which any weathered material has long since been removed. They cannot, therefore, produce drift in quantities comparable with that produced by the earlier Pleistocene ice-sheets which frequently advanced over deeply weathered areas. Thus the clayey till is often more bulky in the deposits of the Pleistocene ice-sheets than it is in those of contemporary ice-sheets. Nevertheless, a close study of present-day glaciers and ice-sheets can provide valuable evidence on the processes of glacial deposition, and currently developing forms are described wherever possible in the subsequent chapters of this section.

One of the problems in studying glacial and fluvioglacial deposits is the confusion over terminology. J. K. Charlesworth (1957) for example, considers that the terms 'esker' and 'kame' are more or less synonymous, and he uses the term 'esker' to include features which some authors call kames. Thus, under the heading of 'eskers', he also discusses 'kame-terraces', 'kame-moraines' and 'osar'. In this book, eskers and kames are differentiated. Another term that causes problems is 'moraine', which has been used for a very wide range of features. Some of these are discussed in Chapter 15, but some of them, such as delta-moraines, are in reality fluvioglacial forms and these are considered in Chapter 16.

CLASSIFICATION

Glacial and fluvioglacial deposits have for long been referred to collectively as glacial drift, a term introduced by C. Lyell in 1840. The term dates from the time when it was widely believed that erratics and other glacial deposits were associated with major inundations. Lyell himself in 1841 supported the iceberg theory. It is useful to classify glacial drift into groups of sediments that have important properties in common. Further subdivision within the classification can then be based on the relationship of the deposit to the ice. The basic subdivision of drift is into glacial and fluvioglacial deposits. These groups can normally be differentiated by their sorting, for whereas water-laid fluvioglacial deposits are stratified, glacial deposits, laid down directly by the ice, lack this property. Further subdivision is based on the relationship of the deposit to the ice forming it (Table 12). The basis of such a classification

TABLE 12   *Classification of Glacial and Fluvioglacial Deposits*

Glacial drift

Stratified drift	Unstratified drift
Ice-contact deposits	Lodgement till (ground moraine)
Proglacial deposits	Ablation till (moraine)
Lacustrine deposits	Pressure (squeeze) till
Undermelt deposits	Push till (moraine)

was first suggested by T. C. Chamberlin in 1894. Ice-contact deposits are differentiated from proglacial deposits, which are laid down in front of the ice by meltwater. Lacustrine deposits, laid down in still water, differ significantly from the deposits laid down by rapidly flowing melt-streams. Undermelt deposits contain some characteristics of both lacustrine and ice-contact deposits and can best be recognized by their relationship to unstratified deposits under which they normally occur.

The terms 'till' or 'boulder-clay' are often used to denote the unstratified deposits. A. Geikie used the term 'till' in 1863. The term had been used in Scotland previously to denote a coarse stony soil which forms on the bouldery drift of the north of Britain. It is a better term than boulder-clay, as only occasionally does till consist of boulders and clay. The main distinction in unstratified drift is between lodgement till and ablation till. These have been distinguished since at least 1877. The former is deposited under the ice while the latter is let down from the ice surface on to the ground as the ice wastes away. The distinction between lodgement and pressure till is that between a deposit laid down under relatively dry conditions, often by a plastering-on process, and one laid down either under water or with material in a water-logged state. Lodgement till is sometimes referred to as ground moraine, for example by L. Agassiz in 1837, and ablation till as ablation moraine. The use of the term 'moraine' is particularly confusing. It has been used in glaciological literature from very early days, H. B. de Saussure having employed it as early as 1799, and Agassiz having introduced it to a wider public. It was originally applied to topographical features and its use by Scandinavian workers as a synonym for till has led to confusion.

Another important distinction that can be made is between deposits laid down by active ice and those accumulating from stagnant ice. The importance of this distinction has been appreciated since the beginning of this century, and the interest of stagnant ice conditions is such that a separate chapter has been devoted to the features formed in this way.

OCCURRENCE

The relative importance of these different types of glacial and fluvioglacial deposit will vary with a number of factors. Some of the most important include the nature of the land surface covered by ice, the nature of the climate, which partly determines the glacier régime, and the changes in the climate which cause advance and retreat of ice. The land surface over which the ice moves affects not only the nature of the ice-flow but also the size of the ice mass. In addition, the rocks over which the ice moves determine to a major extent the character of the drift deposited by the ice. Normally, in mountain areas, the ice is near its source and movement may be expected to be fast if the slopes are steep and the snow supply plentiful. In the regions bordering the mountains, the ice often attains its greatest thickness and can still move actively. Ice in these areas may be expected to remain active even during frontal retreat. Thus in these mountain and marginal areas, the deposits are more likely to be those deposited by active rather than stagnant ice even during retreat. Examples of rapidly retreating glaciers which are still moving forward actively at their snouts have been cited in Norway. Observations on

Austerdalsbreen (Chapter 4) have shown that the glacier is moving actively over its bed right up to its rapidly retreating snout. It is producing thin moraine that shows slight fluting, which is evidence of movement at the base of the glacier.

A situation that tends to restrict the amount of lacustrine stratified deposition is that in which the glacier margin is retreating upslope. This allows meltwater to drain away freely without being ponded in front of the glacier. Any stratified deposits that do occur will consist of proglacial out-wash, apart from subglacial forms of deposition. The well-developed valley trains that fill the floors of Alpine valleys both in the European Alps and the Southern Alps of New Zealand are examples of this type of deposit. The easily eroded rock type and heavy precipitation in the latter area also help to increase the amount of material available for deposition. In mountain areas, depositional features associated with active ice are more likely to form. These features include drumlins and various types of moraine.

When climatic conditions allowed the development of very large ice sheets, such as those that covered much of north-west Europe and North America during the major ice advances of the Pleistocene, ice moved very far from its mountain sources down on to the low ground. It was natural, therefore, that when conditions ameliorated and general recession set in, these outer parts of the ice-sheet would be the first to be separated from their accumulation areas and to become stagnant. In fact it appears that in north America, the ice margin at first retreated while the ice remained active, but that an acceleration of climatic amelioration coupled with topographical control led to widespread stagnation. R. F. Flint (1929) has shown that the presence of well-marked terminal moraines and clay-rich lodgement till in the outer parts of the glaciated area of North America indicate active frontal retreat in this area, which is 300 to 500 km wide (see p. 392, Chapter 17). The inner parts, north of the Great Lakes border moraines in north America and the Brandenburg moraines in north Europe, on the other hand, show mainly evidence of deposits laid down in connection with decaying ice, in which eskers, kames and fluvioglacial deposits abound.

The ice near the marginal zones of a large ice-sheet will probably never have been so vigorous as that near the source. This is indicated by the inability of ice masses far from their source to override even fairly low hills in their path in some instances. Where hill masses extend above a stagnant ice mass, as in the Cleveland Hills of eastern England during the last glaciation, the deposits will tend to include many ice-contact meltwater features, deposited by the drainage from the hills escaping into and beneath the ice that is still filling the low ground around the hills. Stagnant ice lobes would be expected to be fairly common on low ground far from the ice source.

The former presence of such dead ice masses can often be deduced to explain phenomena that are difficult to explain in other ways.

The formation of the trench of the Trent between Nottingham and Newark is a good example. This trench, which is about 3 km wide, is cut obliquely across the strike of the strata. It diverts the river from what appears to be an easier route, which it followed pre-glacially, across the Vale of Belvoir eastward to the Ancaster Gap and the Wash. It has been suggested that the route along the trench and through the Lincoln gap was initiated when the former route was blocked by a lobe of dead ice lying in the low ground of the Vale of Belvoir. Only an isolated lobe of dead ice could reasonably account for this diversion of the river to the north. The trench contains some fluvioglacial gravels, now part of the Hilton Terrace, to support this hypothesis. The Vale of Belvoir may have been previously deepened by glacial scouring of its soft clays when the ice was more vigorous.

When an ice-sheet becomes stagnant and decays *in situ*, it is clear that a great deal of meltwater must be produced, although sometimes a relatively large loss may also take place by evaporation. Stagnant ice conditions are favourable to the preservation of fluvioglacial forms and deposits. Such conditions, in which lobes of stagnant ice develop at the edge of retreating ice-sheets, occur at times when the ice source lies on a plateau surface. As long as the snow-line lies below the plateau surface, the ice mass will be reasonably well nourished, but as soon as the snow-line rises above the ice on the plateau, then the supply is very suddenly cut off and the whole ice mass may become virtually dead at once. In these conditions marginal decay will be extensive. These conditions appear to have occurred over part of the Prairie provinces in Canada (C. P. Gravenor, 1955), where dead ice features are very widespread.

Only a small change in climate leading to a slight raising of the snow-line can bring about the situation just described. Other climatic controls are also important to the nature of the glacial deposits. It is important to differentiate between those deposits that are associated with cold glaciers and those that occur in association with temperate ice in which meltwater can move freely. Examples of features formed along the margin of cold glaciers will be described in Chapter 15. Stratified subglacial meltwater deposits can only be associated with temperate ice masses in which it is possible for water to circulate beneath the ice (Chapter 12). For this reason many ice-contact stratified features must be associated with temperate glaciers. Nevertheless, cold-based ice-sheets can and often do have meltwater flowing along their margins and held up in ice-dammed lakes alongside them, because, being frozen to their beds, they make more waterproof dams than temperate glaciers.

Methods of Analysis of Glacial Drift Deposits

A number of techniques has been developed to study the character and origin of glacial deposits: the study of erratics, till fabric analysis, and stone shape analysis. Another method of investigation that is useful is the load test in till, for compressibility tests can often differentiate between lodgement till and ablation till. These tests are designed to assess the weight of ice under which the till was deposited; and lodgement tills show the effects of compression by a heavier load of ice than the ablation tills.

ERRATICS

Erratics were long ago recognized as glacial phenomena. They can provide valuable evidence concerning the movements of now-vanished ice masses, but they can also pose very difficult problems, such as that of the origin of the very large erratics that occur on the north coast of Devon and the south coast of Cornwall. It has been suggested that these rocks reached their present coastal position on icebergs driven ashore when sea-level was at about its present level during one of the earlier glaciations. However, it is the erratics in the tills that provide the most useful evidence of ice-flow.

Erratics vary greatly in size. Some rocks, such as Skiddaw slates, do not yield large recognizable erratics, as the rocks break up too easily. The ability of erratics to survive is proportional to their size. Very large erratics have been recorded: for example, many erratics in Scotland are over 100 tons in weight, and one in Caithness, of Cretaceous rock, is 200 m by 137 m by 8 m. In Northumberland, an erratic raft of Great Limestone is 800 m long; at Great Potton, there is one of Lincolnshire limestone over 300 m long, and near Melton Mowbray, a Lincolnshire Oolite block 275 m by 90 m. There are also Carboniferous Limestone erratics 90 m long in Anglesey and a grit one 180 m long near Abergavenny. These blocks are, however, small compared with the vast 'Schollen' of Germany, the largest of which are 4 km by 2 km by 120 m in thickness. They are formed mainly of Tertiary and Cretaceous clays and sands. Large erratics also occur in North America; for example in south-west Alberta, one quartzite erratic measures 24·5 m by 12 m by 9 m and weighs 18,150 tons.

These vast erratics illustrate clearly the immense transporting power of ice, but there remains the problem of how they got into or on to the ice sheet. Some may have slipped on to the ice from steep rock walls, and some may have been quarried off the lee faces of scarps, for example, those of Marlstone and Lincolnshire Limestone in eastern England. In soft rocks, freezing of the ground is a necessary preliminary and the thin blocks appear to have been sheared off along bedding planes or joints in many instances. A good example of the prising-off of erratics is seen in north-west Yorkshire where Silurian boulders lie on Carboniferous Limestone at Norber Brow. If the erratic train is traced back to the Silurian outcrop, the boulders become more angular and the nearest have clearly only been moved a very short distance from their outcrop. The beginning of the process is seen in enlarged rectangular cracks developed in the Silurian bedrock as the prising process started to operate. The Norber boulders show that ice can also move erratics uphill, as they lie at a higher elevation than their outcrop. This is most likely to occur where the ice-shed does not coincide with the earlier watershed. In western Scotland, for example, Torridonian sandstone has been carried up 450 m west of the Fannich Mountains. Uplifts are sometimes steep, such as 300 m in 4 km east of Loch Maree. Uplifts of up to 1000 m occurred in the Baltic area and 1525 m has been recorded on the stoss side of Mount Washington and 1615 m in the Winnipeg area.

The ability of ice to move uphill is also well demonstrated in the distribution of Shap granite erratics. Their distribution pattern in the neighbourhood of the outcrop of the granite has been studied by S. E. Hollingworth (1931) who has deduced interesting patterns of ice movement based on this distribution and drumlin alignment. He shows that the Shap granite erratics do not extend north of a line from Penrith to Melmerby. The form of the drumlins in this area, which is discussed in Chapter 14, shows however that basal ice was moving north throughout this part of the Eden valley. To explain this anomaly, Hollingworth suggests that the basal ice, forming the drumlins and carrying the Shap granite erratics northward, must have moved up along shear planes in the ice until it reached the upper layers of the ice-sheet. The movement of the upper ice is said to have been determined by the slope of the ice surface. The ice appears to have reached its maximum elevation along a line extending from near Helvellyn to Cross Fell across the Eden valley and to have moved north and south from this position. Thus the Shap granite erratics appear to have been carried first by the basal ice and then to have moved up into the upper layers by the time they reached the area around Penrith. From here they moved southward with the upper ice towards the Stainmore depression. This col was a large funnel into which ice from 15 km around was channelled. The evidence for the passage of ice by this route from the Lake District is very clearly indicated by the large trail of Shap granite boulders that stream away from the outcrop north-eastward to the Stainmore gap. The Shap granite erratics do not penetrate north of the position at which the powerful glacier from the Ullswater valley met the northward-moving ice stream. It is possible that this glacier was one of the agents that forced the northward-moving ice to override it and thus enter the upper ice circulation.

A basal ice-shed exists in the neighbourhood of Appleby from where the southward- and eastward-moving ice flowed uphill towards the Stainmore col (Fig. 13-1). Through the gap, the train of Shap granite boulders can be followed down to the limit of the Newer Drift ice advance that built the York moraine. Shap granite erratics were also moved southward down the Lune gorge near Tebay, although there is evidence here too of northward-flowing basal ice in the form and erratic content of the drumlins between Shap and

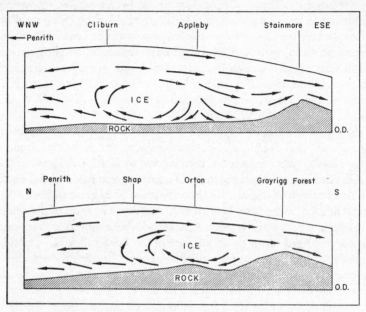

Fig. 13-1 Hypothetical ice flow in north-west England to account for the distribution of Shap granite erratics and drumlins (S. E. Hollingworth, Q. J. geol. Soc. Lond., 1931)

Tebay. Ice moving over Shap from the south must, therefore, have moved up from the basal layers into the upper part of the ice-sheet to follow the slope of the ice surface southward through the Lune gorge in which the ice would once more reach the ground level. Diagrammatic sections, shown in Figure 13·1, illustrate the type of ice movement reconstructed from the evidence of Shap granite erratics and the form of drumlins.

Other erratics indicate very extensive ice movement. The riebeckite eurite of Ailsa Craig off the west coast of Scotland has been found in abundance along the east and south-east coast of Ireland and on the north and west coasts of Wales and around Anglesey, showing clearly the movement of the Irish Sea ice-sheet. The rhomb porphyry stones found on the east coast of England in Yorkshire and Durham could only have come from Norway and give good evidence of the spread of the Scandinavian ice-sheet on to the British coast at at least one stage in the glaciation.

On the west coast of Baffin Island in Arctic Canada, limestone erratics, derived from Foxe Basin, were carried eastward to the mainland of Baffin Island. They give evidence of the existence of a large ice centre over Foxe Basin from which ice spread out in all directions. The maximum extent of this ice sheet occurred about 8000 to 9000 B.P. Over the central western part of Baffin Island, the direction of flow at this time was the reverse of that which set in later. By about 7000 B.P., an enlarged Barnes ice-cap was flowing westward to Foxe Basin in the opposite direction to the earlier ice. This later ice mass only reached the position of the present coast in many parts of the western side of the Island.

TILL FABRIC

The fact that stones in till are not randomly distributed was noted as early as 1859 and was discussed by H. Miller in 1884. The analysis of the direction and dip of elongated stones in till is being increasingly used to study both the direction of former ice-flow and the processes of glacial deposition. It is generally agreed that the elongated stones in till become orientated parallel to the direction of ice flow, although this is not always true and the problem of till fabric analysis is a very complex one. It is necessary to treat the data statistically to obtain a valid idea of the orientation pattern. It is also desirable to obtain several sets of samples from any one site as variations over short distances can be considerable.

Some valuable work on till fabric analysis was carried out by C. D. Holmes (1941) who attempted to analyse the effect of shape on the orientation of stones in till. He divided stones into six shape classes, discoid, ovoid, tabular, wedge-form, rhombohedroid and varihedroid, and four roundness classes, sharply angular to well rounded. The measurements were plotted on the lower hemisphere to indicate both direction and dip of the stones in a 360-degree pattern. A composite plot of all the data collected showed that the greatest number of stones lay parallel to the direction of flow of the ice and they mostly had a low angle of dip, one-third of the stones dipping at less than 10 degrees. In many till fabric diagrams, there is a minor concentration of stones transverse to the flow direction. The stones parallel to the flow move by sliding, but the transverse stones move by rolling about their long axis which is transverse to the ice-flow. Many stones with a steeply dipping intermediate axis become orientated parallel to the flow. The ovoids that are well rounded usually maintain the best orientation parallel to the ice-flow. A useful means of assessing the method of movement of the stones in the ice is to examine the striae on the measured stones. On wedge-shaped stones, they run parallel to the edges and not to the long axis. This suggests that these stones move mainly by sliding. Large, not very elongated, stones often have a strong transverse pattern. More elongated, well-rounded, stones of rhombohedroid shape are particularly adapted to rotational movement, and hence also tend to lie transverse to the flow. The very well-rounded and shorter stones of this form tend to lie parallel to the flow. Because of the wide variety of stone shape and the many factors influencing their emplacement in the till, the fabrics usually show a considerable scatter. From this extensive study, Holmes reached the following general conclusions. He considered that ground moraine accumulated gradually and that stones carried above the glacier bed were subject to rotation, generally about their long axes, but sometimes about the other axes. The stones in contact with the floor moved generally by sliding with both their long and intermediate axes parallel to the floor plane and the long axis parallel to the movement except for wedge-shaped stones. The degree of stone roundness may thus have a very profound effect on the preferred orientation of the stone, while shape also plays a very important part.

More recently, P. W. Harrison (1957) has studied a clay-till fabric near Chicago, Illinois. He measured the maximum projection plane of shale particles 3 to 40 mm in length. Disk- and blade-shaped particles were mainly selected. Harrison came to the conclusion that much of the fabric in this clay-rich till is inherited from the fabric that developed when the ice was moving. The mean dip of the stones seemed to be related to vanished shear-planes in the heavily laden basal ice, the planes dipping upstream. Along these shear planes, ice tends to override the basal ice that has become static. Stones in such planes would become orientated parallel to them. This orientation also occurs in the ice around the plane, as observed in Greenland today. In the till fabric as measured when the ice has melted, the strength of the orientation is probably partly lost by melting but in basal ice which is laden very heavily, the proportion of ice to debris may only be about 25 per cent. The clay also contains a considerable amount of water and consolidation tests on the till showed that the overlying ice was not more than 90 m thick. Thus it is possible for the ice to retain its fabric from the time when it was moving, if it becomes stagnant and decays slowly in situ. End moraines show similar till fabrics which also appear to have been inherited from the time when the ice was moving up thrust-planes as it deformed internally to move over already stagnant ice. Some of the moraine fabrics studied by Harrison showed no up-glacier dip and the stones were dipping with the surface slope of the moraine.

These fabrics had probably been induced by mudflows down the slope under periglacial conditions.

The bulk of the ground moraine does not appear to have been disturbed since its deposition and Harrison considers that the till fabric indicates that the till accumulated slowly as the ice stagnated by melting from below. The till does not appear to have accumulated by the plastering-on of material beneath moving ice. Thus if this pattern is widespread, it suggests that most till deposition takes place during decay of the ice during deglaciation. The till does, however, retain the pattern imposed upon it when the ice was able to move, giving the common up-glacier dip of the stones, related to the upward shear of ice along planes near its margin where the load is high.

Till fabric analysis has been used by R. G. West and J. J. Donner (1956) to differentiate the Lowestoft and Gipping till sheets discussed by D. F. W. Baden-Powell (1948) in East Anglia. West and Donner measured both the dip and direction of 100 stones at each site in undisturbed till. Most of the samples showed a strong preferred orientation when plotted on a symmetrical diagram showing only orientation. The dips on the whole were balanced on either side of the diagrams. Those samples which did not show this balance appeared to have been disturbed by solifluction. Many of the undisturbed stones had a very low dip and many were horizontal. As a result of this detailed till fabric analysis, West and Donner were able to confirm that the ice depositing the Lowestoft till advanced generally from the north-west over the Midlands and from the west over East Anglia, while the succeeding Gipping ice came from the north or the north-east over the Midlands, but east of Cambridge it came from the north or the north-west.

J. W. Glen, Donner, and West (1957) have discussed the processes that lead to the preferred orientation of stones in till. On theoretical grounds, elongated pebbles placed at random in a flowing liquid should rapidly become orientated with their long axes parallel to the flow, but, after a long period, the particles turn so that their long axes are transverse to the flow. This is the position of theoretical minimum energy requirement. They suggest that parallel orientation is produced by free flow in the ice, by collision of oblate stones and by dragging on a stationary layer or over a shear-plane. Transverse orientation should be developed by protracted flow and by collision of prolate stones. Orientation unrelated to ice flow can be induced by the disturbance of the deposit when the ice melts or solifluction occurs.

Observations on currently developing tills in Nordauslandet (Spitsbergen) suggested that thick till which showed a parallel peak could most likely have been formed by the plastering-on process. Narrow bands of till between ice layers showed a transverse peak that could have been produced by long-continued flow or by collisions in suitably shaped stones. This method of analysis, which takes into account only the long axis direction and not the dip of the stones, although justified by these authors, probably omits some valuable evidence concerning the nature of the fabric. The value of the dip measurement was apparent in the analysis by Harrison and it provides a means of eliminating fabrics disturbed by solifluction as indicated by West and Donner.

J. T. Andrews and B. B. Smithson (1966) have shown the value of dip measurements in addition to orientation values in their study of the fabric of the cross-valley moraines of north central Baffin Island (see Chapter 15). These moraines consist of till which is unstratified, but which has been emplaced subaqueously at the margin of the retreating ice-cap by a pressure or squeezing process. The orientation pattern is such that the preferred direction lies normal to the length of the moraine in most types studied. The dip pattern is much weaker on the steeper distal slope of the moraine. It is argued that the strong up-glacier dip, characteristic of the proximal side, indicates that liquid till may have been forced up into ridges at the edge the ice-cap. This forcing-up of the ridges would have occurred when th summer melt produced a large volume of water, which liquefied the sub glacial material and allowed the ice to settle into it and to push up the marginal material into a moraine ridge.

Valuable information may also be obtained much more readily by measuring the orientation of the long axes of surface boulders in a till sheet or moraine. Andrews (1965) has shown that the orientation pattern of the surface stones shows a preferred value that agrees closely with the direction of ice-flow. The orientation strength on the till sheets was consistently strong, and stronger than those found on the surface of moraine deposits. The weakest orientation patterns were found on outwash plains where the boulders tended to show a weak orientation perpendicular to the water-flow direction. Samples of moraine stone orientation measured in west Baffin Island showed a statistically significant orientation at right-angles to the moraine crest in many instances. The stones, therefore, lay parallel to the ice-flow direction.

The study of till fabrics and orientation of surface stones provides a valuable method of reconstructing former ice movement as well as elucidating the processes of till deposition, but it is necessary to analyse the data statistically if valid results are to be obtained. Three-dimensional till fabric analysis is discussed by Andrews and K. Shimizu (1966).

SHAPE ANALYSIS

A study of the roundness of stones can give useful information concerning the nature of the feature in which the stones are incorporated. Some information is also provided about processes of deposition. There are several methods of measuring the shape of a stone, but an analysis of three methods indicates that A. Cailleux's formula (1945) for roundness is the most useful. The first of the methods is based on Cailleux's roundness formula, which is given by $\frac{2R}{a} \times 1000$, where R is the minimum radius of curvature in the principal plane and a is the long axis of the stone. Secondly, W. C. Krumbein (1961) has suggested a sphericity measure $\sqrt[3]{\frac{bc}{a^2}}$, where a, b, and c are the long, intermediate, and short axes respectively. This measure is closely and inversely related to the third measure, Cailleux's flatness formula, given by $\frac{a+b}{c} \times 100$.

Cailleux's roundness formula was applied to samples of 50 stones taken from a variety of different glacial and fluvioglacial features in west Baffin Island. The deposits sampled included material from fluvioglacial outwash deltas, eskers, kames, moraines, and solifluction material. Analysis of variance tests showed that the roundness of stones in the different deposits was significantly different. Pairs of samples were also tested by using the Student's t test to ascertain whether the roundness values were significantly different. The difference was not related to rock type as all the samples tested consisted of granite-gneiss from the basement Complex which outcrops in this part of Baffin Island. Some of the results are shown in Table 13.

TABLE 13   *Analyses of Stone Roundness for Various Glacial Deposits*

Pairs	Means, mm		Differences, mm	t	Degrees of freedom	Significance level
Deltas, eskers	334	332	2	0·14	22	Not significant
Deltas, kames	334	238	96	5·65	16	0·001
Eskers, kames	332	238	94	6·86	13	0·001
Moraines, kames	138	238	100	5·44	10	0·001
Moraines, eskers	138	332	194	12·60	13	0·001
Moraines, deltas	138	334	196	10·90	16	0·001
Moraines, solifluction	138	185	47	1·68	7	Not significant

It is interesting to note that deltas and eskers cannot be differentiated by roundness. This indicates that their stones have undergone a similar amount of rounding by meltwater streams. The eskers must have developed in close proximity to the ice and their stones cannot have been carried very far by meltwater. It is unlikely that meltwater can exist far beneath cold-based ice-caps such as the Barnes ice-cap in Baffin Island. Rounding in these fast-flowing, heavily-laden melt-streams must, therefore, be very effective to produce such a marked degree of rounding in such a short distance. The size of the largest boulders also points to this.

Table 13 also suggests an important difference in the methods of formation of eskers and kames. Some rounding of stones has taken place in the kames, as they are significantly rounder than the moraine stones, but they can only have been carried a very short distance by meltwater before being deposited at the ice margin. The meltwater was probably not concentrated into powerful streams where the kames were formed.

The lowest mean roundness was found in the moraines. This low value suggests that the morainic material was carried either on top of the ice or was frozen beneath the base of the ice. Most of it probably only started to be abraded a short distance and time before it was deposited as moraine. In temperate glaciers, rounding is probably more effective. This is indicated by a significant increase in stone roundness over a distance of about 1 km in recent moraines laid down in front of Blea Water under High Street in the Lake District. More active movement of ice over its bed would be expected in these circumstances and this would produce more effective stone rounding. The material in this case was from the Borrowdale Volcanic Series.

Drift Stratigraphy and its Analysis

A very great deal of work has been accomplished in the analysis of drift stratigraphy. The aim of these studies is to ascertain the nature and sequence of Pleistocene events and to correlate these from place to place. The analysis of glacial deposits is not only concerned with the stratigraphy but the form of the deposits. Their attitude must also be taken into consideration. For example, it is recognized that the deposits of the Newer Drift (Würm, Weichsel, Wisconsinan) period have much fresher forms than the older deposits. Newer Drift deposits show original constructional forms and occur within the valley floors. The deposits of the older glaciations tend to show mainly erosional forms and they are often missing from the valley bottoms because of later river down-cutting. They therefore tend to have interfluve distributions. However, the most valuable evidence used to build up Pleistocene chronology is derived from a study of the drifts themselves. Data may be obtained from exposures, boreholes, and augering. In studies of this type, the sequence of different types of deposits is very important and valuable evidence is obtained by studying the weathering of the drifts (Chapter 1). It is not intended to discuss drift chronology, but merely to consider the possible interpretation of special drift sequences, to comment briefly on weathering of drifts and to give one example of the analysis of a sequence of drifts which reveals a complex pattern of events of lasting importance.

THE UNDERMELT DRIFT SEQUENCE

R. G. Carruthers (1947-8 and 1953) has described a series of drifts which he considers reveals a complex process of deposition, occurring when an ice-sheet becomes stagnant and melts *in situ*. The sequence of drifts contains both stratified and unstratified deposits. The stratified undermelt deposits can be differentiated from lacustrine deposits by certain special characteristics. A complex layer of drift often contains a basal till overlain by stratified deposits and capped by another layer of till. The base of the stratified series frequently contains clay deposits. The term 'shear clay' has been used to describe these clays, which consist of extremely sharply divided but very thin layers of clay, giving a laminated effect, which occurs without rhythm and can be as fine as 0·01 mm. Small particles in the clay show equal flow around them, the underlying and overlying laminae passing under and over the particle respectively. Some of the layers show crumpling or minute faults and some patches are very contorted. There is no trace of organic material in the clays, but they may contain very thin sand partings. The thicker sand layers may show current bedding. The fine definition of the clay layers and the lack of graded bedding suggest deposition by moving water. Where the sand layers show no structure, they are probably secondary, being laid down as the ice slowly melted. The sand usually increases in amount upwards.

The suspension structures are an interesting feature of the deposits. These developed where the depth of flowing water was rather greater. The clay was sometimes held firm until a late stage in the melting process and, when it did melt out, the lower edge remained unbroken, but the upper one developed a flame-like appearance. In this form it became incorporated in the rapidly accumulating sand. At times, roof-falls produced hour-glass-like forms which provide a valuable clue in identifying the nature of the deposit, because they do not appear to form under other conditions.

At the top of the sequence there is often a gravel and upper till layer. The contact between the tills and the sand zone beneath is important in their interpretation. The contact sometimes forms a clean, straight line, while at other times there are wisps of till incorporated into the sand. Sometimes there are thin lenses of sand within the lowest layers of the till. Such forms do not suggest that the upper till was laid down by a readvance of the ice, but that the two deposits are intimately associated at their junction. The upper till is not always ablation moraine, as it may have the character of ground moraine. Carruthers suggests that it could have been introduced by an overriding glacier or be material that has moved up from the base of the ice-sheet. Ice in which a large amount of material was incorporated would resist melting while the sand accumulated beneath it. Where the roof remained firm, a straight contact would result. However, if the till was also melting, parts of it would be incorporated into the sand below. Finally, the till would melt out at the top of the section.

This type of sequence of deposits could be interpreted, therefore, as evidence for the gradual melting from below of a complex and stagnant ice-sheet. Shear clays only occur on the lower ground as the fine clay requires special conditions that are not often present in areas of greater relief. Good examples may be seen in the area between Darlington and Stockton in north Yorkshire. Carruthers suggests that the deposits are formed by shearing action in the lower layers of the ice where it is heavily loaded. The silts and sands are not always present and only occur where meltwater streams were concentrated. Thus in many areas, only till is found. The intermediate sand deposits, which are very extensively developed in the drift area of the Cheshire plain, may have formed in a similar way.

WEATHERING OF TILLS

The study of the weathering of the drift is an important aspect of drift stratigraphy in establishing glacial chronology. Different types of drift will weather in different ways: Two main processes are important in the weathering of tills. Oxidation occurs first and extends deepest. It consists of the alteration of ferrous compounds to oxides, giving the soil a yellow, red or brownish colour. Leaching consists of the solution of carbonates by percolating waters. Ancient soils, sometimes called 'palaeosols', are those that are not related to current processes. They provide a valuable means of establishing glacial chronology. The Sangamon palaeosols in North America provide a good example: they are extensive and well preserved, and vary with different conditions as do modern soils covering the same area. Thus they vary from podsols south of the Great Lakes to chernozems in the Great Plains area, and in suitable conditions gumbotil also occurs.

If the tills are calcareous, they will become progressively more decalcified by leaching with time and the depth of decalcification will be at least a partial measure of their age. In more clayey tills, such as those of parts of the United States, weathering produces special characteristics in the till. These deeply weathered clay tills were defined and called gumbotils by G. F. Kay (1916). Their recognition in a number of sections has enabled the till sheets of Iowa to be differentiated and associated with the main advances of the Nebraskan, Kansan, Illinoian, and Wisconsinan glaciations. Their characteristics have been discussed more recently by J. C. Frye and others (1960) and by A. C. Trowbridge (1961). The latter lists the characteristics of gumbotils by which they can be recognized and their genesis identified. Gumbotil is a compact sticky clay when wet, and hard when dry. It is completely leached and contains no easily weathered materials except at the base. When it is wet, it breaks up into small, shiny-sided polyhedral pellets. It is massive, not stratified, and rarely exceeds 5 m in thickness. The pebbles that it contains are mainly of materials that do not weather readily, such as quartz and chert, and they become more numerous towards the base of the deposit. The base of the gumbotil is not sharply defined, but it grades down through oxidized and leached till, to oxidized and unleached till, and finally to unoxidized and unleached till. Sometimes there are ghost pebbles of crystalline rocks in the lower part of the gumbotil. At its upper limit the layer of gumbotil is overlain by unweathered till or loess. These characteristics indicate that the gumbotil is the product of a long period of weathering of clayey till and its presence indicates an interglacial period. It also provides a method of differentiating the older and newer drift because the Wisconsinan tills never have a gumbotil layer at their upper limit. The recognition of weathering in drifts provides a valuable tool in chronological analysis.

AN EXAMPLE OF DRIFT STRATIGRAPHY ANALYSIS

The example of drift stratigraphy selected for brief review is from the work of F. W. Shotton (1953). The area lies between Coventry, Rugby and Leamington and the deposits of this area provide evidence for a complex series of events that led to a complete change in the relief and drainage. The main sequence of deposits is shown in Table 14. The Bubbenhall clay is very restricted in outcrop and is largely a stoneless, red silty clay, laminated in at least one auger hole and containing pebbles in some places. It contains Bunter pebbles that denote a northerly source. Its deposition was followed by a long period of erosion during which much of it was removed. The overlying Baginton-Lillington gravels were derived almost entirely from the Bunter. The deposit thins towards the south-west. These gravels were deposited on an uneven surface and are overlain, without break in some places, by the Baginton sand. The two deposits show current and deltaic bedding in the lower parts, but in the upper part the bedding is level, which suggests that ponded conditions were beginning. The sand passes undisturbed into the Wolston clay, supporting the view of ponded conditions. The deposits contain a cold steppe fauna. They become thicker to the north-east and were laid down in a valley that drained in this direction. The stones in the gravels indicate that several feeder rivers probably carried them into the area. The

TABLE 14   *The Sequence of Pleistocene Deposits around Coventry, Rugby and Leamington*

Newer Drift	Modern alluvium	
	River terraces	⎧ Avon No. 1 ⎨ Avon No. 2 ⎩ Avon No. 3 Avon No. 4
	Long time interval	
Older Drift	Dunsmore gravel	
	Wolston series	⎧ Upper Wolston clay ⎨ Wolston sand ⎩ Lower Wolston clay
	Baginton sand Baginton-Lillington gravel	
	Long time interval	
	Bubbenhall clay	

reduction in calibre of the deposit upwards shows increasing influence of ponded conditions as ice advanced from the north.

This ice advance is clearly seen in the Wolston series. The deposits are up to 16 m thick. At their base is a stoneless clay laid down in still water. Above the lowest metre, there are occasional pebbles, but the bulk of the deposit is a stoneless red or pink clay, with evidence of varves in places. The clay overlaps and covers a larger area than the Baginton-Lillington deposits and is interpreted as a lake deposit. Nevertheless in the upper part, a small amount of true till also occurs, owing to the spasmodic advance of the ice over its own lake deposits. The rock types contained in the deposit suggest that a northern ice mass was displaced by one coming from the east, as flints increase in number in the upper part, becoming of equal importance with the Bunter pebbles. This tendency continues into the overlying Dunsmore gravels. The Wolston sand between the two clay layers indicates a temporary retreat of the ice with more summer melting carrying coarser deposits. The Dunsmore gravel is the highest of the Older Drift deposits and caps a flat plateau between the Avon and the Leam. It slopes gently to the west-north-west to 98 m O.D. The deposit is waterlaid and contains 8 cm pebbles, mainly of flint. It nowhere occurs in the same place as the eastern till and these two deposits are mutually exclusive. The Dunsmore gravel appears to be the outwash from the eastern till sheet.

The terrace deposits are everywhere at a lower elevation than the Older

Drift Dunsmore gravel. The terraces are related to a river flowing south-westward, while the Older Drift river deposits were laid down in a north-east draining valley. The terrace deposits will not be considered in detail.

In interpreting the events that led to the deposition of these sediments, Shotton recognized the presence of a large lake, called Lake Harrison, in the present Avon valley. The Wolston clays give evidence for this lake. Clays outside the area studied in detail show that the lake was very extensive, reaching from just north of Leicester to Moreton in the Marsh in the south, its maximum length being 90 km (see Chapter 19). The water level reached a maximum height of 125 m O.D. when it overflowed the Jurassic scarp to the east, as its southern outlet was dammed by western ice. The present-day watershed between the Soar and the Avon lies at the site of a deep pre-glacial hollow where 66 m of drift have been proved by boring. The pre-Chalky till surface has been mapped by using the evidence of boreholes and augering. The map shows a broad valley rising to the south-south-west to about 88 m O.D. near Bredon Hill, with a gradient of 12 m in 40 km. The present Soar is, therefore, very much shorter than it was pre-glacially and the Warwick-shire Avon is a very recent river. The Avon was initiated by the outwash streams of the retreating eastern ice that eventually overran the lake deposits and advanced as far as Moreton. The detailed study of the drift stratigraphy in this relatively small area, using borehole data and augering as well as examining the available sections, has revealed a complex and important sequence of events.

Drift Tectonics

Some drift deposits show tectonic disturbance producing either folding and contortion of the strata or faulting. Some of the structures, such as over-thrusting or compressional faults, may be the result of pressure by advancing ice against earlier deposits. Examples of this type of structure are described in Chapter 15, in discussing push moraines. Some small-scale faulting may also be the result of the melting-out of lumps of ice incorporated in deposits, as for example in the kamiform deposits near Carstairs in south Scotland. Other types of fault, particularly normal faults, may require different explanations, such as those put forward by Shotton (1965).

Some of the faulting described by Shotton is on a fairly small scale. At Strett-on-Fosse, a section and auger holes have revealed a succession of normal faults giving rise to small horsts and graben. The displacements are up to about 3 m. At the base of the graben, the underlying Lias clay bulges upward. In this type of fault the vertical force is the dominant one. The pattern of the faulting in relation to the deposits shows that dislocation occurred at several periods. Ice loading does not appear to be a satisfactory explanation for the faulting because it would have to vary over such short distances. The force that Shotton considers could have produced the observed features is the tension induced by the cooling of frozen ground in winter (see Chapter 21). The fault character suggests that the ground was frozen when the movement occurred. The tension resulting from shrinkage in the winter would be reversed by expansion and compression during the summer. This latter movement appears to have been taken up by movements between the grains of the sands and gravels. The orientation pattern of the faults, which appear to meet at 60-degree angles, also suggest that tension resulting from contraction has been important in their formation. They appear to be fairly common in the Older Drift.

Other examples of normal faulting and shear plane development have been described by B. J. Taylor (1958) in the Middle Sands of Lancashire and Cheshire. These are water-laid sands lying between two till sheets, belonging to the Irish Sea Glaciation in the Newer Drift period. The disturbance of the sands is of two types. First, near the top, the sand beds have been folded and tilted and small-scale thrusting has occurred. The movements are thought to have been caused by lateral shear-stress imposed by advancing ice. Below these features, the second type of disturbance consists of reticulate shear planes penetrating several metres into the sands. Some of these shear planes penetrate 10 m into the sands and are inclined at 60 degrees either side of the vertical. Most of the faults are normal. The pressure of the ice, which must have been a few hundred metres thick, compressing the sands beneath, which were probably water-logged and frozen, caused shear planes to develop. These planes shortened the vertical dimension. Most of the planes lie between 30 and 60 degrees to the vertical. These features are not present in the Upper Sands which overlie the upper till. This interpretation differs from that already suggested to explain the character of the Middle Sands (p. 314).

Some of the fault structures described by Shotton are considerably larger in scale. He has examined a section exposed in motorway construction near Kilsby, Northampton. The glacial sequence in this area is 30 m thick at least and is of Saale (Gipping) age. It consists of coarse gravel separating two tills. Normal faults have let down the upper till at least 12 m into the gravels. Similar examples of large-scale normal faulting have been revealed in cuttings made along the M1 motorway near Narborough in Leicestershire. The fault throws in this section are from 7·5 to 21 m, and the faults continue down into the Keuper Marl beneath. It seems likely that these faults are reactivating earlier fault patterns, as those near Narborough are fairly closely connected to the fault pattern in the nearby Charnwood Forest area. The weight of the ice-sheet that deposited the tills is thought to be a possible cause of the reactivation of old faulting. The ice must have been of considerable thickness in this area and it seems likely that it caused isostatic depression of the ground. The faulting took place after the uppermost of the drift deposits was laid down. The updoming as a result of the reduction of pressure when the ice-sheet melted seems to be the most likely cause for these faults. The faulting probably took place quite quickly after the retreat of the ice because there is no indication of the faults at all in the present landscape, neither do they appear to affect the terraces that were formed soon after the ice retreated and in the subsequent interglacial. Tectonic movement in drifts can, therefore, have a variety of origins.

Conclusions

Glacial deposits provide very valuable evidence concerning processes operating in and beneath ice masses and they also play a very important part in modifying the landscape of areas formerly covered by ice-sheets and glaciers. Some of the ways in which these deposits can be analysed have been mentioned and the following chapters enlarge upon some of the specific forms and types of deposit. An example of the analysis of drift stratigraphy has been given and this example shows how profound can be the changes wrought by the ice-sheets and their deposits.

These deposits cover about one-third of Europe, nearly one-quarter of north America, and 8 per cent of the earth's land surface. They have produced much new land in Britain (K. M. Clayton, 1963). The eastern parts of Yorkshire, Lincolnshire, and Norfolk, for example, would be under the sea if it were not for their thick and extensive glacial deposits. Drift thicknesses vary enormously over the glaciated area. Flint (1957) gives the following figures for different parts of North America: Great Lakes area—12 m, Illinois —35 m, Iowa—46 to 66 m, Central Ohio—average 29 m, maximum 232 m, New Hampshire—average 10 m, maximum 122 m, Spokane valley, Idaho-Washington—335 to 442 m. Charlesworth (1957) records a wide variety of estimates of drift thickness. For example, the mean drift thickness in north Germany has been given as 58 m from 470 boreholes, Norrland (Sweden) 4 to 7 m, Denmark 50 m. In some places, exceptionally thick deposits have been recorded, such as 470 m near Lubbendorf in Mecklenburg, 288 m near Rostock, about 400 m near Grenoble, 397 m near Heidelberg, and about 800 m near Imola in the Po valley. In East Anglia, 143 m has been recorded and 175 m in the northern part of the Isle of Man. The greatest thickness recorded in North America is 670 m in the Fraser Delta area, though this consists mostly of outwash.

In all the areas that were once covered by ice, the soils have been modified or completely changed. In some areas, former soils have been swept away, but in many areas new drift soils have been laid down. The character of the drift is usually very important in determining the soil type, which in turn has an influence on vegetation and many other aspects of the geography of the area concerned. Glacial deposits are of interest for their own sake, but also for their widespread influence on many other facets of the landscape. Whole new landscapes have been created by the burial of the solid rocks beneath the thick mantle of glacial deposits. Sometimes this occurs with intriguing surface forms, such as drumlins or eskers. Even the smooth till sheets hide much valuable information within them.

REFERENCES

L. AGASSIZ (1837), 'Des glaciers, des moraines, et des blocs erratiques', *Verh. schweiz. naturf. Ges.* 22, v-xxxii

J. T. ANDREWS (1965), 'Surface boulder orientation studies around the north-western margin of the Barnes ice cap, Baffin Island, Canada', *J. sedim. Petrol.* 35, 753-8

J. T. ANDREWS and K. SHIMIZU (1966), 'Three-dimensional vector technique for analysing till fabrics: discussion and Fortran program', *Geogr. Bull.* 8, 151-65

J. T. ANDREWS and B. B. SMITHSON (1966), 'Till fabrics of the cross-valley moraines of north-central Baffin Island, North West Territories, Canada', *Bull. geol. Soc. Am.* 77, 271-90

D. F. W. BADEN-POWELL (1948), 'The chalky boulder clays of Norfolk and Suffolk', *Geol. Mag.* 85, 279-96

A. CAILLEUX (1945), 'Distinction des galets marins et fluviatiles', *Bull. Soc. géol. Fr.* 15, 375-404

R. G. CARRUTHERS (1947-48), 'The secret of the glacial drifts', *Proc. Yorks. geol. Soc.* 27, 43-57 and 129-72

R. G. CARRUTHERS (1953), *Glacial drifts and the undermelt theory* (Newcastle)

T. C. CHAMBERLIN (1894), 'Proposed genetic classification of Pleistocene glacial formations', *J. Geol.* 2, 517-38

J. K. CHARLESWORTH (1929), 'The South Wales end-moraine', *Q.J. geol. Soc. Lond.* 85, 335-55

J. K. CHARLESWORTH (1957), *The Quaternary Era* (London, 2 vols)

K. M. CLAYTON (1963), 'A map of the drift geology of Great Britain and Northern Ireland', *Geogr. J.* 129, 75-81

R. F. FLINT (1929), 'The stagnation and dissipation of the last ice-sheet', *Geogr. Rev.* 19, 256-89

R. F. FLINT (1957), *Glacial and Pleistocene geology* (New York)

J. C. FRYE, P. R. SHAFFER, H. B. WILLMAN, and G. E. EKLAW (1960), 'Accretion gley and the Gumbotil dilemma', *Am. J. Sci.* 258, 185-90

A. GEIKIE (1863), 'On the phenomena of the glacial drift of Scotland', *Trans. geol. Soc. Glasg.* 1, 1-190

J. W. GLEN, J. J. DONNER, and R. G. WEST (1957), 'On the mechanism by which stones in till become orientated', *Am. J. Sci.* **255**, 194-205

C. P. GRAVENOR (1955), 'The origin and significance of Prairie mounds', *Am. J. Sci.* **253**, 475-81

P. W. HARRISON (1957), 'A clay till fabric: its characteristics and origin', *J. Geol.* **65**, 275-308

S. E. HOLLINGWORTH (1931), 'The glaciation of western Edenside and adjoining areas, and the drumlins of Edenside and the Solway basin', *Q.J. geol. Soc. Lond.* **87**, 281-359

C. D. HOLMES (1941), 'Till fabric', *Bull. geol. Soc. Am.* **52**, 1299-354

G. F. KAY (1916), 'Gumbotil: a new term in Pleistocene geology', *Science* **44**, 637-8

G. F. KAY and J. N. PEARCE (1920), 'The origin of gumbotil', *J. Geol.* **28**, 89-125

W. C. KRUMBEIN (1961), 'The analysis of observational data from natural beaches', *Tech. Memo. Beach Eros. Bd U.S. (Washington)* **130**, 58 p.

C. LYELL (1830-33), *Principles of Geology* (London, 3 vols, 1st Ed.; 1840, 6th Ed.)

H. MILLER (1884), 'On boulder-glaciation', *Proc. R. phys. Soc. Edinb.* **8**, 156-89

H. B. DE SAUSSURE (1779-96), *Voyages dans les Alpes* (Neuchâtel), 4 vols

F. W. SHOTTON (1953), 'The Pleistocene deposits of the area between Coventry, Rugby and Leamington, and their bearing on the topographic development of the Midlands', *Phil. Trans. R. Soc.* B, **237**, 209-60

F. W. SHOTTON (1965), 'Normal faulting in British Pleistocene deposits', *Q.J. geol. Soc. Lond.* **121**, 419-34

B. J. TAYLOR (1958), 'Cemented shear planes in the Pleistocene Middle Sands of Lancashire-Cheshire', *Proc. Yorks. geol. Soc.* **3**, 359-65

A. C. TROWBRIDGE (1961), 'Discussion: accretion gley and the gumbotil dilemma', *Am. J. Sci.* **259**, 154-7

R. G. WEST and J. J. DONNER (1956), 'The glaciations of East Anglia and the East Midlands: a differentiation based on stone-orientation measurements of the tills', *Q.J. geol. Soc. Lond.* **112**, 69-91

Chapter 14

Drumlins

The whole County is remarkable for a Number of small Hills, which are compared with wooden Bowls inverted, or Eggs set in Salt; and from thence it is said to have taken the Name of Down, which signifies a hilly Situation. (C. SMITH, 1744)

The term 'drumlin' is of Gaelic origin, being derived from *druim*, a word for a mound or rounded hill, and has been used in glaciological literature since 1866. It refers to the small round, oval or elongated hills, which diversify some areas of glacial deposition and which are largely, if not entirely, composed of till. The term has been used to describe quite a wide variety of features, from isolated rounded hills up to 60 m in height to low, gentle swells. Drumlins can occur as isolated features, but more often they develop in swarms. Often, thousands or more together cover very considerable stretches of country, as for example in Ireland and New York State.

Normally the drumlins have a length to breadth ratio of about 2·5:1, extending up to 3 or 4:1. But in some instances very long narrow features have also been referred to as drumlins. The term has, therefore, been used with some lack of precision. Until the method of formation of drumlins is known with more certainty, it seems best to use the term for a specific form rather than in a genetic sense. A drumlin could be defined as a low hill, having an oval outline and not exceeding 60 m in height. It is formed mainly of till, but sometimes contains stratified material or a rock core. It is not desirable to call features of similar form, but composed entirely of solid rock, drumlins, as the main essential of a true drumlin is that it should consist of glacial deposits, shaped to create a streamlined form. There are certain intermediate forms that may contain a considerable amount of solid rock. Crag and tail features are well-known examples. These also show the streamlined form associated with a moving medium.

The Shape of Drumlins

Various studies have been made of the shape of drumlins in relation to specific geometrical figures. Drumlins are particularly easily treated in this way as they show more symmetry than many geomorphological forms and occur in large numbers in close proximity, so that a large sample can easily be obtained. They are revealed clearly on aerial photographs and often on good, closely-contoured topographical maps, such as the Ordnance Survey 1:25,000 series and the U.S. Geological Survey topographic maps, such as the 1:24,000 series, some of which have a contour interval of 10 feet (3 m).

B. Reed et al. (1962) have used these maps to study the drumlins of three areas, near Boston (Mass.), central-western New York State, and the Voltaire area of North Dakota. The drumlin form was compared with an ellipsoid, the shape of which can be defined by use of the formula $\frac{x^2}{a^2}+\frac{y^2}{b^2}+\frac{z^2}{c^2} = 1$, where a, b, c are the long, intermediate, and short semi-axes respectively. This gives the x, y and z co-ordinates. Each contour should be an ellipse if this formula is used and the ratio of major to minor axes must be constant for each contour.

The results of plotting the values for major and minor axes of a series of drumlins shows that, at their lower contours, the drumlins approximate closely to the elliptical form. However, at their summits, they depart slightly from the ideal shape. The elongation of the drumlins is also apparent in the steepness of the curve plotting length against breadth. Those in New York are more elongated than those near Boston. A study of the orientation and spacing of the drumlins was also carried out. The spacing both parallel to and normal to the mean direction of orientation was measured, using adjacent drumlins for the measurements. The results show a wide variation in the spacing with a multi-modal curve resulting. There is, however, some evidence for a preferred spacing and some indication of periodicity.

Another shape model has been applied to drumlins by R. J. Chorley (1959). This shape emphasizes the streamlined form of drumlins and their formation by a moving medium. They are compared with the shape of other streamlined forms, such as aeroplane wings and snow drifts. These shapes are such that the air, which is the moving medium causing their formation or to which they are adapted, can flow smoothly over them. The form is not symmetrical as an ellipse about two axes. It is only symmetrical about the longitudinal axis, having a rounded end facing into the flow and an elongated tapering end on the lee side. The profile is also not symmetrical, but has a steeper slope on the exposed side and a gentler lee slope.

It is interesting to see how closely a drumlin corresponds to the ideal streamlined form. If a close correspondence can be shown, then it is necessary to investigate the processes that can cause such a streamlining effect. In comparing the shape of drumlins with that of aircraft wings, Chorley makes the point that the maximum wing width is situated about three-tenths of the length from the leading edge of the wing. A more important point is that the greater the air speed that the wing has to withstand, the greater the length relative to the breadth. This means that the form is more elongated. Greater elongation of drumlins may well be associated with a more powerful ice-flow.

The type of curve that can most suitably be used is the lemniscate loop. It is adaptable to the varying forms of actual drumlins, and also provides the ideal streamlined form. The formula for the lemniscate loop is given by

$$\rho = L \cos k\theta$$

where L is the length of the long axis (this being the value of ρ when $\theta = 0$); k is a dimensionless number expressing the elongation of the lemniscate loop, such that k is equal to unity when the form is circular, and as the loop increases in elongation, so k also increases. The value of k is given by

$$k = \frac{L^2 \pi}{4A}$$

where A is the area of the loop. The only measurements that are needed to fit the loop to the actual outline of the drumlin are the length, L, and the area, A. These values can be used to obtain k in the second equation, and then k can be used in the first equation to plot the form of the loop. The tapering of the drumlin is often not so extreme as that given by the lemniscate loop when the ice-flow is relatively slow, and the lee end becomes slightly rounded relative to the plotted curve. Chorley points out that the smaller drumlins tend to be more symmetrical, having a form approaching more closely to that of an ellipse than a lemniscate loop. As the size of the drumlins increases, so does the asymmetry about the long axis.

An interesting comparison is made between this tendency to become more rounded and the shape of bird's eggs. A field of drumlins is often referred to as 'basket of eggs' relief. This analogy is a useful one as it appears that there may be some genetic connection between the shape of eggs and drumlins in this respect. A hen's egg is always laid with the blunt end first, so that the pressure at this end is greatest and the shape of the egg is adjusted to this. The blunt end of the hen's egg thus corresponds to the stoss end of the drumlin, which also experiences the greatest pressure of the moving ice. This relationship can be carried further, in that there is a relationship between the shape of a bird's egg and the size of the bird laying it. Birds that lay large eggs relative to their size tend to lay the most asymmetrical and pointed or tapering eggs. Birds laying relatively small eggs lay much rounder, symmetrical ones. The less the pressure on the eggs as they are laid, the rounder they should be. By analogy, the more slowly the ice is flowing, and hence the lower the ice pressure, the more rounded the drumlins should be. This

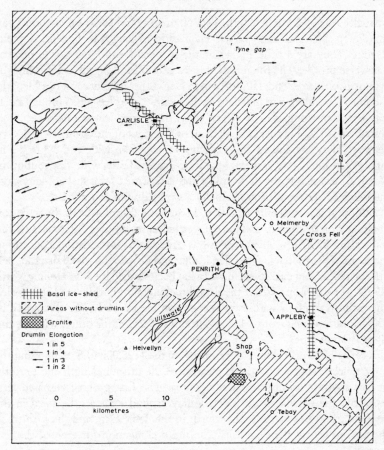

Fig. **14-1** The distribution and elongation of the drumlins in the Eden valley
and Solway lowlands, northern England

analogy should not be carried too far, as the supply of till and other factors
clearly play a part. Nevertheless it does suggest that the shape of drumlins
can be related in part to the pressure exerted by the ice in their formation.
The factor k gives a useful inverse indication of the relative resistance of the
equilibrium form of the drumlin to the moving ice. This resistance depends
both on the consistency of the material and the nature of the force of the
moving ice.

Before leaving the subject of drumlin shapes, it is worth mentioning the
great variety of elongation of the features that have been described as drumlins. At one extreme there are the narrow, linear drumlins described by R. W.
Lemke (1958) in north-central North Dakota near Velva. These features are

normally 1·6 to 4·8 km in length, 46 to 60 m wide and 1·5 to 4·6 m high. They have steep sides and sharp crests. Their ratio of length to breadth is about 60 to 1. They are a little higher at the stoss end and have a symmetrical transverse profile. The longest of this set is 21·5 km long. It is 15 m high at the north-west stoss end, falls to 10 m in height throughout most of its length, and tapers to only 1·5 m in height at the south-east or lee end. This particular feature has an elongation of 240 to 1. Similar shaped features have also been described by J. L. Dyson (1952) in front of the snout of the Grinnel and Sperry glaciers. They are 150 to 180 m long, 0·6 to 3·7 m wide and 0·3 to 0·9 m high. Such long, low features appear to have a fairly wide distribution in North America, but not all drumlins in this area are of this shape. Many have the much lower elongation normally associated with drumlins.

Although it is generally true that, in any one drumlin field, the variation in elongation is limited, there are nevertheless interesting differences within some drumlin fields. These variations give useful information concerning the formative processes. The Eden Valley and Solway Lowland drumlin belt may be taken as an example. This area has been studied by S. E. Hollingworth (1931). A further analysis of the drumlin elongation in this area has revealed interesting variations within it as shown on Figure 14-1. The elongation of 100 drumlins varied from 1·1:1 at one end to 9·7:1 at the other, with a mean elongation of 3·0:1.

A series of 20 drumlins was measured on the 1:25,000 O.S. map, using the lowest closed contour to give a measure of the drumlin length and breadth, in five separate localities within the area. The first group is situated near Appleby, the second in the Eden Valley around Penrith, the third in the neighbourhood of Carlisle, the fourth in the Tyne Gap, and the fifth set in the area to the west of Carlisle. The mean elongation for each of the five areas is:

Area 1. 2·55:1
2. 3·03:1
3. 1·92:1
4. 2·85:1
5. 4·66:1

These figures were then used in an analysis of variance statistical test. The resulting value of F was 22·2, which is highly significant. The F value shows that the variation between the groups of drumlins is considerably greater than that between the drumlins within the individual groups. The change in elongation within the whole drumlin field is, therefore, statistically significant. These differences in elongation must be accounted for in terms of the formative processes in relation to the environment in each of the five groups of drumlins selected. A discussion of the factors that may be responsible for this variation in drumlin elongation will be considered when their genesis is discussed (see p. 336).

The Material of Drumlins

It has been suggested by some authorities that features having the elongated, streamlined form typical of drumlins can consist of a wide variety of material, ranging from almost entirely solid rock to entirely drift. The glacial material of which most drumlins are formed can vary widely. Some drumlins include stratified, water-laid deposits, although many consist mainly of unstratified till, deposited directly by moving ice. The till can vary according to the rocks exposed in the path of the approaching ice. Thus some drumlins consist of mainly sandy material, while others are made up mainly of clay. Some are formed of true boulder clay, consisting of a mixture of subangular and striated boulders set in a clayey matrix, such as many of the drumlins in the north of England. The narrow, linear drumlins of North Dakota that have already been mentioned, consist mainly of stratified sand. This sand includes balls and lenses of till; there are also layers of gravel, and the whole is shrouded in a thick layer of till. The till also underlies at least part of the sand. Finally a layer of wind-blown sand is banked up against one side of the drumlins. This complex structure is found in the larger ridges, while the smaller ones are often composed only of till. The included lenses of till within the sand lie with their long axes parallel to the axis of the ridge.

In some areas, studies have been made of the orientation of stones included in the till of the drumlins. G. Hoppe (1959) has shown that the drumlins of Norrbotten in northern Sweden, which are composed mainly of sandy till, contain stones that show a preferred orientation in the direction of the ridge, except when the stones are too numerous to enable them to move freely. These particular drumlins are irregular, elongated ridges, usually less than 10 m high and up to 500 m long. The stones in them show no preferred plunge and the drumlins show no longitudinal asymmetry as many drumlins do. Hoppe suggests that these particular features may have been formed by the squeezing of water-soaked till into elongated cavities within or under the ice or by direct subglacial deposition. The fact that the stones near the surface of the features also show a preferred orientation suggests that the amount of superglacial till must have been small where the features were formed. Nearly all these ridges have no rock core, although a few do contain a rock core. Their orientation is parallel to the latest striations and thus to the final direction of ice movement in the area. This suggests that they were formed near the front of the ice, when it was receding in the area. Such data conflict with characteristics of drumlins in other areas to be considered, and it may be that these ridges are not true drumlins.

A rather more detailed study of the orientation of stones in till has given some valuable evidence concerning the character of drumlins. This work by H. E. Wright (1957) was based on observations made in the Wadena drumlin field of Minnesota. Measurements of stone orientation in drumlins were made in the area to the west of Lake Superior, where the drumlin field is 100 km long and covers an area of 5000 km². To the west, it becomes buried by outwash, to the south by drift of younger ice, and on the east lies a younger moraine. The younger ice did not destroy the drumlins in the area where it overlapped them in the north-east. In the north of the drumlin field, the drumlins trend south 45 degrees west, and they fan out to south 30 degrees east and north 80 degrees west to the south and west respectively. The drumlins range from an elongated form, such as 12 km by 0·5 km, to oval ones about 0·7 by 0·5 km, and they vary from 5 to 10 m in height. They are formed of yellowish-brown sandy calcareous till.

The stones that were measured had a long axis of 1 to 8 cm and an elongation ratio of at least 3 to 2, fifty stones being measured at each site. The strike and plunge of the long axis were both measured. Nearly all the sites show a major orientation trend parallel to the elongation of the drumlin and to the assumed direction of ice-flow. Some have a minor transverse concentration as well. The evidence suggests that the drumlins were formed by accretion of basal material from the lowest layers of the ice, and that the stones were carried by basal sliding. Stones carried in this way tend to be orientated parallel to the flow, whereas stones carried by rolling are usually arranged with their long axis at right angles to the direction of movement (Chapter 13).

Measurements of stone orientation in drumlins are of value as the actual form of the drumlin gives an independent indication of the direction of ice-flow. It is, therefore, possible to relate stone orientation to direction of ice movement with more confidence in these circumstances. In a smooth till sheet, orientation studies are usually made in order to deduce the direction of ice movement. This is a topic about which there is still much to be learnt and it is significant that in drumlins it is not always true that the stone orientation is parallel to the elongation and flow direction. For example, in a drumlin in Wensleydale (Yorkshire), the preferred orientation was found to be at a considerable angle to the elongation of the drumlin. This drumlin lies in the valley floor and is elongated down-valley and hence presumably parallel to the ice-flow. The mean orientation of 10 samples of 50 stones each, taken from the exposed interior of the drumlin, makes an angle of 33 degrees to the elongation of the drumlins. The maximum deviation is 54 degrees and the minimum 12 degrees. All but two of the ten samples show a statistically significant preferred up-glacier dip of the long axes in a direction perpendicular to the mean orientation of the pebbles. The observations were made on the south side of the drumlin, where till is exposed by river erosion

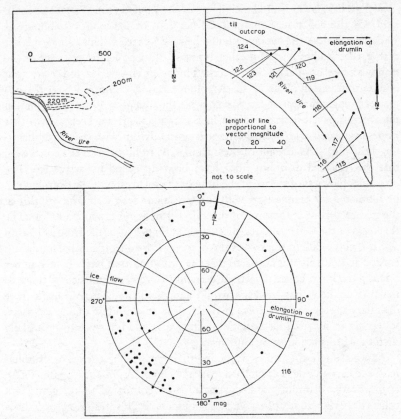

Fig. **14-2** Till fabrics from a drumlin in Wensleydale, Yorkshire

(Fig. 14-2). The orientation, therefore, points in towards the centre line of the drumlin. The two samples that have no significant dip are near the crest of the drumlin. There is a gradual increase in divergence between elongation of the drumlin and direction of the mean vector from the base upwards. The drumlin is elongated in the direction 280-100 degrees and the mean vector directions from the base upwards are 226 degrees, 232 degrees, 237 degrees, 242 degrees and 245 degrees. The values for the summit samples are less consistent and have means of 258 degrees and 268 degrees, but show a much wider scatter. These orientation values suggest that ice was pressing material against the side of the drumlin, with an element of lateral stress, which was missing near the summit of the drumlin. The suggestion that drumlins could occur when the ice-flow has subsidiary lateral stresses may help to explain the occurrence of these features.

Wright has also measured the plunge of the stones in the Wadena drumlin field. Of the 565 stones analysed, 34 per cent of the long axes plunge at 10 degrees or less, the average plunge being 23 degrees. This value is higher than earlier studies made on other types of till deposits. 64 per cent of the stones plunge up-glacier, 27 per cent plunge down-glacier and 9 per cent plunge transversely. The up-glacier plunge can be compared with the up-glacier plunge of shear zones, for example the plunge at 10 to 30 degrees on the margin of the Barnes ice-cap in Baffin Island. It has been shown that stones associated with such shear zones normally dip with them, parallel to the shear plane. As a result of these studies, Wright came to the conclusion that the Wadena drumlins must have been deposited by active ice. The probable thickness of the ice at the time of the formation of the drumlins can be estimated by comparison with the Greenland ice-sheet. The breadth of the marginal zone at the time of drumlin formation was about 140 km. The thickness of the Greenland ice-sheet at this distance from the front is 1000 m and it is about 300 m thick over the outer 40 km zone. If similar conditions existed at Wadena, it is apparent that the drumlins must have formed under a fairly thick ice cover. It would certainly have been sufficiently thick to make it unlikely that crevasses would extend to the base. Although these drumlins appear to have formed fairly close to the limit of the ice-sheet, squeezing of material into cavities in the ice would not seem to be a likely method of formation in this instance.

Another interesting study of the internal structure of a series of drumlins has been carried out near Dollard (Saskatchewan) by W. O. Kupsch (1955). Some of the drumlins studied appear to be composed of till alone, but others have till and stratified sand and boulders. Both forms, however, show clear evidence of having been streamlined by moving ice. The drumlin field lies in a topographical depression with a south-west trend. It is 29 km long and has a maximum width of 10 km and contains 100 drumlins. Their density is therefore much less than in many drumlin fields. They lie about 160 km north of the drift limit. Their average length is 230 m and their width 10·7 m. They vary from 75 to 750 m in length and 23 to 30 m in width in the more elongated ones. Most of them have a length to breadth ratio of about 2·5:1, and a height of 46 m. An interesting point concerning their distribution is that they only occur on the sandy Upper Cretaceous strata and not on shale. The maximum thickness of the drift is 20 m but in places it is as little as 6 m around the drumlins.

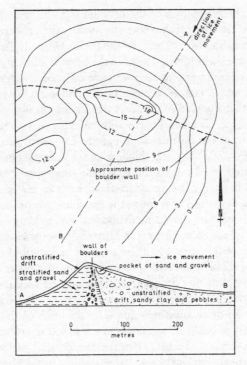

Fig. **14-3** Internal structure of a drumlin near
Dollard, Saskatchewan (W. O.
Kupsch, *Bull. geol. Soc. Am.*, 1955)

The till of which they are partially composed is a silty clay with few
boulders, but containing irregular pockets of sand and gravel. A sand pit
has been dug in one of the drumlins, and this reveals an interesting internal
structure (Fig. 14-3). This drumlin, 490 m long and 335 m wide, consists of
a layer of unstratified till on the gentle distal slope. A line of boulders stands
vertically beneath the highest portion. The steep proximal side consists of
stratified sands and gravels. The whole feature is overlain by about a metre
of unstratified sandy drift. The wall of boulders is perpendicular to the long
axis, and the boulders are rounded and range up to between 0·3 and 1 m in
size. Many of the boulders are fractured and have three sets of joints. These
fractures are clearly related to the formation of the drumlin, as they can be
traced at times through one boulder into a subjacent one. Most of the fracture
planes are nearly vertical. The preferred orientation of the joints is 355 degrees, with a secondary preferred orientation at 92 degrees and a minor set

at 34 degrees. The orientation of the drumlin is 33 degrees. The pattern of the joints can be explained if they were formed by the pressure of ice against the drumlin exceeding the strength of the rocks within it. The joints at 355 and 92 degrees are shear joints. They would occur if the top and bottom of the drumlin were confined and pressure were relieved by lateral movement. The set of joints at 34 degrees represents extension joints. The pattern of the joints suggests that the wall of boulders was already in place and was over-ridden by active ice. The joints agree with an experimental test pattern.

The formation of the drumlin is thought to have been carried out in the following stages. First, the ice thinned and became stagnant. Then the boulders collected in linear crevasses, where they were concentrated by meltwater washing away the finer material. The finer material remained in the rest of the deposits and now forms the proximal side of the drumlin. The ice then readvanced after the kame, formed of the boulders, had frozen into a solid block. The advancing ice streamlined the deposits into drumlin form. The boulders were fractured by the ice pressure and a tail of unstratified till was left on the distal side of the original kame, that is, in the lee of the obstruction. This particular drumlin, in the details of its sedimentary character, shows evidence of both erosion and deposition, the final overall covering being in the form of ablation moraine. The thickness of the ice when the drumlin was shaped could have been as much as 400 m at this point. It was probably much less at the stage when the fluvioglacial material was being deposited to form the original kame, which was later modified by active ice to form the present drumlin.

The variations in the material of which these different drumlins are composed requires a variety of possible methods of formation. Individual drumlins and drumlin fields need not all have been formed by the same mechanism. The presence of both stratified and unstratified material necessitates the consideration of both erosional and depositional processes in drumlin formation.

The Distribution of Drumlins

Drumlins are very widely distributed in those areas that were glaciated during the last major ice advance (Würm, Weichsel, Wisconsinan). Although they are widely scattered, their occurrence is often very localized in any one area. A few are found outside the areas of the last major ice advance: for example, there are drumlins of Riss age in the Danubian part of the Rhine glacier area and of Illinoian age in North America. One of the main problems concerning drumlins is to account for their localized occurrence in any one area. It is also worth noting that, within any one area, the drumlins tend to be of fairly uniform type and dimensions.

Over much of glaciated Europe, drumlins are rare, while in North America they are found infrequently in Pennsylvania, New Jersey, Ohio, Indiana, and south Michigan. However, where they do occur they are often found in very large numbers. In east-central Wisconsin, there are about 5000 drumlins, in south-central New England there are 3000, central-western New York State has 10,000 and Nova Scotia 2300. In Canada, the drumlins tend to be long and narrow in the north. Drumlins also occur in British Columbia.

The North American drumlins are found in five main areas: 1) Manitoba and Athabasca; 2) New England, Ontario, New Brunswick and Nova Scotia; 3) Michigan and Wisconsin; 4) British Colombia; 5) central-western New York State. This last area probably contains the most extensive drumlin belt in the world. The drumlins of this area lie in a belt 56 km broad and 225 km long, between Lake Ontario and the Finger Lakes. Generally, the drumlins of North America tend to occur in belts behind morainic barriers on flat till plains.

In Europe they also occur behind and between morainic ridges, for example, in Switzerland. Drumlins occur in western Pomerania, where they have an elongation of 3·75:1 on average and number over 3000. Many of these contain a fluvioglacial core. Another small but typical drumlin area lies in northern East Prussia between Memel and Nimmerstatt. There are about 100 drumlins in this area and they are formed entirely of till. Drumlins are numerous in parts of the Alps, for example, the drumlin field of the Bodensee and Eberfiner areas (P. Woldstedt, 1954).

In Britain (Fig. 14-4), drumlins are also common in some parts of the Newer Drift areas. The great belt of drumlins, extending across nearly all of the northern half of Ireland, is one of the largest in the world. It extends from County Down in the east across Armagh, Cavan, the Erne area, and into Donegal and Sligo Bay in the west. The drumlins also extend out into the Atlantic in Clew Bay, County Mayo, although this set of drumlins is not directly continuous with the more northerly belt. Drumlins also occur farther south in a great belt in County Clare to the north of the Shannon. These drumlins lie inside the southern Irish end-moraine, which marks the limit of the Newer Drift ice advancing from the north. In the extreme south-west of Ireland, drumlins occur within the limits of the lesser Kerry-Cork glaciation, which also belongs to the Newer Drift period.

In Scotland, drumlins occur in the Midland Valley, in Galloway, and the Merse area of the lower Tweed. Some not very well developed drumlins have also been described in Skye by J. J. Donner and R. G. West (1955). They had previously been described as dead ice features, but examination of stone orientation in them suggests that they are in fact drumlins. The preferred stone orientation is parallel to their elongation. They occur north of

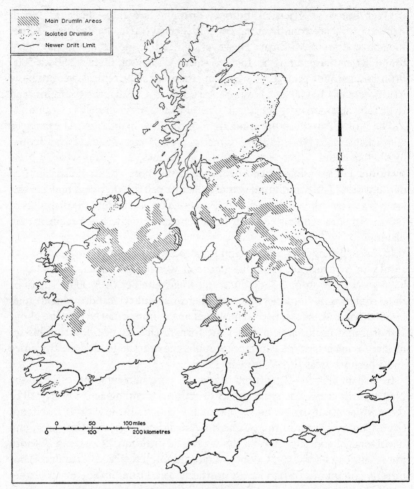

Fig. 14-4 The distribution of drumlins in Great Britain

the Cuillin Hills and are remarkable in that they are small in size, being only 2 to 5 m high and 10 to 20 m long, and their lee side is the steepest.

One of the most extensive drumlin fields in England is that in the Eden Valley and Solway. This belt continues into south-west Scotland, and across the Tyne gap into north-eastern England. It almost links with another set of drumlins to the south of the Lake District around Kendal, Barrow-in-Furness, and Ulverston. To the west of the Lake District, the belt swings round past Wigton and Aspatria towards Maryport. Drumlins also occur in the Ribble valley between Skipton and south of Settle. A very well developed set occurs in the Ribblehead region. This group is almost linked with another set that stretches into Wensleydale. The drumlins in this valley nearly link with the Eden valley set at Kirkby Stephen. The high ground of the Lake District and North Pennines is thus almost entirely surrounded by drumlins. They also occur in Anglesey and in the Welsh Border country near Wrexham.

Other areas in the world with drumlins include central France, the Dinaric Alps, Tienshan, Siberia, Novaya Zemlya, and China.

The more detailed distribution of drumlins in particular areas will be considered. S. Aronow (1959) has discussed the location of drumlins in the Warwick-Tokio area of North Dakota, where drumlins on the whole are not plentiful. The drumlins are confined to the Heimdal moraine and its southern continuation. There are some drumlins on the proximal side of the moraine and on patches of the moraine that extend up through the outwash. Most of these drumlins, however, are not very regular, only twenty being true drumlin-shaped features out of 160. The drumlins are elongated in a north-east direction and are formed of clayey till. Aronow has compared six aspects of the relief and other variables, with and without drumlins, in order to attempt to explain the drumlin distribution. The six factors are: 1) regional slope, 2) topographic grain, 3) character of the drift, 4) configuration of the bed-rock, 5) character of the bedrock, and 6) topography prior to the last glacia-tion. None of these six factors appears to control the distribution of drumlins in the area and several of them are fairly uniform throughout the area. The conclusion is reached that there are no easily discernible differences in the factors enumerated that could account for the variation in drumlin distribu-tion. It is more likely that the cause is to be sought in the character of the now-vanished ice that deposited the drumlins. Where conditions in the ice are suitable for drumlin formation, features showing streamlining will form regardless of the material over which the ice is flowing. True drumlins will occur under circumstances in which till is available for drumlin formation.

THE DRUMLINS OF THE EDEN VALLEY AND SOLWAY AREA

The variation of the elongation of the drumlins in this area has already been noted (see p. 327). The distribution of the drumlins is shown on Figure 14-1. The orientation of the drumlins changes in different parts of the area. This gives an indication of their probable date of origin. Some of the drumlins occur in the area covered by the Scottish Readvance as defined by Hollingworth, but they occur mainly in the area glaciated by the Main Lake District ice-sheet during the main Newer Drift advance. However, there is still controversy over the extent of the Scottish Readvance glaciation. L. F. Penny (1964) adheres to the limit suggested by Hollingworth (1931).

But J. B. Sissons (1964) suggests that most of these drumlins were formed during the Lammermuir advance, which occurred after the maximum of the Newer Drift but before the Perth Advance of Zone I. The ice at this stage according to Sissons, extended well south of the Lake District and the ice front linked with that forming the Kells moraine in Ireland. This moraine separates the major drumlin area to the north in Ireland, from the esker area to the south. Whichever ice-sheet formed the drumlins, it can be shown that most of the drumlins must have formed when the ice-sheet was fairly extensive. The reason for this is that the Lake District ice, when the drumlins were formed, must have moved continuously from the Lake District and Howgill Fells north down the Eden Valley and then it must have come into contact with Scottish ice in the inner part of the Solway Firth. The combined ice masses must have been diverted south-west and south down the Irish Sea along the western flanks of the Lake District.

The pattern of the drumlins conforms exactly to the direction of basal ice movement. This can be inferred from other evidence, such as the distribution of erratics, which include distinctive marker rocks such as the Shap granite (Chapter 13). The ice must have been thick and moving actively during the deposition of the drumlins as their elongation shows that the ice was escaping over every available col during this stage. The drumlins around Appleby are almost circular because a basal ice-shed existed at this point. Ice was diverted northward down the Eden valley in one direction and eastward across the Stainmore pass in another. Around Carlisle, the elongation falls to very low values, and in this area also a basal ice-shed existed (Fig. 14-1). It separated ice of both Scottish and Lake District origin that was moving westward round the edge of the Lake District into the Irish Sea for that moving eastward across the Tyne gap to the low ground east of the Pennines.

The elongation of the drumlins is found to be greatest where the ice was moving fastest and was thickest. This occurred in the area to the north of the Lake District around Wigton and Aspatria. In this area, the drumlins trend east-north-east to west-south-west and their stoss end is to the east. The elongation also increases in the Tyne gap where the ice was being squeezed through the gap and was flowing fast. The greater velocity increased the pressure and the drumlin form became more elongated in response. The evidence of this area points to the fact that these drumlins must have been formed by actively moving ice and not in the sub-marginal zone in which some drumlins seem to occur.

When the Lake District ice parted from the Scottish ice during deglaciation, the ice front to the north of the Lake District lay in an east-west direction almost parallel to the drumlin elongation. Had the drumlins been formed close to the margin of the ice, which would presumably have been moving

PLATE XV   THE ARÊTE OF LLIWEDD.
Seen from Snowdon summit, north Wales, with the rock basin containing Llyn Llydaw
on the left. (C.E.)

PLATE XVI   VIEW FROM GALDHÖPIGGEN (2468 M), NORWAY.
Note the series of arêtes and horns resulting from destruction of the preglacial relief by
cirque enlargement and recession. A bergschrund parallels the cirque walls in the lower
right part of the photograph. (C.E.)

PLATE XVII (*above*)   THE TERMINUS OF THE PASTERZE GLETSCHER, AUSTRIA.
The meltwater stream has carved a gorge through a rock bar. The lake is artificial,
part of a hydro-electric scheme. (C.E.)

PLATE XVIII (*right*)   THE SNOUT OF KJENDALSBREEN, NORWAY.
Showing the emergence of a sub-glacial stream. (C.E.)

northward at this stage, then they should have had an elongation at right angles to their actual one. The drumlins that lie within the Readvance limit as defined by Hollingworth do not show an alignment radial to the ice front, as would be expected had they been formed by this later ice. This advance has left no terminal moraine so that it was probably too short-lived and the ice moving too slowly to form drumlins or seriously to modify the earlier ones. However, there are some drumlins 8 km east of Carlisle that must have been formed by later ice moving down the Tyne gap, as they rest on the Middle Sands and include some of this material in their till. The Middle Sands in this area also contain clay and it is possibly the presence of the clay that allowed drumlins to form in this area. According to Hollingworth, in some parts of the area covered by the Scottish Readvance in northern England and southern Scotland there are two distinct sets of drumlins with different orientations. Most of the drumlins, however, belong to the more vigorous ice-sheets.

Most of the drumlins of the north of England are composed of till, a true boulder-clay with striated boulders set in clay. Some of them contain sand and gravel in a central core, suggesting deposition of material before the shaping by moving ice. Other drumlins have a rock core, while a few consist mainly of rock. It is not possible to distinguish the material from the surface form of the drumlin.

Another point of interest concerning the distribution of drumlins in the Eden valley and neighbouring areas is their restriction to the lower levels. They are found in the valley and lowlands and rarely occur above 300 m. They are common, however, in cols, for example the Tyne gap, Stainmore, the Lune gorge near Tebay, the cols at the head of the Ure valley, Ribblehead, and the broad col linking the Lune, Ribble and Aire valleys near Settle. In areas where the ice comes down into a valley or crosses a col, its thickness or speed will be increased. Both these factors would increase the ice pressure on the bed, and, where suitable material is available, this could well be shaped into drumlin form. It is in these areas of relatively gentle slope that, in the early stages of glaciation, deposition would be most likely to occur.

DRUMLINOID DRIFT TAILS

A rather modified form of drumlin development is well seen in Wensleydale (Fig. 14–5). The ice moving down the valley from the west was of purely local origin and its speed of flow was partially checked by the great width of the valley and the presence of a large ice mass in the Vale of York at the outlet of the dale. Thus a considerable amount of till was deposited along the valley sides. In the valley floor, some well-developed true drumlins occur.

They are elongated down-valley and are composed entirely of a tough boulder-clay. The drumlins along the valley sides have become linked together and are joined on to the solid spurs of the tributary valleys. The spurs are thus elongated down-valley by these series of linked drumlins,

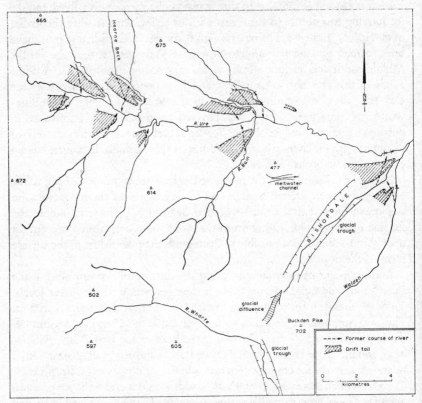

Fig. 14-5 Drumlinoid drift tails and other glacial features in Wensleydale, Yorkshire

which are often arranged en échelon, so that swampy hollows develop between the smooth drumlinoid mounds.

The development of these drumlinoid drift tails can be seen on nearly every spur in the valley. One effect has been to divert all the tributary streams eastward down the valley. The tributaries now flow eastward behind the drift tails in abnormally gentle thalwegs until they can find a low enough route across the drift tail. The stream then rejoins the main river often along an over-steepened course in which waterfalls are common. The best known of these waterfalls is Hardrow Scar, 30 m high. This has been caused by the diversion of Hearne Beck and Fossdale Beck by a long drift tail. Other examples include Whitfield and Mill Gill Forces farther east, both lying on the same stream as the stream finds its way round a double drift tail. The Bain at Bainbridge has been pushed out of its pre-glacial valley by a drift tail, which can be seen to fill the old valley where this is exposed in the flanks of its post-glacial gorge. Part of this gorge is cut through the solid rock (W. B. R. King, 1935).

Theories of Drumlin Formation: Conclusions

In describing the material and distribution of various drumlins, some suggestions concerning the development of drumlins have already been made. It has been shown that drumlins vary considerably in most of their major characteristics, such as elongation, spacing, size and material, but they have one essential feature in common. This feature is their streamlined form, and it is on the basis of this feature that their genesis must be sought. It is not necessary to postulate that all drumlins have been formed in the same way. In fact their great variety points to variations in the formative processes. These depend not only on the nature of the ice forming the drumlins but also on the type and quantity of drift available, the relief and other factors. One point that nearly all those interested in the formation of drumlins agree upon is that they have been formed by moving ice, but here the uniformity of opinion ceases. One of the problems of explaining drumlins is that it is difficult to examine the base of an active and moving ice mass although it is in this situation that the drumlins are formed.

C. P. Gravenor (1953) has assembled and analysed some facts concerning drumlins. He states that drumlins on the whole tend to be sand-rich rather than clay-rich, and clay does not appear to be necessary for their formation, although they sometimes consist of a clay-till. They can also consist of sandy or loamy till, rock or pre-existing drift. They often contain lenses of stratified drift. Rock drumlins may occur with other drumlins. Many glacial areas do not have drumlins. They exist in fields that are often wider than moraine belts. They are streamlined, with the stoss end facing up-glacier. Some drumlins have cores, but most do not. Their long axes are parallel to the ice flow.

Two of the theories put forward to account for drumlins are the erosional and the depositional theories. The erosional theory considers a two-stage formation and can account for stratified material in drumlins. The unstratified material often does not resemble kame deposits and often the drift of which drumlins are composed is the same age as that of the surrounding drift. The basis of the erosional theory is that pre-existing till or stratified glacial deposits are shaped into drumlins by the ice of a readvance. Some evidence for this has been put forward by S. Jewtuchowicz (1956). The drumlins he describes in Poland have a core of ground moraine or till, which has been covered by fluvioglacial deposits. These stratified deposits have suffered cryoturbation and glacial tectonic disturbance. The ice must first have advanced to deposit the till, then it retreated, allowing the fluvioglacial deposits to accumulate on the till. The retreat must have lasted long enough for frost action to produce the cryoturbation features. The ice must then have readvanced, causing the glacial tectonic activity in the deposits already in place and shaping them into the characteristic streamlined drumlin form by partial erosion of the deposits.

R. F. Flint (1957) has put forward an argument in favour of the erosional theory. He points out that there is a complete range of forms from features of pure rock to those composed entirely of drift, all of which have the same drumlin shape. The solid rock features are clearly erosional, so the similarity of form suggests that the others may be as well. It is worth noting, however, that frequently the solid rock features, in the form of roches moutonnées, show a reversal of the normal stoss and lee slope of the typical drift drumlin. The drumlin normally has the steeper slope on the proximal or stoss side, while the roche moutonnée often has a relatively gentle proximal or stoss side and a steeper, more angular and quarried distal or lee slope. This difference suggests that the processes forming drumlins are not identical to those shaping the solid rock. There is also a difference in location. Drumlins usually occur on the lower ground, while smoothed rock features are also characteristic of the steeper slopes and higher areas. Both are, however, essentially streamlined at least in part.

Some authorities consider that drumlins are formed submarginally. There are examples of drumlins associated with moraines. The suggestion that drumlins could form under stagnant conditions has been put forward by F. M. Synge (1952). He pointed out that the drumlins on the north of Ireland lie mainly north of the Kells moraine and they have been destroyed by the Antrim Readvance. He therefore suggests that drumlins may not be able to survive active glaciation. This view, however, does not agree with the streamlined form and orientation of the drumlins.

Gravenor considers that drumlins could form during phases of uneven advance. He suggests that material deposited at the front of a glacier during a halt could be shaped into drumlins when the ice advanced again. The morainic belts would be expected to be wider during ice advance than during retreat when eskers and such forms develop. The bedded material found in drumlins resembles more closely that found in ground moraine than that found in kames. Gravenor contends that this modified erosion theory accounts for many of the characteristics of drumlins.

The drumlins described by Kupsch (1955) in Saskatchewan, which have already been mentioned, show characteristics of both erosion and deposition.

These particular drumlins, consisting of stratified sands and gravels on the steep, proximal slope, with a line of boulders in the centre and a tail of unstratified till, must also have formed in two stages. The stratified sands and gravels and the boulders must have been deposited near the ice margin, under almost stagnant ice conditions, but the tail of till indicates a renewed advance of ice. This advance must have been responsible for the streamlining of the pre-existing drift as well as adding its own contribution on the lee side of the drumlin. Both erosion and deposition must have taken place in the formation of this particular drumlin. The active movement of the ice at the time of the shaping of the drumlin is indicated by the fracturing of the boulders in the centre of the drumlin. This fracturing must have required considerable pressure from thick, fast-moving ice.

The long low tail is characteristic of the streamline form. It is designed to create the minimum of disturbance in the flow of the moving medium, whether it be air, water, or ice. This form is more likely to result from deposition than by erosion because it is this part of the drumlin where the pressure would be least, in the lee of the main body of the feature. Material could therefore accumulate in this position. On a similar argument, the lack of pressure in this position would minimize erosion. It is significant that in roches moutonnées, the lee side is steep and rugged, and is eroded by plucking rather than by grinding and smoothing. In so far as drumlins resemble crag-and-tail features, it seems likely that their shape is partly depositional in character at least, especially the long low tail.

The purely depositional theory has been advocated for drumlins formed entirely of till. It has been suggested that they form around a nucleus of frozen till or rock. A few drumlins show concentric layers, suggesting that they have been built up by accretion of layer upon layer. The depositional view of drumlin formation is supported by J. K. Charlesworth (1957) and Hollingworth (1931). The drumlins that are composed entirely of till show no evidence of two stages of development. The distribution of drumlins on the lower ground suggests that they occur in areas that are predominantly depositional. In detail it has been suggested that the drumlins of northern England form in areas where the pressure of the ice is locally increased by thickening or by increase of velocity. The extra pressure could provide the conditions necessary for drumlin formation in ice in which there is sufficient basal till to form them. The orientation of the stones in the till of drumlins is a strong argument in favour of their deposition by moving ice, which is also responsible for shaping them into drumlin form. Both the drumlin elongation and the stone orientation have clearly been determined by the same ice-flow, and, where this can be shown to be true, those particular drumlins can confidently be stated to be depositional in character.

Some observers have suggested a different method of emplacement of the till in drumlins. Hoppe and V. Schytt (1953) have suggested that elongated hollows occur in the ice as it moves past boulders on the bed and that in relatively thin ice and stagnant ice these cavities would remain in front of the boulder. Water-soaked till would be squeezed up into the hollow to form long linear drumlin features. This explanation cannot, however, be of very wide application because it demands special conditions and does not agree with many of the characteristics of drumlins.

Some of the observations on drumlins suggest that there is some periodicity in the ice flow, which induces zones of deposition. These cores can then be shaped by the moving ice into drumlins. This would apply particularly to the drumlin fields situated on open areas of low relief. The suggestion that variations in the ice flow itself are a factor in the pattern and occurrence of drumlins is worth close investigation. The approximately equal height of the drumlins in any one drumlin field probably reflects the thickness of the debris concentrated in the basal ice.

It would seem reasonable to conclude that drumlins can be formed either by the modification of previously existing drift sheets, the readvancing ice causing erosion and deposition in varying proportions. Alternatively, they can be formed by the direct deposition of till by active ice in areas where till is available and the pressure of the ice is sufficient to produce the streamlining characteristic of drumlins. The precise mechanism of drumlin formation still remains in doubt. The absence of drumlins from most of the areas of Older Drift throughout the world may reflect the great time that has elapsed since this drift was deposited. During this long period, the till surface has been modified and any drumlins that were present have been lost by infilling or erosion. Drumlins remain an intriguing and enigmatic feature of many areas of glacial deposition.

REFERENCES

s. ARANOW (1959), 'Drumlins and related streamline features in the Warwick-Tokio area, North Dakota', *Am. J. Sci.* **257**, 191-203

J. K. CHARLESWORTH (1957), *The Quaternary Era* (London, 2 vols)

R. J. CHORLEY (1959), 'The shape of drumlins', *J. Glaciol.* **3**, 339-44

J. J. DONNER and R. G. WEST (1955), 'Ett drumlinsfält på ön Skye, Skottland', *Eripainos Terresta*, **2**, 45-48

J. L. DYSON (1952), 'Ice-ridged moraines and their relation to glaciers', *Am. J. Sci.* **250**, 204-11

R. F. FLINT (1957), *Glacial and Pleistocene geology* (New York)

C. P. GRAVENOR (1953), 'The origin of drumlins', *Am. J. Sci.* **251**, 674-81

c. p. gravenor and w. a. meneley (1958), 'Glacial flutings in central and northern Alberta', *Am. J. Sci.* **256**, 715-28

s. e. hollingworth (1931), 'The glaciation of western Edenside and adjoining areas, and the drumlins of Edenside and the Solway basin', *Q.J. geol. Soc. Lond.* **87**, 281-359

g. hoppe (1959), 'Glacial morphology and inland ice recession in northern Sweden', *Geogr. Annlr* **41**, 193-212

g. hoppe and v. schytt (1953), 'Some observations on fluted moraine surfaces', *Geogr. Annlr* **35**, 105-15

s. jewtuchowicz (1956), 'Structure des drumlins aux environs de Zbojno', *Acta geogr. Univ. lodz.* **7**, 1-74

w. b. r. king (1935), 'The Upper Wensleydale river system', *Proc. Yorks. geol. Soc.* **23**, 10-24

w. o. kupsch (1955), 'Drumlins with jointed boulders near Dollard, Saskatchewan', *Bull. geol. Soc. Am.* **66**, 327-38

w. o. kupsch (1962), 'Ice-thrust ridges in western Canada', *J. Geol.* **70**, 582-94

r. lebeau (1954), 'Forme mineure du relief sous-glaciaire', *Revue Géogr. Lyon* **29**, 227-56

r. w. lemke (1958), 'Narrow linear drumlins near Velva, North Dakota', *Am. J. Sci.* **256**, 270-84

l. f. penny (1964), 'A review of the Last Glaciation in Great Britain', *Proc. Yorks. geol. Soc.* **34**, 387-411

b. reed, c. j. galvin and j. p. miller (1962), 'Some aspects of drumlin geometry', *Am. J. Sci.* **260**, 200-10

j. b. sissons (1964), 'The glacial period' in *The British Isles, A Systematic Geography* (ed. j. w. watson and j. b. sissons)

f. m. synge (1952), 'Retreat stages of the last ice-sheet in the British Isles', *Ir. Geogr.* **2**, 168-71

f. m. synge and n. stephens (1960), 'The Quaternary period in Ireland—an assessment, 1960', *Ir. Geogr.* **4**, 121-30

p. vernon (1966), 'Drumlins and Pleistocene ice flow over the Ards Peninsula-Strangford Lough area, County Down, Ireland', *J. Glaciol.* **6**, 401-9

p. woldstedt (1954), *Das Eiszeitalter* (Stuttgart)

h. e. wright (1957), 'Stone orientation in the Wadena drumlin field, Minnesota', *Geogr. Annlr* **39**, 19-31

Chapter 15

Moraines

These fragments they [the glaciers] gradually transport to their utmost boundaries, where a formidable wall ascertains the magnitude, and attests the force, of the great engine by which it was erected. (J. PLAYFAIR, 1802)

The term 'moraine' covers a wide range of depositional features associated with glaciers and ice-sheets. Moraines normally consist of unstratified material, although at times stratified deposits laid down in front of ice-sheets have been referred to as moraines. The well-known Salpausselka moraines in Finland, for example, consist mainly of water-laid deposits. This type of feature is best referred to as a 'delta-moraine' and in many respects resembles one type of kame. It will therefore be dealt with when fluvioglacial features are described. This Chapter is concerned with features that have been deposited directly by the ice.

Moraines can be classified in several ways, according to their position, their state of activity, and their method of formation. In the first classification, moraines are divided into terminal, lateral, and median according to their position relative to the glacier. Terminal and lateral moraines are of great interest to the geomorphologist, as they indicate halt or readvance positions, at which the ice margin lay for a considerable time during deglaciation. In the second classification, moraines may be termed 'active' if they are in contact with active ice. Ice-cored moraines constitute a special group in this classification, usually occurring in a marginal position along the side or front of a glacier or ice-sheet. Another group comprises inactive moraines, which have lost all contact with the ice forming them as a result of glacier retreat or decay. It is these moraines that provide evidence for stages of glacier retreat, glacial chronology and correlation.

The third classification subdivides moraines according to their method of formation. Moraines may be subdivided into ablation moraine and ground moraine. Ablation moraine is material that has accumulated on the surface of the glacier by gradual down-wasting of the surface ice. Ablation moraine can be differentiated by its relatively coarse nature, because the fine material has usually been washed out by meltwater. Ground moraine is deposited subglacially and can accumulate in a number of different ways. The second method of classification will be adopted for a primary description of morainic types, although the other classifications will be used for secondary subdivision.

Active Moraines

MEDIAN MORAINES

The median moraine is the most conspicuous form of active moraine in valley glaciers, but it is least likely to form a permanent, stable feature. Median moraines can be seen prolonging a spur between two confluent valley glaciers in a down-glacier direction (Fig. 3-2). Such moraines are sometimes figured as continuing down through the ice to the glacier bed. In many instances this does not, in fact, occur. The median moraine is usually an entirely superficial feature, consisting often of only a metre or so of coarse stony debris lying on the ice. It consists of angular and unsorted material and usually stands out as a conspicuous elevated ridge on the glacier surface. In reality the ridge is mainly formed of ice. In this respect the median moraines of many temperate mountain glaciers resemble dirt cones in having a core of ice. They stand up as ridges because the layer of debris, once it exceeds a critical thickness, protects the ice beneath from melting. The thickness required to prevent or slow down melting is only of the order of 0·5 m.

If the spur from which the debris is derived lies above the firn-line, then the moraine may not appear on the ice surface until lower down the glacier. In other instances, the moraine is hidden under avalanche debris until this melts off as the glacier moves down-valley. Some median moraines show a beaded structure, for example on Austerdalsbreen in Norway (Plate VI). The beading appears to have an annual spacing and can be accounted for by the more active production of waste during the summer season. More of the rock face is exposed at this time to frost and other destructive processes, such as avalanches.

Some median moraines are very straight and orderly, such as those of the Barnard glacier in Alaska. Others, such as those of the Alaska Range glacier, show considerable contortions in the morainic pattern of the glacier surface. The contorted pattern gives evidence of the action of surges, set up by kinematic waves (Chapter 2). The kinematic waves do not affect all the tributary glaciers that join to form the main glacier at the same time. Individual ice streams thrust forward into the main valley at different times, causing the irregular bulges indicated by the median moraine pattern. These moraines, therefore, give useful information concerning the character of the glacier flow.

LATERAL MORAINES

Lateral moraines of present glaciers can vary greatly in character. Some consist of only a thin covering of surface debris resting on the ice and derived from the rock walls of the valley. The lateral moraine can take another form when the glacier is retreating or thinning. Because of the general climatic amelioration of the last few decades, many glaciers are thinning at the present time, resulting in the formation of the trim-lines that are characteristic of many glaciers at present. The trim-line separates an area of bare moraine below from vegetated moraine above. It is often noticeable that the vertical distance between the trim-line on the lateral moraine and the ice surface gradually increases towards the glacier snout. This fact can be explained in some instances by the passage of a kinematic wave down the glacier. The wave frequently reaches its greatest dimensions at the snout of the glacier, owing to the deceleration of flow causing compression in the down-glacier direction. If this factor is greater than the diffusion of the wave, then the change in ice elevation at the snout will be greater than farther up-glacier.

In some areas of temperate valley glaciers, such as south Norway, parts of the Alps, and the Southern Alps of New Zealand, lateral moraines form well defined ridges. They are separated from the valley wall by an ablation valley. Good examples of lateral moraines of this type occur on the flanks of the Tasman glacier and Cameron glacier (Plate XXI) on the eastern side of the New Zealand Alps. These glaciers, draining an area of easily frost-shattered greywacke rocks, have a thick cover of surface and lateral moraine. They are now down-wasting slowly beneath this cover. In order to form the lateral moraine ridge, it must be assumed that at one time the glacier had a higher and more convex cross profile than it has at present. Debris must have accumulated between the valley wall and the higher ice surface. Then as the ice slowly melted down in the valley to form the generally flat or concave surface that it has at present, the valley side of the moraine has maintained its slope away from the ice. The steeper side, facing the glacier, has developed as the ice surface has gradually been lowered. The valley-side slope often lies at the angle of rest of the material, but the glacier side is often much steeper approaching the vertical in places. This very steep slope is unstable and can only be maintained by the presence of an ice core in the moraine, holding the material together. When this ice core finally disappears, the ridge-like form will be considerably reduced in size and will become more symmetrical.

At times, small ridges of lateral moraine can form along the flanks of retreating glaciers. A small example was observed to develop in one year along the lower edge of Austerdalsbreen in Norway. It is thought that, during the winter, the ice thickened and debris moved off the ice to lie against the ice in a pile. Then as the surface was lowered by melting during the summer, the ice-side of the pile collapsed to form a small ridge nearly 1 m high. This type of small ridge had no ice core and was symmetrical in form, each flank lying at the angle of rest.

TERMINAL MORAINES

Few valley glaciers have well-defined terminal moraines in contact with the
ice at present. This is because most glaciers are retreating and terminal
moraines can only form effectively when the glacier is advancing or stationary in position and when the ice is active. An advancing active glacier often
has a fairly steep, high front, against which a bank of material can accumulate.
When the ice thins or retreats a ridge of terminal moraine will form in
the same way as the lateral moraine already mentioned. Terminal moraines
sometimes dam up water to form marginal lakes particularly when the ice
retreat is rapid. Examples of lakes of this type are seen in front of Skeiðarárjökull in Iceland and the Steingletscher and Feegletscher in the Alps. When
the glacier is retreating rapidly, after a period of still-stand and moraine formation, the material released from the melting ice is not sufficient to fill up the
area between the glacier snout and the terminal moraine. Lakes accumulate
temporarily in this position until their outlets are cut down through the
moraine and the lake is drained.

ABLATION MORAINE

Ablation moraine is often a conspicuous form of active moraine on the
surface of a valley glacier. This surface debris usually increases in amount
towards the snout of the glacier, where it is concentrated both by the deceleration of the ice and by the increased ablation near the snout. Where
glacier caves are available, however, it can usually be seen that this layer of
ablation moraine is only a surface layer and the ice beneath is normally very
clean.

A layer of ablation moraine is most likely to be present on a glacier that is
moving slowly and which is slowly shrinking downward while its snout
remains more or less static. In this respect, the contrast between the glaciers
draining to the west of the Southern Alps of New Zealand and those draining
east is of interest. The Tasman, which flows eastward and receives a considerably smaller accumulation as it lies in the lee of the mountains, moves
more slowly and its snout is at 700 m. In an area of easily broken rocks, it
receives a large supraglacial load. This has become concentrated by compression of the slowly moving snout into a thick layer of ablation moraine,
covering the whole of the lower 8 to 10 km of the glacier. Slow down-
wasting of the ice by about 60 m has also helped to increase the thickness of
the ablation moraine. This active ablation moraine merges almost imperceptibly into ice-cored moraine and this again merges into the proglacial
moraine and outwash. The precise location of the glacier snout is, therefore,
difficult to determine.

On the wetter western side of the Southern Alps, the Franz Josef glacier is very different. This glacier flowed actively to within 210 m of sea-level in 1956 and ended in a clean, steep ice-front, despite a rapid retreat which brought its snout to 300 m above sea-level in 1961. The cleanness of the ice is partly the result of the large precipitation which falls as rain in summer when the abundant rain and meltwater can evacuate much of the glacier load. The ice flows fast all the way to the snout and this decreases the accumulation of ablation moraine. The snout of the glacier responds much more readily to changes in accumulation and ablation. This glacier has no deposits comparable with the ablation moraine of the eastern glaciers, such as the Tasman.

ICE-CORED MORAINES

In areas marginal to existing ice masses, some interesting ice-cored moraines have been studied recently, particularly by G. Østrem (1963, 1964). Most of his observations have been carried out in the Jotunheim district of Norway, Kebnekajse (north Sweden), and Baffin Island (arctic Canada). Experiments were carried out to determine the effect of different thicknesses of dirt on the ablation of the ice. The results showed that when the ablation of clean ice was 4·5 cm/day, the ablation rate decreased when the dirt cover exceeded 0·5 cm. It fell to 3 cm/day with a dirt cover of 6 cm and to less than 1 cm/day when the dirt cover exceeded 20 cm. The studies of the effects of varying thicknesses of artificial dirt cover were extended to an examination of the distribution of ice-cored moraines. The study included the nature and age of the ice core and interesting results concerning the length of time the ice remains buried have emerged.

Geophysical techniques were used to investigate the presence and thickness of ice cores. When some ice-cored moraines have been identified in the field, it is possible to continue the study on aerial photographs. Air photographs have been used extensively to locate ice-cored moraines in various parts of Scandinavia. The ice-cored moraines can be identified on the aerial photographs by comparing the size of the moraine and the size of the glacier forming it. When the moraine is large compared with the glacier, it can safely be assumed to contain an ice core and the reverse is also true. This leaves a doubtful middle category that may or may not have an ice core. In all moraines in which ice exists, a snow bank is found on the distal side of the moraine. Many of the moraines consist of a series of small curved ridges of rounded outline. Ice-cored moraines do not normally have a sharp crested ridge.

In the past, it has normally been assumed that the buried ice in the ice-cored moraine was derived from the glacier. From the crystallographic structure of the ice in the core, Østrem showed that it was possible to determine whether the ice had originated as glacier ice or as snow. He found that although some of the buried ice was of glacial origin, in many instances the ice had originated as a snow bank, which had been piled up against the glacier snout.

The results of the study of the distribution of ice-cored moraines show that they are unevenly spread through Scandinavia. They are concentrated in the Jotunheim district, none occurring in the Jostedalsbreen area except for a few on the east. A few occur in Dovre, but the majority are found farther north, beyond the Arctic circle, around Sarek and Kebnekajse. In the extreme north, they are found fairly close to sea-level. The type of bedrock does not appear to affect the distribution of ice-cored moraines.

It is striking that many of the smaller glaciers have ice-cored moraines, while the large ones often do not. This may be accounted for by the relatively small changes in the position of the glacier snout in a small cirque glacier compared with a longer valley glacier. Once a large moraine has formed in front of a small glacier, it will help to slow down its forward movement, thus helping to maintain the relatively stationary position of the ice front. Glaciers with a short, broad tongue will also tend to remain fairly static, and would be expected to have an ice-cored moraine.

Studies of the ice of the core that have been carried out have shown that both glacier ice and snow-bank ice are incorporated in some of the ice cores. On the whole in Scandinavia the snow-bank ice cores are the more common and the presence of snow banks around the outer end of the moraines is also common. Conditions that favour the formation of ice-cored moraines are those where accumulation and ablation are small, that is in areas where the climate tends to be more continental.

A close connection is found between the altitude of the glaciers and the occurrence of ice-cored moraines. The glacial limit in Norway and Sweden gradually rises eastward from 1200 m on the coast near the Lofoten Islands to 2200 m in the Jotunheim area. Two main factors account for the relationship between the glacial limit and the ice-cored moraines. First, the high-lying glaciers in the eastern part of the area are associated with a more continental climate. The effect of this on ice core formation has already been mentioned. The glaciers are at a high altitude and, therefore, ablation will be slow and the ice-front will be static. This will allow time for the formation of ice-cored moraines. Secondly, the winter temperatures will be very low in the areas of the highest glaciers. This factor and the thin snow cover will allow great cooling of the exposed moraine ridge. Only a thin layer will thaw in summer as a result and once the ice core has formed, it will be preserved in the moraine.

Snow-cored moraines are formed of material carried to the end of the glacier and dumped on the frontal snow patch either directly or by slumping. It then becomes mixed with snow, allowing the supersaturation that is sometimes noted in the material of these moraines. When the dirt cover is thick enough, it will protect the snow bank beneath. During this accumulation phase the ice front must remain stationary.

The analysis of the ice from ice-cored moraines can give valuable evidence of the age of the moraine. In order to obtain sufficient organic matter from the snow-bank, it is necessary to obtain a large volume of it. Østrem (1961) has shown that 200 kg of ice may be required to produce sufficient organic matter for radiocarbon dating. The organic particles had blown on to the snow before it was covered by debris and, therefore, the snow must be older than the organic material. It is thus possible to obtain a maximum age for the moraine, as the morainic material must have covered the snow which included the debris. When the organic material had been separated from the ice, it contained 3 gm of carbon, half of which was used for dating purposes. One sample of dated ice was obtained from the outermost moraine ridge of Grausbreen in Jotunheim, Norway. The result of the radiocarbon analysis gave a date of 2600 ± 100 years B.P. The result shows that the moraine is very much older than was formerly thought. K. Faegri (1948) had suggested that these moraines belonged to the great glacier advance in the middle of the eighteenth century in Norway. However, the date given by the carbon-14 analysis agrees with the climatic deterioration that set in at the time of the cool, wet Sub-Atlantic phase. Further observations will show if this was a period of general glacier advance in this part of Scandinavia. This evidence does not necessarily negate Faegri's dates for non-ice-cored moraines, for which there are historical data. This will be mentioned later.

Inactive or Stable Moraines

Moraines no longer in contact with the ice that deposited them adopt a variety of forms, suggesting that they have developed by different mechanisms. They can be broadly divided, first, into those formed marginally to the ice as lateral and terminal moraines, and second, into those laid down over a wider area as ground moraine or ablation moraine. The first category itself includes a range of moraine types. It is important to differentiate between moraines deposited by temperate glaciers and those deposited by cold, polar glaciers. Most of the moraines that have been considered so far belong to temperate valley glaciers. There is also a significant difference between moraines laid down in water (but not *by* water) ponded in front of the ice, and those deposited subaerially. There is, too, a difference between moraines deposited by ice-sheets and those formed by valley and cirque glaciers. Examples of some of these different types will be mentioned. These types include the terminal moraines of valley glaciers in temperate areas, the type of cold ice-sheet moraine that has been designated the Thule-Baffin type, and the cross-valley moraines that have been formed in water.

LATERAL AND TERMINAL MORAINES

Lateral moraines lacking ice cores are usually much less conspicuous than those already discussed adjacent to active glaciers. In mountain areas this type of moraine sometimes only forms an amorphous deposit of debris on the valley side. The lateral moraine usually becomes more distinctive when it leaves the valley side and merges with the terminal moraine to swing across the valley as an arcuate ridge. Examples in the Conway valley of North Wales are given by C. Embleton (1961).

R. W. Galloway (1956) has studied the fabric of lateral and terminal moraines in an area of north Norway and has shown that the preferred orientation of the elongated stones lies at right angles to the lateral moraine ridge crest. This suggests strong radial movements when the moraine was formed. In the terminal moraine the stones lie mainly parallel to the moraine crest. These stones were probably emplaced by rolling and thrusting.

Terminal moraines are also considerably less imposing when they no longer contain an ice core. However, moraines without an ice core can attain considerable heights under suitable conditions. The terminal moraine of the Franz Josef glacier in New Zealand reaches 430 m in height on the narrow coastal plain beyond the mountains. This is an active glacier. Other very high terminal moraines include the arcuate hills that mark the limit of the extension of the Alpine glaciers on the north Italian plain. The ice in this area was also active and built up large moraines from the abundant debris carried out of the mountains. Similar high terminal moraines can be built by ice-sheets under favourable circumstances. In the east of Denmark, for example, the ice-sheet of the Weichsel maximum remained stationary for long enough to build up morainic hills reaching an altitude of 173 m.

Moraines are not always high or continuous. This may result from several possible causes. On the east of the New Zealand Alps, for example, some of the terminal moraines have been considerably eroded by subsequent fluvio-glacial meltwater activity. The lower part of the moraine also has been buried in some areas under the rising mass of fluvioglacial outwash. The absence of terminal moraines does not necessarily mean that they were never laid down. This absence of moraines in some valleys may make the correlation of moraines in neighbouring valleys rather difficult if the number of moraines is used.

A series of well defined lateral and terminal moraine loops in the Walker

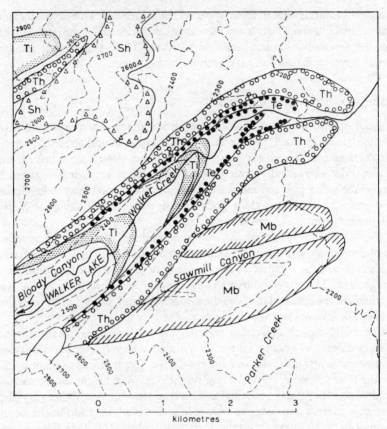

Fig. 15-1 Moraines of Walker Creek (Bloody Canyon) in the Sierra Nevada,
California (R. P. Sharp and J. H. Birman, *Bull. geol. Soc. Am.*,
1963)
Ti—Tioga;   Te—Tenaya;   Th—Tahoe;   Mb—Mono Basin

Creek area of California (in the Sierra Nevada) illustrates stages of glacial
advance and moraine formation. These moraines are quite substantial features
and their distribution is shown on Figure 15-1. The moraines have been
differentiated by studying their position and character. The earliest of the
moraine loops was formed during the Mono Glaciation, probably of Illinoian
age. It now consists of subdued ridges of lateral moraine. The stream changed
course before the deposition of the next set of moraines, the Tahoe series,
which probably belong to an early Wisconsinan glaciation. These are lateral
moraines up to nearly 100 m high and in places they bury the older moraines.
The older moraines were preserved by the change in stream course. This

change suggests a considerable time interval between the two moraines. There are two younger moraines (late Wisconsinan), belonging to the Tenaya and Tioga glaciations, and lying within the arc of the large Tahoe moraines (Plate VIII).

These moraines have been differentiated by interesting semi-quantitative methods (R. P. Sharp and J. H. Birman, 1963). The criteria used were 1) the total number of surface boulders; 2) the ratio of granodiorite to hard meta-morphic boulders (the former are much more susceptible to weathering than the latter); 3) the ratio of weathered to unweathered rocks on the moraine, unweathered being defined as a boulder with a recognizable portion of abraded surface; 4) the proportions of material of different sizes. In the last case, samples are taken from holes dug in the ground. One gallon of material is taken from the upper 15 cm and divided into three size categories, over 5·6 mm, 5·6-0·8 mm, and less than 0·8 mm. A second sample is taken from the next lower 15 cm of material and is similarly sieved in the field. The results shown in Table 15 indicate the striking contrasts between the different moraines.

TABLE 15   *Semi-quantitative Investigations of Moraines of Different Ages in California*

	Mono (oldest glaciation)	Tahoe	Tenaya	Tioga (youngest glaciation)
No. of boulders over 0·3 m diam. in 6 × 30 m strip	60	115	180	300
Granite/Resistate ratio	17/83	29/71	50/50	no count
Weathered to unweathered ratio	95/5	80/20	50/50	10/90
Hole-in-ground data:				
1st Interval—				
Coarse	24	40	67	79
Medium	26	24	7	4
Fine	50	36	26	17
2nd interval—				
Coarse	32	40	53	65
Medium	24	23	13	12
Fine	44	37	34	23

Terminal moraines formed directly of glacial deposits are usually found in front of stationary or advancing glaciers. The material forming these moraines is derived from the valley up-glacier and consists partly of material pushed forward by the glacier in sliding over its bed. Material is also added by the sliding of superficial debris down the steep end-slope of the glacier, which is the characteristic form during advance, and by the movement of debris up shear planes near the snout of the glacier. None of these processes operates so effectively when a glacier is retreating. Thus terminal moraines determine stages of advance during general deglaciation.

The character of the moraine depends on the material available to build it. In general, its slopes will be fairly symmetrical, both sides tending to lie at the angle of rest of the material. The coarser the material, the larger the moraine will tend to be, as meltwater cannot transport the larger blocks. A moraine of very large blocks, about the size of a small house, was formed in the first decade of this century during a short advance of Austerdalsbreen in Norway (Fig. 15-4).

The climate also affects the size of the terminal moraine as it determines to a considerable extent the amount of frost activity around the glacier. If the rocks are suitable, this may be a major source of debris. More important, however, the climate determines the activity of the glacier in large part. In a very cold dry climate, the glaciers are less active than in a wetter, warmer one. On the other hand, the abundant meltwater in the latter type may remove much of the glacial debris. Climatic factors can therefore have opposite effects, in this instance. The active glacier promotes moraine formation but the abundant meltwater either washes away the moraine or buries it beneath fluvioglacial deposits. A cold climate, with active glaciers, should produce the largest moraines.

V. Okko (1955) has discussed the moraines of Iceland and has suggested that moraines be divided into two types, which may be called 'push' and 'dump' moraines. He suggests that most of the moraines of the Vatnajökull glaciers belong to the first type. Evidence for this is seen in the peat layers that have been folded and now underlie the moraines of Breiðamerkurjökull and Kviarjökull. Sometimes these push-moraines are overlain by a layer of ground moraine brought up along thrust planes. The Breiðamerkurjökull moraines are 30 km long and 1 to 2 km broad. Orientation studies of the stones in the moraine give a weak maximum parallel to the last ice-flow and perpendicular to the ridge elongation. In places the moraine consists of several parallel ridges. The fabrics in these double ridges are clearly parallel to the flow in till, which overlies stratified sand, mixed with the peat beds. This succession suggests a double ice advance with sedimentation taking place in the interval between the advances. Only the proximal slope of this moraine is the result of glacial deposition, the distal one being an erosional feature. The Kviarjökull moraines are more than 100 m high and consist mainly of sand with very large boulders. The Morsarjökull moraines show a stone orientation parallel to the ice-flow and this suggests addition of basal till to the moraine. These moraines appear to be of the first type, but they have been subsequently overridden by thicker and more active ice. The end moraines of Hoffellsjökull are mainly dump moraines. The fabric analysis at a depth of 1 m shows no certain orientation and the material is stratified in places. Evidence of the presence of marine shells indicates that the moraines were deposited in water.

PUSH-MORAINES

The term 'push-moraine' has been used to describe features which are not formed in the same way as the push moraines described by Okko in Iceland. Some of these push-moraines occur in front of an advancing ice-sheet, rather than a glacier, and there must be suitable deposits for the ice-sheet to make into the push-moraine. Recently formed moraines of this type can be differentiated from other terminal moraines by their form. Those in Schleswig, for example, are more convex in profile. Their slopes are steeper, especially on the distal side, where the frontal concavity is not developed as in normal terminal moraines. There may be a break of slope between the moraine and the proglacial deposits in front of it.

Push-moraines of this type can be identified best by their material and internal structure. The material shows signs of tectonic disturbance by faulting and thrusting. This causes the development of imbricate structure. A good example of this type of moraine is seen in the Itterback region of north-west Germany. There is also a good example of this type of tectonic structure in the drift exposed in the cliffs near Dinas Dinlle in North Wales. Glacial tectonic structures are well displayed in this coastal section. The material lying in the path of the advancing ice-sheet must have been in a rigid, frozen state in order to fracture abruptly when subjected to very high lateral pressure by the advancing ice. The tectonic features revealed in the structure of push-moraines of this type are formed by this strong pressure. Sometimes waves and folds are formed as well as low angle thrusts and faults.

Where the material shows fracturing throughout a considerable thickness of up to 100 to 150 m, in the Itterback area for example, the depth of permafrost must have been very considerable. It can be inferred, therefore, that the Scandinavian ice-sheet advanced into an area that had previously been deeply frozen. This deep freezing would require a prolonged cold period before the ice advanced to form the push-moraine. Several tens of thousands of years would be required to produce a permafrost layer of 100 to 150 m depth. It can be concluded that the ice only advanced after a period of general cooling of considerable duration. The push-moraines of the last major ice advance were not so extensive as those of the earlier Saale period, to which the Itterback moraine belongs. This conclusion agrees with the known colder conditions in the earlier glacial period.

Similar types of push-moraines have been recorded in North America, for example, in the New York area and in southern Iowa, Alberta, and on the Yukon coast. There are ice-pushed moraines in Holland, which have been described by M. G. Rutten (1960). They consist of early Pleistocene and Tertiary sediments, sand, gravel and clay. The tectonic structures have a dip of 50 to 70 degrees and strike parallel to the length of the push-moraine ridges, dipping away from the outer faces. The ridges are only covered by a veneer of drift. They are 50 to 200 m high and up to 100 km long, arranged in regular sequence following the valleys. It appears that they form best in those areas where the advancing ice ponded the drainage, since this allowed more effective permafrost development. The push-moraines are the result of advancing ice pressure but they are not formed of glacial deposits except for the thin surface veneer.

THULE–BAFFIN MORAINES

The moraines of cold ice-sheets differ in several respects from those of temperate valley glaciers. The areas in which they have been studied most are around the Barnes ice-cap in Baffin Island and in the Thule area of Greenland. J. Weertman (1961) has proposed, therefore, that they be called the Thule-Baffin moraine type. This name is preferable to 'shear' moraine, as this latter term suggests a genetic origin that has not been generally accepted, and for which alternatives exist.

These moraines were described in detail by R. P. Goldthwait (1951). It is on the basis of his explanation of the features that they have been called 'shear' moraines, although he used the term 'end moraine'. The moraines that he described lie around the edge of the Barnes ice-cap, which is a cold ice-sheet, having a mean temperature of about − 10°C. The material forming the supraglacial debris is thought to have come to the surface along shear planes from beneath the glacier, where the ice is thinning near its edge. The ice up to 30 to 60 m above the steep toe of the ice-cap is layered by many thin dirt bands, striking nearly parallel to the ice edge. Although much of the material is fine dirt, there are some pebbles and boulders as well. Goldthwait suggested that the debris moved up along shear planes dipping down towards the bed of the ice-cap. There is good evidence for the existence of shear planes of this type near Flyway Lake. The material accumulates on the surface of the ice near the ice margin to form a dirty zone about 135 m wide. The concentration of dirt on the surface only takes place where the ice thickness is less than 85 m.

The ice-cap surface is clean beyond the 135 m wide strip of dirty ice. Near the edge, there is a black ice surface, which is covered by a thin film of dirt.

This layer is usually not more than 2·5 mm thick and the surface slopes at 15 degrees. The dirt-covered ice melts more rapidly than the white ice above it. The ablation rate measured was 20 cm/week on the clean ice and 32·5 cm/week on the dirty ice. It was estimated that 150-200 cm of the dirty ice was lost in a year's ablation, but only 100-150 cm of the white ice. This difference results in a steepening of the dirty ice slope. Material moving down this steepening slope accumulates at the bottom to form a thicker layer of debris.

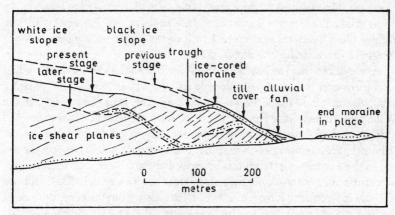

Fig. 15-2 Thule-Baffin cold ice moraine formation at the retreating margin of the Barnes Ice Cap (R. P. Goldthwait, *J. Geol.*, 1951, University of Chicago Press). Debris is transported up inclined shear planes, gathering on the surface as ablation moraine, and finally deposited as end moraine (right)

Thus an accumulation of till 0·6 to 1·2 m thick forms on the last 30 m of the slope, and a depression develops at the junction of the dirty ice slope and the thicker till layer on the outer margin. This is caused by the more rapid melting of the ice under the thin dirt cover than under the thicker till cover. The trough separates the developing ice-cored moraine from the dirty ice beyond. In time, the ice core may melt and a low terminal moraine will be left in place (Fig. 15-2).

Goldthwait estimated that it would take about 25 years to accumulate 0·9 m of debris on the outer ice-cap margin. The ice-cored moraines have a relatively steep slope of 30 degrees and are about 15 m high, but when the ice core has finally melted, the end moraines are no more than 1·5 m to 4·6 m in height and 15 m to 150 m broad. Such features can only form where the ice is moving actively to within 30 m to 60 m of its margin, and they are, therefore, characteristic of active cold ice-caps.

Although the later stage of development of this type of moraine is clear, there is more doubt concerning the method by which the debris reaches the surface. It is this part of Goldthwait's theory that has been modified by the later work of Weertman (1961), who calls these moraines the Thule-Baffin type. The main problem that must be solved is how the debris becomes concentrated in the lower layers of the ice. Weertman points out that the character of the debris does not support the shear hypothesis, as the fine debris can be distributed thinly throughout a considerable thickness of ice. There are also examples of thick layers of debris concentrated in bands up to 0·5 m thick. These layers consist of stones, sand and other debris, while the slightly dirty layers may be up to 1 to 2 m thick. A new suggestion is put forward to account for the dirt in the ice.

It is suggested that the debris becomes frozen into the base of the ice when basal meltwater freezes. A considerable amount of evidence is accumulating to suggest that cold glaciers do not slide over their beds (Chapter 4). In the Thule area, tunnels have shown that there is no movement between ice and bedrock, although there is considerable shearing in the lower layers of ice. Where the ice is temperate, on the other hand, water can exist beneath the ice, and in these conditions debris can be moved as the ice can slide over its bed. Weertman suggests that only the outer part of the ice-cap in Baffin Island and at Thule in Greenland is frozen to its bed. Farther in from the edge, the base of the glacier may be at pressure melting-point. The heat of sliding and geo-thermal heat may exceed the rate at which heat can escape upwards by conduction from the inner part of the ice-cap. The basal ice would melt to form subglacial water. The pressure gradient would force the water out towards the edge of the ice-cap, where it would enter a zone of the ice-cap where the thinner ice can allow the extra heat to escape to the atmosphere. The basal water will then refreeze at the bed of the glacier, while the base of the ice will still be at melting-point owing to the latent heat of freezing. If it can be shown that the base of the ice-cap in its central parts is at the pressure melting-point and yet no water is emerging from the basal margin of the ice-cap, then it may be assumed that basal water is refreezing near the edge of the ice-cap.

An ice-cap in a steady state would not introduce much debris, as distinct layers, into the lower part of the ice. Ice-caps normally, however, change with time. Variations in thickness, and the accompanying change in pressure, will cause variations in sliding velocity. Thus the amount of heat generated at the base of the ice will vary with time, and the point at which the ice re-freezes on to its bed will also vary. If the freezing zone extends outward towards the edge of the ice-cap, the water will freeze at the bottom in an area where before it was liquid. This may occur below a zone into which material had previously been frozen. The overlying debris is thus incorporated into a layer of ice with clean ice above and below. If this cycle is repeated, with freezing following a renewed melting, then the layers of debris can increase in number with each cycle.

Once the debris has been included in the ice, it flows with it and can be carried into the ablation zone where the flow lines must come to the surface. Where the edge of the ice-sheet is in the ablation zone, as in the Barnes ice-cap and the Greenland ice-sheet near Thule, the debris will arrive at the surface and can form terminal moraines in the way already described. At the moment, there are no direct measurements to support the view that the centres of these ice-sheets are at the pressure melting-point. However, from theoretical considerations this appears likely, so that this method of moraine formation, if not proved, is at least possible.

CROSS-VALLEY MORAINES

The Baffin-Thule type of moraine is formed in front of an ice-cap that terminates on dry land. Another type of moraine has been studied that demands that the ice-cap should end in ponded proglacial lakes or the sea. Moraines of this type have been variously named 'wash-board', 'De Geer', and 'cross-valley' moraines. The De Geer moraines, described by G. Hoppe (1959) in Sweden, have also been called 'annual' moraines. However, the suggestion of annual formation cannot be proved, according to Hoppe. Until they have definitely been shown to be annual, this is not a desirable name. The term 'wash-board moraine' is also not appropriate, as it has been used to cover a variety of forms.

The De Geer moraines, described by Hoppe in Norrboten, are about 6 to 7 m high; they have a breadth of 8 to 40 m and may be up to 1 km long. They are asymmetrical, having steeper distal or down-glacier slopes, on which many loose boulders occur. They are usually formed of till, which occasionally shows evidence of pressure structures and occasionally they contain stratified lenses. These features are only found below the highest marine shoreline. They occur below sea-level in places. It is generally agreed that the features were formed at or very near to the ice-front, when this was in the form of an ice-cliff. They were explained by De Geer as a type of push moraine, formed as the glacier advanced in winter. If this were the correct method of formation they would be annual features. However, it has been shown, by comparison with varves, that in fact they are not always annual and more than one can form in one year. G. Lundqvist (1948) showed that the pebbles on the moraines were orientated perpendicular to the ridge direction. Although this direction is often parallel to the last ice movement, it appears to be little related to it. The pebbles remain orientated at right-angles to the ridge direction even when this is not parallel to the ice move-ment. Hoppe suggests that water-soaked moraine under the ice has been squeezed into crevasses near or against the snout, lying perpendicular to the ice front. Irregularities in the pattern can be explained by the calving pattern of the glacier front. The pattern also reflects the rate of retreat of the ice front, which tends to be more rapid where the water is deeper in broken country. The resulting pattern is usually uneven.

The cross-valley moraines, described by J. T. Andrews (1963a and b) from Baffin Island are similar in many ways to the moraines described by Hoppe. The cross-valley moraines were formed in lakes held up in front of the Barnes ice-cap on its north-west margin. Similar lakes now occur on the north-east side of the ice-cap, such as Conn and Bieler Lakes. Andrews undertook a detailed study of the till fabric and the spacing of the moraines. The moraines occur in the valley of the Isortoq River and their spacing and number were studied on aerial photographs. Unlike normal terminal moraines, they tend to be concave rather than convex away from the ice front. In form, the moraines are asymmetrical, having a steep distal face with a slope of 34 degrees and a proximal slope that varies between 15 and 24 degrees. The moraines extend from near the valley bottom up towards the lake shoreline, but are best developed in their middle section, degener-ating towards their lower limit to a rather irregular mass of ridges and hollows. At their largest, the moraines are up to 15 m high and are composed of 20 per cent gravel, 60 per cent sand and 20 per cent silt and clay.

Analysis of the fabric of the till in the moraines has yielded valuable infor-mation concerning their possible method of formation. Both the orientation and the dip of the longer stones in the till were analysed, using pairs of samples from the distal and proximal moraine slopes. The fabrics were found to be very different on the two sides of the moraines. Those on the proximal side of the moraines show a stronger dip and orientation pattern. It is clear, therefore, that the force forming the moraine must have come from the up-glacier or proximal side. The orientation of the pebbles shows that they lie at right-angles to the moraine orientation and not necessarily normal to the direction of the latest ice movement. This is sometimes not at right-angles to the ridge crest. The dip of the pebbles shows a very strong up-glacier plunge on the proximal side. There is little preferred dip on the distal side of the moraines (Fig. 15-3).

The fabric is not associated with the basal movement of the ice, but is imposed by another process of moraine emplacement. In studying the spacing of the moraines, it was shown that there is a correlation between the spacing of the moraines and the distance from the watershed from which the ice is retreating down-valley. Gradually deepening lakes are thus impounded until the exposure of a lower col allows sudden shallowing of the lake by reducing its level. The moraines are closer together farther from the watershed. A graph of moraine spacing against depth shows a striking linear correlation, which is highly significant statistically. The depth of the lake in which the moraines formed varied from 120 to 200 m, the ice thickness being about the same as the water depth. The mean distance between the moraines is 36 m and 33 m in two areas. There is a tendency in part of the valley for two

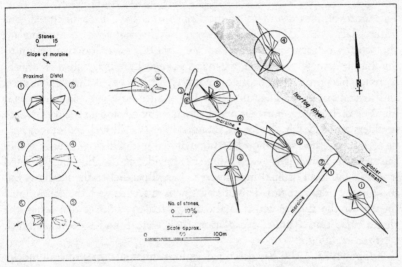

Fig. 15-3 Till fabrics in cross-valley moraines in Baffin Island (J. T. Andrews and B. B. Smithson, *Bull. geol. Soc. Am.*, 1966)

moraines to be closer together, followed by a wider distance before the next pair.

A possible method by which these moraines could form is associated with the hydrostatic pressure present at the base of a high ice-cliff in a lake. The hydrostatic pressure of the water at the base of the cliff would tend to cause seepage under the ice foot. The seepage would tend to thaw the basal ice and the moraine-covered ground on which it was resting. This moraine would then become soaked with water, reducing its strength until it became squeezed up in front of the ice face. The pressure of the ice would induce the till to flow and the stones in it would take up a preferred orientation and dip pattern based on the stresses in the till. The flow on the distal side would not be confined to the same extent as on the proximal side and hence the less strong till fabric on this distal side can be accounted for. The double moraines may form as a result of the ablation and forward movement of the ice acting together to give two periods in the year when the ice front would tend to be stationary for a short period.

There is some evidence that 650 to 800 moraines had formed in a 700-year period. This suggests that they may be annual in character, although this is by no means proved.

Moraine Landscapes

The more distinct examples of morainic forms have been discussed, but features left by the deposition of ground moraine and ablation moraine are also worthy of mention. Hoppe and Schytt (1953) have drawn attention to the fluting of newly exposed, flat-lying ground moraine, which is a fairly common phenomenon. The fluting consists of furrows less than 30 cm wide and several cm deep. They lie parallel to the ice-flow and may be several hundred metres long. Furrows may be formed by erosional processes, where boulders embedded in the ice are dragged over the till surface. Ridges may be formed by the squeezing of plastic material into a hollow groove left in the ice by its movement over a boulder embedded in the till. The elongated stones in the level ground moraine are aligned parallel to the ice-flow, but those in the ridges are not. This observation is in accord with the depositional theory for the ridges, which is also supported by the fact that when the ridges were traced under the glacier, they were found to contain much refrozen meltwater.

Hoppe (1957, 1959) has described various different types of hummocky morainic landscape in Sweden. One of these occurs around Lake Rogen in Härjedalen, and has been referred to as the Rogen type. It covers large areas, stretching from Dalarna in the south to Norrbotn in the north. The other type is known as Veiki moraine, from an area near this place in Gällivare. The Veiki-type moraine consists of three elements: moraine plateaux, hollows, and ridges marking the edges of the plateaux. Stones in these rim ridges show a strong preferred orientation. Hoppe suggests that moraine was deposited unevenly beneath actively moving ice giving the plateaux and hollows. Ice movement then ceased, but water-soaked moraine was squeezed up from the hollows against the plateaux, giving the rim ridges. The presence of drumlins associated with the moraine plateaux and hollows supports the contention that these two elements, unlike the rim ridges, formed beneath active ice.

Ablation moraine, sometimes called kettle moraine, can give rise to very distinctive landscapes under some circumstances. A very good example of an ablation moraine landscape can be seen in the south-east of Ireland in County Wexford. The moraine consists of a thick layer of very sandy till, forming a very hummocky landscape. Amongst the rounded hummocks are small round lakes. The whole landscape is the result of the melting away of an ice-sheet that just impinged on the south-east coast of Ireland from the Irish Sea. Large lumps of ice must have been incorporated in the sandy material and these melted out slowly to form the kettle-holes, now filled with water. The whole landscape appears to be of very recent origin, as the features are very fresh, and it probably represents a late ice advance on to the coast of Ireland. Other features associated with a decaying ice-sheet will be considered when dead-ice features are dealt with in Chapter 17.

Geomorphological Significance of Moraines

It has been shown that different types of moraine require different conditions for their formation. Some only form in water, others only if the ice-sheet is cold and frozen to its bed at least at the margin, others again form at the snout of a glacier when it is stationary or advancing.

It is the latter type of terminal moraine which has been used a great deal to assist chronological analysis of glacial action. The well-known terminology of Alpine glaciations in Europe was partly based by A. Penck and E. Brückner (1909) on the main terminal moraines in the area surrounding the Alps, to which the enlarged glaciers extended. In the same way, the maximum advance of the Newer Drift in Britain is marked over part of its length by terminal moraine formation. The moraines at York and Eskrick in the Vale of York are well-known limits of the Newer Drift maximum as are those of South Wales (for example, at Glais in the Tawe valley). Another very subdued morainic ridge has been identified by A. Straw (1958) on the eastern margin of the ice-sheet where it was pressing up into the Fenland gap near Stickney. This moraine is low partly because of its partial subsequent burial by Fenland deposits.

From the outer moraines of the Newer Drift ice advance, a series of moraines can be traced back towards the ice source in the hills. These moraines were formed by intermittent ice advances during a period of general deglaciation. It is noticeable that the moraines become progressively fresher and more hummocky as they are traced towards the hills. The latest moraines in the sequence are found in the cirques of North Wales, the Lake District, and Scotland, where they date from the post-Allerød cold phase (Zone III of the Late-glacial period). These moraines show very little post-glacial modification, as they were formed only about 11,000 years ago. The ones marking the maximum extent of the Newer Drift ice probably date from between 20,000 and 30,000 years ago.

The moraines in the valleys around the Jostedalsbreen in Norway provide a sequence that continues into the more recent past. The outermost of a series of recent moraines formed by Austerdalsbreen (Fig. 15-4) and Nigardsbreen can be associated with the advance in the mid-eighteenth century. Historical evidence, studied by Faegri (1948) for Nigardsbreen, is valuable in this respect. There is evidence that the glacier advance of this period invaded farm land. A similar moraine in Austerdalsbreen has also been dated to the same period by tree ring analysis. Between this outer moraine, which formed around 1750, and the glacier snout there is a series of low symmetrical moraines. The moraines nearest to the glacier snouts are dated at 1928 and 1930 in Austerdalsbreen and Nigardsbreen respectively. The next nearest moraines in each glacier were deposited around 1909 and 1906 respectively. For this period climatic data are available and also annual measurements of the movement of the glacier snouts. The measurements are complete for Nigardsbreen and nearly complete for Austerdalsbreen. The measurements show quite distinctly that these particular moraines were formed during a period of ice advance and climatic deterioration. Nigardsbreen advanced from 1905 to 1910, when it remained stationary for a few years. It advanced again between 1922 and 1925 and then remained stationary until 1932 when a very rapid retreat started. This retreat has continued uninterruptedly until 1960. This example shows how changes in the position of a glacier margin can be related to the formation of terminal moraines of this particular type, which are low symmetrical moraines.

Conclusions

Moraines can be divided into active moraines, ice-cored moraines and stable moraines. The first type rests on or alongside active glaciers or ice-sheets. This type can be further subdivided into median, lateral, terminal and ablation moraines according to their position relative to the ice mass. The second category of ice-cored moraines can yield useful information when the ice that they contain (partly glacier ice and partly snow-bank ice) has been dated.

Where the ice has retreated, moraines become stable and in this form they provide valuable evidence of past glacier advances. This type can also be subdivided into 1) lateral and terminal, and 2) ground moraine. Another form of terminal moraine is the push-moraine. This moraine is formed of material lying in front of an advancing ice-sheet. It gives evidence of the conditions in front of the advancing ice-sheet, but it is not primarily a glacial deposit, although it may consist of glacial deposits of an earlier glacial period. Moraines formed by cold-based ice-sheets show different characteristics and have been termed Thule-Baffin moraines from the areas in which they have been studied. Another special type of moraine that forms only in proglacial

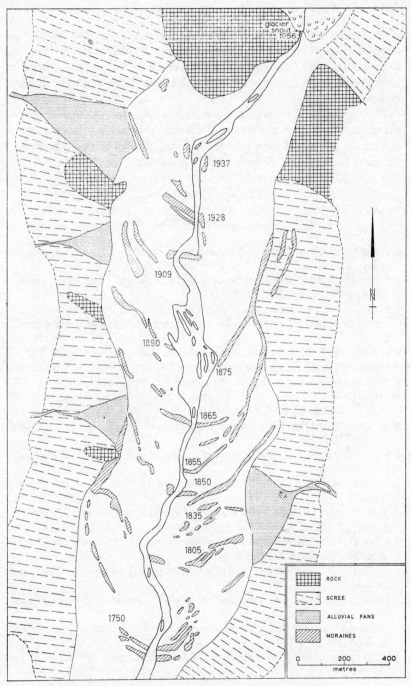

Fig. **15-4** Terminal moraines in Austerdalen, Norway

water bodies, but which nevertheless consists of till, is the cross-valley or De Geer moraine. This type of moraine should be differentiated from the delta-moraine type, which is basically a fluvioglacial form and will therefore be discussed in the next chapter, although it also forms in proglacial water bodies.

The deposition of moraine has given rise in places to distinctive types of landscape, while moraines also give valuable evidence of phases of inter-mittent ice advance during general deglaciation. This is especially true of the more recent phases of the last major glaciation.

REFERENCES

J. T. ANDREWS (1963a), 'Cross-valley moraines of the Rimrock and Isortoq River valleys, Baffin Island, North West Territories', Geogr. Bull. 19, 49-77

J. T. ANDREWS (1963b), 'Cross-valley moraines of north-central Baffin Island: a quantitative analysis', Geogr. Bull. 20, 82-129

J. T. ANDREWS and B. B. SMITHSON (1966), 'Till fabrics of the cross-valley moraines of north-central Baffin Island, North West Territories, Canada', Bull. geol. Soc. Am. 77, 271-90

J. A. ELSON (1957), 'Origin of washboard moraines', Bull. geol. Soc. Am. 68, 1721

C. EMBLETON (1961), 'The geomorphology of the Vale of Conway, North Wales, with particular reference to its deglaciation', Trans. Inst. Br. Geogr. 29, 47-70

K. FAEGRI (1948), 'On the variation of western Norwegian glaciers during the last 200 years', Gen. Assembly int. Un. Geod. Geophys., Oslo (1948), 2, 293-303

R. W. GALLOWAY (1956), 'The structure of moraines in Lyngsdalen, north Norway', J. Glaciol. 2, 730-3

R. P. GOLDTHWAIT (1951), 'Development of end moraines in east-central Baffin Island', J. Geol. 59, 567-77

R. K. GRESSWELL (1962), 'The glaciology of the Coniston basin', Lpool Manchr geol. J. 3, 83-96

H. J. HARRINGTON (1952), 'Glacier wasting and retreat in the Southern Alps of New Zealand', J. Glaciol. 2, 140-4

K. HEWITT (1967), 'Ice-front deposition and the seasonal effect: a Himalayan example', Trans. Inst. Br. Geogr. 42, 93-106

G. HOPPE (1967), 'Problems of glacial geomorphology and the Ice Age', Geogr. Annlr 39, 1-16

G. HOPPE (1959), 'Glacial morphology and inland ice recession in north Sweden', Geogr. Annlr 41, 193-212

G. HOPPE and V. SCHYTT (1953), 'Some observations on fluted moraine surfaces', Geogr. Annlr 35, 105-15

C. A. M. KING (1959), 'Geomorphology in Austerdalen, Norway', Geogr. J. 125, 357-69

G. LUNDQVIST (1949), 'The orientation of block material in certain species of flow earth', *Geogr. Annlr* **31**, 335-47

G. MANLEY (1959), 'The late-glacial climate of north-west England', *Lpool Manchr geol. J.* **2**, 188-215

V. OKKO (1955), 'Glacial drift in Iceland, its origin and morphology', *Bull. Commn géol. Finl.* **170**, 1-133

G. ØSTREM (1959), 'Ice melting under a thin layer of moraine and the existence of ice cores in moraine ridges', *Geogr. Annlr* **41**, 228-30

G. ØSTREM (1961), 'A new approach to end moraine chronology', *Geogr. Annlr* **43**, 418-19

G. ØSTREM (1963), 'Comparative crystallographic studies on ice from ice-cored moraine, snow banks and glaciers', *Geogr. Annlr* **45**, 210-40

G. ØSTREM (1964), 'Ice-cored moraines', *Geogr. Annlr* **46**, 282-337

A. PENCK and E. BRUCKNER (1909), *Die Alpen im Eiszeitalter* (Leipzig)

L. F. PENNY (1964), 'A review of the Last Glaciation in Great Britain', *Proc. Yorks. geol. Soc.* **34**, 387-411

A. RAISTRICK (1927), 'Periodicity in the glacial retreat in west Yorkshire', *Proc. Yorks. geol. Soc.* **21**, 24-28

M. G. RUTTEN (1960), 'Ice-pushed ridges, permafrost and drainage', *Am. J. Sci.* **258**, 293

R. P. SHARP and J. H. BIRMAN (1963), 'Additions to the classical sequence of Pleistocene glaciations, Sierra Nevada, California', *Bull. geol. Soc. Am.* **77**, 1079-86

G. SLATER (1931), 'The structure of the Bride Moraine, Isle of Man', *Proc. Lpool geol. Soc.* **15**, 284-96

A. STRAW (1958), The glacial sequence in Lincolnshire', *E. Midld Geogr.* **2**, 29-40

J. WEERTMAN (1961), 'Mechanism for the formation of inner moraines found near the edge of cold ice caps and ice sheets', *J. Glaciol.* **3**, 965-78

E. M. YATES and F. A. MOSELEY (1967), 'A contribution to the glacial geomorphology of the Cheshire plain', *Trans. Inst. Br. Geogr.* **42**, 107-25

Chapter 16

Fluvioglacial Ice-Contact Features: Eskers and Kames

Eskers consist of washed and sorted, usually water-rolled and stratified materials. . . .
Esker ridges are usually irregular, often very sinuous. . . . (M. H. CLOSE, 1866)

In this chapter some of the features deposited by glacial meltwater in contact with the ice will be considered. One of the problems in discussing these features is that of nomenclature, for terms such as 'esker', 'ose' and 'kame' have not been used systematically. The first two terms have been used to describe the same features by different workers. The word 'esker', of Celtic origin, has been used more frequently in the English literature than the Scandinavian 'ose'. The latter term has been used more in Europe, and is adopted by J. Tricart and A. Cailleux (1962). The term 'esker' will be used in this chapter as it is best known to English-speaking people. The term 'kame' is useful, but again there are other terms that to a certain extent overlap with this one. A method by which eskers and kames can sometimes be differentiated is to use the term 'esker' only for those features that in general run at right-angles to the ice edge (G. Hoppe, 1961). The term 'kame' can be used for features lying parallel to the ice front. Often, however, the features are amorphous and cannot be easily grouped in this way. The term 'delta-moraine' has been suggested for some of the larger-scale features that are probably formed in a similar way to many kames, which are usually fairly small. Kame terraces should be differentiated from kames. Both may lie parallel to the ice edge, but kame terraces usually form along the side of a valley and not at the terminal ice edge. Kame terraces may thus lie at right-angles to those kames that form along the snout of an ice-sheet, and parallel to eskers. They can be differentiated from eskers by their position along the valley side and their different form.

The most important criterion by which fluvioglacial deposits can be differentiated from those laid down directly by the ice is the stratified nature of the meltwater deposits. However, not all stratified drift deposits need occur in one of the forms associated with the fluvioglacial features that will be described. For example, it has already been shown that some drumlins contain stratified deposits. These were laid down by fluvioglacial agencies but

PLATE XIX   MARGIN OF THE BARNES ICE-CAP, BAFFIN ISLAND, SHOWING MARGINAL DRAINAGE AND MORAINE.
(C.A.M.K.)

PLATE XX   LATERAL MORAINE, GLACIER DE LA GRANDE CASSE, FRANCE.
(C.A.M.K.)

PLATE XXI MORAINES BELOW THE TERMINUS OF THE CAMERON GLACIER, NEW ZEALAND.
(C.A.M.K.)

PLATE XXII KETTLE HOLES NEAR GILLIAN LAKE, BAFFIN ISLAND.
(C.A.M.K.)

were later shaped by moving ice into the typical drumlin form. There are also other fluvioglacial deposits that may not have the form normally associated with ice-contact meltwater deposits. These may have been laid down beneath the ice when it was stagnant and melting. Some of these deposits have been referred to as undermelt features and often they show interesting structures. Other features associated with stagnant ice will be considered in Chapter 17.

Eskers and Related Features

An esker is essentially an elongated ridge, formed of stratified fluvioglacial deposits, usually including much sand and gravel. The ridge may be sinuous or straight. Occasionally an esker consists of a number of separated elongated hills of bedded sand and gravel, forming together an elongated, but discontinuous feature. The term 'beaded esker' has been used to describe this rather unusual type. G. De Geer in 1897 used beaded eskers as evidence of still-stands in the ice margin retreat. Eskers also show occasional widening and other irregularities. Esker slopes are variable; at times they are steep, lying at the angle of rest of their material, but sometimes they have been modified to gentler slopes of 5-10 degrees. Some of the largest eskers occur in central Sweden, where they attain hundreds of kilometres in length, with only short gaps. Typical eskers have a fairly constant relationship between their different dimensions. Large eskers, even though they are very long, are never wider than 400-700 m and 40-50 m high. Smaller eskers may be 200-300 m long, 40-50 m wide and 10-20 m high. The material within them is always well worn and sorted. Some contain pebbles 5-20 cm long and although many have smaller material, some in western Baffin Island consist of large stones up to 1 m in size. Many eskers consist of bedded sand, dipping at 10-20 degrees or even more. The dip is normally outward from the centre of the esker and sometimes coincides with the surface slope. An esker fabric measured in an esker in Baffin Island showed a strong preferred stone orientation in the direction of esker elongation. There was also a strong preferred dip in a downstream direction.

OBSERVATIONS OF ESKERS FORMING CURRENTLY

Several eskers have been described in the process of formation. W. V. Lewis (1949) has described an esker forming near the snout of Böverbreen in Norway. The glacier ends on a gentle slope on which much mixed outwash has been deposited. The esker, as it emerged from beneath rapidly retreating ice, was nearly 37 m long and 3·6 m high. It had a flat, almost horizontal top, contrasting markedly with the sloping solifluction area around it. The surface carried a number of small still-active kettle-holes. There were more or less horizontal topset beds on the upper surface. Pits revealed foreset beds beneath 30–45 cm of topset beds, and this bedding suggests that the whole feature must have formed in standing water. It was apparent, however, that no lake could have existed in front of this glacier while the esker was forming. The steep side-slopes of the esker were caused by two meltwater streams that were issuing from the glacier on either side of the esker. The esker still contained an ice core, at least in part, whose melting would disturb the stratification and the form of the esker. Well preserved eskers are therefore only likely to form where ice is unlikely to be incorporated in the deposits. The material forming the esker must have been deposited in subglacially ponded water near the point at which a stream emerged from the glacier snout. The ponded water escaped as the ice front melted back.

J. C. Stokes (1958) has described an esker-like ridge that extends on to dry land in front of the retreating Svartisen ice-cap in north Norway. The ridge emerges from a tunnel, and consists of sand, boulders, and rock flour, with an ice core near the snout, where it is 16 m high. It decreases in height to 1–2 m away from the glacier as its ice core melts. The ice in the vicinity is almost stagnant, although there are signs of thrusting in the dirty ice as it is pressed against a marginal buffer of dead moraine-covered ice. In glacial tunnels beneath the glacier, debris was found piled up against the walls, ready to form ridges as the ice melted. This process would form small unstratified esker-like ridges, similar to genuine eskers in that they form in a subglacial tunnel, but unlike them in their lack of stratification.

Another esker has been studied on the edge of Breiðamerkurjökull in Iceland (P. J. Howarth, 1966). It appears on the ice surface and extends 420 m, running both parallel and at right angles to the ice edge. It is 6 m high near the ice edge and 1 m at the far end, and its width declines from 18 to 5 m. It is ice cored and the melting of the ice core causes the decrease in size along its length. The top of the esker as it melted out of the ice was irregular. The sinuous course and ice core suggest formation in an englacial tunnel. Structural control in the ice is probably exerted on the plan of the esker.

METHODS OF ESKER FORMATION

Two main hypotheses of esker formation have been put forward. First, there is the view that the esker deposits have been laid down by sub- or en-glacial meltwaters as they emerge from the glacier into ponded water. It has been suggested that eskers will form in certain conditions by the annual addition of small mounds of sediment as the ice front retreats. This leads to the formation of the beaded esker type. The summer period of rapid water-flow and abundant material adds each bead, while the winter is represented by the intervening gap. Eskers of this type seem rather rare. The second method of formation is the most favoured. This hypothesis suggests that the eskers are formed subglacially by the deposition of material by meltwater. The fact that meltwater streams can flow uphill under hydrostatic pressure could account for the fact that some eskers do in fact run uphill at times. In general, they are aligned down the ground slope and roughly parallel to the direction of ice-flow.

A subglacial origin of the eskers of the Rondane district of Norway is also supported by the fact that the stratified esker material is overlain by a thin layer of unstratified ablation till, which must have been let down on to the esker as the overlying ice melted away. Such eskers are often more rounded in cross-section. The streams depositing some of these eskers in the Rondane first flowed down the open hillside, and then under a lobe of decaying ice in the valley, before emerging again from beneath the ice. The reduction of gradient under the ice could account for the deposition of material in this instance. The blocking of stream channels by the collapse of decaying ice may at times be instrumental in initiating the deposition of material in the subglacial channels. C. M. Mannerfelt (1945) has described similar eskers in parts of Sweden.

The necessity for the ice to be almost dead in order that true eskers may be formed has restricted their occurrence mainly to areas covered by large ice-caps rather than by valley glaciers. Even rapidly retreating valley glaciers such as Austerdalsbreen move actively right to their snouts and this prevents esker formation. Eskers can also only form in those areas where meltwater can penetrate to the lowest layers of the ice in contact with the bed. Eskers are, therefore, normally found in temperate ice-caps, where the ice is at the pressure melting-point. It has been suggested that a rare type of esker could form by deposition of material in supraglacial meltwater channels. The material could then be let down on to the surface beneath the ice as the surrounding ice melts. Such a feature could form on a cold glacier with surface meltwater drainage, but as most glacier surface streams contain very little debris, eskers formed in this way must be very uncommon.

DESCRIPTIONS OF ESKERS IN DIFFERENT AREAS

R. J. Price (1965) has described eskers left in front of the retreating Casement glacier in Alaska, some of which have been revealed since the glacier became land-ending about 1907. A complex set has been exposed from the retreating ice since 1935. The eskers are ridges 3-9 m high, formed of well rounded gravel, cobbles and boulders. They occur on ground that slopes both away from and towards the glacier. Those formed on the slope down towards the glacier must either have been formed by subglacial streams under hydrostatic pressure or by englacial streams. In the latter event they must have been let down subsequently on to the reverse slope. A complex series of ridges is thought to have developed supraglacially in lakes, which have since drained, leaving shoreline features on the ridges. The largest ridges were probably on the site of meltwater streams and they appear to be true eskers which were underlain by ice at one time. Some near the glacier have ice cores, and photographs showed them extending on to the ice in places. They could either have formed in englacial tunnels or on the ice. Subsequent melting must have produced the uneven crest line. The lakes must have formed after the ice melted out beneath and around the eskers. Some of the eskers were 40 m high and 33 m wide at their base when they were still on the ice. These eskers have been destroyed by meltwater streams subsequently.

A series of north-east to south-west trending eskers has been described by C. Embleton (1964) in Anglesey. They lie to the east of a watershed at 36 m O.D. The eskers nearly all attain a similar altitude of 39·6 to 41·2 m, with an extreme range of 36·6 to 42·8 m at their crests. They were associated with subglacial meltwater streams draining to the south-west across the watershed. Where exposures are available, the internal structure of the eskers reveals current-bedded sand and gravel, capped by coarse ablation moraine 1·2 to 1·5 m thick. The direction of the eskers agrees with observations by J. B. Sissons (1958) that eskers often follow the latest direction of ice movement. Both Embleton and Sissons agree that the concordance in height of the esker crests is caused by the fact that meltwater was flowing freely in open ice-walled channels or englacial courses. The level of the water was controlled by the height of existing cols. This height would control an englacial water-table and deposition would occur up to this level. The eskers in Anglesey cannot have had an ice core or have been formed on ice. The size of the boulders in them suggests active streams. The englacial water-table probably sloped down from 42 m to the col at 36 m.

Hoppe (1961) has described some of the eskers of northern Sweden. These eskers occur both above and below the highest marine limit. They often extend for 5 to 10 km between breaks, which normally occur where the relief is highest. The largest eskers have rather flat, broad crests, supporting the view that the deposits were built up in successive layers in subglacial meltwater channels. Hoppe agrees with V. Tanner (1932) that the sharp-crested narrow variety of esker is the result of subsequent collapse of the sides when the supporting ice walls melt away. The eskers formed above the water-level consist mainly of stratified sand and gravel. Sometimes silt occurs within the esker or on its surface, and occasionally heaps of boulders are found. One particular esker, Hammesharju, shows some unusual features. It is situated south-east of Gällivare and is over 4 km long, 50 to 80 m wide and up to 16 m high, although its crest is broken by gaps. The esker was formed probably in a subglacial tunnel; later, a supraglacial stream was let down across it and eroded the gaps that occur along it. Ridges of till flank the esker on either side. These ridges contain stones preferentially orientated transverse to the ridge elongation. It is thought that they were formed by the squeezing of water-soaked till into the area on either side of the tunnel. There are also examples of a similar type of feature that forms when the till is pressed beneath and into the central core of the esker rather than alongside it.

Another interesting esker is the Vessö esker in southern Finland. It is in part a continuous ridge, but in places it is broken into separate mounds. It is situated about 50 km east of Helsinki, is 8 km long and varies in width from 300 m to 2 km, with a maximum height of 38 m (Fig. 16-1). It runs southeast to north-west radially to the large Salpausselka moraines to which it is linked. Its outer part consists of several islands, while it has been modified by the sea as it has risen isostatically since its emergence from the ice. The esker as a whole is asymmetrical and has a steeper eastern slope with a stony surface, continuing in places below sea-level. In places, there are depressions, which are steep and circular, along the ridge. Along part of its length, the esker forms a broad ridge with associated boulder-covered mounds. Elsewhere, the esker spreads out and merges into an even, sandy plain. The fluvioglacial material of which the esker is composed lies directly on the rock with no intervening till. The bedrock, an easily weathered granite, was eroded to form a depression along which the esker is situated.

The structure of the esker has been revealed in cuttings at a point where the eastern slope is 13 degrees and the western only 3 degrees. The esker is covered with boulders and its internal structure is shown in Figure 16-1. The sloping layer of clay, with a dip of 8 degrees runs almost parallel to the steeper eastern side. This layer probably formed the original esker surface, and the overlying sand and stones consist of later marine additions. Beneath the clay and silt, sand occurs in alternating layers of finer and coarser particles, showing cross-bedding. In the centre, the material is coarser. Some large angular boulders occur in the silt layer and on the upper surface, where the gravel and stone layer is 1–3 m thick.

The form of the esker and its constituent material give indications of its mode of formation. O. Granö (1958) suggests that the fluvioglacial material of the esker was deposited close to the ice margin in deep water. He considers that the character of the esker, particularly its core, agrees with the views of

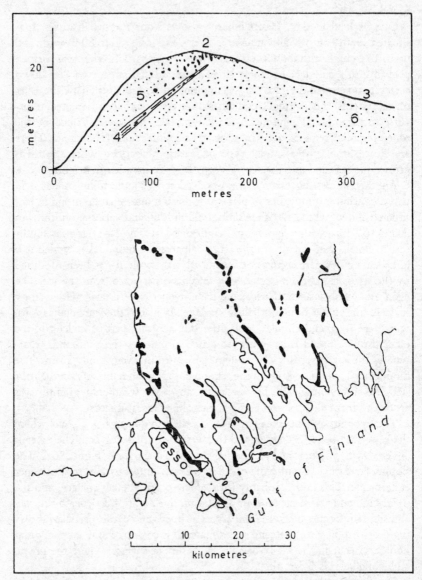

Fig. 16-1 The Vessö esker, Finland (O. Granö, *Fennia*, 1958, by permission of the Johnson Reprint Corporation)

De Geer concerning esker formation in submarine environments. The original core was probably deposited as a small delta at the mouth of a subglacial stream emptying into the sea. A clay and silt layer represents a remnant of the covering of fine material deposited when the esker was still in deep water. It includes large ice-rafted angular blocks. The boulders covering the clay could have been derived from a covering of stony till from which the fine material has since been washed away. The sand above the unconformity on the western side of the esker was deposited by the waves as the esker emerged above the sea. The steeper eastern slope is the result of movement of material to the lee side by the waves whose maximum fetch lay to the west. The feature has been considerably modified by the sea since its formation at the edge of the retreating Scandinavian ice-sheet.

Mention should be made of some of the eskers of Ireland, as it was from this country that the name 'esker' originated. An interesting series of eskers and associated features has been described by F. M. Synge (1950) in the area around Trim, north-west of Dublin. The area lies in the drainage basin of the River Boyne, which flows north-eastward to the north of the main esker area. The area is drained by streams flowing north-westward to join the Boyne (Fig. 16-2). When the ice-sheet associated with the eskers lay over the area, its margin ran roughly from north-east to south-west and was retreating north-west. As it was retreating downslope, it ponded up the drainage in front and lakes were formed which overflowed across the watershed to the south-east. It was in these lakes, according to Synge, that the eskers and associated features were formed. The eskers lie in a strip of land between the Boyne and the elongated ridge of sand and gravel, named by Synge the Galtrim moraine, which will be discussed in more detail on p. 387. The eskers trend from north-west to south-east within this strip. The bedrock is hidden beneath superficial deposits of drift, including sands, gravels, silts and till, which form a gently undulating area. There are twelve eskers in this area, including the impressive Trim esker. They all trend in about the same direction and join the Galtrim moraine at their eastern or south-eastern ends. The eskers are morphologically similar, consisting of long, narrow steep-sided ridges which are sinuous in plan.

All of them consist of bedded sand, gravel and some silt. There appears to be relatively little connection between their internal structure and their external form. One esker of uniform shape may change internally from fine sand to coarse gravel. Structurally also they are varied. Some consist of apparently horizontally bedded sands and gravels, often showing slumping and disturbance by faulting on their flanks. Some show arched bedding. Others consist of unsorted, though washed gravels, containing very large striated erratic blocks of limestone. Some may show delicate deltaic bedding,

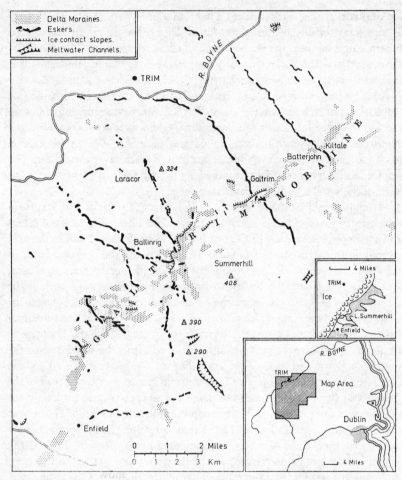

Fig. **16-2** The eskers near Trim, Eire (F. M. Synge, *Proc. R. Ir. Acad.*, 1950)

with some bands of lacustrine clay. The bedding may be graded either from coarse to fine upwards or the reverse. The main impression is one of variety.

The Trim esker is one of the best of the series. Its total length is 14·5 km and it increases in size towards the south-east. It starts as a small ridge 4·6 m high on the west bank of the Boyne. As it is traced to the east of Trim, it runs parallel with the Boyne and increases to 9 m in height, and becomes a well-marked ridge 15 m high for the last 8 km of its length. Where it reaches the Galtrim moraine, it extends up on to it and ends beyond the moraine crest at a height of nearly 100 m. It is unusual in that it does not end in a delta formation as many of these eskers do.

Another rather unusual esker is the Laracor esker, consisting of two parallel ridges which show a tendency to beading along part of their length. It is a small esker only about 7·5-9 m high and 4 km long. Two of the other eskers approach one another, but then swing apart, one ending in the Galtrim moraine and the other ending in a small delta.

Three important generalizations concerning this series of Irish eskers can be made. First, they show winding courses like rivers. Secondly, most of them act as feeders to the Galtrim moraine. Thirdly, many of them show some tendency to beading. The pattern and character of these eskers agree with the hypothesis that they formed under lacustrine conditions against or within walls of stagnant ice. Two of the small eskers continue over the crest of the Galtrim moraine. These probably formed where a cliff of dirty ice overhung the developing Galtrim moraine. Two parallel tunnels formed beneath this ice-cliff, which later collapsed on to the surface of the moraine. The tunnels became choked with gravel, and eventually formed the eskers. Where the eskers show beading, it is suggested that the beads represent stages in the retreat of the ice front, the material being deposited marginally. The absence of eskers north of the Boyne valley shows that ponded water is necessary for their formation. As the ice retreated beyond the valley line, the water could escape freely from in front of the ice margin. However, it has been shown that small eskers can form where water is ponded up temporarily within the marginal zones of a glacier or ice-sheet. It is probable that features of this type can form fairly rapidly. Fluvioglacial streams carry a large load of material and where conditions are suitable, deposits can build up fast. Incision can be equally rapid where the streams are acting erosively.

A model to demonstrate the formation of eskers has been devised by G. F. Hanson (1943). The experiment was designed to simulate a glacier stream discharging into a lake. A pipe, discharging sand and water into standing water, was slowly moved backwards on a platform, representing the ice margin. The sand emerging through the pipe produced an esker in the lake. This deposit remained uncollapsed if the water in the lake was allowed to drain away slowly. The 'esker' formed both with the pipe outlet just above the lake level, and also when the pipe was lowered below the lake level to simulate a subglacial stream. A flat-topped ridge formed when the lake level was lowered below the level to which sand could be deposited. The height and width of the ridge appeared to be related. These experimental eskers illustrate some interesting similarities with genuine eskers. The flat top, in particular, is seen on some of the Trim eskers.

Some interesting observations have also been made of the stone types and mineral grains in eskers. H. A. Lee (1965) has described a long esker in Canada called the Munro esker (latit. 48°N., long. 80°W.). The esker can be traced easily on aerial photographs over a distance of 400 km trending north-south. It consists of a winding plateau 1·6 to 6·4 km wide, with depressions along its axis. Bell-shaped ridges occur between the depressions, and these correspond to subsurface depressions, giving a total thickness of more than 88 m of sediment. The sediment is dominantly sand, but there are layers of coarse gravel or boulders. Sheet bedding is characteristic and cross-bedding suggests sedimentation in braided streams. Foreset beds are not prominent, but deep sections are uncommon. The sides of the esker have been modified by wave action in lake waters.

Samples of the esker material were taken at intervals to assess the distance that different sizes and types of minerals had been carried in the esker. The maximum abundance of any mineral does not occur at the point of outcrop of the rock containing the mineral but at some distance downstream. The distance between the outcrop source of the mineral and its maximum abundance has been called the K distance. This distance is the same for particles that behave in a similar way hydrologically but it differs with the size and density of the particle. The results of the sampling in the Munro esker indicate that dunite in the size range 3·35-8 mm had a K distance of $13 \pm 3·2$ km, while particles in the size range 8-16 mm had travelled less far. Trachyte of 8-16 mm had travelled $4·8 \pm 3·2$ km, and gold grains of 10 microns and over had travelled only $3·2 \pm 3·2$ km. These results show that transport of material in eskers is not very great, although some mineral grains are carried far beyond the K distance, as the curve of percentage number of grains tails off slowly from the K distance peak.

A. Hellaakoski (1931) has made similar observations on an esker at Laitila. The esker is rather sinuous and rarely exceeds 20 m in height. It runs roughly parallel to the striations and is approximately perpendicular to the ice front direction. It is more than 27 km long and is not influenced by the relief, although its shape changes with the altitude. It is ridge-like in the higher parts and forms rows of hills in the lower parts. At one time, it was under 150 m of sea. It runs across an outcrop of Rapakivi granite for 27 km of its length. The stone types were counted both in the drift alongside the esker and in the esker itself from the point where it crossed on to the Rapakivi outcrop. At a distance of 1·5 km from the proximal contact with the Rapakivi granite, this rock gives 50 per cent of the stone count in the drift. However, in the esker the Rapakivi granite only starts to appear 5 to 8 km from the proximal contact. Thereafter the percentage rises steadily and reaches a peak between 15 and 35 km from the proximal contact, where it is twice as high as in the adjacent morainic drift. The percentage of Rapakivi granite stones begins to fall at a similar distance from the distal contact of 5 to 8 km. The morainic drift was the source of the esker material and this was carried about 4 km or more before being deposited. The stones are destroyed in the esker, but preserved in the morainic drift. The stones also become rounded in the esker, but remain angular in the morainic drift. The agency forming the esker could, however, not drag the large boulders, which were lying close to it. The esker material was well worn, mixed and washed. It is suggested that the deposit was laid down by a subglacial river, but the results of the observations indicate that the stones forming the esker have not been carried very far.

Two different types of esker were examined in Baffin Island, where they occur in association with dead-ice phenomena and fluvioglacial outwash deltas. Some of the eskers feed into deltas in the same way as those associated with the Galtrim moraine in Ireland. These short feeding eskers in Baffin Island sometimes contain very large boulders (the mean length of 50 was 89 cm) and merge into the ice-contact slope of the deltas. They must have been formed when the ice front remained stationary for long enough for extensive outwash spreads to build out into the sea or lake as deltas.

The other type of esker is more sinuous and occurs in dead-ice areas among the outwash deposits in association with kettle-holes and kames. Some of the eskers are sinuous and others bifurcate. At least some of these sinuous ridges indicate by their internal structure that they have been largely deposited by flowing water. The fabric of one esker shows a bimodal orientation of the long axes of the stones within it. The main direction is parallel to the esker crest with a marked downstream dip. The secondary mode is at right-angles to the crest with a dip down the flank of the esker and was probably induced by later collapse and slumping of the esker sediment. The variable down-slope dip of the beds also suggests collapse as the retaining ice walls melted away. The sand also shows characteristics of flow in its good sorting ($\sigma_1 = 0.70$).

The form of another esker in the same dead-ice area is shown in Figure 16-3. The ridge is sinuous and rises slowly in crest elevation by about 40 m along its length before it finally merges into a kame terrace. The material is coarse as shown in the Figure. The ridge is sharp crested and flanked on both sides by kettle-holes. Its sides slope at about 25 degrees. These characteristics suggest that the final form of the ridge is the result of slumping. It seems unlikely that the ridge could have been formed entirely by water flowing in a subglacial tunnel owing to the very coarse material and the upward trend of the ridge in the direction of flow. It is more likely that the finer material was deposited by water flowing in ice-walled channels, possibly open, and that the larger stones rolled down into the deposits as the surrounding ice surface melted down. The similarity in roundness value (see p. 311) between delta and esker stones supports the view that most of the smaller esker stones and the finer sediment have been deposited by running water.

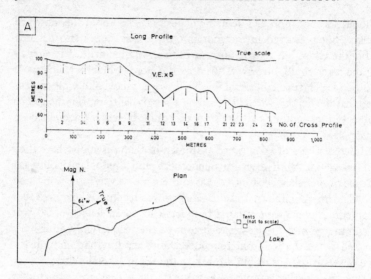

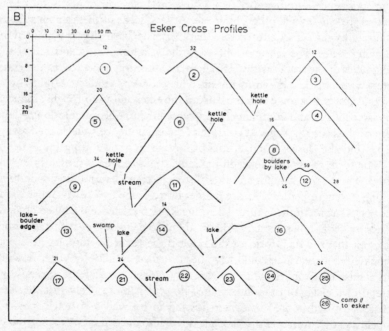

Fig. 16-3 An esker near Reflection Lake, Baffin Island, in profile (A) and section (B). The figures above the cross profiles indicate the mean stone size in cm on the esker surface

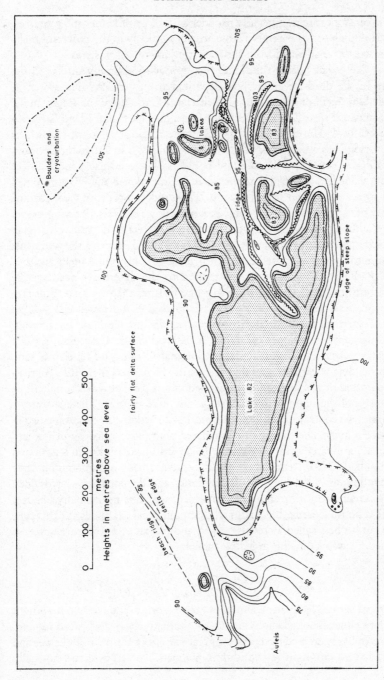

Fig. 16-4 A plane-table survey of a dead ice area near Tikerarsuk Point, west Baffin Island, showing esker-like ridges and kettle holes

The esker stones are significantly rounder than those of kames and moraines. This rounding can apparently take place very quickly in the powerful melt-water streams that issued from the retreating ice-cap in this area.

It is likely that not all the ridges that appear superficially similar to the eskers in the dead-ice areas were deposited mainly by flowing water. A detailed survey of one dead-ice area, with large kettle-holes and ridges, in the large outwash spread near Tikerarsuk Point in west Baffin Island is shown in Figure 16-4. The esker-like ridges bifurcate round small kettle-holes. The largest ridge has a beaded crest, the humps being 2-6 m above the hollows. The ridge is composed of sand and boulders. The sand has a much lower sorting value ($\sigma_1 = 1 \cdot 275$) than the other one mentioned and it is also mixed with stones of a mean size of 4·6 cm. These ridges probably were formed mainly of material washed into crevasses or hollows between decaying masses of ice, although they may have a foundation of water-deposited material. The ice masses finally melt out to form the kettle-holes and the sediments slump into the sharp-crested ridge form as the supporting ice melts away.

R. F. Flint (1928) has drawn attention to the differences between the characteristics of true eskers and esker-like features that are formed in crevasses in dead ice. True eskers may be up to 240 km long, and 4–40 m high. They trend parallel to the ice movement, are discontinuous and sinuous, trend up and down slope, and rarely have level crests. They often show a tributary pattern. Coarse material predominates; fine sand or clay is rare. Their bedding is variable, and they frequently have a coating of till. A cross-section often shows that the marginal bedding is parallel with the side slope as a result of slumping, with an average dip of 20 degrees. These characteristics may be contrasted with those typical of the esker-like ridges formed as crevasse fillings. These ridges are associated with pitted outwash plains, and are small compared with true eskers, being 250 m to 1·6 km in length and 4–4·5 m high. Their trend is variable; they are normally not sinuous but they are continuous. They have no distributaries and do not pass over divides. They contain both fine and coarse material, which is horizontally bedded. There is no till coating, but marginal slumping occurs as in eskers. This type of feature is best called a crevasse filling and not an esker. The true esker is formed by water flowing in glacial tunnels.

Kames

The term 'kame', 'cam', or 'kaim' is of Scottish origin and has two meanings. It means either crooked and winding, or alternatively, a steep-sided mound. The term has been used to describe a great variety of forms. Both kames and eskers have a common relationship to stagnant ice and both are normally composed of sand and gravel. Because kames can form in a variety of ways, the meaning of the term should not be too closely associated with the genesis of the form. A kame can be conveniently defined as follows, after C. D. Holmes (1947), 'A kame is a mound composed chiefly of sand and gravel, whose form has resulted from original deposition modified by any slumping incident to later melting of glacial ice against or upon which the deposit accumulated'. T. F. Jamieson in 1874 was the first person to interpret kames correctly. In 1894, J. Geikie used the term 'kame' to describe both isolated mounds of stratified sand and gravel and also other more complex features of morainic origin. His typical kame consisted of a single hillock, made up of stratified and cross-bedded sand and gravel near the centre, while the marginal parts showed indications of slumping. Kame deposits are sometimes poorly sorted and they may even contain some unstratified material and finer deposits.

While kames consist of more or less isolated mounds (Plate XXIII), kame terraces form more continuous features along a valley side (Plate XXIV). Both kames and kame terraces consist mainly of stratified material. Kame terraces are deposited by meltwater flowing along the edge of an ice mass between the ice and the valley wall (Chapter 12). The side of the deposit near the ice slumps down as the ice melts away to form the steep ice-contact slope that is characteristic of the outer side of the kame terrace. The surface of the kame terrace sometimes slopes gently in towards the hillside, as in the case of some kame terraces in western Baffin Island. Kame terraces may be discontinuous, only forming where the meltwater spreads out and can deposit its load. They may be absent where the water is concentrated in flowing round spurs and other obstructions along the valley side.

J. B. Sissons (1958a) has described the ice-contact deposits of the Eddleston valley near Edinburgh. The ice in this valley gradually decayed *in situ*, providing suitable conditions for kame formation. Kames occur at two main elevations, at 268-274 m and 258-262 m, the latter being the more numerous. These kames were deposited by meltwater, which was flowing freely at the base of decaying ice. They must have formed in large irregular openings in the ice and the presence of kettles indicates the inclusion of lumps of ice among the deposits. Many of the kames have steep ice-contact slopes, which are up to 15 m high on at least one side. Eskers are associated with these kames, and formed as the continuations of subglacial meltwater channels. The heights of the crests of both kames and eskers are similar, suggesting that they were both related to the level of the englacial water-table, which was in turn controlled by a col to the north.

Various different hypotheses have been put forward to explain kames. One view holds that the kames are formed where a meltwater stream, carrying a heavy load, debouches into a proglacial lake. The material is deposited in a deltaic form, and the side of the feature adjacent to the ice later collapses to form a steep ice-contact slope on the proximal side. Where a number of streams enter a lake close together, the small deltaic deposits may merge into a longer feature aligned parallel to the ice front and at right-angles to any eskers present. Normally, however, kames show an unorganized pattern and often exist as isolated mounds, which Synge calls 'lone kames'. These features can originate in a number of ways.

One possibility is the perforation hypothesis. This idea, put forward by J. H. Cook (1946), considers that the debris forming the kame collected in pools on the surface of a stagnant ice-sheet. Sufficient debris must have been exposed on the surface of the ice-sheet by ablation partially to fill the pool. As the pool warmed, it gradually melted its way down to the ground beneath. The material that had accumulated in it then became the kame mound as the surrounding ice slowly melted away. It must be assumed that ablation is largely achieved by evaporation. This process appears to be going on to a certain extent at the present time on stagnant ice masses such as the outer marginal area of the Malaspina glacier. However, the theory raises several problems. If ablation takes place largely by evaporation, there is no means whereby the finer sediment in the till can be dispersed. In some areas where the underlying till is clayey, a layer of fine clay should accumulate at the base of the hole and insulate it from further down-melting. The resulting deposits should contain more fine material than is normally found in kames. This is true of parts of New York State, which was the area discussed by Cook. It also seems unlikely that evaporation was the major means of ablation during the decay of the Pleistocene ice-sheets in areas that are now temperate. Glaciers, such as those in Spitsbergen, do not now lose more than 3 per cent of their ice by evaporation. There is also evidence of abundant meltwater in other features, so that it is reasonable to associate kame formation with meltwater.

An hypothesis put forward by Holmes suggests that kames develop where heavily laden melt-streams flow into small ponds on or within a stagnant ice mass. At the point of inflow the stream would deposit some of its load to form a small deltaic deposit. This would form an isolated mound when the surrounding ice melted away. Glacial meltwater streams often develop moulins (Chapter 6), and as these deepen, the material accumulating in them may be let down on to the glacier bed. The original current bedding will then have a better chance of surviving when eventually the material is exposed as a kame mound. These mounds have some affinity with eskers and sometimes a thin band of deposits linking the kame mounds may be an embryo esker. Kames may at times develop as crevasse fillings in stagnant ice. Such deposits are usually elongated but probably show relatively little internal structure.

One of the best known series of kames is that situated near Carstairs in southern Scotland. In 1926, J. K. Charlesworth interpreted this system of steep-sided fluvioglacial ridges, in places over 15 m high, as an end-moraine formed along the margin of ice from the Scottish Highlands. The ridges extend over a distance of 7 km and consist of coarse sand and gravel much of which shows current bedding. J. B. Sissons (1961) has shown that the features are not consistent with Charlesworth's views, and that they were probably the result of deposition by a complex series of subglacial meltwater streams, flowing beneath decaying Southern Upland ice.

Sissons (1958b) has also described kames and kame terraces that formed during the deglaciation of East Lothian in association with stagnant ice. The kame terraces attain a maximum length of 1·6 km and a width of 200 m, but most of them are only a few hundred metres long and a few tens of metres wide. Their surfaces are mostly flat in transverse profile, but some slope slightly up towards the ice-contact slope, where the supply of debris from the ice must have been greatest. The longitudinal slope varies from almost horizontal to a few metres per kilometre at a maximum. They are mainly formed of sand and gravel, but in some, clay and silt also occur. Some of them show by their bedding that they were built up by streams draining along the edge of the ice or out from the ice. Where the beds are horizontal and the material is fine, the kame terraces were deposited in elongated marginal lakes. At their upstream ends, the kame terraces sometimes merge with thick deposits of valley till, deposited in the tributaries to the main valley in which the ice lay. At their downstream end, they frequently merge into meltwater channels.

As the ice shrank a series of kame terraces developed one above the other. For example, 6 km south of Gifford, a series of four kame terraces has been formed at levels between 244 and 320 m. The uppermost terrace has both a steep longitudinal and front slope, and is composed of coarse gravel. The melting-out of detached blocks of ice has led to this terrace forming a series of separated mounds and ridges.

Most of the kames of this area lie in a belt about 3·2 km wide at 168-228 m on the northern flanks of the Lammermuir Hills. A ring of kames formed around a small lobe of isolated and stagnant ice south-east of Gifford. These kames form flat-topped mounds around a series of marshy hollows about 12 m deep, and formed as material accumulated around the edge of the decaying ice and in crevasses within it. The marshy hollows represent kettle-holes.

The internal structure of one of the isolated kames, situated about 5 km south of Gifford, and 300 m long, is revealed in a good section. This runs from south-south-east to north-north-west along the length of the feature. The lowest beds consist of thick layers of sand, which have been partially eroded before the next layer, consisting of gravel, was laid down. The gravel is overlain by alternating bands of reddish clay and brown silty sand or yellow sand, and then by another thick layer of sand, which merges north-ward into layers of gravel. Similar gravels occur in two lenses within the uppermost layer of alternating sand and silt. These deposits show that, at times, a powerful stream fed material into a small lake, held up in front of the ice, while at other times finer material could accumulate. The flat top of this kame suggests that the lake was filled to the surface at least at the northern end from which the material was derived. In places, the gravels consist of foreset beds dipping at about 20 degrees.

Both kames and eskers have a rather similar distribution. They occur in those areas where the ice-sheets stagnated and melted away slowly and in such a way that much fairly coarse material was available. Much meltwater must also have been present to redistribute the debris and deposit it in and around the margins of the static and decaying ice masses. Kames and eskers are found in large numbers in central Ireland, south of the Kells moraine, and in those parts of Scotland where ice masses became stagnant. They are also very well developed and numerous in the lower parts of Sweden and Finland, and in similar areas relative to the ice-sheets in parts of Canada. In both Canada and Scandinavia, very large ice-sheets wasted away rapidly and released much meltwater into lakes ponded near the ice margins or into the sea, the level of which was then higher owing to isostatic depression. Eskers and kames also occur in western Baffin Island in association with dead-ice phenomena (Plate XXII).

The term 'kame' is restricted by W. Niewiarowski (1963) to forms that occur in association with dead ice; thus he eliminates forms laid down by meltwater around active ice margins. The majority of Polish kames occur in association with drift of the last glaciation between the outermost moraines and the Pomeranian moraines. This same distribution occurs also in Germany. The kames are usually associated with kettle holes and never with drumlins. They occur in four forms: a) kame hillocks, b) flat hills and kame plateaux, c) kame ridges, and d) kame terraces. The first (a) rarely exceed 10 m in height and occur in groups, while those of the second type (b) occur singly, and their flat uniform summit surface is their most notable character. Both types have 5 to 15 degree slopes. One kame plateau covers 12 km². The kame ridges (c) are aligned in uniform directions and range from 8 to 20 m in height. They are transitional between kames and eskers, but differ from the latter in structure and origin, being shorter and straighter. Kame terraces (d) are not common in Poland. The kames are made of clay, silt, sand, gravel and pebbles, but sand predominates. The deposits may reach 20 to 30 m in thickness. Some kames are formed by meltwater streams; others form in still lake-water and are poorly sorted. The lake kames usually consist of finer sediments and may show varves. Sometimes the bedding is disturbed owing to melting of dead ice, which causes subsidence. The disturbance may be in the form of small faults or arcuate bedding, giving an outward dip to the beds along the slopes. The ridges formed in crevasses, and the hillocks in hollows on the dead ice or between blocks of dead ice. There is some doubt whether the kames formed supraglacially, or within or beneath the ice, but the disturbance of bedding supports a supraglacial origin in some examples at least.

Delta-moraines

Many morainic features show evidence of the action of meltwater in their formation, in that they contain stratified drift. Push-moraines, however, which were described in the last chapter, should not be included under the term 'delta-moraine', even though they contain stratified material. They can usually be differentiated from true delta-moraines by their structure. This often shows the effects of intense pressure, producing imbrication and other similar structural features.

Delta-moraines, on the other hand, form where stationary ice fronts remain for some time, ending in a lake or the sea. They have much in common with kames of the type that form at the margin of a stagnant ice-sheet, where this ends in ponded water. The delta-moraines are, however, usually larger features.

A good example of a delta-moraine has already been mentioned in connection with the eskers in the vicinity of Trim in Ireland (Fig. 16-2). Many of the eskers merged into the so-called Galtrim moraine which has all the characteristics of a delta-moraine. It formed adjacent to the stagnant ice-sheet which dammed up the lake into which the deltaic deposits were carried by the meltwater streams. One of the most significant aspects of this delta-moraine is the very well developed ice-contact slope on its northern margin, which gives clear evidence of the former presence of ice here. The moraine trends from north-east to south-west, in its northern section, from the Hill of Tara for a distance of 10 km, and forms a prominent arcuate ridge up to 15 m high. Southward, near Kiltale, the delta-moraine is crossed by an esker. Its typical deltaic structure is revealed in a series of pits near Batterjohn, where foreset beds dip steeply to the south. Overlying the deltaic strata is a layer of coarse unstratified material, which appears to be ablation moraine. The feature has a broad, flat top 1·6 km south-west of this point. At Galtrim, the steep northern slope overlooks flat, marshy ground in which a lone kame 6 m high stands. The slope to the south-east is much gentler. The level top shows that the delta was built up to the lake level during the period when the ice front occupied the position now revealed by the ice-contact slope. As the ice melted, so the marshy hollow developed.

To the west of Galtrim, the delta-moraine splits up into two parallel ridges. The southern one is a rather sharp-crested ridge, indicating that the ice did not stay very long in this position, before it retreated to the ice-contact slope of the north-western ridge. This ridge is wider and was fed by the three eskers that merge into its ice-contact slope on the north-west. One of the eskers climbs up the northern face of the delta-moraine to its top where the esker stops. The esker, therefore, acted as a feeder channel to the delta-moraine. There are kettle-holes, forming enclosed hollows, within the delta-moraine deposits as well as lone kames to the north of it. Some eskers merge with the foot of the ice-contact slope. The largest esker is as high as the 10 m ice-contact slope, and the two merge accordantly together.

At its southern end, the delta-moraine splits up into three separate gravel ridges. All these features have the characteristics of deltas forming where the ice front was stationary for a fairly short period. Synge (1950), who has described these features, believes that they were deposited by heavily laden meltwater streams draining through the ice, the marginal zone of which was stagnant.

The overlying unsorted material on the crest of the ridge is thought to have been derived from an overhanging ice-cliff. This cliff must have contained much englacial material that would collapse into the newly deposited deltaic sediments beneath the water surface. Dirty ice terminating in water frequently forms an overhanging face because the dirty ice has a greater ablation rate in water than in air. Thus the overlying material can be described as ablation moraine. The presence of small eskers on the ridge crest also supports this view.

The two large Salpausselka moraines of Finland are probably the largest examples of delta-moraines. The outer one is 70-80 m high and 2·5 km wide. The inner one lies 20-25 km north of the outer. The Salpausselka moraines form a long arc, stretching right across the southern part of Finland, and making a conspicuous belt of higher ground which marks important halt stages in the deglaciation of the area. The features were laid down as a line of marginal deltas in the sea, which was high at this time since isostatic recovery had barely begun. The material consists of well-washed sand and gravel, which shows deltaic bedding on the distal side. The proximal side shows some evidence of later glacial advance in contortions and overfolds in the structure. Thus the proximal part has some of the characteristics of a push-moraine. Till only occurs along part of the northern margin of the ridge, mainly as a sandy ablation deposit.

Associated with the two major Salpausselka ridges is a shorter ridge known as Jaamankangas, forming an arc about 50 km long and attached to the innermost of the Salpausselka ridges at its northern end. Jaamankangas ridge is associated with many eskers, one of which is 300 km long, the longest in Finland. The ridge consists of three main elements. There is a plateau ridge, an esker ridge and an area of knob-and-kettle relief. The plateau area is 2-4 km broad and 110-120 m high, sloping gently south and including deep kettle-holes and hillocks. It consists of deltaic deposits. In parts of the Jaamankangas ridge, the structure is, however, horizontal rather than deltaic. This may be the result of later resorting by the sea. Another possibility is that part of the feature was formed as an enlarged esker between two lobes of ice. Whatever the precise mechanism by which the feature was formed, it was deposited by heavily laden meltwater streams next to stagnant ice.

Features that have already been described were deposited either within or at the margin of the ice-sheet. It is often assumed that when fluvioglacial deposits underlie a thick layer of unstratified till that this sequence indicates a readvance of the ice. That this need not be so if the till layer is thin and consists only of ablation moraine, has already been pointed out in discussing the Galtrim moraine. It is also probably true that even under fairly thick till, stratified deposits need not necessarily indicate a readvance. The decay of a stagnant ice-sheet can give rise to a complex sequence of deposits, whose chronological significance needs to be very carefully evaluated. The matter will be discussed further in the next Chapter.

Conclusions

Eskers and kames are formed of stratified deposits laid down in contact with slow-moving or stagnant ice by glacial meltwater. Eskers form elongated ridges, some of which originate on the ice and have ice cores. Others are laid down subglacially or englacially and have been built up within the ice to the level of the englacial water-table, or to a uniform lake-level, to give an even crest elevation. Eskers are sometimes beaded, some are straight and some are sinuous in form. They are usually formed of sand and gravel, but may contain large boulders. Most eskers were probably deposited in channels within or beneath the stagnant ice.

Kames are more irregular in form than eskers and can be divided into two types. Those formed at or near the snout of an ice mass can be differentiated from kame terraces formed along the valley-side margin of a glacier. These terraces slope gently down-valley and often have a slight transverse slope away from the hillside. They are formed by deposition from meltwater flowing between the hillside and the ice. The other type of kame may be deposited at the margin of a glacier in a proglacial lake, sometimes forming a short irregular gravelly ridge, orientated parallel to the ice front. Isolated mounds of stratified material, forming lone kames, possibly accumulated in hollows on the decaying ice.

Delta-moraines may be distinguished by their deltaic bedding and their steep ice-contact slope facing the decayed and vanished ice margin. They are formed by the deposition of meltwater load in standing water in front of the ice-sheet and are frequently fed by esker channels. The eskers now merge into the ice-contact slope of the delta-moraine. They mark periods of prolonged halt and decay of stagnant ice-sheets.

REFERENCES

R. G. CARRUTHERS (1947-8), 'The secret of the glacial drifts', *Proc. Yorks. geol. Soc.* **27**, 43-57 and 129-72

R. G. CARRUTHERS (1953), *Glacial drifts and the undermelt theory* (Newcastle)

J. K. CHARLESWORTH (1926), 'The readvance, marginal kame moraine of the south of Scotland, and some later stages of retreat', *Trans. R. Soc. Edinb.* **55**, 25-50

J. H. COOK (1946), 'Kame complexes and perforation deposits', *Am. J. Sci.* **244**, 573-83

C. EMBLETON (1964), 'The deglaciation of Arfon and southern Anglesey, and the origin of the Menai Straits', *Proc. Geol. Ass.* **75**, 407-30

R. F. FLINT (1928), 'Eskers and crevasse fillings', *Am. J. Sci.* **235**, 410-16

R. F. FLINT (1930), 'The origin of the Irish "eskers" ', *Geogr. Rev.* **20**, 615-30

G. DE GEER (1897), 'Om rullstensåsarnas bildnigssatt', *Geol. För. Stockh. Förh.* **19**, 366-88

J. GEIKIE (1894), *The Great Ice Age* (London, 3rd Ed.; 1st Ed., 1874)

O. GRANÖ (1958), 'The Vessö esker of south Finland and its economic importance', *Fennia* **82**, 3-33

G. F. HANSON (1943), 'A contribution to experimental geology: the origin of eskers', *Am. J. Sci.* **241**, 447-52

A. HELLAAKOSKI (1931), 'On the transportation of materials in the esker of Laitila', *Fennia* **52**, 1-41

C. D. HOLMES (1947), 'Kames', *Am. J. Sci.* **245**, 240-9

G. HOPPE (1961), 'The continuation of the Uppsala esker in the Bothnian Sea and ice recession in the Gävle area', *Geogr. Annlr* **43**, 329-35

P. J. HOWARTH (1966), 'An esker, Breiðamerkurjökull, Iceland', *Rep. Br. geomorph. Res. Grp, Symposium at St Andrews, 1966*, 6-9

T. F. JAMIESON (1874), 'On the last stage of the glacial period in North Britain', *Q.J. geol. Soc. Lond.* **30**, 317-38

H. A. LEE (1965), 'Investigation of eskers for mineral exploration', Geol. Surv. Pap. Can. 65-14, 1-17

W. V. LEWIS (1949), 'An esker in process of formation, Böverbreen, Jotunheim, 1947', J. Glaciol. 1, 314-19

C. M. MANNERFELT (1945), 'Några glacialmorfologiska formelement', Geogr. Annlr 27, 1-239

W. NIEWIAROWSKI (1963), 'Types of kames occurring within the area of the last glaciation in Poland compared with kames known from other regions', Rep. 6th Conf. int. Ass. quatern. Res. (Warsaw, 1961), Lodz (1963), 3, 475-85

R. J. PRICE (1965), 'The changing pro-glacial environment of the Casement Glacier, Glacier Bay, Alaska', Trans. Inst. Br. Geogr. 36, 107-16

R. P. SHARP (1953), 'Glacial features of Cook County, Minnesota', Am. J. Sci. 251, 855-83

J. B. SISSONS (1958a), 'Supposed ice-dammed lakes in Britain, with particular reference to the Eddleston valley, southern Scotland', Geogr. Annlr 40, 159-87

J. B. SISSONS (1958b), 'The deglaciation of part of East Lothian', Trans. Inst. Br. Geogr. 25, 59-78

J. B. SISSONS (1961), 'The central and eastern parts of the Lammermuir-Stranraer moraine', Geol. Mag. 98, 380-92

J. C. STOKES (1958), 'An esker-like ridge in process of formation, Flåtisen, Norway', J. Glaciol. 3, 286-90

F. M. SYNGE (1950), 'The glacial deposits around Trim, County Meath', Proc. R. Ir. Acad. B, 53, 99-110

V. TANNER (1932), 'The problem of eskers', Fennia 55, 1-13

J. TRICART and A. CAILLEUX (1962), Le modelé glaciaire et nival (Paris)

F. M. TROTTER (1929), 'The glaciation of eastern Edenside, the Alston Block and the Carlisle Plain', Q.J. geol. Soc. Lond. 85, 549-612

Chapter 17

Ice Stagnation Features

Each glacier loaded with stones from the rocks above it may be regarded as a ship freighted with specimens of its native mountains, which it deposits by thawing in the place where it ultimately rests. (R. BAKEWELL, 1838)

The concept of ice disappearance by stagnation was developed at the turn of the century. F. P. Gulliver hinted at the possibility of ice stagnation in 1899 and in 1904, M. L. Fuller collected evidence in south-eastern Massachusetts that led to the first proposal of the ice stagnation idea for the deglaciation of this area. J. H. Cook in 1924 and R. F. Flint (1929) recognized the very wide extent of the area affected by stagnation of ice during deglaciation. Flint showed that stagnation of the last ice-sheet occurred throughout Connecticut. Other areas affected by stagnant ice include Rhode Island, parts of Massachusetts, New York State east of the Hudson River, northern New Jersey, and all northern New England. Similar features also occur in Yakima Valley, Washington.

When an ice mass ceases to receive an adequate supply to keep it moving, it will become stagnant and waste away by slow down-melting. The ice surface melts down most rapidly where nunataks first appear and lakes form marginal to the rock exposures. These lakes become elongated along the valley side as the ice shrinks down into the valleys. Evidence for their previous existence may be seen in paired terraces of deltaic-bedded outwash deposits. The decaying ice will usually contain crevasses that cannot close up so that a relatively uniform water-table is established in the dead ice mass. For this reason, the terraces tend to be paired at equal heights on either side of the valley, and are often connected with spillways, by which the lakes drained. The terrace margins represent ice-contact slopes that reveal the pattern of the decaying ice mass. The lake deposits also contain varves at times which point to slow rates of down-wasting. The down-wasting becomes extremely slow when the ice surface is thickly covered with debris.

In Connecticut, the debris appears to have been washed on to the ice surface from upstream, and for this reason, the southernmost ice masses were often the last to disappear because they had the thickest debris cover washed on to them by the south-draining streams (Flint, 1929). The horizontality of the terraces formed along the flanks of the Connecticut valley is good evidence that they are of lacustrine origin. The terraces show very little erosion since their formation and this lack of erosion probably reflects the presence of stagnant ice in the valley bottom throughout the whole period of deglaciation. If the ice from the lower reaches of the valley had disappeared before the ice in the upper part had melted, the meltwater from the upper ice would have eroded the terraces in the lower part of the valley. The great extent of the terraces also points to the widespread presence of dead ice throughout the period of deglaciation.

Relief plays an important part in the distribution of stagnant ice. Features of stagnant ice are often more conspicuous in the broader parts of valleys. These parts may be cut off from the ice source by the narrower reaches upstream in which the smaller bulk of ice melted away more rapidly. This local stagnation can only occur, however, when the whole ice supply has become reduced because of climatic factors. Many areas of stagnant ice features occur in areas of considerable relief or in the lee of large relief features. As the ice thins, it becomes unable to surmount the relief barrier and the ice beyond becomes cut off and stagnates (C. M. Mannerfelt, 1945, p. 110).

Various types of ice-contact deposit associated with stagnant ice, such as eskers and kames, have already been discussed. There are, however, other features of interest associated with the decay and disintegration of stagnant ice. Ice stagnation occurred very extensively in some areas, for example in the Canadian Prairies and parts of Scandinavia. The characteristics of an ice mass that is stagnant and decaying will first be described, and then other features deriving their origin from stagnant ice will be mentioned. These features include Prairie Mounds, ice disintegration features, crevasse fillings, ice-pressed forms, and Kalixpinnmo hills.

GLACIER KARST FEATURES

The processes by which some of the stagnant ice features have formed have been likened to those that operate in karst landscapes. In karst country, the limestone is dissolved in a manner analogous to the melting of a stagnant ice-sheet. This idea has been pursued by L. Clayton (1964) in describing the features of a present-day stagnant ice mass. Although the removal of ice is by melting, rather than chemical solution as in true karst, the resulting features have much in common. There are not many glaciers that today have the necessary conditions for the development of good examples of glacier karst features. The stagnant and heavily drift-covered parts of the Malaspina glacier provide such conditions. Another nearby glacier, the Martin River glacier, 200 km west of the Malaspina, also illustrates these features clearly.

The cover of till is necessary to produce the karst-like features, because it prevents too rapid ablation which would destroy the developing forms. These forms require more localized ablation which is provided by the thick and uneven cover of till. In the Martin River glacier, the outer 1-3 km are stagnant, while the terminal 4 km are covered with debris varying from several centimetres to more than 3 m in thickness along the outer margin. Much of this material appears to have reached the surface of the glacier from subglacial sources along thrust planes, where the more active ice has been overthrusting the stagnant ice. The stagnant part of the glacier has gradually thinned. The glacial karst features are most mature near the edge of the glacier, where they have reached a stage of old age. There is an intermediate zone with mature karst forms, and an inner youthful zone.

The funnel-shaped sink-holes are the most striking features. They are almost circular and are mostly between 60 and 370 m across. They form with a density of about 30 to the square kilometre. Many hold small lakes and are 15 to 90 m deep. At times, the sink-holes coalesce to form compound features. They also enlarge and collapse to reveal the meltwater stream below, forming a type of glacial uvala. Even polje-sized features can form. For example, there is a lake 2·5 km long and 0·75 km wide near the margin of the glacier. The floors of these features may be the solid ground below the glacier. They are ice-walled outwash plains, and resemble features that have been described in North Dakota. The normal sink-holes are probably formed mainly by collapse. However, some are formed by 'solution' or melting by the operation of glacier moulins. The moulins are vertical-walled, cylindrical holes only a few metres in diameter, but probably up to 30 or more metres deep. These moulins, however, only appear in abundance in the active part of the glacier (Chapter 6). Ice caves and tunnels are also very common as is indicated by the large number of collapse features. The tunnels were about 3 m high and up to 2 km long. New ones were continually developing as old ones were abandoned. Some of these tunnels were floored with gravel, while the minor ones were often completely clogged with sediment. These would form eskers when the ice melted.

The ablation till on the surface, which is the equivalent of the residual soil on true karst, supports vegetation when it exceeds half a metre in thickness. Where the ablation till is 1·5-3 m thick it is stable, and spruce and hemlock trees over 100 years old are growing on it. In the youthful stage, where the ice-flow is still active, there are moulins, little surface moraine, except median moraine, and no vegetation. In the mature zone, there are funnel-shaped sink-holes, little or no movement, and unstable sediment up to 1 m thick and some vegetation. In the old age zone, the ice is stagnant, and the till cover is up to and over 3 m in thickness. The karst features include uvalas, poljes, and hums. The drainage comes to the surface again and a dense forest can flourish on the deeply buried ice, which only melts away very slowly.

Features of Ice Stagnation

Features of stagnant ice in eastern central Alberta have been described by C. P. Gravenor (1955). These are the Prairie Mounds, consisting of ground moraine and covering an area of about 64 km² about 22 km east of Hemaruksa The mounds are nearly circular and often have a central depression up to 1 m deep. They are about 100 m in diameter and 4½ m high. Molt of the mounds are composed of clayey till, but some contain stratified sit. and clay. The mounds are underlain by at least 12 m of clayey till.

There are two possible methods whereby these mounds could have formed according to Gravenor. First, he suggests that pits could occur where the debris on the stagnant ice was thinnest. Secondly, it is suggested that the ice would tend to collapse into caves formed by undermelting or into enlarged crevasses. Once the pits had formed in the stagnant ice, material would be washed into them from the surrounding ice. Then, as the surrounding ice became cleaner and the cover in the pit thicker, the surrounding area would melt away faster than the ice under the pit, so that eventually the pits would become inverted to form the prairie mounds.

This hypothesis accounts better for the characteristics of these features than that which suggests that they have a periglacial origin (Chapter 21). There is no evidence of ice-wedge fill. In this area, the wastage of the ice was probably largely by evaporation so that meltwater would not be so likely to disperse the sediment. The lack of outwash in this area supports the view that evaporation was important and it also accounts for the availability of englacial moraine to form the mounds.

The ice disintegration features of western Canada have also been described by Gravenor and W. O. Kupsch (1959). Most of the material left by the Wisconsinan ice in this area was till and there is relatively little stratified material, in contrast to eastern North America where stratified deposits predominate. The features are also well preserved and easily visible, owing to the dry climate and lack of thick vegetation. The term 'disintegration' is used to describe the way in which the ice-sheet decayed and disappeared in this region. It appears that the ice must have broken up into a number of separate small blocks. The breaking-up of an ice-sheet may take place under uncontrolled conditions, in which case there is no evidence of alignment of the resulting features; alternatively, the ice may break up along aligned or lobate weaknesses within the ice, and these weaknesses may give a pattern to the resulting deposits. This gives rise to the controlled type of disintegration deposit. The patterns that result from this process are inherited from a more active phase of glacier motion, when the ice developed a pattern of crevasses and thrust planes associated with its movement. The linear zones along which the decaying ice fractured are normally aligned parallel or perpendicular to the ice movement. Sometimes the crevasse pattern in the lower part of an ice-sheet or glacier is aligned at an angle of 45 degrees to the ice movement (Chapter 4). It also sometimes occurs that dead ice features are superimposed on features formed by the ice when it was still active. The patterns can be particularly clearly seen on aerial photographs.

A variety of features occurs in ice disintegration areas. Where the disintegration was uncontrolled, there is no pattern apparent. Ridges, hollows, and mounds are distributed indiscriminately. The major elements of the relief are knobs and kettles, moraine plateaux, and beaded linear ridges. The moraine plateaux are relatively high areas consisting of clay till, sometimes covered by thin lake clays. Some of the moraine plateaux have kettles and poorly developed rims. The features described by Gravenor and Kupsch in Canada seem to be very similar to those described by G. Hoppe (1952, 1957, 1959) in northern Sweden. Hoppe considers that the moraine plateaux, which he calls the Veiki moraine type, have originated subglacially as described in Chapter 15. The evidence for this is the preferred orientation of the pebbles in the raised ridges around the plateaux, the fact that such features are often associated with drumlins, and that some are cut by meltwater channels. Gravenor and Kupsch do not give any information concerning pebble orientation, so that it is possible that the features they described, although similar in relief to those described by Hoppe, are in fact of different origin. Other possible methods of formation will be mentioned later in the chapter.

Another interesting feature of the uncontrolled disintegration conditions is the closed disintegration ridge. The circular features of this type are similar to one of the types of prairie mound, but others are not so symmetrical. They range in height up to 6 m and from 6 to 600 m in diameter. The base of the central depression in some of them lies below the level of the surroundings. In this they differ from the prairie mounds whose central depression is always shallow. Occasionally a small mound is found in the centre rather than a depression. Most of the ridges are formed of till, but some contain stratified material.

The controlled disintegration features are widespread in western Canada and the United States. They are composed chiefly of till with a thin covering of gravel, but some are of stratified material. Their dimensions vary from 1 to 11 m in height, 7·5 to 90 m in width and from a few metres to 13 km in length. They are usually fairly straight or arcuate. Two sets can intersect at right-angles, or at an acute angle to form a diamond pattern. Their alignment is controlled by the last effective ice movement, which caused lines of weakness to develop along which the ice later broke up. Many different shaped ridges occur, from hairpin shapes to those like a shepherd's crook. When two ridges meet or cross, one may lie above the other or they may merge together.

The term 'wash-board moraine' has been used to describe the pattern that the ridges make when they are parallel and close together. Some of the features of this type that occur in Alberta were deposited on land and differ in this respect from the cross-valley moraines that were described in Chapter 15. J. A. Elson (1957) has described the so-called wash-board moraines of Manitoba, where they are formed of sandy till. They are up to 4·5 m high and about 90 to 150 m apart. He suggested that they were formed beneath the ice, at the base of thrust planes in the ice. Gravenor and Kupsch agree with this suggestion, and go on to point out that the ridges owe their preservation to ice disintegration. They also point out that the ridges show evidence of former ice movement and are thus controlled patterns. Their pattern tends to be lobate and in this they differ from the linear disintegration ridges. The linear features, which are sometimes even-crested, probably formed in more open crevasses. They often show a composite pattern, with several different directions superimposed. Erosional features, formed originally in ice-walled channels, also occur. The term 'ice-walled channel' can be applied both to open channels with ice walls and to closed tunnels within the ice (Chapter 12).

In considering the origin of these features, two major processes appear to have operated. First, some of the material was probably let down by the ablation of the ice. Secondly, some may have been squeezed into the cracks and hollows within the ice as it gradually broke up into stagnating blocks. The disintegration features formed late in the glacial sequence; sometimes they overlie drumlins and other features of active ice movement, and must have formed later than these.

Another possibility is that the features that show alignment and controlled patterns may have received this pattern subsequent to their original formation. If the decayed ice that originally formed them had been rejuvenated, by thickening, to become active again, it might have pushed the material into the pattern in which it is now found. This process, however, appears less likely than the view that the pattern was inherited from a period of more active flow prior to the formation of the disintegration features. Thus inherited flow control, rather than regenerated flow control, appears more likely.

The process of ablation must clearly play an important part in the latest stages of the disappearance of a dead ice mass. The superimposition of ridges one above another can best be accounted for by down-wasting, where one crevasse was filled with material, and later, another was opened on top of it, the material in the second being let down on top of the first. The moraine plateaux could form by the accumulation of material in a low part of the glacier, which then becomes a plateau by ablation of the surrounding cleaner ice. This would account for the lake deposits that sometimes overlie the till on the tops of the plateaux. The moraine-filled depressions would have been small lakes on the ice as they were developing. Support for the ablation view is also found in the collapse structures that are common in this type of disintegration deposit.

One of the difficulties of the ablation theory is that ablation moraine is normally loosely packed. It does not contain much fine material, which has been washed away in the meltwater. Many of the features described, however, consist of compact silty till. Another point is that many existing glaciers only carry a relatively small area of supraglacial till close to their snouts. It is difficult to envisage a process whereby huge areas of decaying ice could become thickly covered by supraglacial material. It is known from a study of modern ice-sheets that till can be brought up to the surface near the edge of the ice-sheet. But this only occurs in live ice or at the junction of live and dead ice. Till ridges parallel to the ice front can form in this way.

ICE-PRESSED FEATURES

A. M. Stalker (1960) has described similar features in Alberta and has produced much evidence to show that they probably originated as ice-pressed features. He therefore supports the second major hypothesis. Stalker shows that the same process of ice-pressure or squeezing can account for the wide variety of forms that exists in this area. The forms can be classified into the following types: a) rim ridges of moraine plateaux and plains plateaux, b) ice-pressed moraine ridges, c) ice-pressed plains ridges, d) till-cored eskers and eskers made wholly of till, e) ice-pressed minor moraine ridges, f) ice-pressed long drumlinoid ridges. Some of these features have already been mentioned in connection with moraines, eskers, and drumlins. There is no need to describe all these features in detail, but the processes that may have operated in their formation are worthy of comment. Stalker considers that during deglaciation there were holes and crevasses at the base of the disintegrating ice. The till at the base of the ice was either not frozen or only partially frozen as the ice was disappearing. The melting of the overlying ice produced much meltwater, so that the till beneath the ice was full of water, commonly being completely saturated. The weight of the overlying ice would press the saturated till towards the holes and crevasses within the ice. Where the holes were very large, more material would accumulate around their margins than in the central part of the hole. When the ice finally melted, this material round the margins would form the marginal ridges, while the hollows around the ridges occur in those parts from which the material was pressed. A strong argument in support of this theory is that the till forming the ridges around the moraine plateaux is of basal type, as it contains much fine material.

One of the essentials for the operation of the ice-pressing process is an abundance of water. The water was provided in this part of Alberta because the ice was more active to the east and north and would thus prevent the flow of water in this direction, which is the natural drainage direction because of the general slope of the land. Thus there was plenty of water available to soak the subglacial till and fill the holes and crevasses in the basal ice. The long ridges could form in basal tunnels cut by streams, but it is less easy to account for the much larger holes that formed the moraine plateaux. Many of these features are more than 1 km in diameter. Smaller holes probably occurred at the junction of two or more crevasses.

A process that could account for the enlargement of a hole once it had formed is calving. Where the holes were water-filled, the ice would tend to be undermined at the water level, and this would cause collapse of the ice-walls by calving and the enlargement of the hole. This process of calving had probably ceased before the formation of the rims by ice-pressure. This latter process requires a certain stability of the ice-walls. The cessation of calving would take place as increased melting opened more channels and allowed the water, filling the holes, to drain away. Water in the holes is not essential to the squeezing process. Another advantage of the preliminary calving before the formation of the ridges is that calving would provide steep ice-walls and these assist the pressing process. Calving cannot, however, explain the origin of the larger holes, although it probably played an important part in the smaller holes.

A sufficient ice thickness is necessary to provide sufficient pressure to produce the ice-pressed forms. However, the ice thickness must not be too great or the ice will flow actively and destroy the disintegration forms. Under thick ice, the tunnels and holes would also close up and the process could not operate. There is thus an optimum ice thickness required to produce the features by ice-pressure. Stalker suggests that under favourable conditions, with ice 60 m thick, ice density of 0·8, and till density of 2·0, a ridge about 24 m high could be built up. The observed maximum height of the ridges agrees fairly well with this estimate, as the highest ridges are about 21 m high. A greater height could be achieved if a great thickness of supraglacial till were also available to add to the height of the ridge. However, many of the ridges are much lower than these values, which suggests that the ice thickness generally must have been less than 60 m. Another common limiting factor in the growth of the ridge was the supply of plastic till.

H. A. Winters (1961) has described ice disintegration features on the Missouri Coteau in North Dakota. The landforms of this area resemble the hummocky moraine of western Canada described by Gravenor, Kupsch, and Stalker. Perched lacustrine plains are a conspicuous element of the landforms of North Dakota. They must have formed in hollows on the ice, and were left perched as the surrounding ice melted. Some of the perched features are surrounded by rims of till or fluvioglacial gravel. The rims are not conspicuous where the former lakes were almost completely filled with lacustrine sediment, but where the lake deposits were thin, the ridges stand up like the edges of upturned saucers around the plateau edge (Fig. 17-1).

STAGNANT AND ACTIVE ICE DISINTEGRATION

Many features show evidence of the influence of both active and dead ice. The Kalixpinnmo hills, described by Hoppe (1959), come into this category. These hills are formed of fine sand and coarse silt. They are up to 25 m high and are elongated, aligned parallel to the ice front. They are thought to have originated sub-marginally in areas where the meltwater streams slowed down. The slowing down could be the result either of loss of hydrostatic head caused by thinning, or of the widening of the channel in which the streams were flowing. The material deposited in this way was mixed with till falling from the tunnel roof. The deposits seem to have been thrust together and formed into ridges by the ice, which appears to have been still in motion. The features, therefore, demonstrate characteristics both of dead and live ice sedimentation.

It is difficult to establish the precise nature of the formative processes merely from the external shape of the deposits. Some hummocky drift is the result of live ice deposition. A study of the internal structure and the fabric of the material may be necessary to reach a conclusion concerning the origin of the features. However, more must be learnt about the material fabrics before this technique can provide an infallible method of differentiation. The superimposition of live ice and dead ice features also leads to problems of interpretation, although the different types can frequently be recognized by studying the form of the features and the deposits of which they are composed. Dead ice features must also not be confused with permafrost and periglacial features, which may resemble them in some respects. Rimmed depressions could in fact be either disintegration features or collapsed pingos (Chapter 21).

Features that have been interpreted as the result of stagnating ice by M. M.

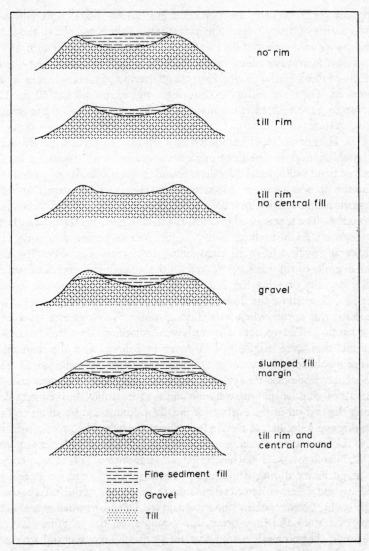

no¯ rim

till rim

till rim
no central fill

gravel

slumped fill
margin

till rim and
central mound

Fine sediment fill

Gravel

Till

Fig. **17-1** Landforms associated with ice disintegration: diagrammatic cross
sections (H. A. Winters, *Prof. Geogr.*, 1961, Association of
American Geographers)

Note: The symbols for gravel and till should be interchanged.

Leighton (1959) occur in the Illinoian drift of south-west Illinois, where the drift consists of many ridges. The ridges trend mainly parallel to the south-westerly ice-movement of the Illinoian lobe. The ridges are thought to have originated as crevasse fillings and are 25–30 m high in places. They consist mainly of fluvioglacial gravel. Features referred to as moulin kames are also described. These are mainly 6–12 m high, but at times attain 60 m. They appear to have formed from material choking moulins at a late stage of disintegration.

J. H. Hartshorn (1958) has shown how till may come to overlie stratified outwash deposits as a result of processes associated with stagnating ice. Till derived from supraglacial deposits is found to overlie sands and gravels near Taunton in south-eastern Massachussetts. Hartshorn envisaged lumps of stagnant ice, covered with a layer of till, separated by zones of fluvioglacial deposition. The supraglacial till must rise to a higher level than the surface of the outwash. As the ice lump melts, the till on its steepening slopes may move off laterally by flow on to the surrounding lower areas of outwash (Fig. 17-2). The deposits of till that have flowed off the wasting ice mass can form 6 m thick deposits of what has been called flow till on top of the stratified deposits. Sometimes the flow till becomes interstratified with the outwash deposits; this occurs when later fluvioglacial activity covers a layer of till that has flowed down over a lower layer of stratified deposits. The outwash deposits that were originally laid down in contact with the stagnant ice lumps would develop ice-contact slopes as the ice slowly melted away. These ice-contact slopes are always associated with the presence of flow tills. The association of flow till with outwash and the formation of ice-contact slopes along the margin of the outwash is usually accompanied by an inversion of the original relief before the ice lumps start to melt. The process by which flow till forms can be seen in operation on the Malaspina glacier at present.

W. Niewiarowski (1963) has shown that large areas of Poland were covered by stagnant ice during deglaciation. These zones of dead ice were up to 250 km wide. Kames, described in Chapter 16, are one of the features characteristic of these zones of decaying stagnant ice. Dead-ice moraines are another type of feature and they occur, for example, on the moraine plateau of Chelmno. These moraines are underlain by ground moraine and are characterized by irregular mounds and ridges and an abundance of kettle-holes. Much of the material is stratified and it is very mixed in size from silt to pebbles. The dip of the pebbles usually conforms to the direction of slope, but the orientation is often varied. Small faults cut the stratified sediment and occur as a result of subsidence as the ice melts. There is a cover of unstratified drift 1–2 m thick. The stones in this drift do not show any preferred orientation and the material is a type of ablation moraine although it differs from

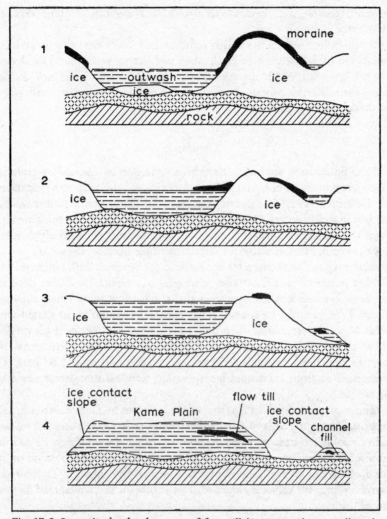

Fig. **17-2** Stages in the development of flow till (J. H. Hartshorn, *Bull. geol. Soc. Am.*, 1958)

the ablation moraine of active glaciers. Dead-ice ablation moraine contains much fine material and many small fault structures resulting from displacement as the dead ice melts. Another important point concerning the dead-ice moraine is that it often lies at a lower level by a few metres than the outwash produced by the ice when it was still active. Marginal valleys also sometimes occur in dead-ice areas and it has been suggested that the very flat areas of ground moraine may be associated with decaying masses of uncrevassed dead ice.

Kettle-holes need not necessarily indicate dead ice. Some kettle-holes form as a result of the freezing of small lakes, melt-streams or ground ice during the accumulation of the deposits. Then when these ice masses melt, kettle-holes form. Kettles associated with dead ice normally occur in combination with kames and eskers.

Conclusion

Ice disintegration appears to have been an important method of deglaciation over wide areas in western North America and also parts of Scandinavia. G. Holmssen (1963) has shown that dead-ice features and deposits occur widely in south-eastern Norway. Most of the morainic drift of this area has been left by stagnant ice, for only terminal moraines and sparsely covered rock occur in areas of active ice retreat. Altitude plays an important part in determining the areas characterized by dead-ice deposits. Both Holmssen and Stalker point out that the higher areas tend to produce the earlier dead-ice deposits, because it is in these areas that the ice first thins where they are situated far from the ice source. If ice-pressed forms can be associated with areas of ice disintegration, then they provide a useful means of identifying those areas in which the ice disappeared in this way. It is particularly important that dead-ice hummocky moraine landscapes should be correctly differentiated from landscapes incorporating terminal moraines formed by active ice.

The stagnation and disintegration of the ice seems to have occurred largely in broad, marginal belts of the ice-sheet, and particularly where the ice was thin over higher areas. Because the zone of disintegration was wide and the hummocky areas extensive, it is not possible to recognize the position of the ice front by studying these deposits. The ice disintegration features do, however, provide valuable evidence of an important and widespread method of ice dispersal.

REFERENCES

L. CLAYTON (1964), 'Karst topography on stagnant glaciers', J. Glaciol. 5, 107-12
J. H. COOK (1924), 'The disappearance of the last glacial ice-sheet from eastern New York', Bull. N.Y. St. Mus. 251, 158-76
J. L. DYSON (1952), 'Ice-ridged moraines and their relation to glaciers', Am. J. Sci. 250, 204-11

J. A. ELSON (1957), 'Origin of washboard moraines', *Bull. geol. Soc. Am.* **68**, 1721

R. F. FLINT (1929), 'The stagnation and dissipation of the last ice sheet', *Geogr. Rev.* **19**, 256-89

M. L. FULLER (1904), 'Ice retreat in Glacial Lake Neponsat and south-east Massachusetts', *J. Geol.* **12**, 181-97

C. P. GRAVENOR (1955), 'The origin and significance of Prairie Mounds', *Am. J. Sci.* **253**, 475-81

C. P. GRAVENOR and W. O. KUPSCH (1959), 'Ice disintegration features in western Canada', *J. Geol.* **67**, 48-64

F. P. GULLIVER (1899), 'Thames River Terraces in Connecticut', *Bull. geol. Soc. Am.* **10**, 492-5

P. W. HARRISON (1957), 'A clay till fabric: its character and origin', *J. Geol.* **65**, 275-308

J. H. HARTSHORN (1958), 'Flow till in south-eastern Massachusetts', *Bull. geol. Soc. Am.* **69**, 477-82

G. HOLMSSEN (1963), 'Glacial deposits of south-eastern Norway', *Am. J. Sci.* **261**, 880-9

G. HOPPE (1952), 'Hummocky moraine regions with special reference to the interior of Norrbottn', *Geogr. Annlr* **34**, 1-71

G. HOPPE (1957), 'Problems of glacial morphology and the Ice Age', *Geogr. Annlr* **39**, 1-18

G. HOPPE (1959), 'Glacial morphology and Inland Ice recession in north Sweden', *Geogr. Annlr* **41**, 193-212

M. M. LEIGHTON (1959), 'Stagnancy of the Illinoian glacial lobe east of the Illinois and Mississippi rivers', *J. Geol.* **67**, 337-44

C. M. MANNERFELT (1945), 'Några glacialmorfologiska formelement', *Geogr. Annlr* **27**, 1-239 (with English summary)

W. NIEWIAROWSKI (1963), 'Some problems concerning deglaciation by stagnation and wastage of large portions of the ice-sheet within the area of the last glaciation in Poland', *Rep. 6th Conf. int. Ass. quatern. Res.* (*Warsaw, 1961*), Lodz (1963), **3**, 245-56

A. STALKER (1960), 'Ice-pressed drift forms and associated deposits in Alberta', *Bull. geol. Surv. Can.* **57**, 1-38

R. VIVIAN (1965), 'Glaces mortes et morphologie glaciaire', *Revue Géogr. alp.* **53**, 371-401

W. A. WINTERS (1961), 'Landforms associated with stagnant ice', *Prof. Geogr.* **13**, 19-23

Chapter 18

Proglacial Features

The first impression of Breidamerkursandur is that of a desert, . . . an almost smooth plain of dark, blackish gravel which from the end-moraines of the glaciers, high as the ramparts of a fort, slopes gently down towards the sea. The water from the ice for ever changes its course; . . . the swiftest rivers change their course almost daily on account of the huge quantities of gravel which they carry with them. (H. W. AHLMANN, 1938)

Where glaciers are active and can erode their beds effectively the meltwater streams issuing from their snouts are very heavily laden. Their load usually consists of coarse waste as well as fine silt, the latter giving the meltwater streams their characteristic milky appearance. This coarse material is normally deposited in front of the glacier margin to form proglacial features that consist of fluvioglacial deposits. There are two major landform types resulting from this deposition, depending on the character of the local relief. If the glacier is confined within steep valley walls, the outwash must be similarly confined and the resulting deposit is generally known as a valley train. But where there is a wide ice margin, terminating on a broad lowland area, the melt-streams can spread widely and build up an extensive surface. This is usually called an outwash plain, also known by the Icelandic term 'sandur' (Plate XXVI). Valley trains are associated essentially with valley glaciers and the outwash plains with ice-sheets. Both landforms, but particularly the outwash plains, can cover a large area.

Before considering these two major types of proglacial landform separately, some points concerning the general characteristics of these features will be mentioned. The debris provided by the ice is normally very poorly sorted and consists of a wide range of size of material. Normally the coarser material is deposited near the ice margin, and the finer, sandy sediment is carried farther. This material gives rise to the Icelandic name 'sandur', which means sandy area.

Because of their large load, the meltwater rivers usually aggrade their valleys and follow extremely braided courses that are continually changing in pattern (Plate XXV). Another characteristic of outwash streams is their variable flow. The variation near the glacier snout may be diurnal in character in summer, but it is the seasonal variation that is more important. It is associated with the summer ablation season when the flow increases very markedly for a few months, diminishing to very low values during the winter.

Some glacial rivers are also liable to more violent and short-lived variations in the discharge rate. These result from glacier bursts or jökulhlaups, as they are called in Iceland where they are particularly common. These sudden outbursts are caused by the rapid drainage of lakes dammed up by the ice, as described in Chapter 19. The largest sandur in Iceland, Skeiðarársandur, is

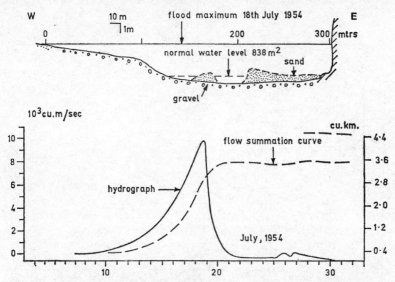

Fig. 18-1 Hydrograph for the Skeiðará jökulhlaup of July, 1954. A cross section of the main channel, showing water levels, is included (S. Rist, *Jökull*, 1955, Iceland Glaciological Society)

particularly liable to experience jökulhlaups owing to the periodic draining of Grimsvotn, a lake on Vatnajökull. From the point of view of the sandur development, it is the volume of water and the load it carries that is of significance. The usual discharge pattern of a jökulhlaup consists of a fairly gradual rise to a peak in the water level, followed by a short crest height, and a sudden rapid fall to normal level. For a short time, the velocity of flow may reach figures of 7-8 m/sec. In addition to the very large volumes of water suddenly discharged on to the sandur, a great deal of material is also carried down by the water. Near the glacier snout, large lumps of ice may be broken off and left stranded on the sandur as the flood subsides, eventually melting out to form kettle-holes. The kettle-holes often occur within the large channels, which are cut by the powerful streams before they start depositing their load.

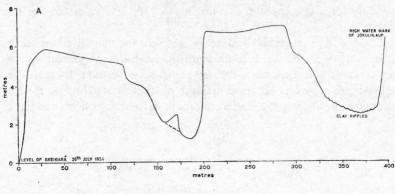

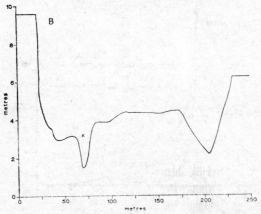

Fig. 18-2 Jökulhlaup channels on Skeiðarársandur, Iceland
A surveyed after the 1954 jökulhlaup
B older channel with kettle hole (K)

The hydrograph of the 1954 jökulhlaup of the Skeiðará shows that the water rose gradually at an accelerating rate from 7 July to reach a peak on 18 July and then fell rapidly to reach its normal level by 21 July (Fig. 18-1). During this period, S. Rist (1955) estimated that a total of $28,930 \times 10^6$ kg of silt were carried by the river. The load of silt reached a maximum of 9·5 gm/litre on 18 July, the day of the flood peak. During this day, a total load of 7800×10^6 kg of silt was carried on to the sandur. The total volume for the whole period, about 28·5 million tons of silt, must clearly represent very rapid removal of material from beneath the glacier. The much larger jökulhlaup of 1934 produced about 150 million tons of silt. If this were spread over the area drained by the Skeiðará it would amount to a loss of a layer 2·8 mm thick each year over the whole area (Chapter 11).

When these vigorous flood streams emerge on to the sandur, they at first erode deep, wide channels. One of these measured 200 m in width and varied from 5 to 7·5 m in depth. There were deeper troughs on the margins of a flat central section (Fig. 18-2). Very soon their heavy load causes the rivers to spread out into the typical braided pattern, the accompanying deposition causing very rapid aggradation of the sandur.

One result of the observed very rapid fall in the discharge at the end of a jökulhlaup is that the river suddenly loses its transporting power and much of its load will be deposited. Such sudden variations in the discharge will help to account for the aggradation that goes on under these conditions on the sandur. The same process will also operate to a lesser degree as a result of the annual variations of flow, with the relatively rapid fall off of discharge in later summer.

Valley Trains

Valley trains owe their characteristics to the glacial meltwater streams that form them and the fact that these streams are confined by valley walls. The valley train in Morsárdalur, Iceland, is a good example. This valley is tributary to the Skeiðará, which has already been mentioned in connection with jökulhlaups. Morsárjökull is a small outlet glacier from Vatnajökull and it ends in a steep-sided valley about 7 km from Skeiðará. The valley is 870 m

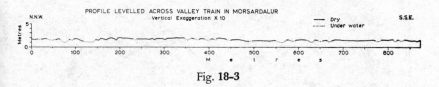

Fig. 18-3

wide and a profile levelled across it, and shown in Figure 18-3, does not vary more than 0·3 m on either side of the mean height, apart from one or two deeper channels that are about 0·5 m deep. The slope down-valley is, however, fairly steep, at about 9·7 m/km to 6·25 m/km, decreasing down-valley. The floor of the valley consists of coarse gravel with little or no vegetation over most of its width. This lack of vegetation testifies to the unstable and variable nature of the channels of the braided river.

The rate at which aggradation takes place depends mainly on the discharge and load of the meltwater rivers. Morsá aggrades less rapidly than Skeiðará, into which it drains. The difference in level was apparent during the jökulhlaup of 1954. During the outburst, Morsárdalur became flooded for a long distance upstream of the confluence as it lay at a lower level. The difference

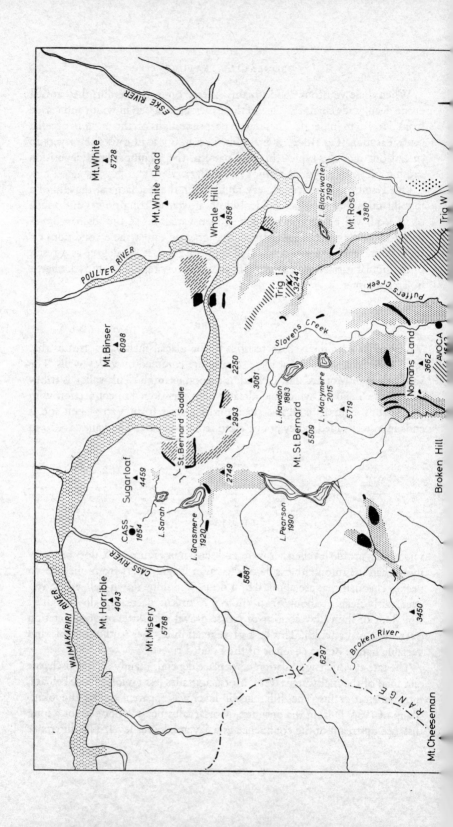

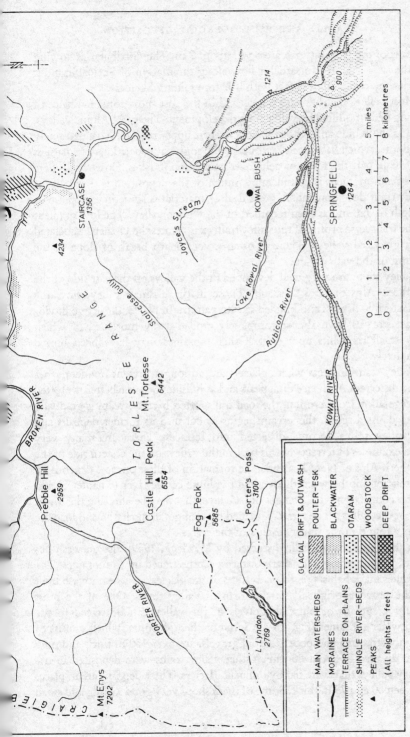

Fig. 18-4 Terrace deposits and moraine in the Waimakariri valley, New Zealand (M. Gage, N.Z. Jl Geol. Geophys., 1958, DSIR, R. E. Owen, Govt. Pr., Wellington)

in level of the two outwash areas was about 2 m. The distributaries flowing on the west side of Morsárdalur flow along the margin of Skeiðarársandur when they come into contact with it before joining Skeiðará.

Most valleys that have been glaciated in the past show some evidence of having been modified by valley train development. The great Alpine valleys, such as the Rhine and Rhône valleys in their upper reaches, have extremely flat floors, which have been built up by fluvioglacial deposition. The flatness may be in part the result of deltaic sedimentation in lakes. However, where the stream gradient is considerable, much of the deposition is the result of valley train development. A great deal of material is sometimes deposited: a depth of 120 m has been recorded in the Rhône valley. The low gradient of the transverse profile of the valley train exaggerates the U-shape associated with glaciated valleys. There is often a very sharp break of slope at the margin of the valley train.

Valley trains are very well developed in the valleys on the east side of the Southern Alps of New Zealand. Those in the Tasman valley show the characteristic flatness and braided stream pattern. In places, the active fluvioglacial streams have almost completely eroded away moraines, of which only small remnants now exist. Other moraines may have been buried completely.

Valley trains in areas where glaciers are no longer present frequently become dissected. As the ice disappears and as hillsides become better protected by vegetation from erosion, the load transported by the stream is gradually reduced. As a result, the stream begins to cut into its former deposits and the valley trains become dissected into terraces. There are many very good examples of terraces of this type in the valleys on the eastern side of the Southern Alps of New Zealand. The formation of these terraces depends on the relationship between the load and volume of the river (Chapter 5) and has no connection with changes of base level that may be affecting the lower part of the valleys. The terraces formed by changes in discharge and load are mainly found in the upper reaches of the rivers.

In the Waimakariri valley, studied by M. Gage (1959), the outwash deposits, now forming very sharp terraces, can be traced up to moraines. The moraines mark halt stages or readvances of the glaciers, whose outwash built up the terraces, originally in the form of valley trains. One of the most extensive spreads of outwash gravels in the valley can be traced to the Blackwater moraines (Fig. 18-4). Some of these deposits, which consist of roughly stratified and poorly sorted gravels, are over 150 m thick and now cover an area of about 25 km². Other valley trains were deposited in tributary valleys that were fed by diffluent glaciers. These deposits are in places between 60 and 75 m thick. Some of them show very good sorting and small sized gravel. The periods of aggradation when the moraines and valley trains were being deposited and the glaciers were active were followed by down-cutting and terrace formation when the ice had retreated.

It is worth pointing out that very similar occurrences, but not usually so clearly displayed, are well known in Britain. The greater age of the deposits has rendered them less conspicuous in parts of England. The Main Terrace of the Severn, of Würm age, is a good example. The upper Hilton terrace of the Trent, which was aggraded at the beginning of the previous inter-glacial, shows many of the stratification characteristics of glacial outwash deposits. The stratification is crude and patches of till are included in the deposits. This suggests that dirty lumps of ice were incorporated and that the current, laying down the deposit, was strong and tumultuous. The deposits have been mainly removed by later incision to form terraces, during periods when the loads of the rivers were very much reduced after the glaciers had retreated. The case of the Thames floodplain and terraces is mentioned briefly in Chapter 5.

Outwash Plains or Sandar

The outwash plains or sandar of Iceland have already been mentioned in connection with jökulhlaups. They are a very good example of the contem-porary development of these features and thus provide a model by which older features, no longer actively developing, can be studied and recognized.

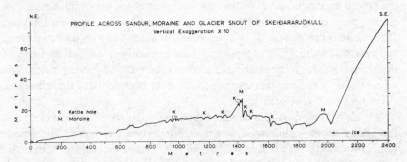

Fig. 18-5 Profile levelled across the upper part of Skeiðarársandur, crossing the moraines to the glacier margin

Skeiðarársandur is the largest outwash plain in Iceland. It stretches down to the sea from the snout of the large piedmont glacier Skeiðarárjökull, a dis-tance of 20 to 30 km. Over this distance, the Skeiðará, the major outlet river, has a gradient of about 5 m/km. The river follows a braided course and only during the maximum of a jökulhlaup is the whole sandur flooded. The slope of the sandur close to the terminal moraines, which were formed in the mid-nineteenth century, was levelled and found to be about 1 in 77 as shown in Figure 18-5. This sandur is aggrading rapidly at the present time. The whole sandur has been built up by the slow migration of the major meltwater streams across it from one side to the other. At the present time, the river is at the eastern extremity of the sandur; in the future, it may start migrating westward again across it. Skeiðará is not the only stream to emerge on to the sandur, although it is the largest. Towards the seaward limit of the sandur, the material becomes much finer. Instead of gravel and boulders, black basaltic sand becomes the dominant material and a shallow, flat lagoon separates the coastal sand barrier from the dry sandur. At the eastern end of the sandur this lagoon is about 6 km wide and less than 0·3 m deep throughout this width. The gradient of the sandur, therefore, falls off as the material becomes finer and the slope required to carry it is reduced.

Detailed work has been carried out on Hoffellssandur, the outwash of another glacier draining from Vatnajökull (F. Hjulström, 1955). It lies to the east of Skeiðarársandur. The work included a detailed study of the hydrology of the fluvioglacial rivers that have built up the sandur and a study of the climate. The character of the sediments on the sandur was studied by boring through the gravel and sand. Studies were also made of the marginal lakes that are associated with the sandur. Repeated surveys have given some indication of the variation of the stream positions with time. During the period 1820 to 1873, the river Austerfljot moved gradually north-eastward across the sandur and then by 1937 it had moved most of the way back again. The river appears to have remained in about the same position between 1903 and 1938. During this period, the glacier gradually started to split into two parts. Between 1945 and 1951, a proglacial lake gradually developed in front of the glacier, although the position of its snout did not change very much. The actual width of the area through which the river has fluctuated is fairly small, covering less than 400 m.

A study was made of the depth of the proglacial lake which developed as the glacier snout withdrew slowly from the terminal moraine. The sandur slopes down very steeply into the lake and the almost vertical ice front in the lake extends far below the level to which the sandur surface has been built up. The greatest depth of the lake, 23 m, was found along the ice margin. This deep point occurs where the meltwater stream emerges from the glacier. The floor of the river slopes up steeply to a depth of 2 m and then more gradually to the sill at a depth of 20 cm. The sandur level is about 4–8 m above the river level, the difference in height decreasing downstream. A horizontal line from the bottom of the lake does not cut the sandur surface until 3 km from the ice front. The lake floor character suggests that, at present, sedimentation of fine material is in progress. The lake also acts as a trap for the coarser material brought down by the glacier streams. The river, under these conditions, starts with almost no load and erodes its bed as indicated by the depth of the stream below the sandur surface. Sandar cannot, therefore, be built up near the ice front where proglacial lakes exist.

One problem is to account for the process by which the meltwater streams can build up the level of the sandur above the lower limit of the ice. This aggradation must take place before the formation of the proglacial lake. Such lakes are common features in front of many of the outlet glaciers of Vatna-jökull (Chapter 19). They appear to form most readily when the glacier snout is retreating and thinning after a period of advance, during which frontal moraines are formed.

The profile surveyed from the margin of Skeiðarárjökull (Fig. 18-5) shows how the level of the sandur behind the major moraine has not been built up to the level in front of it. In this vicinity, proglacial lakes have formed during the rapid retreat of the last few decades. J. Jonsson (1955) has suggested that the proglacial lakes form at points where large meltwater streams issue from a retreating ice front. He suggests that the lake is deepened by the powerful erosive activity of these streams, which flow mainly beneath rather than within the ice. This process can only apply to temperate glaciers. Three pro-glacial lakes in front of Breiðamerkurjökull were found to be 60 m, 48·5 m and 34·5 m deep. The river issuing from beneath the ice erodes downward. Then, when the river moves its position, the ice settles into the hollow so formed. The old channel thus becomes blocked by ice and erosive activity is transferred elsewhere. Slow sedimentation reduces the depth at the old river site, although the ice blocking the old channel prevents rapid sedimentation at this point.

The level of the sandur frequently seems to be linked with the position of a moraine, to which the outwash is graded. It appears, therefore, that aggradation of outwash material can take place most effectively, if not only, when the ice front is advancing and then remains stationary for long enough for a terminal moraine to form. Under these conditions, the meltwater streams must flow upward from beneath the ice to reach the level of the developing sandur. It is possible to observe that this does in fact take place. The melt-rivers can be seen to be emerging under strong hydrostatic pressure from the glacier. The source of Skaptafellsá illustrates this point well.

Because of the gravelly and sandy nature of the sandur sediments, much meltwater percolates underground near the ice margin where the material is coarser. The water emerges from the sediment farther from the ice front where the material is finer and less permeable. The emerging water is clearer than glacial water, but still carries some fine sediment. Nevertheless much of the fine sediment must have been deposited amongst the coarser material. Some of this ground-water emerges as springs, and some forms shallow lakes in old channels. The ground water-table slopes at almost the same gradient as the sandur surface, which is from 1 in 66·7 to 1 in 108·6. The sandur material tends to become finer vertically downward. This increase in fine material also helps to prevent water penetration, so that ground-water movement is probably restricted to the uppermost 10 m of the sediment. An estimate suggests that the ground-water discharge is 0·4 per cent of the total water discharge (Hjulström, 1955).

Most of the fluvioglacial outwash probably accumulates during advancing stages, when the erosional capacity of the glacier is also enhanced, so that the meltwater streams carry a larger load. The sandur reaches its maximum elevation when the ice front becomes stationary and terminal moraines are forming. The relationship between terminal moraines and outwash spreads is clearly seen in the valleys on the eastern side of the Southern Alps of New Zealand. The features described by J. G. Speight (1963) in connection with the Pukaki stage of the glaciation of the area around Lake Pukaki illustrate these points well. The Pukaki moraine consists of various ridges with steep ice-contact slopes facing up-valley. A series of outwash surfaces form broad channels through an area of knob-and-kettle formation in front of the moraines. These outwash surfaces still show clearly the pattern of braided streams that formed them. The outwash is in the form of a broad fan with a gradient of about 9 m/km. This outwash surface stands about 2 m below that of the former stage, which was formed in relation to moraines a little farther down the valley. The ice-contact slope, facing up-valley and the Lake Pukaki basin, indicates that the glacier must have eroded down well below the level to which the outwash was built up when the ice occupied the position indicated by the moraine. The streams depositing the outwash must have been flowing under hydrostatic pressure if they were flowing under and not on or within the ice. The ice depositing the moraines must have been 300 m thick in the main basin behind the moraine. This gives some indication of the upward component of flow of the streams under the ice.

Some of the areas of older outwash formation can usefully be compared with these recent examples. The Canterbury Plains in New Zealand were at one time thought to be formed by fluvioglacial deposition. This area of smooth but gently sloping land is formed of thick layers of coarse gravel on the east of the Southern Alps. It is now thought to be only partially of fluvioglacial origin. Its surface slope is very similar to that of the sandar of Iceland. R. P. Suggate (1958, 1963) has shown that the Plains consist of three major fans of material brought out from the mountains by heavily laden streams. It was thought that the two lower fans were built up as outwash plains in association with two of the major moraines recognized in the foot-hill area, the Otarama and the Blackwater moraines. It was recognized that the upper fan was formed in post-glacial times. It lies on the seaward side of remnants of the earlier fans that form the surface near the mountain area. It is now thought that only the oldest fan is in fact of glacial outwash material and that this fan correlates with the Blackwater stage in the Waimakariri. This stage probably belongs to part of the last glaciation in the Northern Hemisphere.

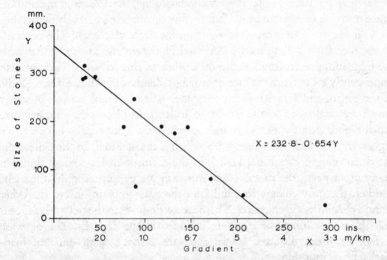

Fig. 18-6 The relationship between the size of the material and the gradient of an outwash delta in western Baffin Island

Fluvioglacial outwash deltas were formed in Baffin Island as the ice retreated from its readvance position near the present central west coast about 7000 B.P. These proglacial deltas link the ice-front position, which is marked by a steep ice-contact slope, with the sea-level or lake-level controlling the deltaic deposition. The deltas consist of very large boulders near the ice-contact face. The mean value for seven deltas, taking the mean of 50 stones on each, was 44 cm, with a standard deviation of 25 cm. The size of the material and the gradient falls off from the ice-contact face outwards. There appears to be a linear relationship between the stone size and slope as shown in Figure 18-6. The two variables correlate significantly at the 99 per cent level, r, the co-efficient of correlation, being -0.858. The coarser deltas usually have large dry channels, which show that they were built up by braided streams. At their front faces, the deltas are much finer and consist of sand and finer sediment. The sand layers show the steeper dip of foreset beds beneath a layer of horizontal topset beds in some instances. Many of the large deltas in this area are 4–5 km across. They indicate halt stages in the recession of the ice, and are also associated with dead-ice features, some of which have already been noted in Chapter 16.

Outwash plains are not very conspicuous in Britain, although examples of valley trains have already been mentioned. In continental Europe, on the other hand, they are widely distributed and of great scenic importance. In Denmark for example, there are large areas of outwash associated with the ice front at the time of the Main Stationary Line, the Weichsel maximum. These outwash areas form broad fans cutting through and partially destroying the Old Morainic landscape, dating from the Saale glaciation. The outwash plains, therefore, belong to the Weichsel glaciation and are associated with the high moraines formed at the maximum of this advance. The outwash slopes gently away from the moraines at a gradient of 1 in 700 near the middle of the plains, and 1 in 1000 near the outer margin. These spreads of sandy gravel are analogous to the sandar of Iceland. In places, the moraines are much higher than the outwash, but elsewhere the outwash has built up to a higher level than the moraines with which it is associated. In these areas the proximal margin of the outwash is in the form of an ice-contact slope. In many other parts, the moraine and outwash merge imperceptibly into one another, the only change that indicates the passage from one to the other being a change in soil character. There are kettle-holes in the outwash deposits that show that lumps of ice were included in the deposits. The outwash surface has been dissected by later stream activity, with different load-discharge relationship.

Another classic area where the outwash fans are associated with terminal moraines is in the northern Alpine foothills. This is the area where the original glacial sequence of A. Penck and E. Bruckner (1909) was worked out by studying the moraines and associated outwash deposits. The outwash of the Alpine valleys is called 'Schotter', which means gravels. The so-called 'Deckenschotter', which belong to the two older glaciations (Gunz and Mindel), were true outwash plains, spreading extensively over the country in front of the moraines. In the later glaciations (Riss and Würm), however, rejuvenation during the Mindel-Riss interglacial restricted the outwash to the valleys. Thus valley trains were formed in the later glaciations. The Munich plain is covered by outwash from the Alps, brought down by the Inn and Isar valleys. The gradients of these deposits, varying from 1 to 1·2 per cent, are steeper where the outwash is confined within valleys than where it can spread more widely as in the North European plain. Outwash deposits are not confined to the northern fringe of the Alps. They also occur in the Po valley where they reach a thickness of 237 m at Cremona. The outwash is now buried by deposits owing to subsidence in the area.

The outwash plains of North Germany are very extensive. They form wide sandy heaths, such as the Lüneburger Heide. The material gradually changes from gravel near the moraines (Fig. 12-5) to sand and then to finer clay deposits southward. Other areas where there are sandar of large dimensions and considerable thickness are found to the east of the Baltic. In Russia, they occur in the Dnieper valley. There are also very extensive outwash deposits, or modified drift, in North America. The effects of meltwater on the drift are apparent in the New England states westward to Minnesota, the Dakotas and Manitoba. The Mississippi River at one time carried a large load of glacially derived material. Its valley drained a 3000 km stretch of the Laurentian ice margin at a time when the precipitation was higher, during the maximum of the Wisconsinan glaciation. This coarser material can be traced to the delta of the river 1120 km from the nearest ice edge.

REFERENCES

L. ARNBORG (1955), 'Hydrology of the glacial river Austerfljot', *Geogr. Annlr* **37**, 185-201

L. ARNBORG (1955), 'Ice-marginal lakes at Hoffellsjökull', *Geogr. Annlr* **37**, 202-28

M. GAGE (1958), 'Late Pleistocene glaciation of the Waimakariri valley, Canterbury, New Zealand', *N.Z. Jl Geol. Geophys.* **1**, 123-55

F. HJULSTROM (1955), 'The ground water', *Geogr. Annlr* **37**, 234-45

J. JONSSON (1955), 'On the formation of frontal glacial lakes', *Geogr. Annlr* **37**, 229-33

A. PENCK and E. BRUCKNER (1909), *Die Alpen im Eiszeitalter* (Leipzig)

S. RIST (1955), 'Skeiðarárhlaup 1954', *Jökull* **5**, 30-36

J. G. SPEIGHT (1963), 'Late Glacial historical geomorphology of the Lake Pukaki area, New Zealand', *N.Z. Jl Geol. Geophys.* **6**, 160-88

R. P. SUGGATE (1958), 'Late Quaternary deposits of the Christchurch metropolitan area, New Zealand', *N.Z. Jl Geol. Geophys.* **1**, 103-22

R. P. SUGGATE (1963), 'The fan surfaces of the central Canterbury Plains', *N.Z. Jl Geol. Geophys.* **6**, 281-7

Chapter 19

Glacial Lakes and Lacustrine Deposits

The great glacier of Aletsch gives rise ... to a small lake called the Märjelen See, which ... is periodically drained by changes which take place in the internal structure of the glacier. Rents or 'crevasses' in the ice open and give passage to the waters, which escape in a few hours, producing destructive inundations in the country below. The Märjelen See was about two miles in circumference when I visited it in August 1865, and about forty feet below its normal level. ... Such a state of things gave me an opportunity of examining a point of great geological interest, namely, the ... large terrace or line of beach which encircles the lake basin all round its margin, and which constitutes its shore when full, and when its surplus waters flow over to the Viesch valley. I satisfied myself that this terrace is a counterpart of one of those ancient shelves or parallel roads, as they are called, of Glen Roy in Scotland. ... (C. LYELL, 1866)

Glacial lakes are not today so common, nor do they attain such dimensions as in the Pleistocene period. Nevertheless, enough survive and in a sufficient variety of circumstances to throw much light on the nature and effects of the Pleistocene lakes. Present and former glacial lakes are responsible for distinctive landforms and deposits, including shoreline features such as strandlines, terracettes, deltas, and bottom deposits such as laminated clays. The term 'glacial lake' really embraces a whole family of lake forms. There are both lakes actually dammed by glacier ice, and lakes where glacier ice forms part of the lake margin though not the actual barrier. Relationships with the ice are as varied as in the case of meltwater channels. Some glacial lakes are supraglacial, some are marginal in various ways, and some exist subglacially or englacially although direct study of lakes in this latter category is rarely possible. The methods by which glacial lakes drain is equally diverse, but it is well established that the majority periodically empty, often catastrophically, by subglacial or englacial routes.

It is logical to begin by looking at present-day glacial lakes, and then to review the morphological and depositional features relating to such lakes and their Pleistocene equivalents.

Modern Glacial Lakes

LAKES IN LATERAL VALLEYS DAMMED BY ICE IN THE MAIN VALLEY

Probably the biggest concentration of glacial lakes in the world today is to be found in southern Alaska, especially in the coastal strip (K. H. Stone, 1963). Nearly all lie in ice-free tributary valleys that have been blocked by glacier ice in the trunk valley, the trunk glacier often sending a short distributory arm up into the side valley to form the actual dam. The largest lake, covering 124 km², is Lake George; in 1963, there were 52 other lakes in an area 1200 × 150 km. All drain under or into the ice—no instance of the lake waters overtopping the ice dam has been recorded. Until 1963, and since 1918 at least, Lake George emptied annually. As in all such regions, the number of lakes is slowly diminishing, the result largely of the current glacier recession, but the former lake sites are easy to recognize. R. P. Sharp in the St. Elias Range of Yukon Territory, Canada, describes the site of North Fork Lake, in a tributary valley dammed by Wolf Creek Glacier, where the wave-cut shoreline terrace and the delta built by the North Fork River indicate a maximum water depth of 50 m. Depth data for existing Alaskan lakes are few; the deepest measurement is 73 m for Lake Tulsequah in 1958, but this lake is known to have reached a depth of 195 m in 1910.

Another area where glacial lakes are still relatively numerous is that of Vatnajökull, Iceland. Here again the majority lie in tributary valleys blocked at their lower ends by glaciers. Such is the case with Graenalón, the largest, covering 18 km² in 1935 and 150 m deep, dammed by Skeiðarárjökull on the east. Until recently, Graenalón emptied subglacially every four years, but later every second year. Its maximum possible level has been determined by the col at 635 m leading south to the Nupsvötn River, the lake overflowing through this col in May 1939, for example. Farther east along the margin of Vatnajökull, another outlet glacier, Heinabergsjökull, has also been responsible for blocking some small tributary valleys (Fig. 19-1). Vatnsdalur, one of the latter, contained a lake up to a level of 464 m in the 1870s, which drained over a col at that height into a lower ice-dammed lake, Dalvatn. In 1898, Vatnsdalur lake emptied subglacially in a sudden and violent manner, and the overflow col was abandoned. Until the 1940s, it drained each year thereafter in a similar manner though the violence of the bursts decreased with the gradual shrinkage of Heinabergsjökull which became less and less capable of impounding any sizeable lake. A further result of the diminishing efficiency of the ice dam was that the date of drainage each year became earlier—from late September in 1899, late August in 1910, to July in the 1930s. In the 1950s, Vatnsdalur was draining two or three times each year, its level not exceeding 400 m.

In Europe, the most celebrated ice-dammed lake, also occupying a tributary valley, has undoubtedly been the Märjelensee, held up by the Aletsch glacier (Fig. 3-3). It attained its greatest dimensions in July 1878, when the water rose to 2366 m and the lake was 1600 m long. Its depth may have been

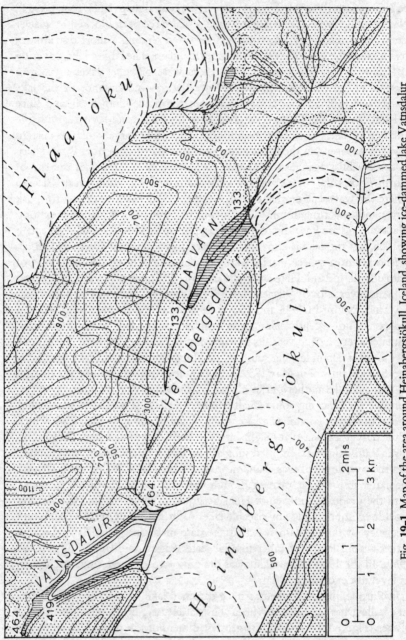

Fig. **19-1** Map of the area around Heinabergsjökull, Iceland, showing ice-dammed lake Vatnsdalur and the site of formerly ice-dammed Dalvatn (S. Thorarinsson, 1956)

as much as 80 m. It has long been known as a peculiarly variable lake, changing greatly in size and shape from one year to the next. When full, it has been on rare occasions observed to spill over the ice, but usually the ice is too crevassed to permit this, and its normal mode of emptying is into crevasses in the impounding ice wall, or beneath the ice. The last method empties the lake particularly rapidly. From 1813 to 1900, the lake emptied subglacially nineteen times. In 1873, 10 million m^3 in 8 hours were said to have been released, causing a flood-wave on the Rhône. As the lake continued to attain unprecedented levels in the 1870s, and its bursting created flood problems in the valley below the Aletsch snout, construction of a tunnel from the lake into the Seebach was put in hand. After completion in 1896, it soon became superfluous, for slow shrinkage of the Aletsch glacier has since that date gradually lowered the maximum lake level. The 1957 map (Switzerland, 1:10,000) of the Aletschgletscher showed that the lake had all but disappeared, a small pond at 2300 m encased by moraine and draining into the ice at a level of 2288 m being the only survivor. As the edge of the Aletsch glacier does not exceed 2314 m at any point south of this, the possibilities of lake formation are now very limited. N. E. Odell (1966) recorded the Märjelensee as 'empty' in July 1965.

LAKES IN MAIN VALLEYS DAMMED BY ICE FROM LATERAL VALLEYS

A second category of ice marginal lake concerns situations, often causing highly dangerous lake bursts, where a side glacier advances into and blocks a main valley. Examples of Pleistocene age in western North America were described in Chapter 12. Large-scale present-day examples are known from the Karakoram (K. Mason, 1935), especially among the upper Shyok glaciers. In 1924-5, for instance, the Chong Kumdun glacier suddenly advanced to form a complete block across the Shyok river, the lake bursting in October 1926 with catastrophic effects. The process was repeated several times subsequently, until the glacier ceased to advance in the 1930s. Another glacier, the Biafo, has been responsible for similar temporary lakes at times when it has advanced out of its side valley to block the main valley of the Braldu (K. Hewitt, 1964). The lake bursts cause severe floods and damage in the upper Indus valley. In Switzerland, the Mattmark See was, until the 1920s, an example on a smaller scale (R. Keller, 1950). The steeply sloping tongue of the Allalin glacier extending across the Saaser Visp valley impounded the Mattmark lake in the latter to a depth of 30 m or more at times; in 1834 the lake was 2·5 km long. It has had a history of violent outbursts in the past— 26 since 1859—but by 1909, the Allalin glacier had retreated sufficiently to unblock the main valley, and its lateral moraine was being cut through by the lake overflow, and the lake itself was rapidly silting up.

A slightly different example is provided by the Moreno glacier in Patagonia. This is one of the rare glaciers in the world that has been steadily advancing this century (R. L. Nichols and M. M. Miller, 1952). In 1917 it entered the main valley in which lies Lago Argentino and divided the latter. The part of the lake behind the Moreno ice barrier is known as Lago Rico; the water level of the latter was raised 17 m above Lago Argentino in 1942. The ice dam normally does not survive more than a year before rupturing, and after the burst, which commences subglacially and later follows an open ice-walled channel when the subglacial tunnel has collapsed, the water level settles down again to that of Lago Argentino.

FRONTAL LAKES

The term was employed by L. Arnborg (1955) to describe lakes formed between the ice terminus (or glacier snout) and a barrier such as moraine downstream. Glacial ice thus contacts the water but the lake is not actually ice-dammed. The glacial Great Lakes of North America (up to the Algonquin stage) were examples on a huge scale of frontal lakes. Well-known minor examples today occur at the foot of Brixdalsbreen (Norway) and the Steingletscher (Switzerland). Arnborg has investigated two frontal lakes at Hoffellsjökull (Iceland): Svinafellslón and Hoffellslón. The former is held by moraines dating from 1890 and 1903, and lies across minor terminal ridges dating from the 1930s. Such lakes are deepest next to the ice front, but soundings here are hazardous because of calving ice. J. Jonsson (1955) has suggested that some frontal lakes may occupy basins or a series of furrows eroded by subglacial streams when ice covered the site of the lake (Chapter 18). Ice collapsing into the furrows prevented them from being infilled with sediment, and when the ice melted back, meltwater filled the hollows. This may explain how the floors of some frontal lakes come to lie at a lower altitude than that of the outwash plain surface downstream, and how some frontal lakes form without morainic barriers.

MINOR MARGINAL LAKES

Small lakes may develop beside nunataks (owing to reflected heat from the rock), in marginal channels blocked by slides or alluvial cones, or in the angle between confluent glaciers. J. D. Forbes (1843) showed how, at times when melting is adequate, a small lake forms by the Tacul promontory between the Géant and Leschaux glaciers, north-east of Mont Blanc, and noted that marked variations in lake level suggested a subglacial outlet.

LAKES ON GLACIERS AND ICE-SHEETS

Factors involved in the formation of supraglacial lakes include: 1. Local reversed slopes of the ice surface; 2. collapse of subglacial or englacial cavities; 3. differential ablation; 4. blocking of crevasses or moulins. The Martin River glacier, Alaska, possesses in its marginal portions numerous roughly circular sink-holes with standing water, up to 90 m deep and 300 m in diameter (J. R. Reid and L. Clayton, 1963). The lakes are apparently connected with subglacial tunnel systems, and their periodic draining and refilling results from ice movement opening or closing the tunnels. R. P. Sharp (1947) noted similar features on the Wolf Creek glaciers; he also described 'dust basins' on the ice surface formed by localized melting under thin layers of dark silt, some as much as 5 m deep. Once such surface ponds are formed, the relative warmth of the water quickly enlarges them.

SUBGLACIAL AND ENGLACIAL LAKES

The existence of such water bodies at depths within glacier ice is indisputable. Sudden increases in the flow of subglacial streams emerging from glacier tongues (as preceded the Allalin glacier catastrophe of 1965: see Chapter 4) are likely to result from the tapping of such water bodies; and local subsidence of the ice surface over presumed subglacial or englacial cavities accompanied again by sudden increases in meltwater discharge from the glacier also points to the existence of internal lakes. The most impressive example is that of Grimsvötn in the heart of Vatnajökull, Iceland. This is an ice-filled depression, 300 km² in area and normally 400 m deep; beneath a thickness of ice, a lake accumulates periodically in this depression owing to ablation and to slow subglacial melting related to solfatara activity. When approximately 7 km³ of water have accumulated subglacially, the lake bursts, escaping by tunnels 55 km long to the edge of Skeiðarárjökull. The bursts occurred at almost regular 10-year intervals until 1934 (10 years' annual snow accumulation at Grimsvötn totals about 7·5 km³) and each was accompanied by a volcanic eruption at Grimsvötn. S. Thorarinsson (1953) argues that the eruptions are triggered off by the sudden release of water pressure when the lake bursts, for the maximum discharge of water just precedes the earth tremors accompanying the eruption. Since 1934, the bursts have tended to occur at intervals of 4 to 6 years. During and after the burst, the ice surface of Grimsvötn sinks more than 200 m into the drained cavity beneath.

The Problem of Glacial Lake Drainage

All observers agree that the usual mode of glacial lake drainage is subglacial or englacial. T. G. Bonney emphasized this as long ago as 1896. A few lakes, as already noted, drain occasionally by overflow cols: this certainly happened in the case of many Pleistocene ice-dammed lakes (see Chapter 12) and no further discussion is required on this point. Drainage of a glacial lake over the ice barrier is rare except in the case of sub-polar glaciers where sub-zero ice temperatures preclude subglacial or englacial escape routes. V. Schytt (1956) mentions two large ice-dammed lakes in the Thule area of north-west Greenland which normally drain supraglacially.

There is considerable controversy over the mechanism by which glacial lakes drain subglacially, especially in regard to the sudden bursts or 'jökul-hlaups'. The lake water undoubtedly finds its way into crevasses, into gaps between the ice and bedrock, or perhaps soaks through subglacial gravels, and may travel for great distances beneath the ice—note the 55 km length of the Grimsvötn escape route. The problems involved are several: 1) Why do glacial lakes often drain regularly and periodically? 2) How are subglacial outlets first opened and how are they kept open until the lake has emptied? 3) How are such long escape routes opened in such a short time? 4) What factors control the actual time when the burst commences?

S. Thorarinsson (1939) contended that when the depth of water in the lake reached a critical figure (about nine-tenths the height of the ice barrier), the ice barrier would begin to float. The height (h) of the ice barrier would thus control the maximum possible depth of the lake, and can be computed from the formula:

$$h = \frac{x - m}{1 - D}$$

where x = the height of the barrier above the lake surface, m = an allowance for crevassing and unevenness of the barrier surface, and D = ice density (about 0·9). Applied to the Vatnsdalur lake levels of 1903 and 1938, this gives very reasonable results; and Thorarinsson claims that during the Grimsvötn jökulhlaups, a large part of the edge of Skeiðarárjökull floats for a brief period. But it has also been pointed out that, on this hypothesis, one would expect the outflow of water to be intermittent and not a sudden rush, for once lake drainage had begun, lowering of lake level would reduce the pressure on the ice barrier which would then cease to float, until the lake once more filled to the critical level. Moreover, it is very difficult to conceive of several kilometres, let alone 55 km, of ice floating simultaneously, and the hypothesis could not apply where the ice was frozen to bedrock. J. W. Glen

(1954) has put forward a different hypothesis. Since the density of water is greater than that of ice, water of more than about 200 m depth resting against an ice dam or filling a hole in a glacier will exert a stress on the ice sufficient to cause deformation of the ice at the base of the water mass. This hypothesis, however, also encounters problems. The depths of glacial lakes are often nowhere near 200 m when bursts occur—the head of water preceding the 1958 burst of Lake Tulsequah, British Columbia, was no more than 73 m (M. G. Marcus, 1960). In this case, too, it appears that tunnels in or beneath the ice and 7 km long were opened up in the space of a few days, which is difficult to equate with the slow rate at which ice deforms (for instance, compare known rates of contraction of man-made tunnels at depths of 50 m or more in glaciers: Chapter 4). Furthermore, there is the same problem encountered by Thorarinsson's hypothesis that as soon as lake depth was reduced, pressure on the ice tending to enlarge an escape route would also diminish.

O. Liestøl (1955) thinks that enlargement of the escape route by melting is a most important factor. Initial opening of a subglacial passage is achieved by ice flotation, following Thorarinsson; the passage is then kept open and enlarged by melting until the lake has emptied. Calculations show that water at 1°C. and flowing at only 1 m³/sec, can theoretically melt over 270 m³ of ice in 24 hours. Another possibility for maintenance of an open escape route during emptying is suggested by N. Aitkenhead (1960). After the ice barrier has been lifted by flotation, and lake level has dropped, the ice will never fall back exactly in its former position on a bedrock floor which is likely to be uneven; and moreover, icebergs wedged in the tunnels might help to prevent collapse. Marcus (1960) makes similar suggestions. Thorarinsson's flotation hypothesis seems adequate to explain the timing of the initial lifting of the barrier, and does not demand excessive water depths as does Glen's hypothesis. Liestøl's and Aitkenhead's suggestions for maintenance and enlargement of tunnels to allow complete lake draining in one burst seem reasonable additions to Thorarinsson's hypothesis.

THE NATURE OF JÖKULHLAUPS

The following account of the Grimsvötn burst of 1934 by S. Thorarinsson (1953) is typical of the large Icelandic jökulhlaups:

The river Skeiðará, which in March and April normally has its smallest discharge, started to rise on March 22nd. The rise was slow at first, but on the 24th it had reached approximately the normal summer high-water level. On the 28th the water started forcing its way out from under the glacier at several places, breaking up its border. On the morning of March 31st, the glacier burst reached its climax. Forty to fifty thousand cubic metres of muddy grey water plunged forth every second from under the glacier border bringing with it icebergs as big as three-storeyed houses. Almost the whole of the *sandur*, some 1000 sq. km in area, was flooded. At 17.30 hr the same day the burst suddenly started to abate, and by the following morning the discharge of the Skeiðará was normal.

It is impressive to recall that the source of this water escaping from the edge of Skeiðarárjökull is Grimsvötn, 55 km to the north. Jökulhlaups are associated with other glaciers in Iceland. The most intense and dangerous bursts are the 'Katlahlaups' from Myrdalsjökull (Thorarinsson, 1957) where the peak discharge may exceed 100,000 m³/sec. The sudden release of such quantities of ice-impounded water has far-reaching influences on the morphology of the valley or floodplain over which the water escapes; vast quantities of sediment are carried away or redistributed, and in the course of a few days the whole form of the outwash plain may be radically altered (Chapter 18). Although present-day jökulhlaups have been studied in Iceland, Norway, North America, and elsewhere, recognition of possible Pleistocene jökulhlaups and their effects has advanced little. The bursts of Pleistocene Lake Missoula (Chapter 12) are an exception, and also the bursting of glacial Lake Tahoe in Nevada, postulated by P. Birkeland (1964). Birkeland describes the evidence for such floods in the Truckee valley below Reno, where 3-metre granitic boulders lie stranded many kilometres beyond the limits of any glaciation. In Britain, T. F. Jamieson (1892) tentatively suggested that the glacial lakes of Glen Roy, Scotland, suffered a sudden burst when the lake fell from the higher to the middle level (see p. 432).

Glacial Lakes as Indicators of Glacier Oscillations

As Thorarinsson (1939) demonstrated in the case of Icelandic ice-dammed lakes, they can be of the highest value in recording changes in ice thickness. If the flotation hypothesis is accepted, the lake level prior to draining provides a means of accurately determining the ice thickness in the barrier. The lake shorelines provide records of former lake levels, and historical records can often provide the date of lake emptying. On the evidence from lake fluctuations alone, Thorarinsson reconstructs the recent glacial history of Iceland: 1700-50 general glacier advance; 1750-1800 stagnation or slight recession; readvance, probably to maximum extent in historical times between 1800 and 1850; and beginning about 1890, a general recession, which has gradually accelerated and threatens extinction of many glacial lakes.

Features Associated with Present and Former Glacial Lakes

STRANDLINES AND BEACH DEPOSITS

Studies of modern glacial lakes show that within surprisingly short periods of time, lake waters standing at one particular level are able to create a strandline provided that there is sufficient unconsolidated sediment available. From a distance, strandlines appear as horizontal markings (except where later isostatic movements have caused them to be tilted or warped), often at successive positions on the hill slopes to record changes in the lake levels. Closer examination reveals that the strandlines are features caused both by erosion by lake wave attack and by deposition from the lake waters. It is common to find that strandlines, particularly those of Pleistocene age, become less and less distinct as one approaches them; it may be exceedingly difficult to recognize and locate them on the ground over which one is walking, especially in wooded country. Sometimes they are represented only by slight changes in vegetation, related to differences of soil and drainage at the actual shoreline, and such vegetational changes, though easily discernible over a large area viewed from a distance, may be almost undetectable and inseparable from other variations in ground cover at close quarters. Tracing of former shorelines is further complicated by their discontinuity, by isostatic tilting, and by the fact that their sequence on the ground is not necessarily a chronological one.

The general morphological and depositional features that together form a strandline are those normally associated with the margins of any water body, and will vary in scale with the dimensions of the lake.

The Pleistocene Great Lakes of North America offer the most impressive lake-shore beaches to be found anywhere in the world. Recognized as old lake beaches as long ago as the 1830s, J. W. Spencer in 1890 described them thus: 'In ascending from the modern lakes to the highlands, several old shores must be crossed. The country may be described as a series of terraces or steps, whose frontal margins are moulded into hills, and whose surfaces are plains, most commonly of clay, although sometimes of gravel or sand, at the back of which there may be found the beach in some form. These gently rising terrace plains may each be several miles in width . . .' (p. 76). In places, the beaches consist of boulder pavements up to 1 km wide; the larger boulders were probably deposited by lake ice. The strandlines and beaches have been traced continuously for hundreds of kilometres in some cases (see Chapter 5, p. 129 and Chapter 12, p. 268); as expected, they terminate abruptly at the former ice margins, though successively lower strandlines extend farther and farther north and east, and the lowest completely enclose the present lakes showing how the ice gradually wasted away.

Even in the case of such huge lakes, wave-cut notches and cliffs are usually only present where the lake washes against unconsolidated materials such as glacial drift, or other soft formations. R. P. Sharp (1947) states that the wave-cut terrace of the former North Fork Lake in the Yukon Territory is 23 m wide in till and up to 5 m in bedrock, but the nature of the latter is unspecified (p. 51), nor is the duration of the lake known. Much more important in the definition of strandlines are the associated deposits, built into lake shore terraces. The deposits range from very fine silts, such as composed the small terraces recorded by G. de Boer (1949) for a former glacial lake beside Leirbreen, Norway, to coarse material including cobbles and boulders in the case of some of the larger Pleistocene glacial lakes. Much depends on the nature of the material locally available, on the size of the lake, determining strength of wave action, and its duration. If morainic debris is available, wave action may wash out the finer fraction from one place, leaving a lag deposit of boulders, and redeposit it elsewhere. Changes in lake level will also have important effects, as R. F. Flint (1957, p. 145) suggests. The more massive accumulations of cobbles and boulders may be related to a rising lake level and advancing shoreline, concentrating and reworking the beach material, whereas the smaller pebble ridges are perhaps to be associated with stable or even receding shorelines.

Glacial lakes which have recently drained often reveal on the surrounding slopes a series of minor shoreline features or 'terracettes'. Lake George, Alaska, in 1951 just after emptying showed more than twenty terracettes below the high-water mark of that year. All were evenly spaced and horizontal; some were no more than 0·5 m apart vertically (K. H. Stone, 1963). Gjanupsvatn beside Hoffellsjökull in Iceland drained suddenly in June 1951 and its sides presented terracettes at 193, 184, 175, 170, 153, 136, and 132 m (L. Arnborg, 1955). De Boer (1949) noted traces of about twenty terracettes below the two main shorelines of the Leirbreen marginal lake. The terracettes are clearly formed by the lake waters, and they are depositional features mainly. Yet it has not been explained why they occur in series, unless they represent an intermittently falling lake level (which itself requires explanation), nor why they are found in some drained lakes but not in others. Some have been recorded which are only 5 cm high and 5 cm wide. Further investigations are needed on all these points.

Other features associated with glacial lake strandlines include spits and bars' lake ramparts, and dune systems. These are to be found on the margins of the large Pleistocene lakes only. The finest examples of spits and bars ever described for Pleistocene lakes are along the shores of Lake Bonneville, Utah (G. K. Gilbert, 1890), which although not usually regarded as a glacial lake, did in fact receive glacial ice at certain points from the Wasatch Mountains.

Gilbert's monograph depicts on numerous close-contoured maps magnificent examples of spits, tombolos, bay bars, and shore terraces.

Lake ramparts require rather more consideration. Such features were recorded by F. Leverett and F. B. Taylor (1915) in their classic monograph on the Pleistocene history of the Great Lakes. Glacial Lake Maumee, for instance, is margined in places by peculiarly crooked shore ridges of coarse material 3-5 m high, which are almost impossible to interpret as normal beach ridges. Leverett and Taylor suggest their formation by lake ice which, after beginning to break up in a thaw period, has been driven on-shore by strong winds. If the shore gradient is gentle, the ice will slide up out of the water, pushing beach material in front. The suggestion has been amply confirmed by many observers on Arctic coasts today (for instance, J. D. Hume and M. Schalk, 1964). In the case of Arctic lake shores, A. L. Washburn (1947) described at first hand the way in which floating ice, driven by winds or currents, can push up ridges of angular and some rounded boulders to heights of 1·5 m. One such ridge was 20 m long and several metres wide. The beach material may be displaced several hundred metres inland, and, most important, up to 7-10 m vertically above water level. Hence, Pleistocene lake boulder ridges should be used with caution in estimating former lake levels.

Dune systems beside glacial lake shores need little comment. They are well known from the margins of such large Pleistocene lakes as glacial Lake Chicago, where they attain heights of up to 20 m above the Glenwood strandline, for instance.

In Britain, few examples of glacial lake strandlines and beach deposits exist. Those of the Glen Roy area in Scotland are unique as regards their clarity and extent and deserve some detailed consideration (Fig. 19-2). L. Agassiz (1842) correctly identified the horizontal markings or 'Parallel Roads' as old glacial lake shorelines; he claimed that the series of lakes was impounded by a 'lateral glacier projecting across the glen near Bridge Roy and another across the valley of Glen Speane' (p. 332). T. F. Jamieson (1863, 1892) was responsible for the first detailed field examination and interpretation. The Roads lie at three distinct principal levels. In Glen Roy itself, the three may be traced along a 10-km section of the Glen, while elsewhere in the Glen and in neighbouring Glens Gloy and Spean, only one or two of the strandlines may be observed. Commenting on the nature of the strandlines, Jamieson remarks that 'in the narrower land-locked parts, the little terraces jut sharply out from the hill, almost perfectly flat . . .; where however the valley is wide, . . . the terraces are broader, ruder, and more shelving' (1863, p. 240). The strandlines tend to die out where they pass over resistant rock outcrops, but elsewhere their beaches may be 12-20 m broad. They usually slope valleywards at 5 to 30 degrees (1892, p. 17). Rounded beach pebbles are found on them in places. Levelling by the Ordnance Survey, extended by Jamieson himself, gave the mean heights of the strandlines as 376·6, 350·1, and 280·5 m, the actual range of measurements being, respectively, 375·3–378·9, 348·5–353·3,

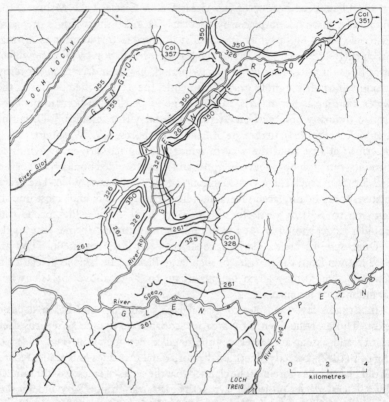

Fig. 19-2 The parallel roads of Glen Roy and adjacent valleys, Scotland (*Mem. geol. Surv. Scotland*, 1935, Crown Copyright reserved, reprinted by permission of the controller, H.M.S.O.)

and 278·9–282·8 m. The cols through which the lakes drained (circled on Fig. 19-2) and which determined the lake levels differed by 1–2 m from the mean heights of the respective strandlines. The variations have no significance, however, for as Jamieson pointed out (1892, p.17), the Ordnance Survey attempted to level along the median lines of the beaches, a procedure which would be expected to give slightly varying results for their levels, considering the ill-defined nature of the beaches in places and their transverse slopes. Allowance must also be made for accumulation of peat in the cols since lake waters drained through them. Lower lake levels of 138 m and 98 m were recognized later in Glen Spean by G. Wilson (1899).

There are many possible reasons why so few examples of glacial lake strandlines are known in Britain or in several other parts of the world. As already noted, modern glacial lakes are characterized by sudden and frequent changes of level, precluding the development of strongly marked strandlines. Hard rock will resist wave action, which will itself be limited in power if the lake was of small dimensions. The possibility of the lake being frozen for considerable periods must not be forgotten, nor the ease with which strandlines consisting mostly of soft unconsolidated deposits may be effaced by superficial movements or mass wasting, especially if subjected to periglacial climatic conditions. The length of time needed to form effective strandlines is a quantity variable with lake size, constancy of level, and shore materials, and few data are available. Vatnsdalur in Iceland (Fig. 19-1) provides one example, though, for in 1898 it first established a water level of 419 m. When visited by S. Thorarinsson (1939) forty years later, the lake had drained and a well-marked shoreline at 419 m was apparent. The lake area was no more than 1 km².

The most important factor determining the clarity of glacial lake strandlines is undoubtedly their age. In this respect, the Glen Roy strandlines are particularly significant, for it is now known that they date from the Late-glacial (Zone III) episode, terminating less than 10,000 years ago. The strandlines of glacial lakes formed in areas outside the limited range of Zone III glaciation in Britain are considerably older and this is the prime reason for their indistinctness or, in some cases, complete effacement. At the same time, great caution must be exercised in postulating the existence of such glacial lakes in the absence of any evidence in the form of strandlines.

DELTAS

The sudden checking of velocity where streams enter glacial lakes results in the formation of deltas. Deltas will be particularly large where they are deposited by meltwater flowing either directly (for instance, subglacially) from ice into the lake, or indirectly by a normal river channel, for the rate of discharge and the load carried will be high. The typical form of a delta not laid down in contact with the ice bordering the lake will consist of a relatively steep and lobate frontal slope, and a top surface gently inclined lakewards. The junction of these two surfaces will approximate closely to lake level, though sub-aqueous extensions are known, and the lake level is liable to fluctuate. Positive changes of lake level will result in submergence and burial of the former delta by new deposits; negative changes will cause emergence of the whole or part of the delta, the incision of the stream feeding it (resulting in varying degrees of dissection), and the construction of a new delta beyond. Repeated changes of water level will thus produce complex forms.

The internal structure of a simple delta comprises the well-known sequence of bottom, foreset, and topset beds, but such a simple arrangement is by no means common. Changes of lake level already mentioned will produce complex structures, slumping and settling of the sediments will result in contorted bedding, and further complications will arise in the case of ice-contact deltas (see below).

North America holds the largest glacial lake deltas. Warren Upham's monograph (1896) on glacial Lake Agassiz records the evidence for the existence of that vast body of water (see Chapter 12) into which immense deltas were built out by meltwater. The Assiniboine delta, for instance, covers no less than 5000 km² (one-fifth as large as the Nile delta) and has an estimated volume of 80 km³. Its front rises 100 m, while wind-blown dunes up to 25 m high diversify its top surface. The foreset beds include waterworn cobbles up to 20 cm in diameter, testifying to the velocity of the currents which swept the material into the lake. From what we know of the duration of Lake Agassiz, the average annual increment of sediment for this delta must have been about one-fifth that of the present Mississippi delta.

Deltaic accumulations support the interpretation of the Parallel Roads of Glen Roy as glacial lake shorelines. Jamieson described one such delta on the eastern side of Glen Roy, 'bulging out like an artificial mound, compact and clear in outline as when first formed, save that the stream . . . has now cut a great gash through its midst' (1863, p. 241). The delta surface in this case is related to the level of the lowest principal strandline. The largest of the Glen Roy deltas lies at the western end of Loch Laggan, a clearly defined mass of silt and gravel whose front rises 6 m above the present loch, and whose top surface slopes gently backwards to a height of about 15 m. Jamieson draws attention to the fact that deltas are mostly associated with the lower of the three principal shorelines; he suggests that at the earlier and higher levels, more of the surrounding land was ice or snow covered and hence less debris might have been available to enter the lakes.

Figure 19-3 shows a fine suite of glacial lake deltas near Mold in north-east Wales. Their surfaces rise steplike at levels of 131-134, 151-154, and 170-174 m; sections reveal delta-bedded sand and gravel. The meltwater stream carrying these materials entered through and partly excavated the Bellan gorge upstream. The lakes may not have been extensive, for there are no strandlines evident and the deltas prove only the existence of standing water at the places where they accumulated (C. Embleton, 1964).

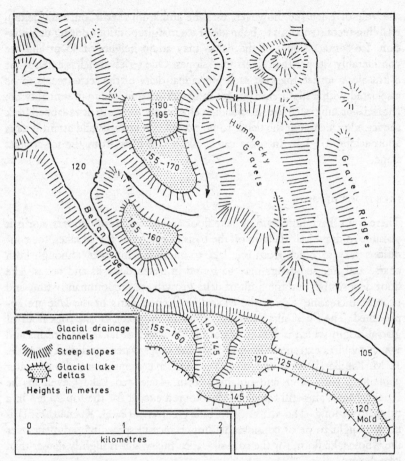

Fig. 19-3 Glacial lake deltas in the Alyn valley, north Wales. The lakes were impounded between high ground on the west and south-west, and decaying ice on the east and in the Alyn valley

Of somewhat different form are ice-contact deltas, sometimes termed esker deltas or delta moraines and already mentioned in Chapter 16. Figure 16-4 provides an example. These features possess an ice-contact slope at the back, although the frontal portion building out into the lake is normal. Owing to the proximity of the ice, kettle-holes often appear on the top surface and along the rear portions; slump structures characterize the bedding as the ice withdrew its support; and eskers may be present at the rear, representing the courses of the subglacial streams which injected the material into the lake. Fine examples from Britain have been mapped by F. M. Trotter (1929) in the Vale of Eden (he, however, uses the ambiguous term 'outwash delta'), enabling reconstructions to be made of ice margin positions during deglaciation. Ice-contact slopes of the deltas may attain heights of 45 m and are considerably steeper than the frontal slopes. One series of deltas extends for 8 km along an ice margin; each individual delta of the series seems to be associated with a slight embayment in the ice margin where presumably the englacial or subglacial stream issued. Such streams appear to have sapped back the ice edge, enabling the deltas to be extended rearwards and attain widths approaching 1·5 km. In some cases, eskers run back from the ice-contact slope.

LAKE FLOOR DEPOSITS

There is no sharp line dividing lake floor and lake margin deposits, nor is it possible to draw a distinction on the basis of grain size for all lakes, for conditions vary so much from one lake to another. However, although both coarse and fine sediments may be found in beach deposits and deltas, lake floor deposits beyond the limit of delta growth are predominantly fine, and become increasingly so as the central or deepest parts of the lake are approached. The finest silts and clays represent the water-borne rock flour of glacial origin which is seen to discolour the water of lakes and streams and which requires extremely long periods and quiet water conditions to settle. E. M. Kindle (1930) found that even after five months, samples of water containing rock flour and obtained from glacier-fed Lake Cavell in the Rocky Mountains still remained discoloured except for the top 25 cm in a tube 1·6 m long. The action of turbidity currents (P. H. Kuenen, 1951) is now thought to be of considerable importance in spreading sediment over the whole lake floor, for the cold silt-laden meltwater is slightly denser than lake waters.

Lake depth is an important factor affecting the nature of the deposits. Deep lakes provide the best opportunities for the trapping and deposition of the finest material. In the case of shallow lakes fed by meltwater, the existence of more powerful currents will prevent the settling of fine material and often cause it to be washed towards and through the lake outlet resulting in its loss from the lake. This is the reason why the infill of former lakes frequently shows a gradation towards coarse material at the top, apart from the fact that the last stages of infilling will be carried out by a braided and rapidly aggrading stream.

Occasionally in the fine lake bottom deposits, isolated pebbles or boulders appear. R. S. Tarr (1897, p. 151) described the exposed bed of a drained lake beside the Cornell glacier, Greenland, which showed fine clays and scattered blocks of rock. The boulders are clearly ice-rafted or dropped into the lake from the ice margin. Their weight and impact are responsible for folds in the lake floor sediments (R. P. Sharp, 1947).

A very common and highly significant characteristic of glacial lake floor deposits is their laminated nature. The laminations are caused by sudden changes of grain size between successive thin layers, from finest mud to slightly coarser silt; there is also an accompanying colour change, the finest laminae being darker (and also thinner) than the coarser. These regularly banded deposits have been termed 'rhythmites'; an individual pair consisting of one fine and one slightly coarser lamina is known as a 'couplet'. The thin dark layer of a couplet which consists of very fine and partly colloidal material represents a phase of slow deposition under very quiet water conditions, such as would be expected if the lake were frozen at its surface and little or no meltwater were entering. Kindle's work (1930) already quoted showed that rock flour will remain in suspension in still water long enough to allow deposition to continue for many months even if no further sediment enters the lake during this time. The light-coloured silt comprising the coarser member of a couplet represents more rapid accumulation of lake floor sediment under more disturbed conditions, as when meltwater was entering the lake and lake currents, especially turbidity currents (Kuenen, 1951, p. 82) were strong enough to spread the silt over the whole lake floor. It seems that rhythmites will readily develop in cold freshwater lakes into which meltwater streams intermittently flow; they are rarely found in salt or even brackish water where the salt aids flocculation of the fine particles and an unbanded deposit results. Most rhythmites of Pleistocene age indicate the nearby presence of glacial ice at the time they were laid down. Non-glacial rhythmites are known in some Pleistocene sequences and in many pre-Pleistocene geological formations, but the following discussion will be limited to glacial lake rhythmites.

Couplets may vary in thickness from less than 1 mm to as much as several dm, but 1 to 5 cm is the normal range. Many local factors will affect the thickness—the load and volume of the meltwater stream entering the lake, the depth of the lake and the relief of its floor, the strength of the currents in the lake and the distance from the point of entry of the meltwater stream. It is therefore impossible to correlate rhythmites in one lake with those in another on the basis of their absolute thickness.

Since the first half of the nineteenth century, a number of workers had suspected that the couplets in rhythmites were seasonal. A. Smith (1832) hinted at this possibility in a very early study of laminated clays in the Connecticut valley, the couplets here being 0·8-1·2 cm thick. G. de Geer from 1882 onwards developed the theory of the annual nature of couplets, and introduced the term 'varves' to describe such annual couplets in 1912. There is indeed strong evidence that many couplets are annual, though it is also now recognized that not all are, and great care must be taken to prove the point one way or the other if rhythmites are to be employed as a means of dating, as De Geer did. De Geer regarded the fine lamina of a couplet as the result of deposition in winter when the lake was frozen and meltwater entry limited or non-existent. The abrupt break at the top of the fine lamina represents the spring thaw when new coarser silt once again re-entered the lake. Confirmation of this notion has come from study of the pollen found in rhythmites, which provides evidence of seasonal changes in vegetation surrounding the lake during the time interval represented by one couplet. Further confirmation has been provided by studies of modern glacial lake-floor deposits, such as that made by W. A. Johnston (1922) in the case of Lake Louise in Alberta. This lake, about 0·8 km square and up to 70 m deep, receives meltwater from the Victoria glacier which then terminated about 1·6 km from the lake. The lake freezes for about seven months of the year, but some meltwater continues to flow into the lake. The meltwater has a temperature close to 0°C., and is appreciably colder than the lake in summer when water temperatures vary from a minimum of 2°C. at the delta to as much as 4°C. higher at the far end of the lake. Johnston calculated the approximate average thickness of the annual sediment layer to be expected on the lake floor from the known discharge and load of the meltwater stream as 4 mm. A core from the deepest part of the lake showed faint banding, the spacing being about 4-5 mm, therefore strongly suggesting a seasonal rhythm.

Non-annual rhythmites (which should not be termed varves) are known to occur. They may arise from sudden fluctuations of discharge and load on the part of the meltwater stream, sometimes caused by the bursting of glacial water-bodies upstream; from warm and cold spells of a non-annual nature; and from the action of periodic storms stirring up the lake waters at an otherwise quiet period. But if, as seems likely in much of southern Sweden, non-annual rhythmites are rare, then the laminations provide a valuable time-scale for the late Pleistocene.

De Geer was the first to attempt to make regional correlations of varve sequences (see his accounts published in 1912 and 1940), by measuring varve thicknesses on different sections, plotting these thicknesses on a time-scale assuming each varve to be annual, and visually matching the curves so obtained. By this means, localities up to 1 km apart were correlated, a varve sequence established over a distance of almost 1000 km, and a chronology established over nearly 17,000 years. The method of correlating different sections is unreliable if the varves are thin or too regular in thickness—there is a limit to the possible variations in the form of the graphs—but de Geer's work is now supported by some radiocarbon dates. The idea behind the correlation based on relative varve thicknesses is that such variations reflect fluctuations of climate (and hence rates of summer ablation) which should similarly affect all localities in a region of limited size.

De Geer's method was applied in eastern Canada and the United States by E. Antevs (1925). A useful summary of the varve chronology gradually built up over many years was given by Antevs in 1957. It is based on discontinuous varve series measured between New York City and Cochrane (240 km south of James Bay) and on estimates of the gaps in the varve series, the total chronology extending back over the last 19,000 years. With the advent of radiocarbon dating, doubt has been cast on the validity of this record, suggesting that some of the varves may not be annual or that false correlations between different varve sections have been made. Antevs (1957, 1962), however, maintains that the radiocarbon dates are more likely to be in error than those based on varve counts. An example of the discrepancy is given by the varve dating of the Two Creeks Forest Bed, a critical horizon in the late-glacial chronology of North America, as 19,000 B.P., compared with a radiocarbon age of 11,000-12,000 B.P. for this formation. Most workers now prefer to rely on the radiocarbon data.

One of the principal weaknesses of the traditional method of varve correlation has always lain in the process of visually matching varve sequences from different areas. Application of statistical techniques here holds considerable promise. R. Y. Anderson and D. W. Kirkland (1966) have recently applied statistical correlation techniques to some non-glacial varved sediments, finding that thickness variations of varve laminae have correlation coefficients higher than +0·8 over the area of varve formation at any given time.

Extensive deposits of rhythmites are not known in Britain, but the sites of some late Pleistocene glacial lakes of limited extent are occupied by laminated clays. Trotter (1929) noted 2·4 m of such clays in the Carlisle district; K. M. Clayton and J. C. Brown (1958) recorded up to 6 m of laminated silts and clays near Ware which were laid down in Lake Hertford impounded by the last ice to reach the London Basin. It has been claimed by R. G. Carruthers (1939, 1940) that many of the laminated clays said to have been laid down in extensive ice-dammed lakes, particularly in northern and eastern England, are not lacustrine or water-laid deposits at all, but the banded dirts of englacial detritus, released by rising bottom melt (Chapter 13). There has been much opposition to Carruthers' interpretation of the laminated clays as 'pressed clays' or 'pressed melts', though he was undoubtedly correct in maintaining that Pleistocene ice-dammed lakes in Britain were not so common nor so large as has been suggested. Examining the specimens of laminated clay obtained by Carruthers, E. B. Bailey, and S. E. Hollingworth (1940) were equally convinced that they were true lacustrine deposits; and that the contortions found in the laminae recorded flow distortion of bedded mobile mud that had accumulated on the floor of a glacial lake, or that they resulted from the settling of boulders released from floating ice.

Conclusion

Glacial lakes include lakes dammed by ice and lakes in which glacier ice forms part of the lake margin though not the actual barrier. They occur today in lateral valleys dammed by trunk glaciers, in main valleys blocked by ice emerging from a tributary valley, and in other marginal, supraglacial, subglacial, or englacial positions. Most glacial lakes drain subglacially or englacially, though a few sometimes overflow through cols cut in rock or drift if the lake-level is high enough, and a few lakes dammed by sub-polar glacier ice drain over the ice barrier. The method by which lakes drain subglacially is not fully understood, especially in connection with the occurrence of sudden lake bursts or jökulhlaups. Thorarinsson's hypothesis (1939) which involves temporary flotation of the ice barrier seems more likely to apply, with certain modifications, than Glen's hypothesis (1954) which demands water of sufficient depth (more than 200 m) to cause ice deformation.

Features associated with present and former glacial lakes include strandlines, beach formations, terracettes, lake ramparts (formed by floating ice impinging on the shore), deltas, and lake-floor deposits. Strandlines of former lakes may be absent or indistinct if the lake level was variable, if the lake was too small to permit effective wave action, or if sufficient time has subsequently elapsed to allow their effacement by mass wasting. Deltas may be preserved for much longer periods and are particularly useful in determining former lake levels. Ice-contact deltas or delta moraines are valuable indicators of ice margin positions.

Glacial lake floor deposits frequently include rhythmites (varves), comprising a succession of couplets which are in most cases of an annual nature. On this basis late-glacial chronologies in Scandinavia and eastern North America have been established, though in the latter area, radiocarbon dating has shown some major disagreements. Rhythmites must be carefully distinguished from laminated deposits resulting from the melting-out of englacial or subglacial silt layers.

REFERENCES

L. AGASSIZ (1840-41), 'On glaciers, and the evidence of their having once existed in Scotland, Ireland, and England', *Proc. geol. Soc. Lond.* **3**, 327-32

N. AITKENHEAD (1960), 'Observations on the drainage of a glacier-dammed lake in Norway', *J. Glaciol.* **3**, 607-9

R. Y. ANDERSON and D. W. KIRKLAND (1966), 'Intrabasin varve correlation', *Bull. geol. Soc. Am.* **77**, 241-55

E. ANTEVS (1925), 'Retreat of the last ice-sheet in eastern Canada', *Mem. geol. Surv. Can.* **146**

E. ANTEVS (1957), 'Geological tests of the varve and radiocarbon chronologies', *J. Geol.* **65**, 129-48

E. ANTEVS (1962), 'Transatlantic climatic agreement versus C 14 dates', *J. Geol.* **70**, 194-205

L. ARNBORG (1955), 'Ice-marginal lakes at Hoffellsjökull', *Geogr. Annlr* **37**, 202-28

E. B. BAILEY and S. E. HOLLINGWORTH (1940) in discussion on 'Northern glacial drifts', *Q.J. geol. Soc. Lond.* **96**, 249-69

P. BIRKELAND (1964), 'Pleistocene glaciation of the northern Sierra Nevada, north of Lake Tahoe, California', *J. Geol.* **72**, 810-25

T. G. BONNEY (1896), *Ice work present and past* (London)

R. G. CARRUTHERS (1939), 'On northern glacial drifts: some peculiarities and their significance', *Q.J. geol. Soc. Lond.* **95**, 299-333

K. M. CLAYTON and J. C. BROWN (1958), 'The glacial deposits around Hertford', *Proc. Geol. Ass.* **69**, 103-19

L. W. COLLET (1926), 'The lakes of Scotland and Switzerland', *Geogr. J.* **67**, 193-213

G. DE BOER (1949), 'Ice margin features, Leirbreen, Norway', *J. Glaciol.* **1**, 332-6

G. DE GEER (1912), 'A geochronology of the last 12,000 years', *C. r. 11th int. geol. Congr.* (*Stockholm, 1910*), **1**, 241-58

G. DE GEER (1940), 'Geochronologia suecica principles', *K. svenska Vetensk-Akad. Handl.* Ser. 3, **18**, 6. Text and Atlas

C. EMBLETON (1964), 'Sub-glacial drainage and supposed ice-dammed lakes in north-east Wales', *Proc. Geol. Ass.* **75**, 31-38

R. F. FLINT (1957), *Glacial and Pleistocene geology* (New York)

J. D. FORBES (1843), *Travels through the Alps of Savoy* (Edinburgh)

R. J. FULTON (1965), 'Silt deposition in late-glacial lakes of southern British Columbia', *Am. J. Sci.* **263**, 553-70

G. K. GILBERT (1890), 'Lake Bonneville', *U.S. geol. Surv. Monogr.* **1**, 1-438

J. W. GLEN (1954), 'The stability of ice-dammed lakes and other water-filled holes in glaciers', *J. Glaciol.* **2**, 316-18

K. HEWITT (1964), 'The great ice dam', *Indus* **5**, 18-30

J. D. HUME and M. SCHALK (1964), 'The effects of ice-push on Arctic beaches', *Am. J. Sci.* **262**, 267-73

T. F. JAMIESON (1863), 'On the parallel roads of Glen Roy, and their place in the history of the glacial period', *Q.J. geol. Soc. Lond.* **19**, 235-59

T. F. JAMIESON (1892), 'Supplementary remarks on Glen Roy', Q.J. geol. Soc. Lond. 48, 5-28

W. A. JOHNSTON (1922), 'Sedimentation in Lake Louise, Alberta, Canada', Am. J. Sci. 204, 376-86

J. JONSSON (1955), 'On the formation of frontal glacial lakes', Geogr. Annlr 37, 229-33

R. KELLER (1950), 'Niederschlag, Abfluss und Verdunstung in Schweizer Hochgebirge', Erdkunde 4, 54-67

F. A. KERR (1934), 'The ice dam and floods of the Talsekwe, British Columbia', Geogr. Rev. 24, 643-5

E. M. KINDLE (1930), 'Sedimentation in a glacial lake', J. Geol. 38, 81-87

P. H. KUENEN (1951), 'Mechanics of varve formation and the action of turbidity currents', Geol. För. Stockh. Förh. 6, 149-62

F. LEVERETT and F. B. TAYLOR (1915), 'The Pleistocene of Indiana and Michigan, and the history of the Great Lakes', U.S. geol. Surv. Monogr. 53

O. LIESTØL (1955), 'Glacier-dammed lakes in Norway', Norsk Geogr. Tidsskr. 15, 122-49

C. M. MANNERFELT (1945), 'Några glacialmorfologiska formelement', Geogr. Annlr 27, 1-239 (with English summary)

M. G. MARCUS (1960), 'Periodic drainage of glacier-dammed Tulsequah Lake, British Columbia', Geogr. Rev. 50, 89-106

K. MASON (1935), 'The study of threatening glaciers', Geogr. J. 85, 24-41

R. L. NICHOLS and M. M. MILLER (1952), 'The Moreno glacier, Lago Argentino, Patagonia', J. Glaciol. 2, 41-50

N. E. ODELL (1966), 'The Märjelen See and its fluctuations', Ice 20, 27

J. R. REID and L. CLAYTON (1963), 'Observations of rapid water-level fluctuations in ice sink-hole lakes, Martin River glacier, Alaska', J. Glaciol. 4, 650-2

V. SCHYTT (1956), 'Lateral drainage channels along the northern side of the Moltke glacier, north-west Greenland', Geogr. Annlr 38, 64-77

R. P. SHARP (1947), 'The Wolf Creek glaciers, St Elias Range, Yukon Territory', Geogr. Rev. 37, 26-52

J. B. SISSONS (1958), 'Supposed ice-dammed lakes in Britain, with particular reference to the Eddleston valley, southern Scotland', Geogr. Annlr 40, 150-87

A. SMITH (1832), 'On the water courses and the alluvial and rock formations of the Connecticut River valley', Am. J. Sci. 22, 205-31

J. W. SPENCER (1890), 'Ancient shores, boulder pavements, and high-level gravel deposits in the region of the Great Lakes', Bull. geol. Soc. Am. 1, 71-86

K. H. STONE (1963), 'Alaskan ice-dammed lakes', Ann. Ass. Am. Geogr. 53, 332-49

R. S. TARR (1897), 'The margin of the Cornell glacier', Am. Geol. 20, 139-56

S. THORARINSSON (1939), 'The ice-dammed lakes of Iceland, with particular reference to their values as indicators of glacier oscillations', Geogr. Annlr 21, 216-42

S. THORARINSSON (1953), 'Some new aspects of the Grimsvötn problem', J. Glaciol. 2, 267-74

S. THORARINSSON (1957), 'The jökulhlaups from the Katla area in 1955 compared with other jökulhlaups in Iceland', Jökull 7, 21-25

F. M. TROTTER (1929), 'The glaciation of eastern Edenside, the Alston Block, and the Carlisle Plain', *Q.J. geol. Soc. Lond.* **85,** 549-612

W. UPHAM (1896), 'The glacial lake Agassiz', *U.S. geol. Surv. Monogr.* **25**

A. L. WASHBURN (1947), 'Reconnaissance geology of portions of Victoria Island and adjacent regions, Arctic Canada', *Am. geol. Soc. Mem.* **22**

G. WILSON (1899), work described in *Mem. geol. Surv. Summ. Prog.* (1899), 159-62

PART IV

Periglacial Geomorphology

The 'subglacial' climate must be looked on as an optimum of destructive action. (J. G. ANDERSSON, 1906)

Chapter 20

The Periglacial Environment: An Introductory Survey

Freezing is the most important factor in a periglacial climate. (W. LOZINSKI, 1912)

The idea that distinctive phenomena were associated with a cold climatic environment induced by the presence of nearby glacial ice in Pleistocene times was formulated in the nineteenth century. Crystallization of thought on the matter was encouraged by J. Geikie's volume *Great Ice Age* (1874) in which were summarized the known occurrences of 'rubble drift' or 'head' in Britain. Geikie and others favoured an origin for these deposits under former cold conditions. In 1900, F. E. Matthes drew attention to a process, nivation, demanding a freeze-thaw snow climate but not actually involving glaciation. Six years later, J. G. Andersson reported on his visit to Bear Island in the Arctic, describing a hitherto unrecognized form of mass wasting which he termed 'solifluction', a process active under 'sub-glacial' climatic conditions. The adjective 'sub-glacial' was clearly ambiguous, and in 1909, W. Lozinski proposed the term 'periglacial'. Lozinski had studied mechanical weathering of sandstones in the Carpathians which seemed to require a former frost climate. His ideas, presented to the Eleventh International Geological Congress meeting at Stockholm in 1910 (Lozinski, 1912), stimulated discussion and awoke considerable interest in the subject of periglacial geomorphology, which was further intensified among the participants of the subsequent excursion to Spitsbergen. Unfortunately, Lozinski failed to define the term 'periglacial' explicitly; he, moreover, was mostly concerned with frost weathering and gave little consideration to the distinctive processes of nivation and solifluction recognized by Matthes and Andersson. Lozinski used the term periglacial to refer to areas lying near the margins of Pleistocene ice, and to the supposedly distinctive climatic conditions of these areas. By extension, the term came to be applied also to the phenomena induced by such climatic conditions. To the range of periglacial phenomena so far recognized were added, in the next few years, equiplanation (D. D. Cairnes, 1912), altiplanation (H. M. Eakin, 1916), and frost-heaving (B. Högbom, 1914). Högbom's monograph on frost phenomena, with particular reference to Scandinavia and Spitsbergen, marked a milestone in the progress of periglacial studies, and is still a standard reference. The term periglacial is widely used today, though its usage and varied application is far from satisfactory. The most acceptable application of the term is to a zone, albeit of indefinite width, peripheral to the glacial ice of today or of any phases of the Pleistocene. To speak of a 'periglacial climate', unless it is arbitrarily defined, is much less satisfactory, for conditions in the periglacial zone of present and Pleistocene ice-sheets vary enormously from place to place: the severely cold dry climate of the Weichsel maximum in southern European Russia, for example, when great spreads of wind-blown loess accumulated beyond the ice margin, must have been very different from the more maritime and probably 'Icelandic' type of climate found then in ice-free areas of southern Britain. Although large areas of the Pleistocene periglacial zone were certainly characterized by continuous or semi-continuous permafrost, there were also undoubtedly other parts whose climate was not too different from that of the present, as in the Wisconsinan ice-marginal areas of Illinois, Indiana, and Ohio where tree growth may have continued close to the actual ice front (see p. 575). Moreover, the periglacial zone of the last glaciation in middle latitudes is not simply to be equated in terms of its range of climate with present-day high-latitude arctic regions, for as J. Büdel (1959) has pointed out, the former was characterized by higher angle sun and greater insolation, thus promoting more rapid snow melt, a greater depth of seasonal ground thaw, and possibly a greater drying-out of the ground in the late summer. Büdel attempts to distinguish such a régime as 'temperate periglacial', and the concept of differing periglacial climates is to be encouraged, as opposed to the often-expressed notion that there could be one periglacial climate experiencing low temperatures, many freeze-thaw fluctuations, and occasional strong winds blowing from a supposed glacial anticyclone.

L. C. Peltier (1950) attempted to define a periglacial morphogenetic region, which he took to be an area whose approximate range of mean annual temperature was $-15°$ to $-1°C.$, whose mean annual precipitation was from 120 to 1400 mm, where strong mass movements and occasionally strong winds prevailed, and where the action of running water was relatively minor. Such a generalized concept is not a particularly useful one, nor are the figures chosen the most suitable; moreover the positions and climates of the periglacial zone through the Pleistocene up to the present have changed radically. The periglacial zone is not a static phenomenon, nor are its characteristics everywhere the same. Nevertheless, the term periglacial remains a convenient one under which to group certain features and processes related to a range of climates characteristic of a broad periglacial zone, either now or in the past.

The terms applied to periglacial phenomena have often been confusing, inadequately defined, or developed with a particular theory in mind. In an attempt to remedy the situation, K. Bryan in 1946 proposed an entirely new set of terms, grouped under the heading of 'cryopedology', the name proposed for a sub-discipline dealing with frozen ground phenomena. Some of these terms will be introduced in subsequent chapters, and a useful review of their European language equivalents is given by A. Dylikowa and J. Olchowik-Kolasinska (1956). Many of Bryan's terms have never been widely accepted, for there are often simpler and unambiguous alternatives: 'congelifraction' is a cumbersome replacement for 'frost action'; 'permafrost', though etymologically unsound, is too well established to be supplanted by 'pergelisol'; and the term 'active layer' is much more intelligible than 'mollisol'. While it is always desirable that terminology should be logical and scientifically based, there is an equally great need to avoid unnecessary proliferation of jargon.

Periglacial Weathering Processes

Freeze-thaw action is undoubtedly the most important process of rock weathering in the periglacial zone; it is the primary agent responsible for such features as talus accumulations and blockfields, and the breakdown of debris into particles fine enough to be handled by running water and the wind. The process, however, is by no means a simple one, nor is it fully understood: O. R. Grawe (1936) was one of the first to cast doubt on the simple idea that expansion of water to form ice would inevitably result in rock shattering. Many complex pressure-temperature relationships are involved in the water-ice change. If water is present in a completely closed cavity within rock, and if there are no impurities (including air) in the water, then a fall of temperature of the water to $-22°C$. accompanied by its conversion to crystalline ice will produce a theoretical maximum force of about 2100 kg/cm^2, though in fact practically no rocks could withstand a tensile stress of even one-tenth of this amount. In practice, one is never dealing with a completely closed cavity, otherwise there would be no means of entry for the water in the first place. It is possible, however, that in a wedge-shaped water-filled crack, rapid cooling from outside would cap the crack with ice and convert it into a closed system, though the expansive force of ice crystal growth might be relieved by deformation of the ice rather than further rupturing of the rock. The temperatures required to produce freezing of moisture in rocks are often much lower than expected, for the water contains air and impurities, latent heat must be removed, and water may be super-cooled in some circumstances. Other factors that must be considered in connection with freeze-thaw rock weathering are the intensity of freezing, its duration, and the rate of freezing.

Many workers have investigated the relationships between the effectiveness of freeze-thaw as an agent of weathering in a given region, and the climatic régime of that region. Unfortunately, climatic data are often inadequate: for instance, most of the temperature records normally available are of air temperatures measured in a standard screen at intervals which may be too infrequent, whereas much more relevant to freeze-thaw action would be detailed records of ground temperature fluctuations. Air temperatures provide no more than a rough guide to ground temperatures, for ground temperatures are affected by insolation, the presence of water, ice, or snow, and rates of heat absorption and radiation. Hillslopes in sunshine will experience melting long before the air temperature rises above freezing; alternate sunshine and cloud may even themselves induce freeze-thaw in critical conditions, while dark-coloured rocks may retain heat to melt ice even when the air temperature has fallen below freezing. However, many workers have used air temperature records as a first approximation in studying the freeze-thaw process, and taken temperatures a few degrees above and below 0°C. as the points of effective melt and freeze respectively. An example is J. K. Fraser's work (1959) on freeze-thaw cycles in Canada, which confirmed the suggestions made long ago by Högbom (1914) and E. K. Leffingwell (1919, p. 176) that high-latitude climates are usually unfavourable for freeze-thaw weathering. F. A. Cook and V. G. Raiche (1962) made an intensive study of freeze-thaw cycles at Resolute, N.W.T. (latitude 740°N.), which, lying some 250 km from Devon Island ice-cap, may fairly be said to lie in the present-day periglacial zone. Table 16 enumerates some of their findings.

TABLE 16   Number of Freeze-thaw Cycles at Resolute, 1 May–30 September 1960

Level of temperature record	Range −2 to +1°C.	Range −4 to +2°C.	Frost-change days (at least one crossing of the freezing line)	Number of crossings of the freezing line
Stevenson screen	9	3	27	194
Stevenson screen, average for 1948–59	5·5	1·4	37·8	
Ground	18	7	44	170
−2·5 cm	1	0	13	16
−10 cm	0	0	8	24
−20 cm	0	0	9	8

The small number of freeze-thaw cycles is striking (nearly all occur in May and June), as is the large number of mostly small amplitude crossings of the freezing line which may be too small to be significant in the weathering process. As expected, the ground surface experiences the most freeze-thaw cycles. Cook and Raiche contend that freeze-thaw cycles in this and probably many other parts of the Arctic are far less important than was once thought, a conclusion which has been recently supported by other investigators such as J. T. Andrews (1961, 1963a). Fraser (1959) found it difficult to avoid the conclusion that the abundance of apparently frost-riven rock in northern

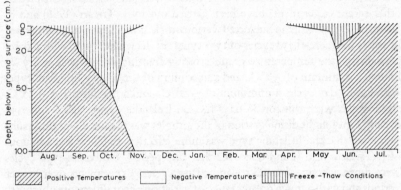

| | Positive Temperatures | | Negative Temperatures | | Freeze -Thaw Conditions |

Fig. 20-1 Annual temperature régime recorded at Hornsund, Spitsbergen, for the ground surface and depths up to 100 cm (Z. Czeppe, *Bull. Acad. pol. Sci.*, 1960)

Canada is related not to the present climate but to conditions in the Pleistocene more favourable here for freeze-thaw action.

The data just presented for Resolute are probably typical of a present-day continental arctic climate. For comparison, Fig. 20-1, compiled from observations by Z. Czeppe (1960), illustrates a maritime arctic climate at Hornsund in Vestspitsbergen. The number of freeze-thaw cycles is greater than in the case of Resolute, and at ground level, the total length of the two seasons when temperatures are frequently changing across the freezing line is over $5\frac{1}{2}$ months. This type of temperature régime is clearly much more favourable for mechanical frost weathering, but even so, it should not be thought that such conditions represent the optimum. C. Troll's world survey (1944), though based on air temperature records, showed that the greatest frequency of freeze-thaw cycles is associated with low-latitude high-altitude areas, as in the Andes of Peru, and not, as Peltier (1950) thought, in the humid subarctic regions. However, air temperatures frequently crossing the freezing line are not the sole requirement for effective freeze-thaw action; moisture must also be available, and must be able to enter crevices or pore-spaces of suitable shape and size in the rocks. Availability of moisture is not adequately reflected in precipitation data from rain or snow gauges, as snow melt, fog, and dew may be just as important as measurable precipitation in wetting the rocks.

Little is known as to whether rapid freezing of moisture in rocks is more effective than slow freezing, or whether intense cold during the period of freezing is more potent than temperatures just a few degrees below freezing. Experiments on freeze-thaw action under laboratory conditions are rather inconclusive. W. R. B. Battle's experiments were mentioned in Chapter 8; other recent experiments have been carried out by J. Tricart (1956) and S. Wiman (1963). Wiman subjected various rock samples partially immersed in water to freeze-thaw cycles of two types: a) 'Icelandic' type, with a daily periodicity (the temperature graph crossing freezing-point twice every 24 hours), a minimum of − 7°C., and a maximum of + 6°C.; b) 'Siberian' type, with a four-day cycle, a minimum of − 30°C., and a maximum of + 15°C. Both cycles were run for 36 days (i.e. 36 Icelandic cycles, and 9 Siberian cycles). Only slight disintegration of the samples was obtained, but the results suggested that the Icelandic type was more effective, unlike the conclusions of J. Tricart, who found that over the same period, 8 Siberian-type cycles were more effective than 25 Icelandic-type cycles. Experiments over longer periods and with a much greater range of variation of conditions are obviously required.

Little work has been done on other processes of rock weathering in the periglacial zone. S. Taber (1943) and others acknowledge that sudden large falls of temperature will cause contraction cracking (sometimes audible) of ice-rich sediments, but low temperatures in themselves are not significant except in that they slow down rates of chemical action and also result in the elimination of free water to take part in chemical reactions. Chemical weathering under periglacial conditions has received scant attention apart from the work of A. Cailleux (1962) and Z. Czeppe (1964), though a large number of writers such as C. Troll (1944) have remarked that ever since Lozinski linked the term periglacial with intensive frost weathering, there has been too much concentration on the latter at the expense of other possible processes. Czeppe has examined sandstone outcrops in Spitsbergen which appear to be exfoliating in concentric shells to leave rounded core stones. His mineral analyses of the rocks and their shells prove successive chemical changes, and the splitting-off of the shells may be partly caused by expansion resulting from hydration or oxidation, though frost has undoubtedly aided the process once started. Cailleux described efflorescence and coatings of mirabilite (hydrated sodium sulphate), calcareous, ferruginous, and silicic compounds, on rocks in the McMurdo Sound region of the Antarctic, and considered that alternate solution and recrystallization of mirabilite might be an important agent of rock weathering. Further studies are clearly essential before a more general assessment of chemical weathering in the periglacial zone can be attempted.

Stream Action and Valley Development in the Periglacial Zone

River flow in the periglacial zone is characterized by irregularity and sudden fluctuations. Prolonged periods of sub-zero temperatures will result in complete cessation of flow and therefore of all fluvial activity, but during spring or early summer thaw, river ice suddenly begins to break up and may release floods of catastrophic dimensions for a brief period, when considerable erosion and movement of debris occurs. Rivers with such a régime may develop braided sections and aggrade their beds with the debris moved by the spring flood.

A good example is provided by the Mecham River, near Resolute, N.W.T. (F. A. Cook, 1960). In 1959, it remained frozen until 20 June. Flow then gradually commenced and accelerated steadily to a peak of 4·5 million m³ a day by 2 July, and then diminished rapidly to 100,000 m³ a day by 20 July. Its flow remained at about this level until shortly before it froze completely on 10 September. Eighty per cent of its annual flow was concentrated in a ten-day period, when the river was wild and impassible, but for the rest of the year it carried little sediment and could be crossed practically anywhere. Another example, that of the behaviour of the Colville River, Alaska, is given by L. Arnborg et al. (1967).

Some writers believe that fluvial action is relatively unimportant in the periglacial zone; L. C. Peltier allotted it a significant part only in removing debris supplied by frost weathering and solifluction. Little firm evidence is available to support or refute such generalizations. A. Rapp's study (1960) of the Kärkevagge region in northern Sweden showed that the most important transport process was that of running water carrying salts in solution (removing 26 tons/km²/year), quantitatively much more important than solifluction, but Rapp considered only processes acting on the valley slopes, not the main streams themselves. It would be an impossible task to make similar quantitative evaluations of processes over areas vastly larger than that of Kärkevagge.

Asymmetrical valleys Some observations on the common asymmetry of certain valleys in the present or former periglacial zone have been summarized by H. T. U. Smith (1949), and there is no space here to deal with the subject fully. H. Poser (1947) thought, with others, that valley asymmetry was peculiar to the periglacial zone and that the occurrence of such valleys might even afford a useful clue to the former extent of frozen ground. The best recent discussion of the subject is by D. R. Currey (1964), who has studied asymmetric valleys in western Alaska, from Cape Lisburne to the Arctic Circle. Table 17 summarizes some of the data obtained. In this area,

TABLE 17　*Analysis of Asymmetric Valleys in Part of Alaska*

Azimuth of stream flow	Number of valleys with steeper left-bank slopes	Number of valleys with steeper right-bank slopes	Valleys showing little or no asymmetry	Totals
1-180°	8	24	15	47
181-360°	104	15	66	185
	112	39	81	232

wherever streams flow roughly due north or south, the numbers of valleys with steeper left-bank slopes is approximately equal to the number with steeper right-bank slopes, but as the direction of stream flow departs from a north-south azimuth, valley asymmetry becomes strongly marked, with a definite preponderance of steeper north-facing sides. The effect is not related to bedrock, nor can the situation be explained in terms of stream deflection by the Coriolis force. Currey considers and eliminates hypotheses used to explain steeper south-facing sides (e.g., C. D. Ollier and A. J. Thomasson, 1957, in the Chiltern Hills); of four remaining hypotheses, he argues that the most likely is that postulating less freeze-thaw action and solifluction on north-facing slopes which thus retain more of their initial steepness caused by valley incision. If he is correct, there is a definite relationship between regularly occurring non-structurally controlled asymmetric valleys and the periglacial zone.

Vallons de gélivation　M. Boyé (1950) reported the occurrence in Greenland of valleys ('ravins de gélivation') believed to be formed not by stream action but by the widening of lines of structural weakness by frost action and the evacuation of this debris by solifluction. A similar explanation for some valleys near Knob Lake, central Labrador, was proposed by C. R. Twidale (1956). These valleys are cut in the side of an argillite ridge, some ending abruptly in vertical rock walls but opening out in the other direction on to a shale lowland. In cross section, the valleys show flanking talus slopes beneath rocky scarps; the largest valley is about ten metres deep. Protalus ramparts (see p. 556) mark their mouths and were considered by Twidale to mark the soliflual movement of material down-valley to their exits. J. T. Andrews (1963b) has criticized some of Twidale's findings. Although he

accepts that the valleys may have been initiated by joint-guided frost shat-
ering, he proposes a pre-Wisconsinan rather than a post-glacial age for this,
and produces evidence to suggest that the valleys have been much modified
by subglacial meltwater erosion during the Wisconsinan glaciation—a large
ice-emplaced erratic at the mouth of one valley bears witness to an ice-cover
since the valley was formed. Andrews deplores the tendency to assume that
freeze-thaw action is necessarily important in a region of cold climate; as
already noted, conditions in many high-latitude arctic regions today are
distinctly unfavourable for freeze-thaw action, with temperatures continuously below freezing in the winter and continuously above freezing in
the summer.

Boyé's hypothesis remains a possible one for the formation of certain
structurally-guided small-scale valleys in present or former periglacial regions, but it should be applied with caution and with alternative modes of
origin carefully considered.

Conclusion

This chapter has introduced the subject of periglacial geomorphology and
considered certain aspects of the periglacial environment. Weathering processes and the part played by running water have been outlined. The following Chapters will deal with other features and processes connected with the
periglacial zone: frozen ground phenomena, patterned ground, mass movements and slope deposits, the action of snow both mobile and stationary,
and the action of wind.

REFERENCES

J. G. ANDERSSON (1906), 'Solifluction, a component of sub-aerial denudation', *J. Geol.*
14, 91-112
J. T. ANDREWS (1961), 'Vallons degélivation in central Labrador-Ungava: a reap-
praisal', *Can. Geogr.* 5, 1-9
J. T. ANDREWS (1963a), 'The analysis of frost-heave data collected by B. H. J. Hay-
wood from Schefferville, Labrador-Ungave', *Can. Geogr.* 7, 163-73
J. T. ANDREWS (1963b), 'So-called "vallons de gélivation" in central Labrador-
Ungava', *Biul. Peryglac.* 12, 137-43
L. ARNBORG, H. J. WALKER and J. PEIPPO (1967), 'Suspended load in the Colville River,
Alaska, 1962', *Geogr. Annlr* 49, 131-44.
M. BOYÉ (1950), *Glaciaire et périglaciaire de l'Ata Sund nord-oriental (Groenland)* (Paris;
Expéditions polaires françaises, 1)

K. BRYAN (1946), 'Cryopedology—the study of frozen ground and intensive frost action with suggestions on nomenclature', *Am. J. Sci.* **244**, 622-42

J. BÜDEL (1959), 'Periodische und episodische Solifluktion im Rahmen der Klimatischen Solifluktionstypen', *Erdkunde* **13**, 297-314

A. CAILLEUX (1962), *Études de géologie au détroit de McMurdo* (*Antarctique*), *Com. natn. fr. Rech. Antarct.* **1**, 41 p.

D. D. CAIRNES (1912), 'Differential erosion and equiplanation in portions of Yukon and Alaska', *Bull. geol. Soc. Am.* **23**, 333-48

F. A. COOK (1960), 'Periglacial-geomorphological investigations at Resolute, 1959', *Arctic* **13**, 132-5

F. A. COOK and V. G. RAICHE (1962), 'Freeze-thaw cycles at Resolute, N.W.T.', *Geogr. Bull.* **18**, 64-78

D. R. CURREY (1964), 'A preliminary study of valley asymmetry in the Ogotoruk Creek area, north-western Alaska', *Arctic* **17**, 84-98

Z. CZEPPE (1960), 'Thermic differentiation of the active layer and its influence upon the frost heave in periglacial regions (Spitsbergen)', *Bull. Acad. pol. Sci. classe III*, **8**, 149-52

Z. CZEPPE (1964), 'Exfoliation in a periglacial climate', *Geogr. Polonica* **2**, 5-10

A. DYLIKOWA and J. OLCHOWIK-KOLASINSKA (1956), 'Processes and structures in the active zone of perennially frozen ground', *Biul. Peryglac.* **3**, 119-24

H. M. EAKIN (1916), 'The Yukon-Koyukuk region, Alaska', *U.S. geol. Surv. Bull.* **631**, 1-88

J. K. FRASER (1959), 'Freeze-thaw frequencies and mechanical weathering in Canada', *Arctic* **12**, 40-53

J. GEIKIE (1874), *The Great Ice Age* (London, 1st Ed.)

O. R. GRAWE (1936), 'Ice as an agent of rock weathering', *J. Geol.* **44**, 173-82

B. HÖGBOM (1914), 'Über die geologische Bedeutung des Frostes', *Bull. geol. Instn Univ. Upsala* **12**, 257-389

A. JAHN (1954), 'Walery Lozinski's merits for the advancement of periglacial studies', *Biul. Peryglac.* **1**, 117-24

E. K. LEFFINGWELL (1919), 'The Canning River region, northern Alaska', *U.S. geol. Surv. Prof. Pap.* **109**, 1-251

W. LOZINSKI (1912), 'Die periglaziale Fazies der mechanischen Verwitterung', *C. r. 11th int. geol. Congr.* (*Stockholm, 1910*), **2**, 1039-53

F. E. MATTHES (1900), 'Glacial sculpture of the Big Horn Mountains, Wyoming', *U.S. geol. Surv. 21st A. Rep.* (1899-1900), 167-90

C. D. OLLIER and A. J. THOMASSON (1957), 'Asymmetrical valleys of the Chiltern Hills', *Geogr. J.* **123**, 71-80

L. C. PELTIER (1950), 'The geographic cycle in periglacial regions as it is related to climatic geomorphology', *Ann. Ass. Am. Geogr.* **40**, 214-36

H. POSER (1947), 'Dauerfrostböden und Temperaturverhältnisse während der Würm-Eiszeit im nicht vereisten Mittel- und Westeuropa', *Naturwissenschaften, Berlin* **34**, 1-9

A. RAPP (1960), 'Recent development of mountain slopes in Kärkevagge and surroundings, northern Scandinavia', *Geogr. Annlr* **42**, 65-200

THE PERIGLACIAL ENVIRONMENT

H. T. U. SMITH (1949), 'Physical effects of Pleistocene climatic changes in non-glaciated areas: eolian phenomena, frost action and stream terracing', *Bull. geol. Soc. Am.* **60**, 1485-516

S. TABER (1943), 'Perennially frozen ground in Alaska: its origin and history', *Bull. geol. Soc. Am.* **54**, 1433-548

J. TRICART (1956), 'Étude expérimentale du problème del a gélivation', *Biul. Peryglac.* **4**, 285-318

C. TROLL (1944), 'Strukturböden, Solifluktion, und Frostklimate der Erde', *Geol. Rdsch.* **34**, 545-694 (English translation, *Snow Ice Permafrost Res. Establ.*)

C. R. TWIDALE (1956), 'Vallons de gélivation dans le centre du Labrador', *Revue Géomorph. dyn.* **7**, 17-23

S. WIMAN (1963), 'A preliminary study of experimental frost weathering', *Geogr. Annlr* **45**, 113-21

Chapter 21

Frozen Ground Phenomena

... banks, which are about six feet above the surface of the water, display a face of solid ice, intermixed with veins of black earth, and as the heat of the sun melts the ice, the trees frequently fall into the river. (SIR ALEXANDER MACKENZIE, *Voyages to the Arctic*, 1789)

As early as the sixteenth century, explorers were bringing back reports of frozen ground from Arctic regions, and in the seventeenth century, news of the discovery of the frozen carcases of mammoth and woolly rhinoceros in Siberia reached Europe. The first intelligent description of ground ice is ascribed to M. F. Adams, reporting from the Lena delta in 1806, while in North America, O. von Kotzebue's expedition of 1816 recorded ground ice at Elephant Point, Alaska. The earliest detailed studies of frozen ground were made by A. T. von Middendorff, who collected data from wells dug in search of water in Yakutsk, Siberia (G. B. Cressey, 1939). Interest in frozen ground was greatly stimulated during the construction of sections of the Trans-Siberian Railway, for entirely new problems of engineering were presented by such terrain, while in Alaska and the Yukon, the discovery and exploitation of alluvial gold also brought about closer acquaintance with the characteristics of frozen ground. A landmark in the growing literature on frozen ground was the publication in 1919 of E. K. Leffingwell's monograph on the Canning River region in Alaska, which included a major section on such topics as the depths and temperature gradients of the frozen ground, the forms, age, and origins of ground ice, and on the related landforms. Since then, the investigation of frozen ground has accelerated to such an extent that its study can now almost be regarded as a separate scientific discipline, claiming attention not only from geologists and geographers but also from geophysicists, engineers and all those concerned with land utilization in arctic and sub-arctic regions. Particularly rapid advances were made during World War II for military reasons.

The Present-Day Occurrence of Permafrost

S. W. Muller (1947) proposed the name 'Permafrost' for a 'thickness of soil or other superficial deposit, or even of bedrock, at a variable depth ... in

which a temperature below freezing has existed continually for a long time'
(p. 3). He distinguished between 'dry permafrost', in which moisture was
lacking or insufficient to allow interstitial ice to form and act as a cementing
material, and 'frozen ground' which contained water mostly in the form of
ice. In permafrost, frozen and unfrozen water may exist in equilibrium,
depending on local variations of temperature, pressure, impurities in the
water, and the grain size of the material. The term 'permafrost' has often
been criticized (K. Bryan, 1946, suggested its replacement by 'pergelisol')
but its usage is now well established as a useful and easily intelligible contraction of 'Permanently frozen ground'.*

Above permafrost there may be a zone in which temperature conditions
vary seasonally. This is the 'active layer' (Bryan's mollisol) thawing out when
temperatures rise sufficiently, and refreezing in winter or in cold spells (or
even partly at night time). In this layer, plants are rooted, and movements
can readily occur in the thawed state. The thickness of the active layer
depends on the air temperature régime, the degree of exposure to insolation
(slope aspect is important), the conductivity of the soil, and the degree of
insulation which might be provided by a cover of snow or vegetation. The
active layer may rest directly on the upper surface of permafrost (the 'permafrost table'), but sometimes, an unfrozen layer may intervene (see, for example, J. D. Ives, 1960). Unfrozen layers are also known within permafrost;
they are termed 'taliks'. They may reflect former climatic changes—the
relations between permafrost and climate will be considered later—or they
may represent aquifers containing water highly mineralized or under pressure, for instance. Beneath all permafrost, unfrozen ground exists, for
permafrost is simply the result of negative heat balance at the earth's surface,
either now or in the past, replacing the normal positive temperature gradient
caused by outflow of heat from the earth's interior.

Conventionally, three spatial zones of permafrost occurrence are recognized (L. L. Ray, 1951): 1) *continuous permafrost*, mainly related to areas
where the climate now is severe enough to form permafrost; in this zone,
the only unfrozen areas are those lying below certain deep and wide lakes,
rivers, or beneath the sea; 2) *discontinuous permafrost*, in which small scattered
unfrozen areas appear; and 3) *sporadic permafrost*, where small islands of
permafrost exist in a generally unfrozen area. It is likely that many such
islands are relics of a former colder climate, and are steadily diminishing in
size. Continuous or discontinuous permafrost underlies no less than one-fifth
of the earth's land areas, including 47 per cent of the U.S.S.R., possibly as
much as 50 per cent of Canada and Alaska, and most of Greenland and
Antarctica.

* Strictly, one should speak of *perennially* rather than *permanently* frozen ground.

Fig. 21-1a shows the distribution of permafrost in the U.S.S.R., where sufficiently detailed knowledge is available to plot isopleths of its thickness: no comparable map can yet be drawn for Canada and Alaska. It should be noted how permafrost underlies parts of the Arctic Ocean floor; these submarine areas possess temperatures below 0°C. and are therefore permafrost by definition, but they generally contain no ice because of the salinity conditions. Siberia includes the coldest area of the world outside Antarctica; at Oymyakon on the Indigirka River (latitude 66°N.), an absolute minimum temperature of − 71°C. has been recorded (I. P. Gerasimov, 1961).

In Fig. 21-1b, the conventional zones of permafrost in Canada are indicated, but the map is far from being a precise representation owing to inadequate data. An indication of this was provided during iron ore company excavations and drillings in central Labrador (Ives, 1960; J. T. Andrews, 1961), for 60-metre thicknesses of permafrost were discovered 220 km south of the limit of discontinuous permafrost depicted by R. J. E. Brown in 1960.

Permafrost has been proved to a depth of over 600 m at Nordvik in N. Siberia, and to 500 m in the adjacent Taymyr Peninsula. 314 m is known near Cape Simpson in N. Alaska, and 390 m estimated from temperature gradients in a drill hole at Resolute, N.W.T (A. D. Misener, 1955), while preliminary measurements in a deep bore at Winter Harbour, Melville Island, N.W.T , suggest that sub-zero temperatures persist to a depth of 450 m. At Braganza Bay, Spitsbergen, ground is frozen in colliery workings to depths of about 320 m, and it extends out beneath the sea for about 100 m from the coast (W. Werenskjold, 1953). Many factors affect the depth of ground freezing; the theory is simple, but precise analysis is extremely complex, as Leffingwell showed as early as 1919. The temperature régime and insolation received at the ground surface (long-term changes must be taken into account), the rate of escape of geothermal heat, and the thermal properties of the ground material (which vary with moisture content) are basic factors. Also to be considered are changes in the thermal properties when ground ice begins to form; the conductivity of dry sand, for instance, can be trebled by formation of interstitial ice. It has long been noted, too, that ground freezes most deeply where the surface is not protected by a cover of vegetation, water, or snow, for all these have a much smaller conductivity than soil or rock. The great depths of frozen ground quoted above are all in the coldest areas of the world, where vegetation is minimal, where summers are short and cool, and where snow does not cover the ground at least until very late in the year.

Temperatures in permafrost are generally lowest near the ground surface, and are influenced by air temperature cycles. Diurnal temperature changes

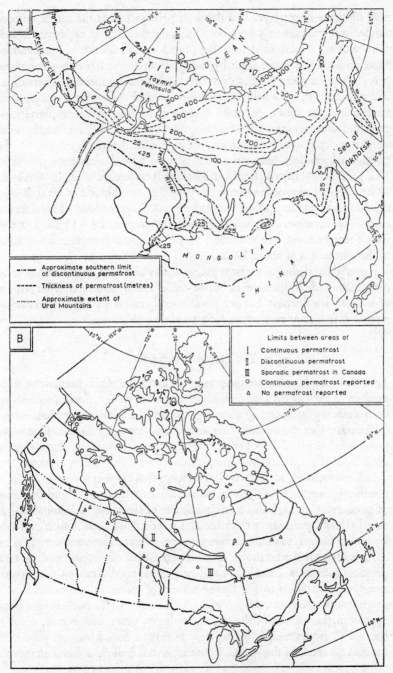

Fig. 21-1 A Distribution and thickness of permafrost in the U.S.S.R. (R. J. E. Brown, *Arctic*, 1960, Arctic Institute of North America)
B Permafrost zones in Canada (R. J. E. Brown, *Geol. Surv. Canada*, 1967)

may affect the upper layer to a depth of perhaps 1 m at the most; seasonal temperature changes may be picked up at depths of 15 m or more, though there will be a substantial time lag (up to several months). Temperature fluctuations in the active layer by definition cross the freezing-point regularly; below the permafrost table, however, seasonal temperature fluctuations will take place entirely below freezing-point, and will become smaller with increasing depth until they vanish at the level of 'zero annual amplitude'. At greater depths, temperatures will not fluctuate and will rise steadily with increasing depth until unfrozen ground is reached. An average figure for the geothermal gradient below the level of zero annual amplitude would be 1°C. per 30 or 40 metres. In the drill hole at Resolute, N.W.T. (F. A. Cook, 1958), seasonal changes amounted to 0·4°C. at a depth of 17·6 m (compare the mean annual range of air temperatures: 42°C.); the level of zero annual amplitude was about 30 m below the surface (temperature – 13·2°C.), below which temperatures rose at about 1°C. per 25·5 m suggesting a lower limit to the permafrost at about 390 m.

Reversals in the mean thermal gradient may be found, sometimes of sufficient magnitude to give rise to buried 'taliks'; many factors may be responsible for such a situation but one possible explanation that has always excited interest is that they may reflect long-term climatic changes.

The Origin of Permafrost

There has long been controversy as to the extent to which permafrost is the product of present climatic conditions, and as to its age. Such questions are far from being satisfactorily answered, though the balance of evidence seems to indicate that permafrost first came into existence in the Pleistocene (whether it survived any of the interglacials is unknown) and that today the great bulk of permafrost is in rough balance with the present climate.

In Canada, R. J. E. Brown (1960) claims that there is a very broad relationship between present-day mean annual air temperatures and the zones of permafrost, though there are important discrepancies in some areas. As noted already, permafrost thickness and temperature conditions are affected by several factors, not only air temperature, so that a close relationship is not to be expected; furthermore, data on permafrost occurrence and air temperature records in Canada are still very inadequate. The – 4°C. isotherm roughly corresponds to the southern limit of discontinuous permafrost in several parts of Canada, though in the Yukon, the – 1°C. isotherm is a better approximation. South of the – 1°C. isotherm, permafrost is rare; north of the – 4°C. isotherm, it is widespread. In eastern Asia, however, permafrost extends far south of the – 2°C. isotherm, and it is thought likely that in this region it is not in equilibrium with the present climate but a relic in part of former cold conditions in the Pleistocene. A number of workers have also tried to correlate permafrost occurrence with other climatic variables, especially indices of freezing or thawing. The results of such analyses do not as yet suggest any precise relationship, though J. A. Pihlainen (1962) considered that mean annual air temperature and a thaw index (= degree-days, Fahrenheit, above freezing-point) provided a first approximation to the likely occurrence of permafrost in Canada. Table 18 gives data for selected stations in Canada. Further investigations are clearly needed, particularly on the exact occurrence of permafrost at these and other places.

TABLE 18   *Comparison of Present Climate and Permafrost Occurrence in Canada*

Place	Mean annual air temperature °F.	Permafrost occurrence	Thaw index °F. degree-days
Resolute, N.W. T.	3	Continuous (up to 350 m)	458
Churchill, Man.	19	? Continuous (up to 30 m)	1975
Norman Wells, N.W.T	21	? Continuous (up to 70 m)	3027
Fond du Lac, Sask.	22	None	2789
Great Whale River, P.Q.	23	None	1941
Wabowden, Man.	28	Discontinuous	3373

There is good evidence in some places that the present climate is adequate not only to sustain permafrost but actively to form it. Permafrost in new floodplain silts has been frequently observed (for instance, S. Taber, 1943, p. 1504). At Port Nelson, Manitoba, in 1929, a swamp was drained (J. L. Jenness, 1949), its bed being then unfrozen. Within three winters, frost penetrated to a depth of 10 m and there, at a level coinciding with the base of the permafrost in adjacent areas, it stopped, suggesting strongly that the permafrost in this area at least was intimately related to the present climate. Also in the area marginal to Hudson Bay, permafrost now exists in areas that before the post-glacial uplift were beneath the sea and were unlikely then to have been frozen.

On the other hand, the great thickness, location, and intense cold of some southerly permafrost areas in the U.S.S.R. has led to the view that the permafrost in such areas is essentially 'old' and is now slowly degrading with the amelioration of climate since the last glacial period. Russian investigators suggest that there is an important contrast between the southerly zone of degrading permafrost and the far northern zone of active or aggrading permafrost; these two zones are separated by an intermediate belt in which permafrost is being neither actively formed nor actively destroyed by the present climate. At one time it was thought that permafrost distribution in

Siberia (and also in Alaska) could be regarded as complementary to the glacierized regions, on the grounds that permafrost developed during Pleistocene cold periods and was unable to form in ice-covered areas because of the insulating properties of the ice. This view is rendered untenable by more recent investigations (see R. F. Flint and H. G. Dorsey, 1945) which show that present permafrost and Pleistocene glaciation are not mutually exclusive.

It will be apparent that the relationships of permafrost to present-day climate are very complex and far from being fully evaluated. Some permafrost may be a late Pleistocene relic, but the greater part has probably developed in response to the climate of the last few thousand years or even of the present day. More precise conclusions must await the collection of much more information on the depths and thermal gradients of permafrost.

Ground Ice

Most ground ice exists in the permafrost of polar and sub-polar lands, and ranges from pore-fillings in sedimentary rocks to masses of clear ice 30 m or more across. Up to 80 per cent of permafrost by volume may consist of ice. R. F. Black (1954) makes a useful distinction between rocks which are super-saturated with ice (they contain more ice than pore space), rocks which are saturated (ice volume approximately equal to the pore space: the sediment is cemented by ice, but the ice does not form visible veins or granules), and rocks which are under-saturated. The forms which ground ice takes have been classified by P. A. Shumskiy (1959); a modified version of his classification is given in Table 19. Categories 5 and 6 demand no further consideration; needle ice (1a) will be mentioned in a later chapter (p. 510). Segregated ice, vein ice, intrusive ice and extrusive ice will be examined in this chapter.

TABLE 19   *Classification of Ground Ice Forms*

1. SOIL ICE	a) Needle ice (pipkrake)
	b) Segregated ice
	c) Ice filling pore spaces
2. VEIN ICE	a) Single veins
	b) Ice wedges
3. INTRUSIVE ICE	a) Pingo ice
	b) Sheet ice
4. EXTRUSIVE ICE	formed subaerially, as in the case of ice formed on river flood-plains (Aufeisen).
5. SUBLIMATION ICE	formed in cavities by crystallization from water vapour.
6. BURIED ICE	buried icebergs, buried glacial ice, etc.

PLATE XXIII   SECTION IN A KAME, PENETRAETH, NORTH WALES.
Ablation moraine rests on water-bedded sands and gravels. (C.E.)

PLATE XXIV   KAME TERRACES (LEFT) OF TIOGA AGE AT SONORA PASS JUNCTION, SIERRA NEVADA,
CALIFORNIA.
(C.E.)

PLATE XXV   OUTWASH PLAIN OF THE HVITÁ, ICELAND.
About 5 km from the margin of the Langjökull. Scale given by person standing in centre foreground. (C.E.)

PLATE XXVI   SKEIÐARÁ SANDUR AND (IN DISTANCE) TERMINUS OF SKEIÐARÁRJÖKULL, SOUTH ICELAND.
(C.A.M.K.)

SEGREGATED ICE

Segregated ice is a phenomenon of super-saturated rocks. It has been suggested that it should be named 'Taber ice' in honour of its discoverer, S. Taber, who has done so much in the field and in the laboratory to elucidate

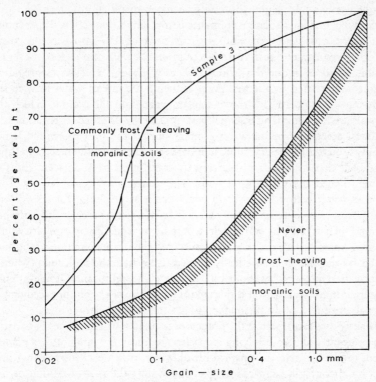

Fig. **21-2** Grain size composition of morainic soils in relation to their susceptibility to frost-heaving (P. J. Williams, *Geogr. J.*, 1957, Royal Geographical Society)

the mechanisms of ground ice formation. Taber was the first to demonstrate that surface uplift following freezing from the surface downward of certain materials containing water ('frost heaving') was not caused by expansion of water on cooling to form ice, for such expansion of water amounts to no more than 10 per cent whereas heaving needing an expansion of over 100 per cent has often been observed. Taber's experiments (1929, 1930) showed that simple freezing of interstitial water causes practically no uplift, and in any case, not all the water normally freezes. His experiments also showed that frost-heaving occurred readily on silts, and less readily on clays, but never on sands. Examination of the frost-heaved materials revealed that uplift was caused by the formation of bands or layers of clear ice ('segregated ice') within the materials; these layers were formed of ice crystals growing within the material in the direction in which heat was being most rapidly conducted away. Ice segregation only occurs in materials now termed 'frost-susceptible', having a suitable grain-size composition (Fig. 21-2). Taber found that quartz dust with a maximum particle size of 0·07 mm was too coarse for ice to segregate; but in materials of grain size 0·01 mm or less, ice segregation could be induced without difficulty, though in clays, segregation might be restricted by the fine pores reducing the rate of water flow to the growing ice crystals. This movement of water into the zone in which the crystals are growing is an essential part of the process of ice segregation: the material actually gains in water content during the process, which is not simply a matter of the crystallization of water already in the material. Taber suggested that molecules of water were transferred from adjacent films of water to feed the growing crystals. The water must come largely from ground-water below the layer in which ice is segregating, for the ice and permafrost will effectively prevent any downward percolation from the ground surface; and the water is drawn up by tension, not by capillarity since no free water surface exists. Taber estimated the ultimate tensile strength of water by the energy theoretically needed to separate the molecules (when it is converted into a gas) as about 1400 kg/cm^2. The significance of fine-grained materials in ice segregation is that extremely thin water columns feeding the crystals are more easily maintained under tension than larger ones.

Many other factors affect the process of segregation. The shapes of particles are relevant (tabular flat-lying particles are most favourable); very small voids, while assisting the upward lift of water, may delay the freezing process by encouraging supercooling of the water; the rate of cooling is important (very rapid cooling is unfavourable for ice segregation and may even prevent it in some materials); and surface loading may have to be considered, though mainly in artificial conditions (such as civil engineering works), for growing ice crystals can exert a pressure of over 14 kg/cm^2 (150 tons/m^2). All these principles were deduced by Taber, and have been later confirmed and elaborated by investigators such as G. Beskow (1935) and E. Penner (1960). A recent theoretical study of ice segregation from the viewpoint of soil moisture flow and heat conduction at a frost line is by K. Arakawa (1966).

The commonest form of segregated ice is that of the ice lens or layer formed parallel with the ground surface. Under favourable conditions, such as slow freezing of water-saturated frost-susceptible materials, ice lenses may grow to great size. Taber (1943) quotes examples 4 m thick from silts in

Alaska; J. R. Mackay (1966) in the Mackenzie delta area states that the segregated ice in typical sections may amount to ten times the weight of dry material in the section. G. B. Cressey (1939) observed ice lenses in shafts at Igarka on the Yenisei river, totalling as much as 7·6 m in thickness out of a section 30 m deep. It is uncertain why lenses commonly form in series, or what controls their exact location in the ground. An ice lens will cease to grow when the source of water supply is shut off for some reason, as for instance when ice begins to fill in the voids of the layer beneath the ice lens.

The principal morphological effect of ice segregation is differential ground heaving, to be followed, if the ice lenses melt, by differential collapse (see p. 479). The thicknesses of segregated ice vary greatly according to such factors as the amount of available moisture, the rates of cooling (affected by vegetation and snow cover if any), and the grain-size composition. Hummocky relief is typical, consisting of blisters and earth mounds. Differential heaving has serious consequences for any construction works, and it is not confined to permafrost areas—W. H. Ward (1948) recorded 4 cm of frost-heave at Saffron Walden, Essex, in Chalk frozen to a depth of only 23 cm in the winter of 1947. Excessive heaving often results from a sudden temperature drop after a spring thaw, for the water content of the soil will be high following melting of segregated ice formed in the winter, and freezing will promote even greater segregation of ice (Taber, 1943, p. 1452). Heaving can affect not only the surface but also individual stones. Suitably shaped stones (e.g., tabular or wedge-shaped downward) may be gripped by an ice lens and slowly lifted, leaving voids beneath them. If on thawing these voids fill with slumped soil, the stones will be unable to drop back into place. A succession of thaw-freeze cycles may eventually lift the stones to the ground surface, a possible basis for the legend common in some cold countries that 'stones grow in the soil'.

ICE VEINS AND WEDGES

Vein ice occurs as vertical or near-vertical sheets of ice from less than one to several millimetres thick and sometimes penetrating 10 m or more below the ground surface. Thicker sheets usually taper downward and are known as ice wedges. The largest ice wedges, whose thickness at the top may reach 10 m, represent the most massive of all forms of accumulation of ground ice. Wedges and veins are commonly parts of a polygonal network of ground ice enclosing polygons of frozen ground from 1 to 30 m in diameter (Chapter 22).

The earliest scientific study of ice wedges was over a period of some eight years by Leffingwell (1915) on the north shore of Alaska, and it is a remark-

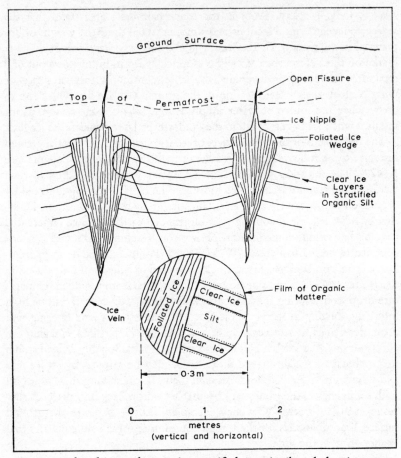

Fig. 21-3 Foliated ice wedges cutting stratified organic silt and clear ice seams, Galena, Alaska (T. L. Péwé, *Biul. Peryglac.*, 1962)

able tribute to the keenness of his observation that his theory of their origin is still widely accepted today. In the Arctic coastal lowlands and the Yukon river lowlands of Alaska, the ice wedges are formed in 'muck formations', the miners' term for the extensive and thick organic-rich silts. The climate of these regions is characterized by intense winter cold and a large temperature range (for example, Galena: mean January temperature – 24°C., absolute minimum – 53°C., absolute maximum 32°C.). The active layer of the muck freezes in winter and contracts on cooling, both in the active layer and the permafrost beneath. The resulting ground cracking may be audible (Leffingwell, 1915, p. 638). Leffingwell thought that subsequent filling of

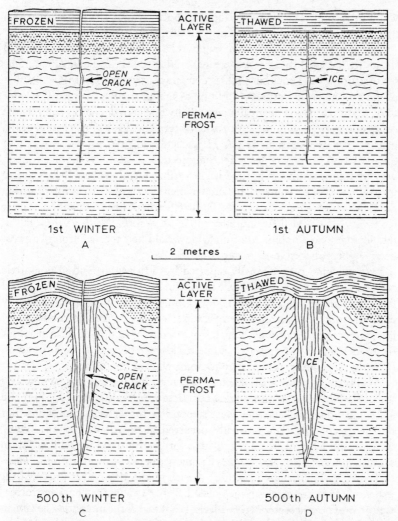

Fig. **21-4** Evolution of an ice wedge according to the contraction crack theory (A. H. Lachenbruch, *Special Paper* 60, 1962, Geological Society of America)

these contraction cracks with water (later freezing) or ice crystals produced the ice veins, and further that the cracks provided planes of weakness which would reopen under stress next winter, allowing deposition of a further film of ice. The maximum width of cracking in any one winter, he observed, was 8 to 10 mm. The veins developed into ice wedges, the largest of which might take about 1000 years to form on this basis.

Leffingwell's theory has received support from many later studies. Fig. 21-3 shows the detailed form of ice-wedges studied by T. L. Péwé (1962). The stratification of the enclosing silts is deformed owing to summer expansion of the ground, as it warms up and pushes against the ice wedges. The latter show vertical foliation or streaking caused by fine particles of dirt entering the contraction cracks with the spring meltwater. The detailed mechanics of the process have been examined by A. H. Lachenbruch (1962) whose diagram of the evolution of ice wedges is reproduced in Fig. 21-4. As the silts contain very high proportions of ground ice in various forms (Taber, 1943, claimed up to 80 per cent ice content), the rates of contraction and expansion probably differ little from those of clear ice. The coefficient of linear expansion of clear ice is $52 \cdot 7 \times 10^{-6}$ at $0°C$. and $50 \cdot 5 \times 10^{-6}$ at $-30°C$. With a $30°C$. fall in temperature, cracks about $1 \cdot 5$ mm wide would be expected around a prospective polygon 1 m in diameter.

Frost cracking is occasionally reported from frozen ground in non-arctic regions. A. L. Washburn, D. D. Smith, and R. H. Goddard (1963) noticed frost-cracking of a golf-course in New Hampshire following an exceptionally cold December (1958; mean temperature $-15°C$.) with little or no snow cover to protect the ground, which froze to depths of up to 2 m. A network of intersecting cracks appeared, some enclosing polygons 6 to 30 m across (Chapter 22); the cracks were up to 5mm wide at the surface.

A different hypothesis of origin for ice veins and wedges was favoured by Taber (1943). While admitting that contraction cracks do form (pp. 1447 and 1521), he did not believe they were responsible for ice wedge initiation, nor did he accept that they normally occurred in polygonal patterns. Veins, wedges, and lenses were all regarded as forms of ground ice resulting from segregation, the veins and wedges having grown downward from ice lenses or layers (p. 1525). The growth of the ice masses themselves provided the expansive force for cracking the ground. Taber's views on ice wedge formation are not now accepted. R. F. Black (1963) makes the interesting comment that whereas Leffingwell worked in an area of continuous permafrost, Taber studied ice wedges in discontinuous permafrost where contraction cracking would be less effective and less evident; the presence of ice wedges in an area of discontinuous permafrost might even imply its degradation from a former state of continuous permafrost. T. L. Péwé (1966) argues that ice wedges only form when the mean annual air temperature is $-6°C$. or colder over a period of many years.

Wedges become extinct when winter stresses at the top of the crack cease to exceed the strength of the wedge ice. This may come about for several reasons. Snow, water, or solifluction debris may bury the wedge and insulate the ground from large changes of temperature; adjacent wedges may develop to relieve the overall stresses; but if these factors can be eliminated, it is possible that the tops of the wedges buried by later deposits may mark horizons at which important climatic or geomorphological changes took place.

After the ice in a wedge melts, debris will be washed or will slump into the wedge, whose form will be preserved in varying degrees. Fossil ice wedges have been widely recognized in areas that once experienced a severely cold climate. They are often best preserved in silts or clays, whose bedding they intersect cleanly. In sands and gravels, slumping may partially destroy them. G. Johnsson (1959), in a beautifully illustrated paper, gives the following as features useful in the recognition of fossil ice wedges: the filling of sediment must have come from above or the sides, never from below; the sides should converge with depth; stones may stand on end in the wedge in contrast to stones in the surrounding material; and the wedge normally stands more or less vertically. Care must be taken to distinguish fossil ice wedges from pseudo wedges (for example, chalk pipes, features resulting from glacio-tectonic disturbance, clay squeezed upward by differential loading, or irregular collapses owing to melting of ice lenses). Johnsson (1962) has studied fossil ice wedges in southern Sweden, where the largest measure 6 m in depth and 75 cm wide at the top. R. W. Galloway (1961) lists over 100 wedges in Scotland between the Moray Firth and the Solway, including ice veins up to 5 m deep, and notes how they occur mostly in lowlands outside the area covered by Zone III (post-Allerød) ice. The marked concentration in the north-east is related to this area having been ice free in the las-Würm glaciation (F. M. Synge, 1956). Ice wedges of probable Würm age are known from southernmost Britain; very fine examples were to be seen in clays and sands during the cutting of the M1 motorway near Watford, Hertfordshire.

Some Features Associated with Ground Ice

INVOLUTIONS

Unconsolidated stratified deposits sometimes display contortions of bedding and interpenetration of one layer by another, to which the term 'involutions' has been applied in a purely descriptive and non-genetic way. Other terms used in the past include 'Brodelböden', introduced by K. Gripp in 1927, but this term has been used in connection with a particular theory now not generally accepted (see p. 502). Involutions may arise in a variety of ways: ice shove, mass movements, turbidity currents (cf. P. H. Kuenen's (1953) load-cast structures), differential volume change of certain minerals, differential loading, undermelt (see p. 313), and tectonic causes are all possibilities, but there still remains a large number of occurrences that cannot be explained in any of these ways, and that are now generally attributed to processes connected with former frozen ground.

Periglacial involutions were first described in North America by C. S. Denny (1936) studying contorted sands and gravels in southern Connecticut. R. P. Sharp (1942a) took up the problem when opencast coal mining south-west of Chicago exposed involutions in glacial and fluvioglacial deposits that could not be explained other than in terms of periglacial action. Originally horizontal bedding had been violently deformed where alternations of sand,

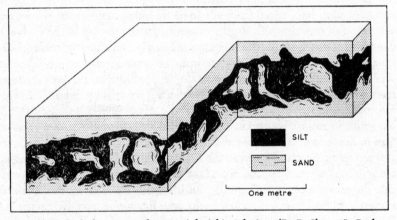

Fig. **21-5** Block diagram to show periglacial involutions (R. P. Sharp, *J. Geol.*, 1942, University of Chicago Press)

silt, and clay occurred, but not where the sections consisted only of sand. The sand appears to have been intruded in all directions by masses of silt and clay, though downward intrusions predominate. Masses of sand or silt appearing isolated in a section face were in all cases found to be connected in the third dimension (Fig. 21-5); the deformations lacked any linear continuity, each individual fold being quickly replaced by wholly unrelated structures. At their base, the involutions abruptly ceased. Festoons of stones whose long axes follow the contorted bedding often picked out the plugs and tongues of the involutions. Similar descriptions of the phenomenon were given by J. P. Schafer (1949). A. Jahn (1956) classified involutions into fold, pillar, and amorphous structures: in the last type, the original structure has become so deformed that it is unrecognizable.

Although a periglacial origin is generally accepted for involutions not related to the other causes enumerated above—and particularly in the case of involutions occurring in association with fossil ice wedges (G. Johnsson, 1962)—there is still some uncertainty as to how they were formed. Two principal hypotheses are advocated. One involves the squeezing of a moist plastic layer between rigid permafrost beneath and a newly frozen crust at the ground surface. The permafrost table would mark the downward limit to involutions, which then formed essentially in, and provide a means of determining the presence and thickness of, the active layer. A second possibility, favoured by Sharp, is connected with the process of ice segregation. Frost-susceptible materials, especially silts and clays, expand rapidly in volume on freezing as already explained, with the segregation of ice. Adjacent materials, if still unfrozen, would then be squeezed and intruded by these active centres. Melting of the ice might induce further deformation.

Galloway (1961) has studied the distribution of periglacial involutions in Scotland. Twenty-eight sites are recorded, some of which involve contortion of Allerød (Zone II) deposits and the deformations in such cases are therefore likely to be Late-glacial (Zone III). The involutions generally extend down to depths of 0·8 to 1·8 m, which may be taken as a rough indication of the permafrost table. E. Watson (1965) has compiled information on involutions in west central Wales, south of Towyn, taken to have been ice-free in the last Würm glaciation but subjected to a periglacial climate. Some, he notes, are covered by undisturbed beds and are therefore unlikely to have had any surface expression; but others occur just below present soil level and might have been related to patterned ground at the surface, now destroyed.

EARTH AND TURF HUMMOCKS

Earth hummocks, sometimes turf-covered, were described by B. Högbom in 1914, and have locally been termed palsen, thufurs, and buttes gazonnées. Their regional occurrence and periglacial relationships have been summarized by C. Troll (1944). R. P. Sharp (1942b) gives a good account of them in the St. Elias Range, Yukon. In this area, they are up to 0·6 m high and 1·5 m in diameter, they are limited to slopes less than 20 degrees (on slopes over 5 degrees, they become elongated across the slope, tending to form small terracettes), and they have earth cores covered with 8-15 cm mats of humus, roots, moss, and grass. No sorting of material seems to have occurred. Sharp follows G. Beskow (1935) in suggesting that expanding patches of frozen ground squeeze earth into small knobs. Vegetation in the hollows around and between these knobs tends to be wet and not such a good insulator as the thicker air-filled vegetation on the tops of the mounds which remain unfrozen. Thus the original patchy ground freezing is perpetuated. The influence of other possible factors on the shape and development of the hummocks, such as varying depths of snow cover and snow melt causing swelling of clayey soils, has been discussed for part of the Jura by A. Reffay (1964), who found that the hummocks ('buttes gazonnées') were most commonly situated on slopes where freeze-thaw cycles will be most frequent and effective.

Taber (1952) argued that the hummocks were elevated by frost heaving following segregation of ice lenses beneath. This hypothesis has been strongly refuted by R. S. Sigafoos and D. M. Hopkins (1951, 1954) who agree with Sharp that it is the areas between the hummocks which freeze first, that there is evidence of lateral thrusting or squeezing in the injection of tongues of silt into or beneath the hummocks, and that the fragile nature of some plant roots growing in the hummocks does not accord with a hypothesis of frost heaving.

A similar phenomenon in parts of western North America is provided by the 'mima mounds', though in many areas these attain somewhat larger dimensions than the hummocks just described. The most famous locality for mima sounds is in the outwash valley plains of the southern Puget Sound region, Washington State. They attain heights of as much as 2 m and may be from 3 to 20 m in diameter; they are built of a mixture of materials including silt, sand, and pebbles. Ever since they were first reported in 1845, there has been speculation and uncertainty as to their origin. The problem became an embarrassment to geological science since all hypotheses prior to about 1940 failed under field tests; some geologists then turned to an origin by gophers, which led to lengthy and unprofitable discussion as to what a gopher might or might not do. It now seems almost certain that the mounds have a periglacial origin, though agreement is by no means reached on the exact processes. A. M. Ritchie (1953), suggests the following sequence of events: a) freezing of the outwash deposits and the formation of ice-wedge polygons; b) thawing of the polygons to a stage where a rounded frozen core still remained surrounded by thawed inter-polygon zones; c) a brief period of erosion by running water, stripping out some of the thawed material.

It may be concluded that there is probably more than one type of earth hummock formed under periglacial conditions, and that a great deal of further work is needed before satisfactory generalizations can be formulated.

PINGOS

'Pingo' is an Eskimo term for scattered, isolated, dome-shaped hills, and was introduced into geomorphology by A. E. Porsild (1938). Pingos are conspicuous features in the Arctic regions of Alaska, Canada, Greenland, and Siberia, where there is continuous permafrost. Many have also recently been identified in the sub-arctic boreal forest region of North America, in the zone of discontinuous permafrost. They may be from less than a metre to over 60 m in height, and up to 300 m in diameter. The smaller ones have rounded tops; the larger ones are often ruptured and broken open at the top which has a crater-like appearance as a result. Internally, they usually show outward dipping or deformed beds of stratified silts or sands, though some pingos have developed in bedrock which has been similarly deformed (J. G. Cruickshank and E. A. Colhoun, 1965). In some, an ice core is inferred or visible (R. P. Sharp, 1942c, therefore termed them 'ground-ice mounds').

A good example of a pingo is described by F. Müller (1962). Ibyuk Pingo is situated in the Mackenzie delta, N.W.T. It rises 41 m above the surrounding flat marsh, with a basal circumference of 900 m. The summit is in the form of a large crater in which a small lake has collected. It possesses an ice core about 40 m thick, overlain by about 15 m of frozen sediments which have cracked open at the crater. Pingos such as this are, in fact, the most striking relief features of the Mackenzie delta in whose 25,000 km² there occur about 1400, the greatest known concentration of these features. In this area, each occurs in a former lake basin or in a shallow lake at present.

From a series of recent studies, it seems probable that there are two main types of pingo, one of which is so well exemplified in the Mackenzie delta region that it is often referred to as the 'Mackenzie type'. In East Greenland, central Alaska, and certain other areas, however, pingos of different origin are termed the East Greenland type. Müller (1959) suggested that the Mackenzie type be termed 'closed-system' pingos, the East Greenland type 'open-system' pingos. It is convenient to discuss their origins separately.

i) *Mackenzie type* A comprehensive investigation of pingos in the Mackenzie delta has been undertaken by J. R. Mackay (1962). They are not simply the result of ice-lens heaving since they are formed in material which is 95 per cent fine to medium sand (0·1–0·5 mm), with very little silt or clay, and therefore not significantly frost-susceptible. On the other hand, the sand allows rapid movement of pore-water. Fig. 21-6 illustrates Mackay's theory of their formation. The initial situation comprises a relatively large deep ice-covered lake surrounded by permafrost, while beneath the central position of the lake there is unfrozen ground. As the lake is slowly infilled with sediment, there comes a point when the lake ice becomes frozen to the bottom, and the bottom sediments in turn freeze, so that a layer of ice and permafrost extends over the site of the lake. Thus there is created a 'closed-system' in the unfrozen ground beneath this new cap of permafrost from which water

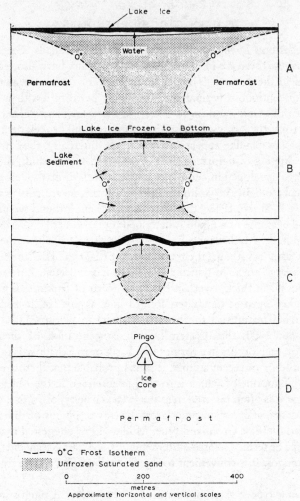

Fig. 21-6 Pingo formation in the Mackenzie Delta area, according to J. R. Mackay (*Geogr. Bull.*, *Ottawa*, 1962, by permission of the Queen's Printer, Ottawa). Depth of lake ice and water exaggerated

A Broad shallow lake over frost-free ground
B Accumulation of sediment in lake causes lake ice to freeze to bottom in winter. Edge of permafrost advancing, causing expulsion of pore-water in the saturated sands and giving upward pressure beneath lake centre
C Further advance of permafrost; up-doming of lake ice and sediment
D Permafrost continuous; up-doming has ceased, and ice-cored pingo remains

cannot escape. Inward growth of permafrost around the unfrozen core increases water pressure in the latter, for pore-water is expelled from the unfrozen sands by the advancing frost plane. To relieve the pressure, the surface layers are bulged up. Eventually, all the water in the closed system is converted to ice and the excess water forms a core of clear ice under the bulge.

The rate of formation of such pingos is generally slow since the ice cores require large volumes of unfrozen sediments from which the water is slowly expelled. Radiocarbon dates for two pingos suggest average growth rates of 0·3 to 0·5 m per thousand years. They are thought to be stable features of permafrost areas so long as the surface remains intact, but the up-bulging helps to cause cracking of the surface, allowing exposure of the ice core and its eventual disappearance by melting when air temperatures permit. The centre of the pingo then collapses to form a crater.

A completely different view of the mode of formation of the Mackenzie pingos has been put forward by R. C. Bostrom (1967). He regards them as being of the East Greenland type, which is discussed in the next section.

i) *East Greenland type*   Leffingwell in 1919 suggested that isolated mounds on the gently sloping coastal plain of north Alaska were formed by hydraulic forces. Progressive downward freezing of water-saturated sediments might interfere with normal ground-water flow and cause sub-surface pressure to build up. In extreme cases, the ground would be bulged up and sometimes even ruptured, giving an outburst of water and debris. The general requirements for 'open-system' pingos are that there should be permafrost, and an aquifer beneath the permafrost by which water flows to the pingo site.

Bostrom (1967), as noted above, has recently reinterpreted the Mackenzie pingos as being of this hydraulic type, and if his hypothesis is substantiated, the term 'Mackenzie type' will clearly have to be replaced. He adduces evidence to show that the Mackenzie delta is an area of contemporary subsidence. Sediment is being added and frozen on to the permafrost layer while at the base of the latter, sediment is thawing on being carried by subsidence through the 0°C. isothermal surface. Thus a rigid frozen layer, possibly 90 m thick, rests on a newly thawed and water-saturated zone below; the resulting hydraulic pressure will be 11·7 kg/cm^2 if the mean density of the frozen layer is taken to be 1·3. In places where the frozen layer is cracked (for instance, as a result of flexure) or otherwise interrupted, water (density 1·0) will rise from the thawed zone by artesian pressure to a theoretical height of 27 m above the ground surface. The figure of 27 m is approximately in accordance with the mean height attained by the Mackenzie pingos. As the water moves upward through the permafrost, its relative warmth maintains and enlarges a thawed passage and forms a pond at the ground surface. When the pond shallows sufficiently by silting, it can become frozen to its bottom in winter, and continued upward movement of artesian water will dome up the ice to form a pingo. The ponds or lakes associated with the Mackenzie pingos are thus explained not as initial features (Mackay, 1962) but as consequent on the upward movement of water under hydraulic pressure.

G. W. Holmes, D. M. Hopkins, and H. L. Foster (1963) believe that many pingos in central Alaska belong to the East Greenland category. Most lie on gentle forested slopes in unglaciated silt-filled valleys; large ice lenses within them are sometimes exposed. Their ages are thought to vary from some decades up to about 7500 years. Further examples in north-east Greenland are given by Cruickshank and Colhoun (1965).

Fossil pingos Few traces of pingos in former periglacial regions have been identified with any certainty. A. Pissart (1963) claims that some circular or semi-circular ramparts in parts of Wales and Belgium are Pleistocene pingo remnants. Diameters of the features are up to 120 m; a human origin is discounted. On slopes, the ramparts tend to be elongated and interrupted on the up-slope side. In general, the features are not incompatible with the development and subsequent decay of ice-cored pingos. G. Wiegand (1965) has discussed similar fossil pingos in parts of south and central Germany.

EXTRUSIVE ICE

A. T. von Middendorff in 1859 (Leffingwell, 1919) first named and explained interesting forms of ground ice that he called 'Aufeisen'. The ones that he discussed were associated with river floodplains. In winter, river flow is impeded by freezing in the form of anchor ice at the bottom and frazil ice at the surface of the river, and shallow stretches may freeze solidly. Water still flowing from upstream floods the area as the flow becomes increasingly impeded. Freezing then forms candle ice which consists of long vertical ice crystals. Alternate flooding and freezing, continuing all winter, may build up layers of candle ice 4 m thick and covering up to 8 km². Domes and ridges form and these may crack as they are forced up by water pressure from below.

Aufeisen have been found in west Baffin Island in association with large kettle-holes. Where kettle-holes exceed 0·7 km in diameter and are deep, the water cannot freeze from top to bottom. The kettle-holes associated with the Aufeisen have no visible outlets and the water must escape through the coarse bouldery beds to feed springs which occur near the Aufeisen. The spring water escapes at intervals during the winter when the pressure exceeds the strength of the ice forming at its margin. The water then floods out over the Aufeisen. Some layers of candle ice consist of candles 42 cm long. One Aufeis was fed by a small kettle-hole with no visible inflow but a strong outflow of water. The small kettle-hole was fed in turn by a much larger one a short distance away and at a higher level. This Aufeis occupied a small gully and was about 1 km long with a gradient of 3·5 degrees; it was at least 3 m thick. It consisted of horizontal layers of candle ice interbedded with compressed snow which must have fallen between outbursts. The most distinctive feature of Aufeisen is the candle ice of which they mainly consist, and the essential requirement for their formation is the periodic availability of water throughout the winter.

THERMOKARST

In karst areas, a distinctive set of landforms is related to processes of solution acting upon limestone or dolomite bedrock. 'Thermokarst' is a term that also applies to a peculiar group of landforms and the action of a particular process in producing them, but in this case it is not a chemical but a physical process. However, there are some striking points of resemblance between karst and thermokarst landscapes. The term 'thermokarst' was introduced by a Russian, M. M. Ermolaev, in 1932, to describe the features resulting from the melting of ground ice in permafrost regions. The principal effect of ground-ice melting is essentially collapse, which may give rise to surface pits and basins, funnel-shaped sinks, dry valleys, or ravines; before collapse occurs, caves may even be found where ice has begun to melt away. Conical hillocks ('baydjarakhs') which may be disposed in a regular pattern, mark the cores of ice-wedge polygons after the ice wedges have melted away (compare 'mima mounds', p. 474). For its best development, thermokarst needs numerous and thick ground-ice masses in the upper parts of the permafrost. General discussions of the phenomena are given by A. I. Popov (1956) and J. Dylik (1964). S. P. Kachurin (1962) recognizes three zones of thermokarst in the U.S.S.R., in which the southernmost represents areas where degradation of the permafrost is today active and where thermokarst is extensively developed, and the northernmost represents a region where there is no degradation of permafrost but where the presence of abundant surface water at times (for example, spring floods) leads to locally intense thawing of ground ice. He and other writers also note how human activities such as ploughing, ditch digging, or building operations in permafrost may greatly accelerate thermokarst development.

In Alaska, some thermokarst forms have been well described by R. E. Wallace (1948) and D. M. Hopkins (1949). Both discuss the development of 'thaw lakes' and collapse basins. Thaw lakes ('cave-in lakes') develop when, for instance, an insulating vegetation cover is broken by frost heave; patchy thawing of ice-rich sediments will then give rise to scattered hollows containing pools of water. Once formed, such pools steadily enlarge themselves to roughly circular forms. Thaw proceeds rapidly at water level, while the banks above may remain frozen and thus become undercut (Hopkins notes 5-metre undercuts). In the early stages, lakes are small and scattered, sometimes even occurring on flat hill-tops, and drainage is extremely haphazard. Subsequently, the lakes begin to coalesce as they grow in size, and integration of drainage commences. Wallace finds evidence for a still later stage when through-flowing streams begin to build levees and separate thaw lakes into segments. The lakes also enlarge along ice wedges, giving thaw gullies extending a hundred metres or more from the lake shore. Sometimes lakes may be drained underground giving 'thaw sinks'. Rates of growth of thaw lakes from 5 to 20 cm a year were determined by Wallace by noting the progressive changes in the trees affected by bank collapse—tilted, bent, and S-shaped trees occur. A lake 100 m in diameter might need only 500-1500 years to form.

Fossil thermokarst basins have been described from some Pleistocene periglacial areas. P. E. Wolfe (1953) mapped about 2500 closed depressions up to 6 m deep and 2·5 km² in size on the coastal plain of New Jersey. The hollows occur on various rock types from clays to gravels. All hypotheses of origin, other than one of thaw-lake development when the ground was perennially frozen, were found unsatisfactory. Very similar forms have been located in the Paris Basin by Pissart (1958). The circular depressions here are normally found on impermeable rocks, and neither karstic nor weathering processes can adequately explain their occurrence. Pissart suggests that they are thermokarst basins, developed when sub-surface ice lenses melted away accompanied by slow surface subsidence.

Conclusion

Permafrost is defined as a thickness of bedrock or superficial material in which temperatures are below freezing over a period of years. Zones of continuous permafrost are mainly related to areas of present-day severely cold climate, though some parts may be relict from the Pleistocene. Maximum known depths of permafrost approach 600 m in the northern U.S.S.R. and 450 m in northern Canada, regions of intense winter cold, short cool summers, minimal vegetation cover, and small snowfall. Above the permafrost lies the active layer which experiences summer or short-period thawing. Buried thawed layers (taliks) within permafrost may reflect long-term climatic changes, or merely the influence of local factors such as pockets of highly mineralized ground water.

Ground ice in permafrost ranges from pore-fillings in sedimentary rocks to large ice lenses and wedges. Segregated ice has been studied in detail by S. Taber and others, and it has been shown that it forms in frost-susceptible materials, having a suitable grain-size composition, when ground water is drawn in to feed growing ice crystals at a frost line. Given slow freezing of water-saturated frost-susceptible materials, lenses or layers of segregated ice formed parallel to the surface may reach several metres in thickness, resulting in localized heaving of the ground surface. Veins and ice wedges form in semi-vertical cracks which, as E. K. Leffingwell postulated fifty years ago, probably result from thermal contraction of frozen ground during cooling. They probably require mean annual air temperatures below – 6°C. After the ice in a wedge melts (for a variety of possible reasons), debris may slump in and preserve the structure of the wedge in fossil form. Other features associated with ground ice include involutions, probably caused by processes of ice segregation, and pingos, which are dome-shaped mounds up to 60 m high produced by freezing of ground water trapped beneath permafrost or lake ice. Another form of ground ice is classified as extrusive ice and includes Aufeisen formed by periodic escape of spring or river water in winter.

The decay of ground ice, resulting mainly from climatic amelioration, gives rise to thermokarst features, such as pits and basins caused by collapse of layers once supported by ground ice. Thaw lakes are another manifestation of decaying ground ice.

REFERENCES

G. S. ANDERSON and K. M. HUSSEY (1963), 'Preliminary investigations of thermokarst development on the North Slope, Alaska', *Proc. Iowa Acad. Sci.* **70**, 306-20

J. T. ANDREWS (1961), 'Permafrost in southern Labrador-Ungava', *Can. Geogr.* **5**, 34-35

K. ARAKAWA (1966), 'Theoretical studies of ice segregation in soil', *J. Glaciol.* **6**, 255-60

G. BESKOW (1935), 'Tjälbildningen och tjällyftningen med särskild hänsyn till vägar och järnvägar', *Sver. geol. Unders. Afh.*, Ser. *C*, **375** (English translation by J. O. OSTERBERG (1947), *NWest. Univ. tech. Inst.*, 145 p.)

R. F. BLACK (1954), 'Permafrost—a review', *Bull. geol. Soc. Am.* **65**, 839-55

R. F. BLACK (1963), 'Les coins de glace et le gel permanent dans le Nord de l'Alaska', *Annls Géogr.* **72**, 257-71

R. F. BLACK (1964), 'Periglacial studies in the United States, 1959-63', *Biul. Peryglac.* **14**, 4-29

R. C. BOSTROM (1967), 'Water expulsion and pingo formation in a region affected by subsidence', *J. Glaciol.* **6**, 568-72

R. J. E. BROWN (1960), 'The distribution of permafrost and its relation to air temperature in Canada and the U.S.S.R.', *Arctic* 13, 163-77

K. BRYAN (1946), 'Cryopedology—the study of frozen ground and intensive frost-action with suggestions on nomenclature', *Am. J. Sci.* 244, 622-42

F. A. COOK (1958), 'Temperatures in permafrost at Resolute, N.W.T', *Geogr. Bull.* 12, 5-18

G. B. CRESSEY (1939), 'Frozen ground in Siberia', *J. Geol.* 47, 472-88

J. G. CRUICKSHANK and E. A. COLHOUN (1965), 'Observations on pingos and other landforms in Schuchertdal, north-east Greenland', *Geogr. Annlr* 47, 224-36

C. S. DENNY (1936), 'Periglacial phenomena in southern Connecticut', *Am. J. Sci.* 32, 322-42

J. DYLIK (1964), 'Le thermokarst, phénomène négligé dans les études du Pleistocene', *Annls Géogr.* 73, 513-23

R. F. FLINT and H. G. DORSEY (1945), 'Glaciation of Siberia', *Bull. geol. Soc. Am.* 56, 89-106

R. W. GALLOWAY (1961), 'Ice wedges and involutions in Scotland', *Biul. Peryglac.* 10, 169-93

I. P. GERASIMOV (1961), 'The recent nature of the Siberian pole of cold', *J. Glaciol.* 3, 1089-96

K. GRIPP (1927), 'Beiträge zur Geologie von Spitsbergen', *Abh. Geb. Naturw.*, Hamburg 21, 1-38

B. HÖGBOM (1914), 'Über die geologische Bedeutung des Frostes', *Bull. geol. Instn Univ. Upsala* 12, 257-389

G. W. HOLMES, D. M. HOPKINS and H. L. FOSTER (1963), 'Pingos in central Alaska', *Geol. Soc. Am. Spec. Pap.* 73, 173

D. M. HOPKINS (1949), 'Thaw lakes and thaw sinks in the Imuruk Lake area, Seward Peninsula, Alaska', *J. Geol.* 57, 119-31

D. M. HOPKINS and R. S. SIGAFOOS (1954), 'Role of frost thrusting in the formation of tussocks', *Am J. Sci.* 252, 55-59

J. D. IVES (1960), 'Permafrost in central Labrador-Ungava', *J. Glaciol.* 3, 789-90

A. JAHN (1956), 'Some periglacial problems in Poland', *Biul. Peryglac.* 4, 169-83

A. JAHN (1958), 'Periglacial microrelief in the Tatras and on the Babia Góra', *Biul. Peryglac.* 6, 227-49

J. L. JENNESS (1949), 'Permafrost in Canada', *Arctic* 2, 13-27

G. JOHNSSON (1959), 'True and false ice-wedges in southern Sweden', *Geogr. Annlr* 41, 15-33

G. JOHNSSON (1962), 'Periglacial phenomena in southern Sweden', *Geogr. Annlr* 44, 378-404

S. P. KACHURIN (1962), 'Thermokarst within the territory of the U.S.S.R.', *Biul. Peryglac.* 11, 49-55

P. H. KUENEN (1953), 'Significant features of graded bedding', *Bull. Am. Ass. Petrol. Geol.* 37, 1044-66

A. H. LACHENBRUCH (1962), 'Mechanics of thermal contraction cracks and ice-wedge polygons in permafrost', *Geol. Soc. Am. Spec. Pap.* 70, 69 p.

E. K. LEFFINGWELL (1915), 'Ground ice wedges', *J. Geol.* 23, 635-54

E. K. LEFFINGWELL (1919), 'The Canning River region, northern Alaska', *U.S. geol. Surv. Prof. Pap.* **109**, 1-251

J. R. MACKAY (1962), 'Pingos of the Pleistocene Mackenzie River delta', *Geogr. Bull.* **18**, 21-63

J. R. MACKAY (1966), 'Segregated epigenetic ice and slumps in permafrost: Mackenzie delta area, N.W.T.', *Geogr. Bull.* **8**, 59-80

A. D. MISENER (1955), 'Heat flow and depth of permafrost at Resolute, Cornwallis Island, N. W. T., Canada', *Trans. Am. geophys. Un.* **36**, 1055-60

F. MÜLLER (1959), 'Beobachtungen über Pingos. Detailuntersuchungen in Ostgrönland und in der Kanadischen Arktis', *Meddr Grönland* **153**, 3, 127 p.

F. MÜLLER (1962), 'Analysis of some stratigraphic observations and radiocarbon dates from two pingos in the Mackenzie Delta area, N.W.T.', *Arctic* **15**, 278-88

S. W. MULLER (1947), *Permafrost or permanently frozen ground and related engineering problems* (Ann Arbor), 231 p.

E. PENNER (1960), 'The importance of freezing rates in frost action in soils', *Proc. Am. Soc. Test. Mater.* **60**, 1151-65

T. L. PÉWÉ (1962), 'Ice wedges in permafrost, lower Yukon river area, near Galena, Alaska', *Biul. Peryglac.* **11**, 65-76

T. L. PÉWÉ (1966), 'Paleoclimatic significance of fossil ice wedges', *Biul. Peryglac.* **15**, 65-73

J. A. PIHLAINEN (1962), 'An approximation of probable permafrost occurrence', *Arctic* **15**, 151-4

A. PISSART (1958), 'Les depressions fermées de la région parisienne: le problème de leur origine', *Revue Geómorph. dyn.* **9**, 73-83

A. PISSART (1963), 'Les traces de "pingos" du Pays de Galles (Grande-Bretagne) et du Plateau des Hautes Fagnes (Belgique)', *Z. Geomorph.* NF **7**, 147-65

A. I. POPOV (1956), 'Le thermokarst', *Biul. Peryglac.* **4**, 319-30

A. E. PORSILD (1938), 'Earth mounds in unglaciated Arctic northwestern America', *Geogr. Rev.* **28**, 46-58

L. L. RAY (1951), 'Permafrost', *Arctic* **4**, 196-203

A. REFFAY (1964), 'Quelques observations sur les buttes gazonnées des pâturages de la région de Saint-Claude (Jura)', *Revue Géogr. alp.* **52**, 315-23

A. M. RITCHIE (1953), 'The erosional origin of the Mima Mounds of south-west Washington', *J. Geol.* **61**, 41-50

J. P. SCHAFER (1949), 'Some periglacial features in central Montana', *J. Geol.* **57**, 154-74

R. P. SHARP (1942a), 'Periglacial involutions in north-eastern Illinois', *J. Geol.* **50**, 113-33

R. P. SHARP (1942b), 'Soil structures in the St. Elias Range, Yukon Territory', *J. Geomorph.* **5**, 274-301

R. P. SHARP (1942c), 'Ground-ice mounds in tundra', *Geogr. Rev.* **32**, 417-23

P. A. SHUMSKIY (1959), quoted by J. R. MACKAY (1966)

R. S. SIGAFOOS and D. M. HOPKINS (1951), 'Frost-heaved tussocks in Massachusetts', *Am. J. Sci.* **252**, 55-59

F. M. SYNGE (1956), 'The glaciation of north-east Scotland', *Scott. geogr. Mag.* **72**, 129-43

484 PERIGLACIAL GEOMORPHOLOGY

S. TABER (1929), 'Frost heaving', *J. Geol.* **37**, 428-61

S. TABER (1930), 'The mechanics of frost heaving', *J. Geol.* **38**, 303-17

S. TABER (1943), 'Perennially frozen ground in Alaska: its origin and history', *Bull. geol. Soc. Am.* **54**, 1433-548

S. TABER (1952), 'Geology, soil mechanics and botany', *Science* **115**, 713-14

C. TROLL (1944), 'Strukturböden, Solifluktion, und Frostklimate der Erde', *Geol. Rdsch.* **34**, 545-694 (English translation, *Snow Ice Permafrost Res. Establ.*)

R. E. WALLACE (1948), 'Cave-in lakes in the Nabesna, Chisana, and Tanana river valleys, eastern Alaska', *J. Geol.* **56**, 171-81

W. H. WARD (1948), *J. Glaciol.* **1**, 146

A. L. WASHBURN, D. D. SMITH, and R. H. GODDARD (1963), 'Frost cracking in a middle-latitude climate', *Biul. Peryglac.* **12**, 175-89

E. WATSON (1965), 'Periglacial structures in the Aberystwyth region of central Wales', *Proc. Geol. Ass.* **76**, 443-62

W. WERENSKJOLD (1953), 'The extent of frozen ground under the sea bottom and glacier beds', *J. Glaciol.* **2**, 197-200

G. WIEGAND (1965), 'Fossile Pingos in Mitteleuropa', *Würzb. geogr. Arb.* **16**, 1-152

P. E. WOLFE (1953), 'Periglacial frost-thaw basins in New Jersey', *J. Geol.* **61**, 133-41

Proc., Permafrost Int. Conf., Indiana (Lafayette, 1963). *Nat. Acad. Sci.—Nat. Res. Council Publ.* **1287**, 563 p.

Chapter 22

Patterned Ground

When one steps upon the tundra almost anywhere in the (Lena) delta, it appears divided into countless irregular polygons of differing size, whose edges are higher than the middle. Between the edges of two such polygons, a small furrow is found, which is used as a pathway by lemming. (A. VON BUNGE, 1884)

One of the most conspicuous and characteristic features of periglacial regions is patterned ground, first described over a century ago. It occurs almost entirely within the cold-climate zones, although some minor forms of patterned ground are exceptional in this respect. The features that show marked sorting are certainly restricted to a climate where temperatures frequently fall below freezing-point, and both frost and the formation of ground ice are fundamental factors in the origin of patterned ground. A wide variety of forms comes under the heading of patterned ground; it is necessary, therefore, to classify the forms before their characteristics can be adequately described and the processes of formation are discussed.

Patterned ground occurs in both an active state and a fossil state. In the latter case, it provides a useful indication of past climate. Fossil patterned ground has been recognized ever more widely as aerial photographs have become available for increasingly large areas, and such photographs are often the best means whereby it can be located. Patterned ground has been recognized on aerial photographs of parts of the south and Midlands of England (F. W. Shotton, 1960). It is revealed through minor changes in the soil often not apparent from ground level, which affect the density of crops or type of vegetation.

The two most important aspects of the patterns are their shape and the degree of sorting that has developed. Classifications must, therefore, recognize these two elements, and they are the basis of the classification suggested by A. L. Washburn (1956). Until more is known of the origin of the features, it is best to avoid a classification based on genetic terms. A purely descriptive classification has advantages in the present state of knowledge concerning the origin of many of the features. Washburn bases the primary division of his classification on the shape of the pattern. A subdivision of each shape is then based on the character of the sorting, so that the complete classification is as follows:

a) Circles   — sorted and unsorted
b) Nets   — sorted and unsorted
c) Polygons   — sorted and unsorted
d) Steps   — sorted and unsorted
e) Stripes   — sorted and unsorted

The features may grade into one another. Elongated steps or circles may be called garlands, where they are outlined by larger stones, and these merge into stripes as they become increasingly elongated. The circles form where the operating processes can work in isolation from one centre. But where two or more centres can react with one another, the net or polygonal form will occur. The irregular mesh pattern of the net is intermediate between the circles and polygons, which have angular and often regular boundaries. Some of the unsorted forms are made conspicuous on the surface by changes in vegetation or elevation. Earth hummocks, described in Chapter 21, are a common form of unsorted nets. At times, these features are in the form of bare patches surrounded by vegetation.

The gradient of the surface is the essential factor that determines whether a circle, net, or polygon will form as opposed to steps and stripes. The circles, nets, and polygons are restricted to flat or nearly flat surfaces. The steps and particularly the stripes occur on ground with a considerable degree of slope, varying between 6 degrees and about 15-30 degrees. Slopes of 2-6 degrees have forms that are intermediate between the truly circular or polygonal forms and the true stripes. Slopes steeper than 30 degrees are usually too steep to develop patterns.

Description of Patterns

Each of the main types will be briefly described and then some of the more important forms will be examined in greater detail in an attempt to arrive at conclusions concerning their method of formation.

Circles can occur singly or in groups. In the sorted type there is normally a border of coarser stones around an area of finer material (Plate XXVII). Circles of this type are very well developed in Spitsbergen, where they attain dimensions of 0·8 to 3 m in diameter. The stones in the border appear to get larger as the size and thickness of the border increase. There are also good examples in west Baffin Island, particularly where limestone outcrops and forms a litter of small, flat stones, or on raised beach material. These features form in many polar and high mountain areas, and are not restricted to areas of permafrost. They occur in Iceland, for instance, in areas where there is no permafrost. Unsorted circles can also occur in environments where no frost occurs, for example in parts of Australia.

Nets occur where the pattern is less circular but not yet polygonal. Earth hummocks of unsorted material come in this category. Again, permafrost is not necessary. Nets occur in parts of Scandinavia, Siberia, Iceland and Canada, and are most common in sub-arctic and Alpine areas. The features are usually fairly small, those in Iceland being 1 to 2 m in diameter.

Polygons are the best known type of patterned ground and their description and classification has given rise to a wide variety of terms. Polygons can form either in permafrost areas or in areas having seasonal frosts. Indeed, some types of polygon can form in warm semi-arid environments, where they are the result of desiccation. If sorting is present in this instance, it may be the result of stones being washed into the cracks by seasonal floods.

There are types of polygon that only form where there is permafrost. These forms are particularly valuable for palaeoclimatic studies. The type of polygon requiring permafrost must be differentiated from the much smaller features that can be seen at higher levels on the hills of Britain, for example, on the top of Helvellyn. These small polygons have been shown to be growing actively under present climatic conditions in Britain (S. E. Hollingworth, 1934). In the Tinto Hills in Scotland (R. Miller et al., 1954), they have been destroyed purposely, yet have reformed in a few years.

The truly arctic features, which only form where permafrost is present, have been called tundra polygons or ice-wedge polygons. The former term is preferable as it does not imply a specific formative mechanism. However, the association of the large tundra polygons with ice wedges is now very clearly established, and their method of formation is fairly well agreed upon, as outlined in the last chapter.

T. L. Péwé (1963) has discussed the climatic conditions necessary to generate the ice wedges that form the polygonal pattern. He divides the ice wedges into three categories and the same classification can be applied to the polygonal forms that are the surface manifestation of the ice wedges. Active ice wedges are associated with the zone of continuous permafrost and temperatures falling below $-6°C$. and sometimes to $-12°C$. The mean annual number of degree-days of freezing (°C.) ranges from 2800 to 4500. In the areas where inactive ice wedges occur, the temperature conditions are less severe, with mean annual values ranging from $-2°C$. to about $-6°C$ to $-8°C$, and 1700 to 4000 degree-days of freezing. The zone where these conditions occur lies to the south of the active ice-wedge zone. Permafrost is discontinuous, while frost cracking rarely occurs and ice is not added to the wedges, although the climate is cold enough to preserve them. Péwé's third category comprises fossil ice wedges, which occur where the ice has melted and been replaced by sediment. Those found in central Britain imply a fall of temperature of 16° to 18°C. below the present.

The true ice-wedge type of tundra polygon is considerably larger than the small type which develops by frost sorting of material. The tundra type may be over 10 m across, and those in the Avon valley in England (Shotton, 1960) range from a minimum of 4·5-6 m to a maximum of 60 m. The fossil ice-wedge polygons on the North Yorkshire Moors near Scarborough are between 15 m and 30-40 m across. The smaller forms on the other hand are often not larger than about 3 m and those described by Hollingworth (1934) in the Lake District are only about 0·6 m across. Size provides a fairly reliable method of distinguishing the two types of very different polygons. The smaller ones are also nearly always of the sorted variety. The larger ones, on the other hand, are basically unsorted, although in the fossil type the pattern is made by the sedimentary fill of the fossil ice wedges. The polygonal patterns are on the whole more common than the circles or nets, as where conditions are suitable for the formation of one, they usually allow many to form and thus the polygonal pattern develops. The tundra polygons sometimes develop a tetragonal or rectangular pattern and sometimes the five- or six-sided polygonal form. The polygons are nearly always formed in sediments but on rare occasions polygonal patterns have arisen where intense frost activity, working along polygonal jointing in bedrock, has given rise to these patterns in bedrock. Examples of frost cracks in the Palaeozoic limestone on Foley Island in Foxe Basin (Baffin Island) illustrate this point. The limestone in this area shows extreme frost shattering.

Steps form where the slope is rather steeper. In the sorted type, the step is usually bordered by embankments of stones larger than the rest of the material. The steps sometimes form parallel with the slope contours, but at times they become drawn out into lobate form, elongated down-slope, forms which are referred to as stone garlands. Good examples occur on the slopes of the Glyders in north Wales. The term stone-banked terrace has also been applied to these features. They do not normally have the regularity characteristic of polygons although the sorted steps and garlands normally occur in groups. Tabular stones at the margins of the features are often arranged vertically in the lateral edges, but dip 60 to 70 degrees upslope at the lobate front.

In the unsorted variety of this form, vegetation changes often show up the position of the features. Generally the riser of the step is well vegetated and the tread is bare. For this reason, the terms 'turf-banked terrace' or 'turf garlands' have been used to describe the features. In shape they resemble the sorted varieties, but they are often of smaller dimensions.

Neither of these forms need necessarily be associated with permafrost. Since their features are also strongly affected by solifluction, they will be considered in more detail in Chapter 23.

The somewhat irregular step and garland forms merge into *stripes* as the slope increases. The sorted stripes are elongated down the slope and consist of alternating strips of coarser and finer stones. Again the dimensions vary very much; at one end of the scale is the small type that occurs in areas of seasonal frost activity. R. Miller *et al.* (1954) have described very well developed small stripes in the Tinto Hills where the slope is about 20 degrees. These stripes are forming actively under present climatic conditions. The coarse stripe is from 20 cm to 35 cm in width. Coarse material collects in a furrow while finer material forms a slight ridge. The depth of the furrow is about 7·5 cm. In better developed examples the coarse stripe may be 1·5 m wide and the finer two to four times the width. In places the features are quite extensive and have been traced for over 100 m down the slope under favourable conditions. The depth to which the stripes penetrate correlates on the whole with their size, the two increasing together. Because the stripes form on fairly steep slopes, they have less chance of survival when conditions suitable to their formation cease. They are not, therefore, often preserved in a fossil form.

In the unsorted type, vegetation again outlines the pattern and they have been termed 'vegetation stripes'. One report of these features from Arctic Canada describes vegetation in slight troughs 0·3-0·6 m wide and 3-4·5 m apart. But in other areas the widths of the vegetated and bare strips are about equal. This form is not necessarily confined to cold climates and has been reported from parts of Australia.

The Processes Forming Patterned Ground

In discussing the method of formation of patterned ground, the polygonal pattern will be considered in most detail. Circles, nets, and polygons are formed by similar mechanisms and the same processes operate on steeper slopes to produce steps and stripes. Many different possible processes have been suggested. In order to assess their validity, detailed observations of the features and the changes occurring within them, as well as measurements of the processes and controlling factors, are necessary. Detailed studies of the polygonal features will be mentioned first, and then experiments, both in the field and laboratory, on the movements of material and the forces involved will be described. The matter is complicated by the fact that the features can form in a variety of environments by different processes and yet possess many of the same characteristics.

Definitions In discussing the formation of patterned ground the concept of the *active layer* is a useful one. As explained in the last chapter, this is the layer which melts during the summer season in an area where the ground is permanently frozen at depth. It is, therefore, in this layer that material can be moved and the patterned ground can form as the material is subject to alternating freezing and thawing. The term *frost-susceptible* is applied to material of particular grain size composition (see p. 466 and Figure 21-2) that is subject to frost heaving. A *non-frost-susceptible* soil is one which consists of coarser sediments with less than one per cent fines, the soil not being subject to frost heaving. The term *fines* is applied to particles of less than 0·074 mm diameter, or passing through sieve 200 ASA. This includes silt and clay particles. The coarser material consists of sand, gravel and stones.

OBSERVATIONS OF POLYGONS

As already mentioned, the polygonal patterns include two very different types of feature. The ice-wedge polygons are not necessarily associated with sorting of material. On the other hand, some polygons are essentially sorted features and the movement of the material that causes the sorting is an essential process in their formation. In dealing with the development of this type of feature, it is essential to study the pattern both on the surface and in depth. The study in depth is not easy in an area of permafrost and A. E. Corte (1963) considers that a bulldozer is an essential piece of equipment in the study of these features.

Corte has described the sorted polygons and other forms of patterned ground in the Thule area of Greenland, where they are formed on outwash. The features vary from very small centres of fines, only a few centimetres across, to large features up to 4 m in diameter. Some of the centres are raised, others are depressed; some are isolated and others are closely grouped. The fine material in the centres of the polygons consists of fines and sand in places, but elsewhere it is composed of small pebbles. The sorted features do not occur in the same areas as the tundra polygons. The tundra polygons tend to disappear or thin to very narrow cracks as they enter an area of sorted polygonal features.

The area chosen for close study by Corte included both types of polygon and also the junction zone between them. It was 80 by 60 m in size and at the northern and southern margins contained tundra polygons. A depression of washed coarser material in the centre contained the sorted features. The ice wedges were apparent as troughs into which some of the large material had been washed from the immediate vicinity. Trenches were bulldozed across the whole area to reveal the structure of the active layer and to study the nature of the ice wedges. In the area of the ice wedges, the active layer showed no disturbance and the horizontal stratification of the original deposits was preserved. Beneath the active layer, the sand and gravel was frozen and solid ice was found only in the ice wedges. When the profile was traced into the sorted area, the stratification was disturbed and amorphous ice masses appeared within the material. The proportion of fines increased and the surface became sorted. Where the surface material was finer, a plug of fine material was found beneath the surface. Within the plug the elongated stones had their long axes vertical, while in the permafrost beneath there were ice masses containing layers of fine particles and odd stones. The ice masses were covered by layers of fines. It appears, therefore, that there is a difference in the character of material in the undisturbed active layer in the area of the ice wedges and in the contorted active layer in the area of sorted polygons.

In order to establish the reason for this difference, soil samples were taken from the disturbed, slightly disturbed and undisturbed parts of the active layer. The samples were taken from the top, middle and bottom of the active layer, while those from the disturbed section came from the plug of finer material. The results of the analysis showed that the disturbed layer consisted of considerably finer material than that in the undisturbed layer. There was only 2 per cent fines in the undisturbed layer and little difference between the top, middle and bottom samples. In the disturbed layer, there was 5 per cent fines at the top, 11 per cent fines in the middle and 14 per cent fines at the bottom. This distribution shows the effect of vertical sorting. An evaporite deposit was found on the lower side of the stones in the undisturbed active layer, but this was missing from the large stones in the disturbed layer.

Corte also studied the character of the ice associated with the polygonal patterns. In the ice-wedge or tundra polygons there were ice wedges below the troughs on the surface. The fabric of the ice revealed thermal contraction cracks along the axial plane of the wedge (Chapter 21). In these cracks, small crystals of ice, 1-3 mm in size, were orientated horizontally. On either side of the more recent crystals were older ones that were orientated vertically and were 2-3 mm long, dipping steeply towards the axial plane of the wedge. The crystals must grow and become reorientated therefore, as the contraction stresses occur.

The ground ice conditions under the sorted polygons were different. There were amorphous masses of ice, containing some contorted layers of fine material beneath the centres of fines. Ice wedges were much smaller, if hey were present at all. The amorphous ice masses consisted of large-grained transparent ice, with crystals up to 10 cm² in size, where there were no silt layers. But where there were dirt or silt layers, the crystals were very much smaller. The C-axes of the crystals tended to lie at 25 to 45 degrees to the vertical. The contact of the mass ice and the wedge ice showed that the wedge ice was the younger of the two types.

Corte describes another type of feature, consisting on the surface of mounds and depressions of low relief, formed in fine grained outwash. This pattern does not show surface sorting, but sorting was well developed vertically. Plugs of finer material occurred, containing at times as much as 28 per cent fines at the bottom and 5 per cent at the top. The active layer showed considerable disturbance and only small remnants of stratification survived between the plugs of fine material. The soil in these intervening bands only contained 2 per cent fines. The ice in these areas consisted almost entirely of ice masses, which were covered by layers of fine material. There appeared to be no systematic relation between the ice features of the active layer and the pattern on the surface, which showed no sorting. This lack of sorting may have been related to the absence of coarser material in this area, which contained no boulders or cobbles.

It appears that the nature of the material plays an important part in the character and development of sorting. Corte indicates three types of material, each of which produces a different pattern. The coarsest material, in which the percentage of fines is less than 2 per cent, produces ice-wedge polygons, and there is little disturbance of the bedding or surface sorting. The intermediate material produces uneven ground with sorted polygons and fine centres, in which ice masses occur below the surface and the bedding is lost. In the finest material, dense ice masses form and patterns of elevation and depressions develop, with relief up to 30 cm. There is little sorting by size. In the last two conditions, ice wedges do not form conspicuously. The sorting of the surface layer takes place most readily when the percentage of fines in the plugs is between 3 and 8 (Fig. 22-1).

J. V. Drew and J. C. F. Tedrow (1962) have classified arctic soils and related the soils to patterned ground features. In Lithosols and Arctic Brown soils, both of which are well drained, the sorted type of polygon occurs. Non-sorted tundra or ice-wedge polygons occur in Arctic Brown and impeded drainage soils, which include Upland Tundra, Meadow Tundra and Bog soils. Tundra polygons are divided into types A to F, in order of increasing relief. Type F forms mounds or ridges, type E have fairly high raised rims, type D have lower raised rims, type C have slightly raised rims, type B have troughs only, and type A have very small troughs only. In Arctic Brown soils only types A and B occur. In Upland Tundra soils, types A, B and F occur and in Meadow Tundra soils, types B, C, D, E, and F occur, while in Bog Tundra soils types C and F occur. The widths of the ice wedges increase in types D to F but are more or less constant in types A to C, although in these types the depth of the ice wedges increase from A to C. Thus the greatest relief of the ridges is associated with the largest ice wedges, which occur in the soils in which drainage is most impeded.

A detailed study has been made by K. Philberth (1963) of the polygonal soils of part of the French Pyrenees. This is a middle latitude area without true permafrost, but one where the smaller, sorted features of the high

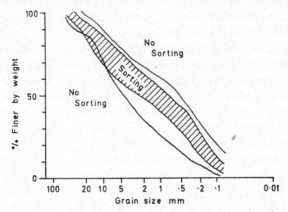

Fig. 22-1 Relationship between size of material and degree of sorting in periglacial deposits (A. E. Corte, *Biul. Peryglac.*, 1963)

mountain environment are well developed. Nearly all high mountains have examples of this type of polygonal pattern. The area studied included mountains with summits over 3000 m, which is the altitude of the snow-line. Most of the polygons were small and showed marked sorting, the edges of the polygons consisting of a mixture of material including many stones. In studying the distribution of the polygons it was found that they do not occur on the granite outcrops. Their absence could be accounted for by the lack of fine material of size less than 0·02 mm: water could, therefore, drain very readily from the soil. On the other hand, sorted polygons were well developed on schistose rocks, which weather to give a fine soil. The polygons developed best on well weathered outcrops of schist in areas where snow melt can add to the soil moisture. They only occurred above 2700 m at elevations where small patches of soil could freeze from early autumn onwards.

Various factors influenced the size of the polygons, including the altitude, the depth of soil, the size of the stones and the stage of development. The size tends to diminish with increasing altitude and where the soil depth is small. Where soil thickness is not the limiting factor there is a tendency for large stones to form large polygons, but this is not always found. Until the polygons are fully formed and exert an influence on their neighbours their size is determined by the length of time they have been forming. At this stage they may be considered as embryonic forms.

Where the coarse border is formed of flat stones, these often stand on edge. The border of stones does not show any sign of soil on the surface, but by digging it was shown that the stones only extend 2 to 4 cm downwards. The lack of fine material resulted from the washing out of the coarse borders in this instance where the stones were small.

R. Miller *et al.* (1954) in their study of the polygons on the Tinto Hills show that the sorting did not extend to a depth greater than 10-15 cm from the surface. Other observers have shown that the stone bands tend to narrow downward. T. T. Paterson (1940) showed that where the soil thickness over bedrock was 25 cm, the coarse stone borders extended throughout the soil thickness. H. W. Ahlmann on the other hand, found polygons in which the stone borders widen downwards and extended to 60 cm. This depth was the extent of the active layer in the area.

Field and Laboratory Experiments on Polygons

Experiments on the movement of stones and the penetration of frost can give useful information with which the various theories of polygon formation may be tested. A.Pissart (1964) has studied small polygons in the French Alps at Chambeyron. Two polygons, 80 cm in diameter, were destroyed in this area. In the first patch, small polygons, about 20 cm in size, reformed, and the outward movement of the stones was measurable. In the second patch, polygons did not reform. On another polygon, six pebbles were placed 15-25 cm from the centre of a 160 cm wide polygon. These six pebbles had moved 7, 7, 10, 15, 16, and 21 cm between 1947 and 1964. They were all found to be at the edge of small polygons about 10-20 cm across that lay within the larger one. The movement of the pebbles shows that small polygons can form under present-day conditions in this area at 2800 m. The small polygons are probably temporary and do not affect the movement of the pebbles in relation to the larger structure.

The formation of polygons such as these is closely related to the annual and diurnal temperature changes and the rate of frost penetration. In arctic regions, the annual freeze-thaw cycle is the most important (Chapter 20), but in high-altitude middle-latitude areas the daily freeze-thaw cycle is of greater significance. On clear days, the range at the surface in the latter environment may be 30-35°C., but it may be only a few degrees at a depth of 15 cm. In the middle latitude areas, the depth of snow is more important than the mean winter temperature in determining the depth of frost penetration. Years of thin snow allow much deeper frost penetration than unusually cold years. P. J. Williams (1961) has suggested that the nearness of the mean annual temperature to 0°C., rather than the intensity of winter cold, determines the depth of frost penetration. The types of patterned ground that depend on deep frost penetration cannot form if the mean annual temperature is greater than 3°C. However, there are other forms of patterned ground that do not depend on this limit, such as stony earth circles.

Experiments were carried out by Philberth (1964) to study the way in which frost penetrates and the hydrostatic pressure caused by freezing. For the experiments, he used a sample of soil that included both fines and stones, and that showed well developed natural polygons. The largest stones were 20 mm in size. Of the material below 2 mm in size, 60 per cent was below 0·2 mm and 30 per cent below 0·02 mm. Only those particles smaller than 2 mm were used for the experiments. The material was well mixed with 30 per cent water before the experiment started. A layer of the material was put in one vessel. In a second vessel, a thin layer of the material was covered with dry stones. Both vessels were subjected to freezing. The soil beneath the dry stones had frozen throughout, but in the first sample the frost had only penetrated 27 mm. At first, the soil froze in a relatively soft form, in which it could respond to external pressure because of its viscosity. However, as the temperature fell still further, the soil froze hard and solid.

In a second experiment with a lower rate of freezing, the frost penetrated only 6 mm into the first sample of uncovered soil, but the soil beneath the dry stones in the second sample had frozen to a depth of 56 mm. Thus moist, fine material freezes much less rapidly than coarse materials, and dry stony materials have a higher conductivity.

Experiments were carried out to measure the hydrostatic pressure of moist soil surrounded by a freezing border. Damp soil was placed in a cylindrical box open at the top, so that the soil could rise upward. An instrument to measure the hydrostatic pressure was placed in the centre of the box. It consisted of a compressible tube containing salted water, the level of the water indicating the pressure in the soil. The water in the tube started to rise after freezing had proceeded for 2 to 3 hours. During the following 9 hours, the water reached half its final height. Horizontal ice needles formed at the surface during the first few hours before the water started to rise in the tube; the rise took place after the surface appeared to have become impermeable. At first the surface was compressible, but later it became rigid. The maximum increase of water level in the tube indicated a change in volume of 2 c.c. in the box. The change in volume could have resulted from either pressure or expansion. Further experiments, using mercury in the tube, showed that at first the force was caused by pressure, but that in later stages, an increase in volume became important. This occurred when the surface ice layer became too strong and solid to give way to the pressure. The rise of the liquid was more rapid in the second phase.

Experiments were also carried out using lead boundaries between one large and two neighbouring small polygons. The experimental polygons were rectangular compartments filled with soil; the small polygons had a base of 70 × 70 mm and the larger 140 × 140 mm. The lead boundaries between the compartments were 2 mm thick, and were free to move except at their ends. The lead was cut into thin strips so that the movement at different depths could be studied. The damp soil filling the polygons was 70 mm thick, which gave a width to depth ratio of 1:1 in the small polygons and 2:1 in the large one, ratios suggested by field measurements. The rate of freezing was such that 80 mm of soil froze in 12 hours, the rate of frost penetration decreasing with time. As freezing took place, the lead bands between the polygons were pushed out towards the larger polygon from the smaller centres at a maximum rate at the bottom, the distance moved after one freezing period being 0·6 mm. At a depth of 60 mm, the movement was 0·2 mm. The values increased to 1·7 mm and 0·8 mm respectively and extended upwards to a depth of 50 mm (where the movement was 0·4 mm) after two freezing periods. After three freezing periods, the bending of the lead strips had penetrated to within 30 mm of the surface. The amplitude of movement was 2·3 mm at the base and 1·4 mm at 25 mm above the base. These values suggest that hydrostatic pressure plays an important part and that the pressure in the smaller freezing polygons is greater than that in the larger. This means that the smaller polygons will grow at the expense of the larger and this accounts for the uniformity in size that is found in any one small area, such as that studied by Philberth.

Corte (1962) has studied the differential movement of particles of different size in experimental conditions. He found that when freeze-thaw cycles operate in a horizontal plane, coarse particles will move up and fines will move down. Horizontal and vertical sorting will always occur together as a result of this differential movement. Vertical sorting is produced by freezing and thawing from the top, and horizontal sorting by freezing and thawing from the side. The degree of sorting is a function of the rate of freezing and gravitational forces. The finer fraction moves away from the cooling front, the fines moving downward when freezing takes place from the surface and laterally in a parabolic path when freezing is from the side. The finer particles migrate away from the coarser, and this causes the soil under the coarser ones

PLATE XXVII   STONE CIRCLES, FOLEY ISLAND, BAFFIN ISLAND.
  (C.A.M.K.)

PLATE XXVIII   PATTERN OF LARGE ICEWEDGES, FOLEY ISLAND, BAFFIN ISLAND.
  (C.A.M.K.)

PLATE XXIX (*above*) MASS MOVEMENT, UNDER FORMER PERIGLACIAL CONDITIONS, OF LARGE BLOCKS OF NIAGARAN DOLOMITE, NEAR PLATTEVILLE, WISCONSIN. (C.E.)

PLATE XXX (*right*) LOWER PART OF THE ARAPAHO ROCK GLACIER.
Rising above the snow in the foreground, in the Front Range, Colorado.
(C.E.)

to sink. The soil into which the finer particles move will heave. Maximum rates of freezing, measured in the field near Thule in Greenland, indicated a frost penetration of 12 mm/hour. The segregation of particles was greater with slow freezing taking place from the top downward.

Corte (1966) has recently discussed the results of detailed laboratory experiments on the effects of thawing and freezing. In summarizing his results, he states that the degree of sorting decreases as the amount of moisture decreases. The movement of particles depends on the amount of water between the ice-water interface and the particle, the rate of freezing, distribution of particles and the nature and position of the freeze-thaw plane. If a mixed soil is saturated, a large number of sizes can be sorted out if the rate of freezing is slow. For example if a coarse heterogeneous soil is frozen it becomes better sorted if the rate of freezing is very slow, but if it is frozen very quickly, no sorting takes place. Fine particles can be sorted under a wide range of rates of freezing. The results of one experiment are shown in Figure 22-2a. Some sorting occurred in the laboratory with a rapid freezing rate of 30-33 mm/hr, a rate which is higher than natural rates. However, lateral sorting under natural conditions should not be limited to a maximum rate of 12 mm/hr, at least for samples similar to those tested in the laboratory.

There are three mechanisms that allow sorting: 1) sorting by uplift or frost-heaving; 2) sorting by migration in front of a freezing plane; and 3) mechanical sorting. The first type occurs when freezing and thawing take place from the top; the second takes place when freezing and thawing proceed either from the top or sides. The third occurs when freezing is from the bottom and thawing from the top, and the process allows mounds to form. When the freezing and thawing occur from the top, the coarse particles move to the top and the fines move downward to the bottom of the freeze-thaw layer. This results in vertical sorting. When freezing and thawing occur from the sides, the particles move away from the cooling front and the coarsest remain on the cooling side. The result of this process is lateral sorting. Mechanical sorting occurs when mounds are formed and when the material freezes from the bottom upward. The presence of vertical sorting allows lateral sorting to take place if the direction of freezing changes. Pockets of fines can be a starting point in the vertical sorting process. Moreover, during vertical sorting, mounds can form and this assists the process of mechanical sorting as indicated in Figure 22-2b. Both these effects are enhanced by lateral migration, which occurs if side freezing takes place. This can occur, for example, from the coarse stone borders that may penetrate downward to form a vertical plane and that possess a high conductivity.

The importance of the nature of the material in determining the ease with which polygons can form has been mentioned by S. Thorarinsson (1964) in

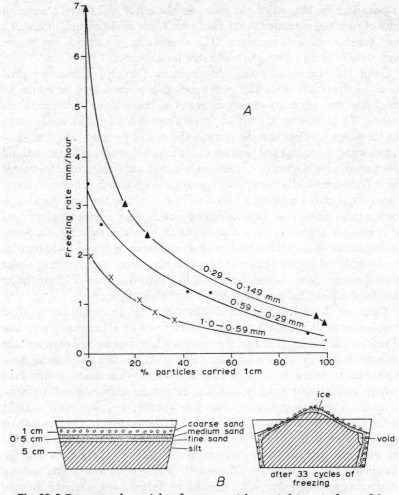

Fig. 22-2 Percentage by weight of quartz particles carried 1 cm in front of the freezing plane under different rates of freezing (A. E. Corte, *Research Report* 85, 1962, U. S. Army C.R.R.E.L.)

describing polygons in Iceland. He has shown that polygons of the ice wedge or tundra type are forming under present conditions in the fine loess soils in the interior of the country. In this type of soil, the polygons are about 20 m in diameter and occur down to levels of about 320 m in the north. Actively growing polygons in this type of soil probably only occur above 420 to 450 m above sea-level. Those formed in coarser outwash soils only occur several hundred metres higher and these are probably fossil. The polygons are about 15 m in diameter, with troughs 30-40 cm wide, which extend down as cracks to 80 cm. Sorted polygons of various dimensions also occur in Iceland.

Fossil Polygons

Although it is more common for fossil ice wedges and involutions to be preserved, fossilized patterned ground features have sometimes been recorded. Those found in parts of Britain, mainly eastern England, provide good examples. They have been studied both on aerial photographs and on the ground by R. B. G. Williams (1964) and by A. S. Watt *et al.* (1966). Their distribution is fairly widespread. Two major types of feature have been identified: first, the fossil ice-wedge polygons, and secondly the *trough* type, consisting of chains of polygons merging into stripes when the slope exceeds 2 degrees. The trough type features make a very large pattern, with a mean polygon diameter of 9·4 m. They show up on the aerial photographs because the soil of the centres differs from that of the margins, leading to a difference of vegetation types. This is especially clear where the patterns are revealed by heather growing amongst grass. This pattern occurs where sandy and chalky patches alternate on Thetford Heath in East Anglia, where this type of trough polygon is well developed.

A trench on Thetford Heath revealed the subsurface pattern. Deep sand underlies the heather, filling ring-like or linear troughs, while the sand is very thin in the centres of polygons or stripes, which consist of very chalky till carrying a grass cover. The sand is probably eolian in origin but the process by which it has reached the troughs is not clear. It is possible that this may have been by small-scale solifluction, moving a uniform sand cover into the troughs. The sand itself is not derived from the surrounding chalky material. The features show no evidence of sorting and can, therefore, be classified as unsorted polygons. They differ from the ice-wedge type in that the troughs are gently rounded and as wide as they are deep. The material beneath the troughs and centres, where it can be seen, is very contorted. The connection between the contortions and the troughs is not clear, however. The disturbed layer extends down to 1·5-1·8 m, indicating the thickness of the active layer during the formation of the features.

The trough polygons and stripes are restricted to the chalk outcrops of eastern England only, and do not occur on the more westerly chalk outcrops. The reason for this distribution is not clear, but it may be connected with the greater continentality of the climate in the east. It is likely that the features date from the last major glaciation, for they are restricted to areas outside its extent, and they appear to have been little modified since.

Black and T. E. Berg (1963) show how patterned ground can be used to record glacial fluctuations in Victoria Land, Antarctica. The patterned ground can be used to date ice advances, for the width of the ice wedges gives an indication of the time since they started to form. In this area, the wedges grow at a rate between 0·3 mm to more than 5 mm each year. All the patterned ground that occurs in this part of Antarctica could have developed during the last 10,000 years. In the older features, the raised rim is double, and this also gives an indication of age.

Processes Forming Polygons

Some of the processes causing polygons to form have already been mentioned in connection with polygon features. The main processes will now be discussed more systematically to provide a summary of present views on polygon formation. It is not possible to list all the hypotheses of polygon formation, for as J. K. Charlesworth (1957) has pointed out, there are almost as many views as there have been observers. All the different types of polygons mentioned cannot have been formed by the same processes. It is, therefore, necessary to divide the features into various categories. The two most important categories are the sorted polygons and the ice-wedge or tundra polygons.

THE FORMATION OF SORTED POLYGONS

Philberth's views (1964) on the origin of sorted polygons have already been described. His theory is based on the effect of freezing pressure from the coarser border. This pressure causes the finer centre to bulge up, and as the pressure increases, the centre becomes smaller. The polygons, therefore, tend to become uniform in size, the smaller growing at the expense of the larger. The coarser border allows the cold to penetrate more readily and the border freezes before the fine plug, which is compressed by hydrostatic pressure. The stony borders are displaced towards the large polygons and the contact between two polygons causes them to become rectilinear. These suggestions are supported by the relationship between the size of the polygon and the depth of the active layer. The first stage in the process is the updoming of any fine centre that may exist in the soil. Permafrost is not required for the process to operate in mountain areas.

Washburn (1956) has assembled many of the earlier theories of polygon formation. He emphasizes the necessity of adopting a polygenetic origin to account for the wide variety of features and even for one type of polygon. Some of these theories will be mentioned briefly.

Expansion caused by freezing was advocated as early as 1910 by B. Högbom. He suggested that the fines attract more water through capillarity. The fines expand more than the coarser parts, which are pushed out by the freezing centre of fines. Fine material is drawn back when the ground thaws and thus the material becomes segregated. Sorted polygons could form where the centres interact with each other.

A. Hamberg (1915) suggested a variation of this theory. He supposed that where stones overlie a fine layer, the fines would be heaved up where they approached the surface most closely. A vacuum would form beneath them into which more fines would be drawn until the fines reached the surface. The stones would then be moved off the upheaved centre to form a marginal circle or polygon.

S. Taber (1943) advocated differential heaving to account for polygons. He considered that differential heaving and gravity could cause sorting in mixed material. Local differential heaving is, however, common where there are no polygons and this process does not account for their even spacing. It has also been suggested that cryostatic pressure, caused by freezing downward towards the permafrost table, could result in heaving. Pockets of unfrozen material would remain and these might be pushed up through the already frozen crust or a coarser surface layer to form centres of fines. Such sorted fine centres are seen on Victoria Island, where they extend downward in an almost cylindrical shape, sometimes resting on stones or gravel at depth.

Weathering has been suggested by some workers as an important process. F. Nansen (1922) stressed the importance of weathering, and suggested that depressions contained more moisture and that this accelerated the weathering, causing a concentration of fines in these areas. Once the fine centres had formed, freezing could push away the larger stones, as the fine centres were thought to freeze first. A similar view has been expressed by C. S. Elton (1927). He emphasized the importance of vertical differences in comminution. This would result in a fine layer underlying a coarser layer. The finer material could then be concentrated by differential upward heaving. Once the polygons had formed, the easier drainage through the coarser borders would make the process self-generating.

Contraction is another process that has been advocated to account for small desiccation polygons of the non-sorted variety. This process may also initiate sorted polygons if the border cracks are filled with stones by flooding. Such polygons are not necessarily associated with a periglacial climate. Contraction at low temperature will be considered later in connection with tundra polygons.

Some other hypotheses that have been put forward will be mentioned very briefly. These include the convection theory of O. Nordenskjold (1909).

He thought that sorted polygons could form as a result of aqueous convection currents, set up in the ground by the difference between the temperature at the surface and at the frost table. Although he later rejected the hypothesis, it was revived by A. R. Low (1925) and K. Gripp (1927). They suggested that convection resulting from the difference of density of water at 4°C. and 0°C. could cause currents. It is not generally thought that these would be capable of moving material.

T. T. Paterson (1940) developed a theory based on a freezing mass that exerted pressure in an outward direction, perpendicular to the centre. This would result in the orientation of rocks parallel to the freezing surface. Rocks would stand vertically around the margin if the freezing centre were at the surface. This theory assumes wrongly that the ice pressure is exerted radially from a centre of fine material.

It will be apparent that many suggestions have been put forward. Most of them are based on some form of pressure generated by frost, caused by expansion of volume on freezing and subsequent reduction of pressure on melting. These changes can produce cryoturbation and sorting of material, where heterogeneous particles are available. Sorted circles, nets or polygons can result. From the experiments that have been described, it seems that the finer particles are moved farther and that these do not freeze so readily as the coarser centres. Thus the freezing centres do not consist of the fines but of the coarser particles. Alternating freeze and thaw does not require a permafrost climate to operate. The rapid temperature changes of high mountains provide suitable conditions in which freeze-thaw processes can operate. This helps to account for the common occurrence of sorted polygons in these areas.

THE FORMATION OF TUNDRA OR ICE-WEDGE POLYGONS

The larger ice-wedge polygons, which do not involve much sorting of material, are formed by contraction caused by extreme cold. This view is now generally accepted. The mechanics of the process have been studied by A. H. Lachenbruch (1962) and described in Chapter 21. It is thought that the polygonal form results from the pattern of the contraction cracks. It is generally found that the diameters of the polygons are about twice or more the depth of the cracks. Lachenbruch has shown that polygonal cracking to form ice-wedge polygons is one example of a phenomenon that takes place in many different ways and in different media. The ice-wedge polygons form as a response to contraction under extreme cold. The thawing of the water that enters the cracks in summer causes troughs to form on the surface. When the tension near the surface is nearly as great as the tensile strength, the stress at a level 3-6 m lower may be compressive. This difference in stress is caused by a time lag in the penetration of the cold wave. The tension crack thus forms at the surface and is propagated downward.

Cracks can form in a variety of patterns. One pattern is the orthogonal one, which occurs in heterogeneous material or plastic media. The cracks do not all form simultaneously and new ones tend to meet old ones at right-angles. The size of the polygon depends on the magnitude of the thermally-induced stress. Some of the orthogonal systems are random, but others are preferentially orientated. The latter normally occur in association with gradually receding water bodies, such as a shifting river channel or receding sea-level. One set of cracks develops normal to the water body and the other at right-angles. This pattern is well exemplified on Foley Island in Foxe Basin. The cracks here are about 35 m apart and the more continuous ones are aligned parallel to the receding shoreline. The cracks are marked by double or triple ridges of limestone pebbles, the total width of the ridges being about 5 m (Plate XXVIII).

The other type of tundra polygons is non-orthogonal and the cracks tend to meet at angles of 120 degrees. These cracks form in homogeneous material, subjected to uniform cooling. The cracks probably propagate laterally until they branch into two at a limiting velocity. The velocity again increases to a limiting value when renewed bifurcation takes place.

THE FORMATION OF THE TROUGH TYPE OF POLYGON

This type of feature, described by Williams (1964) and others in the chalk areas of eastern England, may have a polygonal pattern on flat ground or become elongated on slightly sloping ground. There is an abrupt change of pattern between the upper flat surface and the slope beneath it. This change suggests that the features are not the result of solifluction modifying a poly-gonal pattern on the slope as has been suggested by some workers. It appears that the features cannot be formed by ice wedges as these do not give rise to stripe forms, because even on quite steep slopes, ice wedges can form a polygonal pattern. The origin of the trough type of polygon is still not known.

THE FORMATION OF STRIPES

Stripes are sometimes linked directly with polygonal features, developing from them as the gradient increases downslope. The sorted stripes are par-ticularly likely to form as continuations of sorted polygons, for the same processes can operate, with modifications caused by gravity, on a slope. Like the sorted polygons, stripes are not confined to permafrost areas, and active stripe development is taking place on some of the hills of Britain. T. N. Caine (1963) has described currently developing stripes in the Lake District. He measured an upheaval of a fine stripe amounting to 4 cm after a cold spell, with a freezing period of 3 to 4 days, on Grasmoor at 760 m. Ground ice had formed during this period in thin parallel layers. The differential heave is thought to maintain the stripes. Hollingworth (1934) has also described striped scree on Blencathra in the Lake District. Some of the stripes had developed on old mine tips, and must therefore, be less than 300 years old. They are actively forming at present. This has also been proved by R. Miller *et al.* (1954) by observations on the Tinto Hills. Experiments were carried out in this area by digging over the stripes to a depth of 30 cm. This destroyed the stripes, which had formed on a slope of about 20 degrees. After two years, the stripes were beginning to reform. Differential heaving and better drainage in the coarser elements were thought to cause the sorting of particles and to account for the stripes.

Conclusions

Patterned ground is one of the most conspicuous elements of many peri-glacial landscapes. The patterns can be classified according to their form into circles, nets, and polygons, each of which can be sorted or unsorted in character. These shapes are restricted to fairly flat ground. On steeper slopes of moderate gradient, steps and stripes occur. Polygons are the most wide-spread type of pattern and these can be subdivided into two main types. The smaller type is most characteristic of high mountains in middle latitudes that are subject to freeze-thaw but not necessarily to permafrost. The larger type, called tundra or ice-wedge polygons, only form in areas of active permafrost and low mean annual temperatures of at least $-6°C$.

Many theories have been put forward to explain the first type. Some form of frost action causing cryoturbation and sorting of mixed material is in-volved in their formation, although the exact process is not clearly estab-lished. Observations and experiments have been made to assess the validity of the suggested processes. The nature of the material has been found to play an important part. The tundra polygons are generally agreed to form by contraction cracking caused by intense cold in winter. The cracks fill with ice; later, they may become fossilized by being filled with sediment when the ice melts. The existence of fossil ice wedges and their associated surface polygonal pattern gives valuable evidence on palaeoclimatic conditions. The stripes are mostly formed by similar processes to the smaller, sorted polygons, where these processes act on slopes up to 30 degrees.

REFERENCES

R. F. BLACK (1954), 'Permafrost—a review', *Bull. geol. Soc. Am.* **65**, 839-55

R. F. BLACK (1963), 'Les coins de glace et la gel permanent dans le Nord de l'Alaska', *Annls Géogr.* **72**, 257-71

R. F. BLACK and T. E. BERG (1963), 'Glacier fluctuations recorded by patterned ground, Victoria Land', *Antarct. Geol.* (S.C.A.R. Proc., 1963), **3**, 1, 107-22

T. N. CAINE (1963), 'The origin of sorted stripes in the Lake District, northern England', *Geogr. Annlr* **45**, 172-9

J. CORBEL (1954), 'Les sols polygonaux: observations, expériences, genèse', *Revue Géomorph. dyn.* **5**, 49-68

A. E. CORTE (1962), 'Horizontal sorting. The frost behaviour of soils: laboratory and field data for a new concept', *U.S. Army Cold Reg. Res. Engng Lab., Res. Rep.* **85**, 2, 20 p.

A. E. CORTE (1963), 'Relationship between four ground patterns, structure of the active layer, and type and distribution of ice in the permafrost', *Biul. Peryglac.* **12**, 7-90

A. E. CORTE (1966), 'Particle sorting by repeated freezing and thawing', *Biul. Peryglac.* **15**, 175-240

G. W. DIMBLEBY (1952), 'Pleistocene ice wedges in north-east Yorkshire', *J. Soil Sci.* **3**, 1-19

J. V. DREW and J. C. F. TEDROW (1962), 'Arctic soil classification and patterned ground', *Arctic* **15**, 109-16

M. W. DYBECK (1957), 'An investigation into soil polygons in central Iceland', *J. Glaciol.* **3**, 143-6

C. S. ELTON (1927), 'The nature and origin of soil polygons in Spitsbergen', *Q.J. geol. Soc. Lond.* **83**, 163-94

E. A. FITZPATRICK (1958), 'An introduction to the periglacial geomorphology of Scotland', *Scott. geogr. Mag.* **74**, 28-36

K. GRIPP (1927), 'Beiträge zur Geologie von Spitsbergen', *Abh. Geb. Naturw., Hamburg* **21**, 1-38

A. HAMBERG (1915), 'Zur Kenntnis der Vorgange im Erdböden beim Gefrieren und Auftauen so wie Bemerkungen über die erste Kristallisation des Eises im Wasser', *Geol. För. Stockh. Förh.* **37**, 583-619

B. HÖGBOM (1910), 'Einige Illustrationes zu den geologischen Wirkungen des Frostes auf Spitsbergen', *Bull. geol. Instn Univ. Upsala* **9**, 41-59

B. HÖGBOM (1914), 'Über die geologische Bedeutung des Frostes', *Bull. geol. Instn Univ. Upsala* **12**, 257-389

S. E. HOLLINGWORTH (1934), 'Some solifluction phenomena in the northern part of the Lake District', *Proc. Geol. Ass.* **45**, 167-88

A. H. LACHENBRUCH (1962), 'Mechanics of thermal contraction cracks and ice-wedge polygons in permafrost', *Geol. Soc. Am. Spec. Pap.* **70**, 69 p.

E. K. LEFFINGWELL (1915), 'Ground ice wedges', *J. Geol.* **23**, 635-54

A. R. LOW (1925), 'Instability of viscous fluid motion', *Nature, Lond.* **115**, 299-300

506 PERIGLACIAL GEOMORPHOLOGY

R. MILLER, R. COMMON, and R. W. GALLOWAY (1954), 'Stone stripes and other surface features of Tinto Hill', *Geogr. J.* **120**, 216-19

F. NANSEN (1922), *Spitsbergen* (Leipzig, 3rd Ed.), 327 p.

O. NORDENSKJOLD (1909), *Die Polarwelt* (Leipzig and Berlin), 220 p.

T. T. PATERSON (1940), 'The effects of frost action and solifluction around Baffin Bay and in the Cambridge district', *Q.J. geol. Soc. Lond.* **96**, 99-130

T. L. PÉWÉ (1963), 'Ice wedges in Alaska—classification, distribution and climatic significance', *Geol. Soc. Am. Spec. Pap.* **76**, 129

K. PHILBERTH (1964), 'Recherches sur les sols polygonaux et striés', *Biul. Peryglac.* **13**, 99-198

A. PISSART (1964), 'Vitesse des mouvements du sol au Chambeyron (Basses Alpes)', *Biul. Peryglac.* **14**, 303-10

A. RAPP and S. RUDBERG (1964), 'Studies on periglacial phenomena in Scandinavia', *Biul. Peryglac.* **14**, 75-90

F. W. SHOTTON (1960), 'Large-scale patterned ground in the valley of the Worcestershire Avon', *Geol. Mag.* **97**, 404-8

S. TABER (1930), 'The mechanics of frost heaving', *J. Geol.* **38**, 303-17

S. TABER (1943), 'Perennially frozen ground in Alaska: its origin and history', *Bull. geol. Soc. Am.* **54**, 1433-548

S. THORARINSSON (1964), 'Additional notes on patterned ground in Iceland with a particular reference to ice-wedge polygons', *Biul. Peryglac.* **14**, 327-36

A. L. WASHBURN (1956), 'Classification of patterned ground and review of suggested origins', *Bull. geol. Soc. Am.* **67**, 823-66

A. S. WATT, R. M. S. PERRIN, and R. G. WEST (1966), 'Patterned ground in Breckland: structure and composition', *J. Ecol.* **54**, 239-58

P. J. WILLIAMS (1961), 'Climatic factors controlling frozen ground phenomena', *Geogr. Annlr* **43**, 339-47

R. B. G. WILLIAMS (1964), 'Fossil patterned ground in eastern England', *Biul. Peryglac.* **14**, 337-49

Chapter 23

Periglacial Mass Movements and Slope Deposits

As noon passed, the soil in all the hollows or small watercourses became semifluid and very uncomfortable to walk on or sink into. The entire slope, in consequence of the thaw, had become a fluid moving chute of debris for at least one foot in depth. (SIR EDWARD BELCHER on Buckingham Island, 77°N., quoted by J. Geikie, 1874)

Certain features of the periglacial environment are particularly favourable to mass movement of weathered debris (sometimes loosely described as 'soil' or 'earth', though the existence of a soil profile is not normally implied). Alternations of freezing and thawing will disturb the debris in various ways; meltwater from snow, ice, or ground ice will soak the debris and facilitate its movement; frozen ground at depth prevents downward percolation of moisture in thawed surface layers; and the vegetative cover may be too restricted to prevent mass movements even on relatively gentle slopes. J. G. Andersson's observations (1906) on Bear Island led him to propose the term 'solifluction' for 'the slow flowing from higher to lower ground of masses of waste saturated with water' (p. 95). Andersson recognized, however, that solifluction thus defined was not restricted to areas of any one type of climate; he maintained nevertheless that a sub-arctic climate (Andersson—'subglacial') provided optimum conditions, and that in such regions, solifluction was the chief agent of denudation: 'here, the removal of waste runs on at a rate that may be unsurpassed in other parts of the earth's surface . . .' (p. 112).

Since the literal meaning of 'solifluction' is 'soil flow', it covers a number of different processes of mass movement occurring under a variety of climatic conditions, and also varying rates of movement, from slow downhill creep to relatively rapidly moving mud flows. Opinion is divided as to whether 'solifluction' should be used in such a wide sense. A. Cailleux and J. Tricart (1950) have used the term to include both slow creep and rapid mud slides under climatic conditions varying from periglacial to moist temperate. P. D. Baird and W. V. Lewis (1957) describe as 'solifluction' certain mass movements in the Cairngorms resulting from the heavy rainstorms of the summer of 1956. A. Rapp (1962), however, distinguishes sharply between rapid sporadic movements such as mud flows and debris slides, the slow creep of coarse talus on steep slopes, and the equally slow creep of finer grained material on less steep slopes. Only the latter of these movements he is willing to designate as solifluction. J. Dylik (1951) proposes the term 'congelifluction' to describe earth flow in the presence of permafrost, but excluding frost creep, sheetwash and rapid mud flows, while K. Bryan (1946) had earlier suggested 'congeliturbation' to comprise all mass movements under periglacial conditions, including solifluction.

There is obviously a problem of distinguishing precisely between slow and rapid mass movements if the term solifluction is to be restricted as Rapp advocates, especially as data on rates of movement are still inadequate. As A. L. Washburn points out (1947), there are many cases in which the distinction between mud flow and slow creep is not one of process but the arbitrary one of rate of motion. A. Jahn (1967) supports R. S. Sigafoos and D. M. Hopkins (1952) in their contention that solifluction may take place by simultaneous creep and flow.

In this chapter, the term solifluction will be retained, but discussion limited to its action under cold climatic conditions. The discussion will also be confined to what C. Troll (1944) referred to as 'macro-solifluction', excluding small-scale forms of movement such as occur in stone sorting within polygons and which Troll describes as 'micro-solifluction'.

Processes of Solifluction

Two major groups of solifluction processes have been distinguished: the movements of water-soaked debris akin to mud flows, and movements induced by alternate freezing and thawing of debris resting on a slope (frost creep). In both, gravity is the motive force, and the periglacial environment provides unstable conditions in the slope debris encouraging its downhill movement.

FLOW OF WATER-SOAKED DEBRIS

Permanently or seasonally frozen sub-surface layers prevent downward percolation of moisture. The upper layer affected by seasonal thawing (the 'active layer') becomes soaked with water from melting snow, from melting of any temporary segregations of ice within it, or from rainfall. Washburn (1947) in part of the Canadian Arctic found that the melting of ground ice was the principal source of moisture, since the climate is semi-arid, most rainfall comes in autumn (the active layer is already water-saturated long before this in the spring) and most slopes have lost their snow cover by sublimation or wind action before any melting occurs. S. Taber (1943) also emphasized the importance of ground ice in concentrating excessive amounts of water in the active layer, a process unknown to Andersson. Taber notes how some frozen silts in Alaska contain over 80 per cent of ice by weight and would convert to liquid mud on thawing. In other regions, the contribution of snow-melt seems to be of greater importance. This is borne out by P. J. Williams' study (1957) of solifluction below snow-patches in Rondane, Norway. These appeared to melt substantially from the underside, soaking the mantle of debris beneath and downslope from the snow-patch. The debris in this case was not significantly frost-susceptible (Chapter 21, p. 466) so that melt from ground ice was negligible. However, studying solifluction in frost-susceptible materials in the Dovrefjell area, Williams concluded that even in these, ice-layer melting did not represent a water increase that could by itself have caused flow to occur. It is likely that snow-melt is a more important source of water in some areas, and that in others, ground ice may be a major contributor. Differences of climate and of frost-susceptibility of the materials must both be considered.

The effect of excess water in the active layer is to reduce its shear strength. The general explanation of the process as one of lubrication of soil particles by water is incomplete and misleading: Williams (1959) observes, for instance, that water acting on quartz grains is not a lubricant. The shear strength of a material depends on internal friction and cohesion. The former depends in turn on the pressure at the contacts between individual grains and this varies with the pore-water pressure. The magnitude of cohesion in saturated debris depends on water content and therefore on the ratio of voids to solid. Soaking of debris with water thus reduces both internal friction and cohesion (it will be noted later that frost-heave has a similar effect), and it also increases the weight of the material. Snow patches resting on the active layer may also contribute to instability by their weight (see Chapter 25), increasing the shear stress; on the other hand, when the snow is thickest, the ground may well be frozen up to the surface and easily capable of bearing the load, and at other times of the year, residual frozen ground beneath a surviving snow patch may serve to spread the load.

Vegetation acts as a most important restraining factor. Turf and even moss may be able to hold the semi-fluid layer in place until the load becomes too great. S. R. Capp's descriptions (1919) of the phenomena in Alaska have never been improved on. In an area of permafrost, where seasonal thawing takes place to depths of a metre or so and where melting snow supplies water, movements of debris vary in rate from slow to rapid according to ground slope and the strength of the surface mat of grasses and mosses. At times on steep slopes, the turf ruptures suddenly, liberating rapid mud-flows and leaving scars. The slower and sometimes imperceptible flow movements are more widespread, however, and were considered by Capps to be quantitatively more important.

CREEP INDUCED BY FREEZE-THAW

During freezing of a frost-susceptible material, ice crystals grow normal to the cooling surface and displace particles in this direction. On thaw, the particles re-settle in a direction controlled by gravity. Thus if the cooling surface is inclined, the displaced particles will always re-settle slightly downhill from their original positions. The process has long been known. C. Davison in 1889 successfully reproduced such creep in tilted soil-filled boxes exposed to repeated freezing and thawing. Tricart reported similar but more closely controlled experiments in 1954. The amount of creep can be readily computed: for a 10 degree slope, frost-heaving of 10 cm will give a theoretical maximum downslope movement of 1·76 cm, though exactly vertical consolidation after freezing is unlikely. The amount of movement will also decrease from the surface downward (Williams, 1957) giving a curved velocity profile; it is this which is mainly responsible for the turning of stones and boulders to lie parallel to the surface.

Frost-heave also reduces the shear strength of a material. In the first place, ice layer formation will disrupt frost-susceptible materials, forming discontinuities at the ice surfaces which will be significant planes of weakness when the ice melts. Secondly, frost-heave increases the void ratio and therefore permeability, and in turn reduces cohesion between the particles. These conditions persist during thaw, and allow the material to soak up more water if it is not already saturated.

Williams (1957, 1959) has studied the part played by frost-heave in solifluction in Norway. Below a snow patch, solifluction was measured over periods of several weeks in successive years on a slope averaging 19 degrees. No permafrost was observed. The grain-size composition of the debris (see the continuous line for one sample on Fig. 21-2) indicated a high frost-susceptibility, confirmed by the existence of free ice layers amounting to one-fifth of a 15 cm soil column. In this case, there was strong evidence that frost-heave was the principal process of solifluction. There was no significant change in water content of the debris as between spring and late summer, and little increase in soil weight during the thaw period. Unexpectedly low pore-water pressures were recorded, resulting from increases in permeability consequent on frost-heaving. Rates of downslope movement were small (see p. 511), typical of slow creep caused by frost-heave.

The formation of needle-ice (piprake, mush frost, or efflorescent ice) on slopes gives rise to downhill creep very similar to that caused by frost heave, but it is usually a nightly rather than a seasonal phenomenon. Clusters of ice needles growing at the surface may be capable of lifting pebbles as well as finer debris several centimetres; when the crystals melt or break, the particles drop back to a lower level on the slope. Troll (1944) terms it 'needle-ice solifluction'.

The relative importance in solifluction of 'flow' and 'creep' varies according to local circumstances. Creep will only be significant in frost-susceptible materials. Washburn (1947) considered it not important in comparison with the flow of water-saturated debris. Taber (1943) who excluded frost creep from the term solifluction thought the primary effect of frost-heave was to loosen the soil and thereby render it more susceptible to slope wash and flow. Among the factors that must be considered in assessing the relative importance of creep and flow are the grain-size composition of the material, the availability of water, the depth of frost penetration, and the competence of the vegetative cover.

Rates of Movement

There are various methods of measuring solifluction movements. Painted markings or stakes can record surface displacements (for example, Rapp, 1962). Velocity profiles may be obtained by sinking an iron pipe, filling this with small plastic or wood cylinders one above the other, withdrawing the pipe, and then at some future date carefully exhuming the small cylinders and recording their changed positions (S. Rudberg, 1962). Alternatively, lead cable can be packed into a borehole and exhumed in the same way, or electrical strain gauges can measure the amount of bending of metal strips (Williams, 1957).

Rates of surface movement vary with local conditions as expected. The following are some examples of movements occurring partly or wholly by creep.

A. L. Washburn (1947), Canadian Arctic. Maximum 3·8 cm in one month.

P. J. Williams (1959), Dovrefjell. Slope 19°. Maximum 25 cm in 3 weeks. One movement of 10 cm in 24 hours recorded.

A. Jahn (1960), Spitsbergen. Slope 7–15°. Maximum 12 cm a year.

J. Smith (1960), South Georgia, Slope 21°. Maximum 5 cm a year. Surface pebbles moved up to 70 cm a year.

A. Rapp (1962), Sweden. Slope 15–25°. Maximum 30 cm a year.

S. Rudberg (1962), Sweden. Slope 15°. Maximum 7·5 cm a year (in this case, the bulging front of a solifluction terrace).

J. B. Benedict (1966), Colorado Front Range. Maximum 3 to 4 cm a year.

All observers find that movements are spasmodic (short-term movements must not be extrapolated for longer periods) and are in agreement that rates of movement are less than was previously thought. As Rapp points out, B. Högbom's (1914) suggestion of one or more metres a year for the upper limit of solifluction by *creep* seems likely to be a ten-fold overestimation. On the other hand, local and sporadic *flow* movements can be much more rapid —several metres in a few days under optimum conditions according to A. Jahn (1967, p. 218). Most movement seems to occur in spring.

Sub-surface records of movement show the expected and relatively constant diminution of velocity with depth. Rudberg (1962) gives the following velocity profile for a solifluction tongue at Kärkevagge in Sweden:

Surface	5	cm/year
– 10 cm	3·8	„
– 20 cm	2·8	„
– 30 cm	1·7	„
– 40 cm	0·8	„
– 50 cm	0·3	„

The Climatic Environment

Solifluction may occur in areas of ground frozen permanently, seasonally, or diurnally (Troll, 1944). Permafrost is not essential. The requirements in areas lacking permafrost are fairly deep and rapid frost penetration followed by thawing from the surface downward. In the case of solifluction in areas of seasonally frozen ground, summers must be relatively cool, as otherwise heat stored in the ground from the previous summer will promote thawing from below the level of winter frost penetration; this will be adverse to the preservation of a frozen layer in spring. Solifluction on a large scale is therefore more characteristic of areas not experiencing warm summers and where the mean annual air temperature is not higher than 1°C. (Williams, 1961). It appears to need a more severe climate than certain other periglacial phenomena such as frost hummocks.

At the present day, the lower limits of solifluction in Britain appear to be at about 970 m in the Cairngorms, and possibly as low as 370 m in the Shetlands where windy conditions prevent more than a scanty vegetative cover (R. W. Galloway, 1961). In Central Europe, the lower limit is about 2000 m in the central Alps and 1500 m in the Riesengebirge (J. Büdel, 1944). In the Pleistocene, solifluction was probably ubiquitous in central Europe, for solifluction deposits are widespread and well preserved; Büdel (1937) claimed that solifluction was then active on slopes as low as 2 degrees and that debris had moved as much as 2 km. In Britain, solifluction was active down to sea-level in the last glaciation, even in southern Britain, and in the Late-glacial (Zone III), solifluction deposits are found as low as 230 m on Bodmin Moor (A. P. Conolly et al., 1950).

Solifluction Deposits

Numerous terms have in the past been used to designate such deposits. In 1839, for instance, H. T. de la Beche put forward the term 'head', still familiar to British geologists; in 1887, Clement Reid explained the origin of 'coombe rock' as the natural solifluction deposit of chalk country. Another term used from time to time in the past is 'trail' (H. G. Dines et al., 1940). In 1946, Kirk Bryan proposed the name 'congeliturbate' and advocated the abandonment of local terms, but since congeliturbate as he defined it is the product of all movements resulting from frost heave and the flow of water-saturated debris, it can include other deposits than those affected by down-slope mass movements.

Solifluction deposits are closely related to local rock type since it is rare for material to have travelled farther than 1 or 2 km by the processes described. Thus, unlike glacial drift, erratics are rare (unless the source of the material is drift itself). Angularity of the fragments is characteristic since they have not travelled far, though some reduction in angularity with increasing distance from the source has been occasionally noted (for instance, T. T. Paterson, 1951, p. 11). The thickness of solifluction deposits is extremely variable. Depressions and valley bottoms may collect considerable infills, sometimes several tens of metres thick. The longer axes of stones in the deposit lie in the direction of movement (compare glacial drift) until the point is reached where movement is arrested, as at the edge of a solifluction lobe. Then the blocks are suddenly turned at right-angles to the previous movement (G. Lundqvist, 1949). Reorientation at the border is facilitated by the fact that blocks never reach the front of the moving mass at exactly 90 degrees to it. Obstacles within the moving mass also cause local reorientation of stones. Flow orientation is less persistent and regular in the lower layers of solifluction deposits, unless the latter are derived from glacial drift, when orientations in the lower layers may express the last direction of ice motion.

Contorted structures or 'involutions' are often typical of those solifluction deposits where bedding can be distinguished. Involutions associated with ground-ice formation were considered in Chapter 21. Solifluction can itself produce deformation or drag structures owing to differential lateral movements: K. Bryan (1946) distinguishes these as 'plications'. A. Jahn (1956) shows stages in the development of such forms, the final stage being characterized by roll-like or cylindrical masses (Fig. 23-1).

A number of problems arise in distinguishing solifluction deposits from glacial drift. The problem is particularly intractable if the solifluction deposit

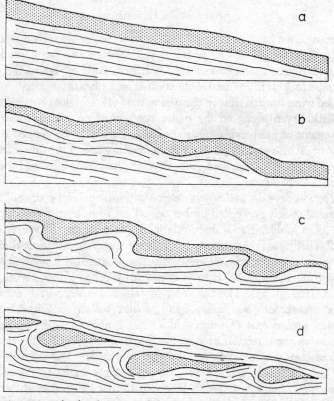

Fig. 23-1 The development of plications in slope deposits by solifluction (A. Jahn, *Biul. Peryglac.*, 1956)

has itself originated from drift. But apart from this, solifluction deposits may sometimes be distinguishable by the downslope orientation of stones, the more angular nature of these stones, the absence of erratics, and the generally looser texture.

Morphological Effects of Solifluction

By stripping off rock waste from higher ground and redepositing it lower down, especially in valley floors, solifluction can slowly reduce the amplitude of relief; at the same time, smoothing of the landscape is effected by burial of minor bedrock irregularities. The concept of landscape levelling under a periglacial climate was termed 'equiplanation' by D. D. Cairnes (1912) though he included processes other than solifluction. Bryan (1946) offered the term 'cryoplanation' to indicate general reduction of the land surface by periglacial weathering and mass movements, and included the work of rivers (often flowing only seasonally, as in parts of east Greenland and Spitsbergen described by H. Poser, 1936) in transporting material delivered to the valley floors by solifluction.

Wherever there are changes in the rates of movement downslope, a number of minor landforms such as steps or banks will result. These have already been briefly described in Chapter 22, since they contribute to the general category of patterned ground. The following is Lundqvist's classification (1949):

	Ground rich in boulders	Few boulders
Rich cover of vegetation	Stone-banked lobes and terraces	Turf-banked lobes and terraces
Little or no vegetation	Stone streams	Earth wrinkles

The difference between lobes and terraces is purely morphological: terrace edges are more linear and may follow or run slightly obliquely to the contour, while lobes have a small extent along the contour in comparison with their downslope dimension. Stone streams will be dealt with in Chapter 24 during consideration of the related forms of blockfields and boulder fields.

STONE-BANKED LOBES AND TERRACES

These comprise soliflual masses confined by crescent-shaped stony embankments (the *Steinguirlanden* of Högbom, 1914, p. 335), and stone-banked terraces. They occur on moderate slopes, usually 10 to 25 degrees. Their size is variable: Jahn (1958) describes examples from the Tatra Mountains ('solifulial tongues') which have frontal banks 1–4 m high; R. P. Sharp (1942) in the St. Elias mountains of Yukon notes stone garlands consisting of stony embankments up to 1 m high and extending up to 8 m downslope (Fig. 23-2); Galloway (1961) quotes stone-banked lobes and terraces from the Scottish Highlands with risers as much as 5 m high and treads extending up to 30 m from front to back (Fig. 23-3). The treads of the terraces and the central parts of the lobes, underlain by relatively fine-grained material, usually slope gently outwards at angles as low as 2 or 3 degrees, with gentle

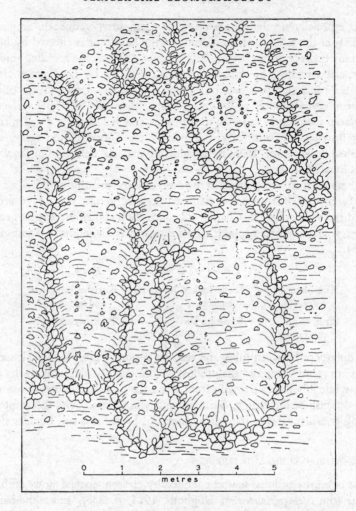

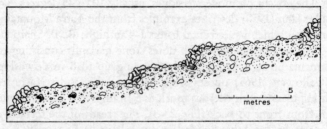

Fig. 23-2 Stone-banked terraces and garlands; plan and section
(R. P. Sharp, *J. Geomorph.*, 1942, by permission of the
author)

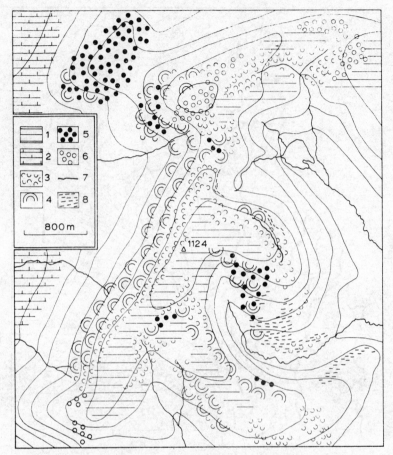

Fig. **23-3** Solifluction phenomena on Ben Wyvis, Scotland (R. W. Galloway,
Scott. geogr. Mag., 1961, Royal Scottish Geographical Society)
1—partly active solifluction sheets on plateau; 2—inactive lower
level solifluction sheets; 3—active turf-banked lobes; 4—inactive
stone-banked lobes; 5—block-fields and stone streams (mostly inactive); 6—ring and stripe patterns; 7—continuous turf-banked
terrace (inactive): 8—discontinuous turf-banked terraces (active)

undulations and sometimes frost hummocks on their surfaces. The formation
of the stony embankments is incompletely understood. Stones may be heaved
to the surface by frost and then move downhill faster than the slow flowage
of the fine materials, according to Sharp, and in support of this, Jahn (1958)
draws attention to the fact that large stones on the soliflual tongue surfaces
often have small turf or earth wrinkles in front of them. Taber (1943) also

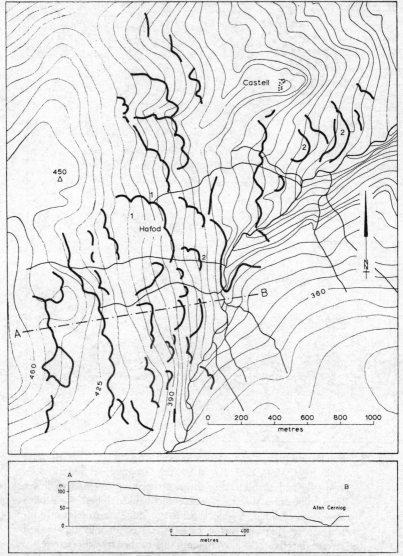

Fig. 23-4 Turf-banked terraces in the Afon Cerniog valley. The profile is located along the line A-B (A. Pissart, *Biul. Peryglac.*, 1963)

thought that large stones move faster on a slope, sometimes by rolling or overturning. The downward movement of the stones is arrested by slight changes in slope, by clumps of vegetation, by bedrock exposures, or by simple accumulation in which several stones catch up with one another and jam to form the nucleus of a growing pile. When the stone bank becomes large enough, little further movement apart from creep will occur. It then grows in height as more material is added. J. B. Benedict (1966) has been able to estimate rates of motion at different periods in the last 2500 years for stone-banked terraces in the Colorado Front Range. The terraces were over-riding humus layers, and radiocarbon dates from different parts of one such layer were obtained. These gave rates of downslope movement of the terrace edge ranging from less than 1·5 mm/year to a maximum of 23 mm/year. Jahn (1967) observes that solifluction lobes and tongues may become stabilized, or their movement retarded, if run-off begins to wash out the fines.

TURF-BANKED LOBES AND TERRACES

In shape and size these differ little from the stone-banked variety, though there seems to be a tendency for them to occur on slightly more gentle slopes

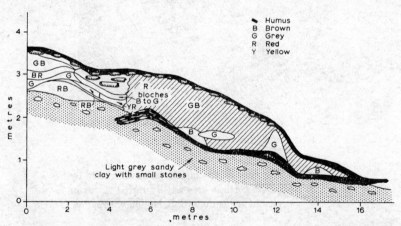

Fig. 23-5 Downslope section of a solifluction terrace on the south slope of Blåho, Trollheimen, Norway (P. J. Williams, *Geogr. J.*, 1957, Royal Geographical Society)

(5 to 20 degrees). A. Pissart (1963) has identified lobate turf-covered terraces in central Wales which he believes have formed by solifluction below trans-verse snow banks persisting on north-east slopes. Preferred orientation of stones in the terraces gives clear evidence of soliflual movement. Terrace

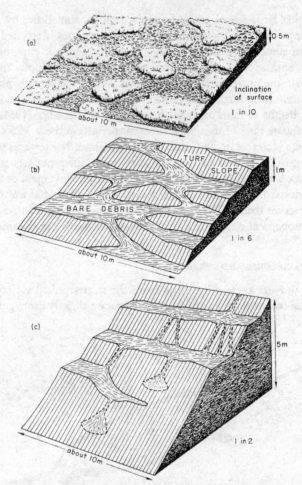

Fig. 23-6 Diagrams illustrating the patterns formed by
moving debris on partially turfed slopes of
various inclinations (S. E. Hollingworth, *Proc.
Geol. Ass.*, 1934)

fronts are from 2 to 15 m high, their treads extend for 50 to 500 m, and they
occur on slopes averaging 2·5 to 8 degrees (Fig. 23-4). Examples described
by Galloway (1961) in the Scottish Highlands are much smaller (risers up to
1·5 m high, treads 2 to 6 m wide). There is clear evidence in some areas that
turf-banked terraces, like the stone-banked variety, are actively in motion.
The turf cover may be ruptured in places, releasing a flow of fine debris
which spreads fan-like on to the next lower terrace. Williams (1957) notes

how the bulging fronts of some turf-banked terraces (Fig. 23-5) on Dovre-fjell, Norway, are overriding vegetation and old humus layers. One such layer was traced beneath an overriding terrace for 13 m, and using rough estimates of the maximum age of the layer, the rate of overriding may have averaged 2 mm a year. In areas where both stone-banked and turf-banked terraces occur (for example, on Niwot Ridge, Colorado: J. B. Benedict, 1966), the former seem to move somewhat faster downslope than the latter, which are overridden by the stones. The materials in the terraces are highly contorted in section, and Troll (1944) shows examples of beautiful recumbent folds in soil horizons of terraces in the Hohe Tauern.

The terraces may develop in areas of either complete or patchy turf cover. Patchiness may be natural, or it may reflect the activities of man or animals, as in the Tatra Mountains (Jahn, 1958). Destruction of the turf enables frost to penetrate more deeply, and facilitates washing of fines by any surface run-off. Terrace surfaces may even support small pools of water at times, encouraging comminution of the sub-surface materials by frost. Patterned ground is commonly observed on them where the vegetative cover is limited.

On the fronts of lobes or terraces, the turf serves to restrain movement as in the case of stony embankments. The variations in its restraining power on different angles of slope are well illustrated by S. E. Hollingworth (1934) (Fig. 23-6).

Periglacial Slope Deposits

It is impossible here even to outline all the various slope deposits and the processes of their accumulation in a periglacial environment. For a complete and precise analysis of the processes, supported by actual measurements over a period of some nine years, reference should be made to Anders Rapp's outstanding study (1960) of the mountain slopes of Kärkevagge, northern Sweden.

In this section, two typical forms of periglacial slope deposit only will be selected: talus, and stratified slope deposits.

TALUS

Talus or scree is familiar in all regions subjected to frost if suitable rock is available. Under a severe periglacial climate, talus accumulations may reach impressive proportions. A good example is in the Devil's Lake region of the so-called Driftless Area of Wisconsin. On the flanks of the depression containing Devil's Lake (see Chapter 12 and Fig. 12-6), talus slopes rise 100 m in places at angles up to 36 degrees, consisting of massive angular blocks of the purple Baraboo quartzite firmly wedged together. Some blocks are 3 m in length; they have clearly fallen from rock ledges above after frost splitting. The talus is now inactive—lichens grow on the rocks, even trees in places; new additions of rock from the outcrops above have virtually ceased, and no injured trees are to be seen. The talus clearly accumulated under much more severe conditions than those of today, and probably most of it at a time when Devil's Lake was formed by late Wisconsinan ice plugging the gap to the south-east. The talus in fact continues at its foot beneath till of Cary age.

In regions of present day periglacial climate, talus accumulation is active. Rapp (1960, 1962) has studied processes of talus movement in Kärkevagge, northern Sweden. On an active talus cone with slopes up to 38 degrees, there is slow downslope creep amounting to 10 cm a year at a maximum, caused by freeze-thaw of interstitial ice, and by settling of the blocks owing to water washing out fines. Creep is active because of the cold humid climate and the rich supply of mobile schistose debris. Fragments also roll and slide down the surface when they fall from above, and small slides occur whenever equilibrium is disturbed. The total displacement of material on talus slopes in the area may amount to as much as 5000 ton/metres per year, though this is small in comparison with other processes of mass movement in Kärkevagge, notably boulder falls, avalanches, and mud slides.

STRATIFIED SLOPE DEPOSITS

Periglacial slope deposits consisting of alternating beds of fine and coarse debris are sometimes known as 'grèzes litées' (Y. Guillien, 1951) or 'éboulis ordonnées' (J. Tricart, 1953, 1956). The fine-grained layers are often silty, while the coarser layers, as well as being thicker, may contain fragments 10 cm or more in length. In Lorraine, grèzes litées attain thicknesses of 40 m. The beds tend to thicken downslope, so that their dip often increases with depth, up to 27 degrees in west-central Wales (E. Watson, 1965) and up to 24 degrees in the southern French Alps (Tricart, 1953). In the Southern Alps of New Zealand, J. M. Soons (1961) has recorded slope and dip angles of 30-35 degrees. Some splitting or joining of beds occurs; cryoturbation structures are rarely seen in the Welsh examples, but are noted, for instance, by T. Czudek (1964) in grèzes litées in Czechoslovakia. J. Dylik (1960) has surveyed the occurrence of grèzes litées in Poland and finds them thinner and containing more fine material than their French equivalents. This difference he relates to the more continental climate of Poland and to the fact that, unlike France, glacial deposits supplied a great deal of fine material.

There is much uncertainty about their origin. The main problem is to account for their stratification. Basically, the deposits as a whole are built out of frost-broken fragments. It is possible that the layers of coarser fragments accumulate under cold conditions (either seasonally or over longer periods) and are spread out over the surface of the scree by sliding on a frozen or snow-covered surface. A. Cailleux (1963) points out that permafrost is not needed—seasonal frost or just snow would suffice, and he notes Jahn's observations in Spitsbergen of coarse gravelly material accumulating on snow slopes. The fine-grained layers may represent either accumulation under different conditions (for instance, slope wash in the presence of snow-melt) or sorting of the material following its descent to the scree. Soons (1961) suggests that needle-ice may play a part in the sorting. Studies have been made of the orientation of stratified screes in the hope that some possible relationships with climatic factors, including the incidence of snow, might be established, but no useful generalizations have yet been established. In Lorraine, grèzes litées occur mostly on east or north-east facing slopes, where most permanent or seasonal snow beds would lie, but in Charente, most grèzes litées face south-east, and of the fourteen sites studied by Watson in west central Wales, seven face south-east, four south-west, and none faces north-east. Further studies are clearly needed on the origin of these deposits.

Rock Glaciers

The term rock glacier was first used by S. R. Capps (1910) in Alaska. In the Kennicott region he investigated thirty or so rock glaciers built of angular talus, extending up to the cirque headwalls, and showing no ice at the surface; but excavations revealed that ice filled the interstices between the blocks at depths within a metre of the surface. In some cases they grade into true glaciers at their upper ends. Microrelief on the surfaces of rock glaciers, especially lobate wrinkles and ridges at their lower ends, often gives a vivid impression of motion, yet it has proved to be exceedingly slow motion, and some rock glaciers are now stagnant, their fronts covered with turf and lichen. Rock glaciers are known from several parts of North and South America and Europe: E. Howe (1909) described their forms in the San Juan mountains (Colorado) but thought, wrongly, that they were landslide debris; A. Chaix (1923) measured their movements in the Engadine (Switzerland); R. L. Ives (1940) in the Front Range of Colorado (Plate XXX), J. E. Kesseli (1941) in the Sierra Nevada, and Mlle A. Faure-Muret (1949) in the French Alps, added to the available observations. Some of the largest recorded by Kesseli attain lengths of 3 km with terminal embankments 60 m high. Component blocks up to 8 m in size are known. There is general agreement that rock glaciers have an upper crust of angular blocks without interstitial debris (though there may be ice) resting on a much thicker lower layer of angular blocks, sand, silt, and possibly mud (again, with or without ice). This differentiation into two layers is caused by downward sifting of fines, and to the greater amount of grinding to which the blocks in the basal layers are subjected.

A recent thorough study of Alaskan rock glaciers by C. Wahrhaftig and A. Cox (1959) makes it unnecessary to consider all the numerous theories of origin of rock glaciers that have been expressed. Their work shows plainly that the motion of rock glaciers is the flow of frozen rubble akin to glacier flow. The first accurate measurements of rock glacier motion were made by Chaix (1923 and 1943), who over a 24-year period showed that, at the terminus of the Val Sassa rock glacier (Upper Engadine), the surface had moved at an average rate of 1-1·5 m/year compared with 0·3-1 m/year at the base of its terminal face. Clearly, the upper layer was moving faster than the lower parts.

Wahrhaftig and Cox measured the surface motion of one Alaskan rock glacier over 8 years (maximum rate 0·76 m/year). In this example, the rock glacier was about 30 m thick and all but the upper 2 or 3 m was perennially frozen. Hence ice must participate in its motion. Since the talus aprons at its front built up to about two-fifths of the height of the front, it was possible to obtain an approximate idea of the vertical distribution of velocity (Fig. 23-7). Such a flow would exist in a material having a strain rate proportional to stress, again suggesting that flow of interstitial ice was taking place. Supporting this was the fact that theoretical calculations of shear stresses in active rock glaciers, assuming a density of 1·8 gm/cm^3, gave figures of 1 to 2 bars. Wahrhaftig and Cox also made approximate calculations of rock glacier viscosity, from surface velocity, flow thickness, slope angle, and assuming that velocity at the base of the flow was zero; for three rock glaciers, viscosity varied between 1·6 and 9·0 poises $\times 10^{14}$. These figures are slightly greater than those normally associated with glacier ice (10^{12} to 10^{14} poises) which is to be expected since rock glaciers contain high proportions of rigid detritus.

The existence of ice in the lower layers of active rock glaciers is therefore almost certain, though obviously difficult to prove by direct observation. Unique in this respect were W. H. Brown's observations (1925) of the internal structure of a rock glacier in Colorado from a mining tunnel dug in from its lower end. This passed first through loose rock, then for some distance through the rock glacier with interstitial ice everywhere, finally passing through a small quantity of glacial ice before entering solid rock.

Microrelief on rock glacier surfaces is analysed comprehensively by Wahrhaftig and Cox. Longitudinal furrows appear to result from wasting of ice-

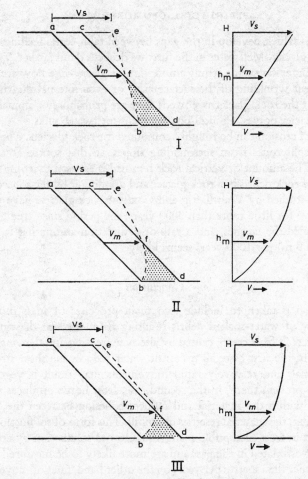

Fig. 23-7 Three possible vertical velocity profiles (on right of diagram) for a rock glacier, as determined from the position of the top of the talus apron at the rock glacier front (C. Wahrhaftig and A. Cox, *Bull. geol. Soc. Am.*, 1959)

ab$\left.\begin{array}{c} \\ \\ \end{array}\right\}$ successive positions of the rock glacier front
cd

V_s surface velocity
V_m velocity of rock debris (at point f, this equals the velocity of the front. Thus the areas cef and bdf are equal in all three cases)
f position of the top of the talus apron

Diagram II is the most common situation (talus apron rising 30–45% of the total height of the front), and diagram III represents the next most common. Conditions of diagram I (h_m at half the total height of the front) are less common. The flow diagram in situation II would characterize a material having a strain rate proportional to stress (*cf.* glacier ice) rich bands which develop in the gaps between talus cones feeding the head of a rock glacier. Meltwater in the furrows deepens them farther. Transverse ridges, conveying such a strong impression of glacier-like flow are thought to represent wrinkling of the surface crust or even internal shearing as the motion of the rock glacier is slowed; they are probably not annual features such as glacier ogives. Conical pits represent meltwater sinks.

Rates of erosion can be roughly computed for rock glaciers. They are fed by frost-split rocks from surrounding slopes, so that source areas can be defined. The amount of vertical rock removal is assessed by computing the volume of rock debris in a rock glacier and dividing it by the source area. In examples studied by Wahrhaftig and Cox, the rock glaciers have probably been active for little more than 3000 years, the period since true glacial ice last occupied the cirques, and a rate of denudation amounting to 3 m (or even 6 to 9 m) per 1000 years seems likely.

Conclusion

Solifluction is taken to include two main processes of mass movement: 1)the flow of water-soaked debris resulting from seasonal thawing of the active layer; 2) frost creep, caused by alternate freezing and thawing of slope deposits. In the first case, an excess of water reduces the shear strength of the material, and relatively rapid movements may result if vegetation is unable to prevent them. In the second case, frost heave produces planes of weakness within the material and reduces cohesion between the particles. On thawing, the material resettles downhill. This form of solifluction is only significant in frost-susceptible materials. Rates of solifluction by creep have been over-estimated in the past and are more likely to be measured in centimetres rather than metres per year; on the other hand, rates of movement by flow, sporadic and localized, may amount to several metres in a few days.

Solifluction deposits show evidence of flow orientation and contorted structures caused by movement. Stone-banked and turf-banked lobes and terraces are common morphological features resulting from differential rates of movement on slopes of 5 to 25 degrees.

Periglacial slope deposits accumulate under a wide variety of conditions, and two forms only are examined in this chapter—talus, and stratified screes. The origin of the stratification in the latter is not clearly understood. A final feature discussed under the general heading of periglacial mass movements is the rock glacier. Rock glaciers are now known to consist of frozen rubble whose flow is akin to that of glaciers. Rates of surface motion are usually about 1-2 m/year, slower than in the case of true glaciers because of the extra rigidity imposed by the high proportion of rock debris. Their surfaces possess a unique microrelief related both to movement and the melting of internal ice.

REFERENCES

J. G. ANDERSSON (1906), 'Solifluction, a component of sub-aerial denudation', *J. Geol.* 14, 91–112

P. D. BAIRD and W. V. LEWIS (1957), 'The Cairngorm floods, 1956: summer solifluction and distributary formation', *Scott. geogr. Mag.* 73, 91–100

J. B. BENEDICT (1966), 'Radiocarbon dates from a stone-baked terrace in the Colorado Rocky Mountains, U.S.A.', *Geogr. Annlr* 48, 24–31

W. H. BROWN (1925), 'A probable fossil glacier', *J. Geol.* 33, 464–6

K. BRYAN (1946), 'Cryopedology—the study of frozen ground and intensive frost action with suggestions on nomenclature', *Am. J. Sci.* 244, 622–42

J. BÜDEL (1937), 'Eiszeitliche und rezente Verwitterung und Abtragung im ehemals nicht vereisten Teil Mitteleuropas', *Petermanns Mitt., Ergänz.* 229, 71 p.

J. BÜDEL (1944), 'Die morphologischen Wirkungen des Eiszeitklimas im gletscher-freien Gebiet', *Geol. Rdsch.* 34, 482–519

A. CAILLEUX (1963), 'Processus supranivaux et grèzes litées', *Biul. Peryglac.* 12, 145

A. CAILLEUX and J. TRICART (1950), 'Un type de solifluction: les coulées boueuses', *Revue Géomorph. dyn.* 1, 4–46

D. D. CAIRNES (1912), 'Differential erosion and equiplanation in portions of Yukon and Alaska', *Bull. geol. Soc. Am.* 23, 333–48

S. R. CAPPS (1910), 'Rock glaciers in Alaska', *J. Geol.* 18, 359–75

S. R. CAPPS (1919), 'The Kantishna region, Alaska', *U.S. geol. Surv. Bull.* 687, 1–116

A. CHAIX (1923), 'Les coulées de blocs du Parc National Suisse d'Engadine', *Le Globe, Genève* 62, 1–38

A. CHAIX (1943), 'Les coulées de blocs du Parc National Suisse—nouvelles mesures et comparison avec les "rock streams" de la Sierra Nevada de Californie', *Le Globe, Genève* 82, 121–8

A. P. CONOLLY, H. GODWIN, and E. M. MEGAW (1950), 'Studies in the Post-glacial history of British vegetation: XI. Late-glacial deposits in Cornwall', *Phil. Trans. R. Soc.* B, 234, 397–469

T. CZUDEK (1964), 'Periglacial slope development in the area of the Bohemian massif in northern Moravia', *Biul. Peryglac.* 14, 169–93

C. DAVISON (1889), 'On the creeping of the soil-cap through the action of frost', *Geol. Mag.* 6, 255–61

H. T. DE LA BECHE (1839), 'Report on the geology of Cornwall, Devon and West Somerset', *Mem. geol. Surv. Gt Br.*

H. G. DINES, S. E. HOLLINGWORTH, W. EDWARDS, S. BUCHAN, and F. B. A. WELCH (1940), 'The mapping of head deposits', *Geol. Mag.* 77, 198–226

J. DYLIK (1951), 'Some periglacial structures in Pleistocene deposits of Middle Poland', *Bull. Soc. Sci. Lettr. Lodz* 3, 2

J. DYLIK (1960), 'Rhythmically stratified slope waste deposits', *Biul. Peryglac.* 8, 31–41

J. DYLIK (1967), 'Solifluxion, congelifluxion and related slope processes', *Geogr. Annlr* **49**, 167-77

A. FAURE-MURET (1949), 'Les "rock streams" ou "pseudo-moraines" du Massif de l'Argentera-Mercantour', *Bull. Soc. géol. Fr.* **5**, 19, 118-20

R. W. GALLOWAY (1961), 'Solifluction in Scotland', *Scott. geogr. Mag.* **77**, 75-87

Y. GUILLIEN (1951), 'Les grèzes litées de Charente', *Revue Géogr. Pyrénées S.-Ouest* **22**, 154-62

B. HÖGBOM (1914), 'Über die geologische Bedeutung des Frostes', *Bull. geol. Instn Univ. Upsala* **12**, 257-389

S. E. HOLLINGWORTH (1934), 'Some solifluction phenomena in the northern part of the Lake District', *Proc. Geol. Ass.* **45**, 167-88

E. HOWE (1909), 'Landslides in the San Juan Mountains, Colorado', *U.S. geol. Surv. Prof. Pap.* **67**

R. L. IVES (1940), 'Rock glaciers in the Colorado Front Range', *Bull. geol. Soc. Am.* **51**, 1271-94

A. JAHN (1956), 'Some periglacial problems in Poland', *Biul. Peryglac.* **4**, 169-94

A. JAHN (1958), 'Periglacial microrelief in the Tatras and on the Babia Góra', *Biul. Peryglac.* **6**, 227-49

A. JAHN (1960), 'Some remarks on the evolution of slopes on Spitsbergen', *Z. Geomorph., Suppl. Band* **1**, 49-58

A. JAHN (1967), 'Some features of mass movement on Spitsbergen slopes', *Geogr. Annlr* **49**, 213-25

J. E. KESSELI (1941), 'Rock streams in the Sierra Nevada, California', *Geogr. Rev.* **31**, 203-27

G. LUNDQVIST (1949), 'The orientation of the block material in certain species of flow earth', *Geogr. Annlr* **31**, 335-47

T. T. PATERSON (1951), 'Physiographic studies in north-west Greenland', *Meddr Grönland* **151**, 4, see pp. 11 and 27

A. PISSART (1963), 'Des replats de cryoturbation au Pays de Galles', *Biul. Peryglac.* **12**, 119-35

H. POSER (1936), 'Talstudien aus Westspitsbergen und Ostgrönland', *Z. Gletscherk.* **24**, 43-98

A. RAPP (1960), 'Recent development of mountain slopes in Kärkevagge and surroundings, northern Scandinavia', *Geogr. Annlr* **42**, 65-200

A. RAPP (1962), 'Kärkevagge: some recordings of mass movements in the northern Scandinavian mountains', *Biul. Peryglac.* **11**, 287-309

C. REID (1887), 'On the origin of the dry chalk valleys and of the coombe rock', *Q.J. geol. Soc. Lond.* **43**, 364-73

S. RUDBERG (1962), 'A report on some field observations concerning periglacial geomorphology and mass movement on slopes in Sweden', *Biul. Peryglac.* **11**, 311-23

R. P. SHARP (1942), 'Soil structures in the St. Elias Range, Yukon Territory', *J. Geomorph.* **5**, 274-301

R. S. SIGAFOOS and D. M. HOPKINS (1952), 'Soil instability on slopes in regions of perennially frozen ground' in 'Frost action in soils: a symposium', *Highway Res. Board* (Washington)

H. T. U. SMITH (1949), 'Periglacial features in the driftless area of southern Wisconsin', *J. Geol.* **57**, 196-215

J. SMITH (1960), 'Cryoturbation data from South Georgia', *Biul. Peryglac.* **8**, 73-79

J. M. SOONS (1962), 'A survey of periglacial features in New Zealand', in *Land and Livelihood: Geographical Essays in honour of George Jobberns* (ed. M. MCCASKILL, N.Z. geogr. Soc., Christchurch), 74-87

S. TABER (1943), 'Perennially frozen ground in Alaska: its origin and history', *Bull. geol. Soc. Am.* **54**, 1433-548

W. F. THOMPSON (1962), 'Preliminary notes on the nature and distribution of rock glaciers relative to true glaciers and other effects of the climate on the ground in North America', *Un. géod. géophys. int., Symposium at Obergurgl, 1962 (Publication No. 58, Int. Ass. scient. Hydrol.)*, 212-19

J. TRICART (1953), 'Les actions périglaciaires du Quaternaire récent dans les Alpes du Sud', *Rep. 4th Conf. int. Ass. quatern. Res. (Rome, 1953)*, 189-97

J. TRICART (1954), 'Premiers résultats d'expériences de solifluxion périglaciaire', *C. r. hebd. Séanc. Acad. Sci., Paris* **238**, 259-61

J. TRICART (1956), *Cartes des phénomènes périglaciaires Quaternaires en France. Mémoires* (Paris)

C. TROLL (1944), 'Strukturböden, Solifluktion, und Frostklimate der Erde', *Geol. Rdsch.* **34**, 545-694 (English translation, *Snow Ice Permafrost Res. Establ.*)

C. TROLL (1947), 'Die Formen der Solifluktion und die periglaziale Bödenabtragung', *Erdkunde* **1**, 162-75

C. WAHRHAFTIG and A. COX (1959), 'Rock glaciers in the Alaska Range', *Bull. geol. Soc. Am.* **70**, 383-436

A. L. WASHBURN (1947), 'Reconnaissance geology of portions of Victoria Island and adjacent regions, Arctic Canada', *Am. geol. Soc. Mem.* **22**

E. WATSON (1965), 'Grèzes litées ou éboulis ordonnés tardiglaciaires dans la région d'Aberystwyth, au centre du Pays de Galles', *Bull. Ass. Géogr. fr.*, **338-9**, 16-25

P. J. WILLIAMS (1957), 'Some investigations into solifluction features in Norway', *Geogr. J.* **123**, 42-58

P. J. WILLIAMS (1959), 'An investigation into processes occurring in solifluction', *Am. J. Sci.* **257**, 481-90

P. J. WILLIAMS (1961), 'Climatic factors controlling the distribution of certain frozen ground phenomena', *Geogr. Annlr* **43**, 339-47

Chapter 24

Altiplanation, Tors, Blockfields, and Stone Streams

At the higher elevations in middle and high latitudes, a range of features reflects the very considerable activity of periglacial processes, particularly frost action, solifluction, and nivation. The formation and even the nature of some of these features are far from being clearly understood, and in some cases a periglacial origin is contested.

Altiplanation

This term was introduced by H. M. Eakin (1916) to signify the processes collectively responsible for producing flattened summits and bench-like features occurring irregularly on spurs and hillsides in part of Alaska, and bounded by scarps built of bare angular talus lying steeply at its angle of repose and including blocks up to 4 m in size. Summit flats and hillside benches admit of many explanations entirely unconnected with periglacial conditions; Eakin was the first to draw attention to the possibility that in some areas, these features might be produced simultaneously at a variety of altitudes irrespective of such considerations as base-level, and that their levels might have no significance in the denudation chronology of a region.

Eakin's account of the formation of altiplanation terraces is confused but it is evident that he considered the terraces mainly as constructional features, built out of rock waste derived by frost from the local bedrock; movement of the rock waste occurred by upward frost heaving and slow transfer to the outer edge of a terrace. On the inner parts of a terrace, comminution of debris by frost was active since here conditions are likely to be wettest, so that gradually all the interstices of the fractured rock debris are filled with fines to give an even surface appearance.

Similar terrace features were described a few years previously by D. D. Cairnes (1912), though he did not use the term altiplanation. He noted that the fronts of some terraces consisted of solid rock but that behind the scarp, the gently sloping terrace surface was one of accumulated debris.

In Russia, altiplanation or 'goletz' terraces have been described from such areas as the Northern Urals. A useful review of the early work is given by G. Jorré (1933), who summarizes the main features of the terraces thus: downslope extent up to several hundred metres, vertical spacing usually 5 to 7 m, scarp slope of 20 to 50 degrees, a gentle terrace surface sloping outward to the scarp edge, a marked lack of relationship with bedrock structure, and a debris cover over bedrock usually less than 3 m thick. Above the highest platforms rise 'stone cities' or huge tor-like rocky outcrops. The terraces do not occur below the lowest Pleistocene snow-line. The major part of their formation is attributed to frost, breaking up the bedrock and comminuting debris to silt-size particles, while snow-melt run-off is held to be responsible for transport.

The first to recognize altiplanation terraces in Britain was A. Guilcher (1950); he described them on Holdstone Down, Great Hangman, and Trentishoe Down in north Devon. Significantly, rock exposures showed that at least one such bench was cut in bedrock; in other cases, a thin debris cover was likely. Nivation may have played a part, particularly in the initiation of the terraces; patterned ground developed on their surfaces, and break-down of material by frost and chemical weathering facilitated its downslope movement. Guilcher's work was extended by Te Punga (1956), and it was confirmed that the terraces were essentially cut in bedrock, with a waste mantle no more than a metre or so thick. Te Punga suggested that the terraces represented the work initially of transverse snow patches (Chapter 25); sub-sequently they were enlarged by frost action and meltwater removed some debris, so that the treads were slowly lowered and the bounding scarps retreated.

A more recent and systematic study of altiplanation terraces in the Dart-moor area of south-west England has been undertaken by R. S. Waters (1962). The following are the main features of the terrace morphology:

Height of scarps	= 2 to 12 m
Dimensions parallel to regional slope	= 10 to 90 m
Dimensions transverse to regional slope	= up to 800 m
Slope of treads	= 3 to 8 degrees
Slope of scarps	= 15 to 22 degrees, with occa-sional vertical rocky outcrops

When traced laterally, few of the terraces maintain a horizontal attitude. Although developed generally on the Dartmoor metamorphic aureole (and mainly in the height range 300–450 m) the benches are in fact cut in a diversity of rock types, but all types have one property in common, namely that they are well jointed. There are also some definite structural relationships: for instance, the terraces are horizontal where they lie along the strike. But their

varied spacing, irregular distribution, the lateral slope of some, and their
association with evidence of periglacial action make it extremely unlikely
that they are the work of 'normal' subaerial, fluvial, or marine action: a
periglacial interpretation raises fewer problems and allows them to be
grouped with morphologically similar features of present-day Arctic regions,
such as Waters describes in Vestspitsbergen.

According to Waters, altiplanation terraces originate as slight depressions
in the land surface, possibly joint-controlled, which become the sites for

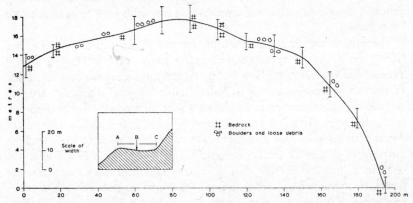

Fig. 24-1 Some features of an altiplanation terrace by Butterfly Lake, west Baffin
Island. The profile follows the median line of the terrace (point B on the
inset.) The width of the terrace is indicated by short vertical lines, representing the distance A-C on the inset (A=outer edge of terrace, C=inner
edge). Symbols above the profile refer to characteristics of the terrace surface
in the inner B-C zone; below the profile they refer to the outer A-B zone.
The altitude of the terrace varies lengthwise through 17 m and may be
partly structurally controlled. The tread of the terrace slopes inward mostly,
as indicated on the inset diagram

collection of moisture and snow, and thus, under a cold climate, the sites of
most intensive frost action. On sloping ground, frost-broken material can be
removed from the downslope edges of any roughly transverse depressions
by solifluction, once sufficient fine material has accumulated by frost comminution. Solifluction can carry away the coarser fragments and allow frost-
sapping to proceed further; the soliflual debris streams down from one terrace
to the next lower one only at selected points in the intervening scarp, so that
the latter is not buried by debris. However, on the lower parts of a hillslope,
the rate of accumulation of solifluction debris will be greater and bedrock
will slowly be covered over completely. The upslope edge of initial trans-

verse depressions is worn back by frost-sapping; at its foot, snowbanks will assist this process by supplying moisture, they will provide meltwater to wash away fines, and will encourage solifluction across the tread below them. Altiplanation terraces are thus envisaged primarily as bedrock-eroded features which at any given time will bear small and varying thicknesses of soliflual debris moving across them. Eakin was probably misled by this debris into regarding the terraces as constructional; the weight of evidence does not now favour this view. Confirmation of Waters' observations and hypothesis is provided by other recent studies, for instance those by J. Demek (1964) and T. Czudek (1964), both of whom recognize altiplanation terraces on slopes in the Bohemian uplands of Czechoslovakia. There are obviously close links between altiplanation terraces and nivation hollows (Chapter 25), and further studies may well suggest that the distinction between these two groups of features should be discontinued.

Figure 24-1 shows some features of a supposed altiplanation terrace in western Baffin Land. The terrace runs along a hillside by Butterfly Lake for about 200 m, its width varying between 7 and 15 m, and its height, measured along a median line, varying from 0 to 18 m above an arbitrary datum. Between points 1 and 12, the terrace surface has a reversed slope (that is, opposite to the general slope of the hillside), its inner edge being up to 1 m lower than its outer edge. The position of the terrace in this case appears to be structurally controlled. Nivation processes were thought to be the most important in the development of this terrace.

Tors

It should be clearly stated at the outset that there are many features which have been or can be described as tors whose formation has no connection with periglacial conditions or processes. Tor was first proposed as a scientific term by D. L. Linton (1955), and geomorphologically is used to designate a residual mass of upward projecting bedrock exposed by differential weathering and removal of weathered debris, though there is considerable variety in their form. They are most commonly associated with intrusive igneous, metamorphic, and some sedimentary rocks such as massively jointed sandstone. Some mark resistant outcrops, others do not; they may occur in groups or singly: they are found on hill summits and on hill slopes, but rarely on valley floors. The joint-blocks of the tor may be loosely piled resembling 'the squared stones of cyclopean masonry' (R. H. Worth, 1930) or they may fit so closely that the blocks cannot have suffered any movement relative to one another. The blocks may be both rounded and angular. The tors frequently rise abruptly from a surrounding relatively smooth surface, which may be of bedrock (such as Linton's (1955) 'basal platforms') or may represent a surface of accumulation: Plate 8 in Eakin (1916) is a fine example of a granite tor surrounded by a smooth solifluction surface.

It is not possible here to review all the various opinions expressed on the origin of tors, for the term has been used to describe a variety of features developing under conditions ranging from tropical to arctic. Attention will be confined to tors in areas which now experience, or have in the Pleistocene experienced, a severe periglacial climate. But even in these areas there is still controversy over whether the tors may be distinctive periglacial landforms, or whether they may be landforms relict from previous periods of totally different climate and later modified by periglacial conditions. Some workers admit both possibilities in the same general area. An example is J. Demek's work (1964) on tors in the Bohemian upland of Czechoslovakia. Two possible models of their evolution are put forward. The first envisages two stages in their formation: (1) a period of deep weathering under warm humid conditions (possibly in the Tertiary) in which a thick regolith developed to 110 m in places. In this regolith, core-stones developed where joint planes were more widely spaced; (2) a period or periods of regolith removal under periglacial conditions in the Pleistocene, when the tors were first exhumed and subjected to frost attack and extensive mass wasting. This hypothesis is applied by Demek to many tors in western Bohemia.

Other tors are considered by Demek to have developed in one stage by rock weathering and simultaneous removal of debris in a periglacial environment. Some hillslopes consist of a series of altiplanation terraces whose edges are marked by tors and frost-riven cliffs; at the foot of these, angular talus has accumulated in places and sometimes gives rise to stone streams. The cliffs retreat by frost attack, intersecting one another to produce tors and other related forms. This scheme of tor evolution is also put forward by T. Czudek (1964) in connection with the Hrubý Jesenik Mountains of Northern Moravia (Fig. 24-2). The tors here, which attain heights of 20 m and sometimes possess overhanging faces, are thought to have formed solely under periglacial conditions, there being no evidence of exhumation of Tertiary weathering forms.

The origin of the tors in south-west England and in the gritstone country of the Pennines has been the subject of prolonged controversy. Linton (1955, 1964) has strongly maintained a non-periglacial origin, invoking deep weathering under warm climatic conditions and subsequent exhumation of tor forms, as in Demek's two-stage hypothesis. Periglacial solifluction has been mainly responsible for removing and redistributing loose debris already in existence and in displacing some of the core-stones. Solifluction and frost action are held not to have assisted in producing tors but in destroying them. On the other hand, J. Palmer and J. Radley (1961), dis-

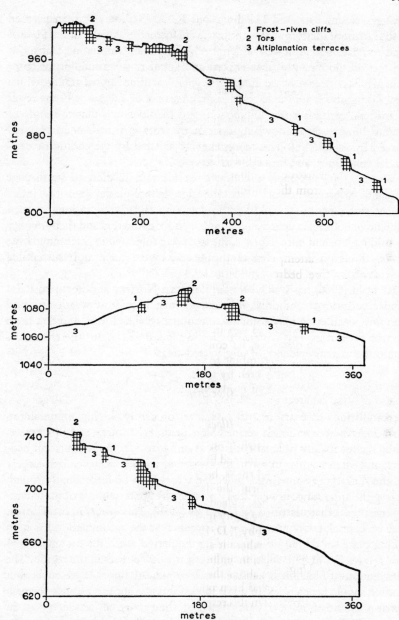

Fig. 24-2 Slope profiles in the Hrubý Jesenik mountains, Czechoslovakia
(T. Czudek, *Biul. Peryglac.*, 1964)

cussing Pennine tors, and J. Palmer and R. A. Neilson (1962), concerned with Dartmoor tors, argue that the tors were formed essentially under Pleistocene periglacial conditions. Palmer and Neilson maintain that on Dartmoor the tors are dominantly frost-broken and angular: any rounding of component blocks is said to be secondary. The *growan* or disintegrated granite, which they distinguish from the kaolinized granite, may be substantially the result of frost pulverization and solifluction, and they contend that the *clitter* of granite blocks spread downslope from the tors is frost-broken material moved by solifluction. The processes were arrested by the ending of periglacial conditions, and thus the tors survive.

Waters (1965) attempts a compromise view in relation to south-west England. Selective bedrock decomposition in pre-glacial (possibly interglacial) warm periods is accepted; the effect of periglacial conditions has been to remove both finer debris and large blocks from the tors and their vicinity by solifluction, and once exposed, the tors were subjected to destructive frost action, though it is very clear that some rocks were much more susceptible to frost action than others.

R. Dahl (1966) has studied tors in northern Norway and northern Italy. Under periglacial conditions, he contends that both frost weathering and chemical weathering may combine to produce tor-like forms, and that these may be present in areas covered by ice in the last glaciation; therefore, tors cannot be used as evidence of unglaciated areas.

Blockfields

Accumulations of coarse detritus on level or gently sloping mountaintop surfaces have been variously termed blockfields, Felsenmeer, and Blockmeer. They consist mainly of local rocks, broken by frost from the subjacent bedrock and often angular in form, the degree of angularity depending largely on the rock-type. Sometimes, indeed, the rocks can be misleadingly round, giving the appearance of very stony moraine. A good example of mountaintop detritus is that described by J. D. Ives (1958). This occurs in the Torngat area of Labrador-Ungava, where it appears that the accumulation has been taking place for a long time, including the period when the mountain tops stood as a series of nunataks above the last major ice-sheet in the region. The distribution of blockfields has been used to indicate limits of ice-cover, but caution should be exercised in this respect, for it also appears possible for such detritus to remain relatively unaffected by the passage of ice-sheets over it. Orientation studies provide a possible method of differentiating between block fields modified by moving ice and those that have not been ice covered. The length of time required for blockfields to form varies considerably with rock-type, and also with climate, especially the number of significant oscillations through freezing point (frost cycles). R. Dahl (1966) finds evidence that some blockfields in Nordland, Norway, have developed in the postglacial, and therefore they cannot be used to indicate possible ice-free areas in the last glaciation.

Stone Streams and Boulder Fields

Stone streams, stone runs, or block cascades are well-known features of some Arctic and Antarctic regions, as are their relict Pleistocene equivalents. In 1846, Darwin gave an account of the 'stone river' at the head of Berkeley Sound in the Falkland Islands, attributing it to some catastrophic earthquake. C. W. Thomson (1877) provided a more realistic explanation, suggesting slow downhill creep of the blocks as water washed out fine debris, and by alternate soaking and drying of the ground beneath the blocks. J. Geikie (1894) compared the stone streams of the Falklands to the 'rubble drift' of southern England formed under conditions of alternate freezing and thawing.

Stone streams consist of accumulations of rock debris usually following valley floors, but sometimes occurring in other topographical locations where the name boulder field is to be preferred. They may begin from blockfields occupying summit areas, extending downslope as narrowing tongues, or from the foot of large talus accumulations. Their margins are sometimes ill-defined, for the surrounding slopes may also possess many boulders, and the distinction then lies solely in the presence of soil and vegetation on the slopes and the almost complete lack of vegetation on the stone streams themselves.

All workers now regard them as of periglacial origin, though the mechanism of their formation and possible movement is by no means clear. One hypothesis that has commanded much support is that of J. G. Andersson (1906), who suggested that the blocks in the stone streams of the Falkland Islands had been borne along down-valley by flowing mud and fine debris at a time of severely cold climate. The Falkland Islands were never ice-covered, but lay at times in the Pleistocene near the margin of the Antarctic ice. Frost action produced the debris, solifluction moved it, and later, running water washed out the fine materials from the upper layers but was itself unable to move the blocks. J. Büdel (1937) adopted a similar view for stone streams in the German Mittelgebirge, and also H. T. U. Smith (1949, 1953) in Wisconsin.

R. P. Sharp (1942) in the Yukon was able to observe mudflows in the thaw season carrying boulders up to 1·5 m in diameter, for mud has a much higher specific gravity than water. On the other hand, A. L. Washburn

(1947) thought the stone streams he examined in northern Canada were not related to the viscous flow of an underlying fine-grained layer but to any factors (such as frost heave or slope wash) which would upset the equilibrium of blocks resting on a slope.

J. R. F. Joyce (1950) has recently re-examined the stone runs of the Falkland Islands, and finds Andersson's solifluction hypothesis untenable. The component blocks are sometimes of enormous size and would have jammed immovably on gentle slopes—some stone rivers here stretch for 4 km on grades of a few percent only. The underlying fine-grained layer demanded by Andersson is not to be found, nor is it at all evident where it could have come from in sufficient volume to convey the vast bulk of huge stones. Furthermore, some of the stone runs lie not in valleys but are really boulder fields across low rounded plateaux. A significant fact is that the stone runs consist of quartzite boulders derived solely from one highly jointed unit of the Devono-Carboniferous series. Joyce argues that the boulders have not in fact travelled any distance but were formed largely *in situ* in the Pleistocene on an outcrop underlain by shale and very susceptible to frost weathering. As the bounding scarps of the quartzite beds retreated under frost attack, so a litter of angular blocks accumulated to carpet the area over which the scarps retreated. Vegetation has since been unable to gain a foothold because of the size of the interstices and thus they remain as barren stone runs. In this case, *boulder field* is a more appropriate term than stone stream, for the latter conveys too strong an impression of extensive downslope movement, which Joyce denies.

A recent study of boulder fields in the Åland Islands off south-west Finland by C. A. M. King and R. A. Hirst (1964) shows that not all such features are of periglacial origin and that care must be taken to distinguish different types; it also throws further light on the origin of periglacial boulder fields and stone runs. In the Åland Islands, two types of boulder field occur in close proximity, one of rounded boulders and the other of angular boulders.

The angular boulder field (Fig. 24-3a) is thought to be of periglacial origin; it is 110 m long and varies in width from 55 m to 21 m, with a mean gradient of 1 in 29. The margins of the boulder field provide useful evidence on the method of its formation. The loose boulders, near the upper edge of the boulder field, can be traced into a zone of greatly weathered and cracked slabs *in situ*. These slabs in turn merge into solid bedrock; this consists of a type of Rapakivi granite, which breaks up very readily along vertical joints and horizontal planes of weakness. At the lower end of the boulder field, the angular boulders of local granite overlie smaller and rounder stones, which include some erratics. This suggests that the angular stones have moved downslope to cover the older deposits.

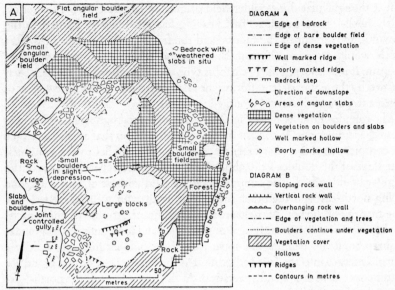

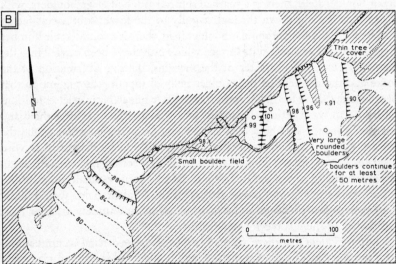

Fig. 24-3 A Map of an angular boulder field, Åland Islands, Finland
B Map of a rounded boulder field, Åland Islands, Finland
(C. A. M. King and R. A. Hirst, *Fennia*, 1964, by permission of the
Johnson Reprint Corporation)

It is thought that this boulder field originated largely by frost shattering of the well-jointed granite. Frost heaving has subsequently assisted in the downhill movement of the blocks, and the fact that the boulder field faces south would facilitate alternate thawing and freezing during spring and autumn. It is considered likely that this type of angular boulder field has been formed by the action of frost during the relatively short period since 4000 B.C. when this part of the island emerged from the sea.

The rounded boulder field (Fig. 24-3b) shows several significant differences from the angular one. It consists of well-rounded stones that are rather larger on conspicuous ridges that extend across the field parallel to the contours. The slope of the boulder field, about 1 in 10, is much steeper than that of the angular boulder field. The rounded boulder field consists of two parts, one facing north-east and the other south-west, both with well-marked ridges. The margins of the boulder field are again significant. Around its upper parts, the boulders rest discordantly against smoothed and solid outcrops of granite. A sharp boundary delimits the upper part of the boulder field, but its lower margin merges into the forested ground as the bare rocks become gradually better covered with soil and vegetation. At the upper limit of the north-east-facing boulder field, there is a marked summit ridge of large boulders, with a backslope leading down through a gully to the lower south-west-facing boulder field. Unlike the angular boulder field, there is a considerable number of well rounded erratics on its surface which must have been carried into the vicinity by ice. The character of the margins, the size distribution of the stones and the gradient of the boulder field all support the suggestion that these rounded boulder fields were formed not under periglacial conditions but by storm wave action. This must have taken place soon after the island emerged from the sea about 6000 B.C., owing to isostatic recovery, and the fact that the boulder field lies at about 100 m in altitude is in agreement with this date. At this time, the fetch would have been considerable in a direction perpendicular to the boulder ridges, which occupy small indentations in the granite relief. The coarse nature of the boulder fields would allow rapid percolation of water and hence the minimum amount of later frost disturbance of these wave-formed features.

It is thus evident that there are many types of coarse detrital accumulations, from summit blockfields to stone runs and boulder fields. Only further field studies can serve to differentiate them more satisfactorily, and the terminology has suffered from confusion in the past (for instance, Kesseli's use of the term 'rock stream' to describe rock glaciers in the Sierra Nevada: see p. 523). Frost action is the fundamental process of derivation of the blocky debris: frost-heave, solifluction, and any other processes that assist gravity by disturbing the equilibrium of large blocks on sloping surfaces, have been responsible for whatever movement of the boulders has taken place. It seems clear that the earlier workers had exaggerated notions of the extent of movement on the more gentle slopes, and solifluction should not be too readily invoked in the absence of any fine-grained material lying beneath the boulders. Another method by which some blockfields and boulder fields may form is by the decay of rock glaciers. Melting of interstitial ice will leave behind spreads of rocky detritus, incapable of further motion.

Conclusion

A number of features related to the work of frost-action, nivation, and solifluction characterizes areas of past and present periglacial activity. Altiplanation terraces represent bedrock-eroded features, often structurally controlled, in whose formation frost action and nivation play a major part. They occur at irregular vertical intervals and commonly vary in altitude as they are traced along a hillslope. Snow-melt and solifluction are important in the removal of debris from them. Their edges are marked by frost-riven cliffs and tors in some cases. Wider studies of their occurrence and formation may necessitate the abandonment of any distinction between altiplanation terraces and transverse nivation hollows. The latter are described in Chapter 25.

The origin of tors is still highly controversial. It is clear, however, that by no means all tors are of periglacial origin, and some workers deny any strong connection between tors and periglacial processes. On the other hand, there is evidence that tors in parts of Britain and Czechoslovakia, for instance, have been extensively modified, and possibly even formed, under Pleistocene periglacial conditions.

Accumulations of coarse debris in the form of blockfields, boulder fields, and stone streams are in the majority of cases related to cold climatic conditions, though rounded boulder fields of marine origin must be carefully differentiated. The angular boulders of the periglacial accumulations are initially derived by frost action, and on sufficiently sloping surfaces may thereafter be slowly displaced by frost-heave, solifluction, or the washing-out of any fine interstitial or subjacent debris which would disturb the equilibrium of the boulders. Large-scale solifluction as envisaged by Andersson, however, is unlikely.

REFERENCES

J. G. ANDERSSON (1906), 'Solifluction, a component of sub-aerial denudation', *J. Geol.* **14**, 91-112

J. BÜDEL (1937), 'Eiszeitliche und rezente Verwitterung und Abtragung im ehemals nicht vereisten Teil Mitteleuropas', *Petermanns Mitt., Ergänz.* **229**, 71 p.

D. D. CAIRNES (1912), 'Differential erosion and equiplanation in portions of Yukon and Alaska', *Bull. geol. Soc. Am.* **23**, 333-48

T. CZUDEK (1964), 'Periglacial slope development in the area of the Bohemian massif in northern Moravia', *Biul. Peryglac.* **14**, 169-93

R. DAHL (1966), 'Block fields, weathering pits and tor-like forms in the Narvik Mountains, Nordland, Norway', *Geogr. Annlr* **48**, 55-85

C. DARWIN (1846), *Geological observations on South America* (London), **7**, 279

J. DEMEK (1964), 'Castle koppies and tors in the Bohemian highland (Czechoslovakia)', *Biul. Peryglac.* **14**, 195-216

H. M. EAKIN (1916), 'The Yukon-Koyukuk region, Alaska', *U.S. geol. Surv. Bull.* **631**, 1-88

J. GEIKIE (1894), *The Great Ice Age* (London, 3rd Ed.)

A. GUILCHER (1950), 'Nivation, cryoplanation et solifluction quaternaires dans les collines de Bretagne occidentale et du Nord de Devonshire', *Revue Géomorph. dyn.* **1**, 53-78

J. D. IVES (1958), 'Mountain-top detritus and the extent of the last glaciation in north-east Labrador-Ungava', *Can. Geogr.* **12**, 25-31

G. JORRÉ (1933), 'Le problème des "terrasses goletz" sibériennes', *Revue Géogr. alp.* **21**, 347-71

J. R. F. JOYCE (1950), 'Stone runs of the Falkland Islands', *Geol. Mag.* **87**, 105-15

C. A. M. KING and R. A. HIRST (1964), 'The boulder fields of the Åland Islands', *Fennia* **89**, 1-41

D. L. LINTON (1955), 'The problem of tors', *Geogr. J.* **121**, 478-87

D. L. LINTON (1964), 'The origin of the Pennine tors—an essay in analysis', *Z. Geomorph.* NF **8**, 5-24

J. PALMER and J. RADLEY (1961), 'Gritstone tors of the English Pennines', *Z. Geomorph.* NF **5**, 37-52

J. PALMER and R. A. NEILSON (1962), 'The origin of granite tors on Dartmoor', *Proc. Yorks. geol. Soc.* **33**, 315-40.

R. P. SHARP (1942), 'Mudflow levees', *J. Geomorph.* **5**, 222-7

H. T. U. SMITH (1949), 'Periglacial features in the driftless area of southern Wisconsin', *J. Geol.* **57**, 196-215

H. T. U. SMITH (1953), 'The Hickory Run boulder field, Carson County, Pennsylvania', *Am. J. Sci.* **251**, 625-42

M. T. TE PUNGA (1956), 'Altiplanation terraces in southern England', *Biul. Peryglac.* **4**, 331-8

C. W. THOMSON (1877), *The Atlantic* (London) **2**, xi, 396

A. L. WASHBURN (1947), 'Reconnaissance geology of portions of Victoria Island and adjacent regions, Arctic Canada', *Am. geol. Soc. Mem.* **22**

R. S. WATERS (1962), 'Altiplanation terraces and slope development in Vest-Spitsbergen and south-west England', *Biul. Peryglac.* **11**, 89-101

R. S. WATERS (1965), 'The geomorphological significance of Pleistocene frost action in south-west England' in *Essays in Geography for Austin Miller* (ed. J. B. WHITTOW and P. D. WOOD), 39-57

R. H. WORTH (1930), 'The physical geography of Dartmoor', *Rep. Trans. Devon. Ass. Advmt Sci.* **62**, 49-115

Chapter 25

The Action of Snow

The daily melting of the snow keeps the rocks around thoroughly soaked with water at zero temperature, so that the action of frost is more than usually effective in splitting them to pieces. (W. B. WRIGHT, 1914)

In considering the geomorphological effects of snow in non-glacierized regions, a distinction may be drawn between snow remaining largely motionless on flat or moderately sloping surfaces, and snow moving rapidly as avalanches on steep gradients. Both aspects will be considered in this chapter.

Nivation

The term 'nivation' was introduced by F. E. Matthes (1900) in his study of the Bighorn Mountains, Wyoming, to describe the erosive effects associated with an immobile and patchy snow cover. Since Matthes drew attention to it, nivation has been studied in both alpine and arctic environments, and its effects widely recognized in areas formerly marginal to the Pleistocene ice-sheets.

The dominant process involved in nivation is freeze-thaw. The first detailed studies of freeze-thaw action associated with snow patches were undertaken by W. V. Lewis (1936, 1939) in an area near the northern edge of Vatnajökull, Iceland and by L. H. McCabe (1939) in Vest Spitsbergen. Both were convinced of the importance of freeze-thaw action around and sometimes beneath snow patches but their work showed that the process is by no means a simple one.

In the first place, it is necessary to distinguish between thin snow patches of limited extent, and thicker snow drifts covering larger areas. No precise distinction can be laid down but the important difference is that the thicker snow drift will insulate the ground surface from atmospheric freeze-thaw cycles; the minimum thickness will therefore depend on their duration and amplitude. An additional consideration is whether or not permafrost underlies the snow patch. There are thus four main situations to be considered:

1) Thick snow over permafrost. The ground surface will be subject to frost attack if meltwater generated on the snow surface seeps beneath the snow and, after melting a thin top layer of the permafrost, refreezes. It should

be noted, however, that permafrost will not usually form or survive beneath thick, extensive, and persistent snow cover (Chapter 21).

2) Thick snow over unfrozen ground. The effect will be protective.

3) Thin snow over permafrost. Atmospheric freeze-thaw cycles may alternately cause melting of the snow and upper permafrost, and refreezing.

4) Thin snow over unfrozen ground. The effect will be as in case 3) except that refreezing, unaided by the presence of permafrost below, may be slower.

The snow itself plays a dual rôle; on the one hand, it is a most important source of meltwater, and on the other it acts as a very efficient insulating medium. Permafrost, if present, acts by providing a reservoir of cold, and by preventing loss by percolation of meltwater from the snow or ground ice.

Since even thin snow will exert some protective influence on the ground beneath in respect of atmospheric freeze-thaw cycles, the zone of maximum destruction by frost action will be around the edges of the snow patch, where meltwater is available but the ground not actually snow covered. Moreover, as the position of the snow margin changes through the year, so this zone of destructive action will migrate with it.

An example of the relationship between air temperature fluctuations and the conditions around and beneath a thin snow patch in Vest Spitsbergen is given by McCabe (1939) for the night of 31 August 1938. Between 8 p.m. and 3 a.m., the air temperature fell from $+2°C.$ to $-3°C.$ By this time, the ground around the snow patch had frozen hard to a depth of 2 cm, though below this depth it remained wet and unfrozen down to the permafrost level. Beneath the snow patch, a thin sheet of ice formed, owing to freezing of the previous day's meltwater. The air temperature rose above freezing-point by 10 a.m. but melting of the snow surface did not commence until 2 p.m. when the air temperature was 4°C.

It should be borne in mind that the average number of days per year when temperatures cross freezing-point is relatively few in Spitsbergen (59 at Green Harbour) and confined to spring and autumn. At high altitudes in tropical regions, however, the number of frost cycles per year is very much greater, exceeding 300 in a few places in the Andes (C. Troll, 1944). Nivation in such areas will be even more potent.

Although it has been generally agreed that thick snow patches overlying unfrozen ground will exert a relatively protective effect, the possibility was considered by Lewis that meltwater tunnels or caves beneath such snow patches, like bergschrunds behind glaciers, might provide access for cold air from outside. One such cave beneath a snow patch possibly 15 m thick was examined by Lewis (1939) and found to extend at least 50 m in from the snow margin. The outside air temperature at the time was 13°C.; the air in the cave stood at 9°C., and meltwater in the cave less than 2°C. above freezing-point. However, the very limited degree of air circulation in caves open only at one end makes it unlikely that freezing conditions in the cave would be re-established simply with each nocturnal fall of outside air temperature below zero. The fact that the base of the snow patch was still frozen in the daytime to the loose rock beneath in the drier inner parts of the cave does not necessarily imply that the cave acts as an avenue for penetration of cold air. The most important function of the cave is, rather, to act as a zone of thawing because of the concentration of meltwater along it.

The occurrence of freeze-thaw action in association with certain snow patches is firmly established. In all cases, however, the thickness of the snow, the presence or absence of permafrost beneath, and the frequency and amplitude of atmospheric freeze-thaw cycles must be taken into account. It will be apparent that the snow itself plays little part except in providing meltwater, and, if sufficiently thick, it may even hinder the process by protecting the ground from sudden temperature changes. The exact extent and significance of temperature changes beneath thick snow patches are difficult to assess, and so far have not been measured, for the digging of any artificial pit or tunnel will obviously destroy the natural temperature régime that existed under completely closed conditions. But we may conclude that the most intensive zone of freeze-thaw action associated with any snow-patch is peripheral to it and extends probably only a short distance inward beneath the thin edge of the snow.

Once the rock has been broken down by freeze-thaw action, its removal is essential if erosion forms are to develop. The main agencies available for transport of finely-broken debris beneath and downslope from snow patches are running water and solifluction, and opinion has supported one or the other at times. W. E. Ekblaw (1918) favoured solifluction moving material beneath the snow towards the lowest point; Lewis in his earlier paper on nivation (1936) refers to downhill creep and solifluction facilitated by the generous supply of moisture beneath and near to the melting snow; T. T. Paterson (1951) also thought solifluction more important than meltwater run-off in removing material from the base of snow patches in north-west Greenland. S. G. Botch (1946) observed that solifluction terraces were frequently present below the lower edges of snow patches, and argued that the weight of the snow would itself encourage the process of solifluction in the lubricated layer of debris beneath the snow patch (see Chapter 23). The evidence in favour of meltwater as a transporting agent is less impressive, though Lewis appealed to it in 1939. He noted how some sixty to eighty runnels emerging from beneath a snow patch on Snaefell (Iceland) were actively removing clay and sand in considerable quantity. McCabe in Spitsbergen was unable to produce such definite evidence in support of meltwater removal, but seems to have thought it the most probable process of transport. It may be that meltwater removal is dominant in relatively wet climatic conditions such as are characteristic of the part of Iceland in which Lewis worked, whereas Botch in the northern Urals and Paterson in north Greenland were concerned with nivation in colder drier conditions where solifluction would prevail. In many cases, both processes probably operate.

Some other processes of weathering and denudation have at times been grouped under the heading of nivation, though their significance is probably slight compared with freeze-thaw action and the accompanying processes of debris removal. J. E. Williams (1949) suggested that chemical weathering might occur beneath snow drifts. It is well known that the optimum temperature for the absorption of oxygen or carbon dioxide by water is near to 0°C., and Williams showed that samples of air taken from inside a snow drift contained more than twice as much CO_2 as normal. The probable reason for this is that, as water freezes, it liberates the CO_2 which, as a relatively heavy gas, then remains in the snow drift ready to be absorbed again by meltwater. But the effectiveness as a weathering agent of water containing CO_2 even in abnormal concentrations is very much to be doubted, except where the bedrock is calcareous (M. Boyé, 1952). Moreover, rates of chemical reaction are reduced with lowering of temperature. Williams failed to consider any processes of chemical weathering other than those that may be connected with water containing O_2 or CO_2.

Another process connected with the presence of snow patches is the possible erosive action of meltwater run-off. W. A. Rockie (1951) in the Palouse observed the development of rills and gullying by meltwater on the ground downslope from snowbanks, while in the very different environment of Antarctica, R. L. Nichols (1963) reported that meltwater dripping off the lower edges of snow patches on to the unprotected ground beneath was a significant eroding agent, as well as producing miniature solifluction forms and fans of fine sand. On the other hand, Lewis (1939) thought the small trickles of water emerging from beneath snow patches in Iceland were not primarily eroding but transporting agents, as already mentioned. The function of the meltwater draining from a snow patch therefore varies according to the local conditions; there may be erosion if the water crosses unconsolidated material on a sufficiently steep slope, there may be mass movement if the debris is saturated, or the water may simply act as an agent of removal.

Although nivation does not include the effects of rapid snow movement in the form of avalanching, some writers have believed that slow mass movement of the snow is a prerequisite of surface erosion by nivation. Matthes (1900), in his original discussion of nivation, found no evidence of such movement, but other writers subsequently were not so sure. Thus, T. C. and R. T. Chamberlin (1911) stated, 'erosion is assumed to be due to the adhesion of the snow-ice mass to the ground on which it rests. . . . A broad patch of soil and loose rock . . . is first dragged away . . .' (p. 199), while I. Bowman (1916) contended that 'in discussing the process of nivation it is necessary to assume a gliding movement on the part of the snow . . .' (p. 286), and that 'compacted snow or névé of sufficient thickness and gradient may actually pluck rock outcrops . . .' (p. 289). At the same time, Bowman denied that snow sliding would cause striation or abrasion. J. L. Dyson's observations (1937) on bedrock striations, which he concluded were the result of rapid snow sliding, have been mentioned in Chapter 6. Lewis (1939), on the other hand, found no evidence to support snow movement as a process contributing to the development of nivation hollows in eastern Iceland, except that, in the case of one snow patch sloping at 15 degrees, slow down-hill movement was seen to be taking place causing the formation of arcuate crevasses.

There is clearly much uncertainty about the occurrence and possible importance of slow mass movements of snow. The movements comprise both basal sliding and internal creep when conditions are appropriate, but there is disagreement as to whether such movements may significantly contribute to surface erosion. A recent contribution by A. B. Costin et al. (1964) suggests that slow movements may lead to some surface abrasion, but not to an excessive degree. In association with a semi-permanent snow-patch in the eastern cirque of Mount Twynam, New South Wales, fresh abrasion marks on granodiorite bedrock were undoubtedly caused by recent movement of stones, some of which were still in position at the ends of their tracks. Both the stones and their tracks of movement sometimes bore traces of fresh rock flour. In this area, moving snow was thought to be the only possible agency capable of impelling the stones; moreover, avalanche action was out of the question for reasons such as the fact that some stones had come to rest in precarious positions and that avalanches are virtually unknown in the area. Slow mass sliding of the snow was therefore postulated. But although striation of bedrock surfaces was occurring, the resulting amount of erosion was probably quite small (and probably minor in comparison with other nivation processes). Patches of peat still adhered to the scratched bedrock surfaces, though it is possible that these peat masses would be protected by being sheathed with ice in the late autumn before the winter snowfall, and that the snow could then slip easily over them.

In areas of unconsolidated rocks, the rapidity with which nivation processes may work has often been remarked on. Following a severe snowfall on the Wasatch Plateau, Utah, in 1952, large snowbanks here persisted throughout the summer. In one season, the turf had been cut away to create nivation hollows one metre deep and up to 500 m across (R. F. Flint, 1957). Similar rates of action were recorded by W. A. Rockie (1951) in the Palouse of Washington State. In neither case was snow movement likely to have made any significant contribution to surface erosion.

On a longer time-scale, the nivation hollows studied by E. P. Henderson (1956) in Quebec province may have developed mainly in the Little Ice Age or at the most within the last 3000 years. The largest of them represent individually the removal of as much as 5 million m^3 of material. The figures are very approximate, especially the time factor, but together with those of Flint and Rockie above, they suggest that nivation must be ranked with glacial erosion as a competent and important process of denudation.

EROSION FEATURES ATTRIBUTABLE TO NIVATION

The most important erosion feature is the nivation hollow. In south-east Iceland, 'the mountain slopes are everywhere scalloped by small hemispherical hollows irregularly spaced and showing all variations in size from a few hundred yards to over a mile across' (Lewis, 1936, p. 431). Lewis (1939) recognized three types of snow patch and corresponding nivation hollow: transverse, longitudinal, and circular. This classification by shape and position agrees well with and includes other workers' groupings.

Transverse nivation hollows lie along the contour of a slope and are frequently structurally determined. They include the hollows produced by Ekblaw's 'piedmont drifts' at the foot of steep lee slopes. Lewis notes that they may individually attain lengths of over a kilometre measured along the slope, and several hundred metres in a downslope direction; a downslope section through one is shown in Figure 25-1a. This hollow fills up with snow completely in winter, and the snow may entirely disappear in summer. Vegetation ceases abruptly at the edges of the hollow. The ground at the lower part is almost flat, covered with stones and fine mud, across which meltwater trickles, while the upper part is steeper (up to 30 degrees) and drier. It is suggested that the nivation processes already described result in the recession of the steeper upper slope into the hillside and the extension of the flat floor. The backslope thus grows in height, and the capacity of the hollow to hold snow is increased, until eventually enough snow is held to survive the summer melt period.

Transverse nivation hollows may thus give rise to a form of hillslope terracing, identical with the altiplanation terraces discussed in Chapter 24. In the area described by Lewis, the initial location of the hollows was determined by the sub-horizontally bedded basalts and tuffs, but slight depressions of non-structural origin might also be adequate to start the process, to which

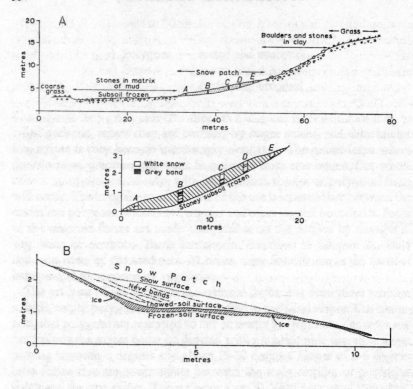

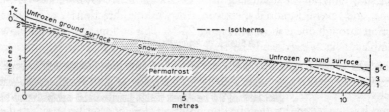

Fig. 25-1 **A** Downslope section through a transverse nivation hollow in Iceland
(W. V. Lewis, *Geogr. J.*, 1939, Royal Geographical Society). The lower
diagram gives details of the pits dug in the snow patch
B Downslope sections through transverse nivation hollows in Spitsbergen
(L. H. McCabe, *Geogr. J.*, 1939, Royal Geographical Society). On the
lower section, isotherms (°C) for the unfrozen ground around the snow-
patch are shown. Beneath this snow patch permafrost rises to the ground
surface

Matthes drew attention in 1900 (p. 182). H. W. Ahlmann (1919) also commented on the way in which snow patches might form and widen notches or shelves on mountain slopes (p. 229), and T. T. Paterson (1957) describes the long lines of snow-bank terraces picking out the strike of the Archaean sediments in north-western Greenland. These also seem to be very similar to the altiplanation terraces described by R. S. Waters (1962) in Spitsbergen.

A special case of miniature transverse nivation hollow is described by M. Boyé (1952) from the Pyrenees (Fig. 25-2). Snow lies at the foot of small limestone scarps in summer, covering them completely in winter. In summer,

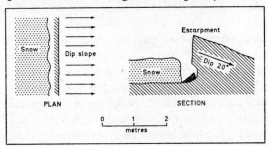

Fig. 25-2 Diagram to show the action of nivation,
undercutting a miniature scarp in limestone;
central Pyrenees (M. Boyé, *Revue Géomorph.
dyn.*, 1952)

a gap appears between the head of a snow patch and its bounding scarp, in which frost-broken debris from the scarp accumulates, similar to a glacier randkluft. At the base of the scarp Boyé found a groove of semi-circular cross-section which he suggested marked the zone of most intense freeze-thaw action in the periglacial microclimate next to the snow patch; possibly also chemical weathering of the limestone by cold meltwater might play a part.

Longitudinal nivation hollows are those which follow the direction of maximum ground slope. The snow patches which give rise to them often accumulate initially in water-eroded valleys or gullies (Ekblaw thus termed these patches 'wedge drifts') and the major problem is that of separating the effects of water erosion from the possible effects of nivation (D. L. Dineley, 1954). In some it appears that nivation has widened the gullies, rounded their heads, and emphasized any structural ledges flanking the gullies. McCabe describes examples which possess a sharp cliff at their head over which a stream pours to plunge beneath the snow patch below the cliff. The streams, however, do not notch these cliffs and appear not to have formed them. In one particular case, the cliff overhangs, and the accumulation of angular boulders fallen down from the overhang into the gap between the rock wall and the snow patch could only have come from the overhang by freeze-thaw action. Water was seen to be trickling down the cliff and refreezing in the gap behind the snow patch.

Thus nivation modifies the original water-cut form. A later stage in the development of longitudinal nivation hollows in Vest Spitsbergen is represented by the establishment of 'niche glaciers' (G. E. Groom, 1959). As the hollow is enlarged by nivation, so the snow drift can extend itself and thicken until ice begins to form by compaction in the lower layers. Such snow-ice masses are seen to form on some exceptionally steep slopes (42 degrees noted by Groom); the characteristic landform is a rounded half-funnel shaped hollow.

Rather similar forms are the 'rasskars' or 'cirque-like ravines' on some precipitous rock slopes in Norway (Ahlmann, 1919). The head is funnel-shaped and the backwall may approach the vertical; downward, the funnel narrows into a V-shaped gorge. Running water is seldom seen in them. Their origin is not well understood, but snow collects in the funnel in winter, probably widens it by nivation together with frost action, and in spring the mass of snow and rock debris, lubricated by meltwater, glides down through the funnel outlet as an avalanche (see p. 566).

The third form of nivation hollow, related to roughly circular snow patches, differs basically from the transverse and longitudinal varieties in being largely independent of structural or water-eroded features. Varying from a few tens of metres in diameter to as much as 1 km, the rounded form is quite unrelated to stream action and is most perfectly developed on gently sloping surfaces where no pronounced variation in geological structure intervenes (R. J. Russell 1933, Lewis 1939). Figure 25-1b shows a section through one measured by McCabe (1939), which should be compared with Lewis's section of a transverse nivation hollow (Fig. 25-1a). McCabe's section was measured in the month of July, before the snow patch dwindled away in August. In July, the ground around and away from the snow patch had thawed out to some depth, but beneath the snow patch, even though this was less than 0·6 m thick, the permafrost extended up to the ground surface, emphasizing the insulating effects of a snow cover mentioned earlier. Wherever permafrost is present at shallow depths, it has the extremely important effect of confining snow meltwater to the topmost layers of the ground. These areas are not only then most susceptible to frost action if refreezing later occurs, but also to solifluction, a major process of debris removal. Away from the snow patches where the ground has thawed out to greater depths, what meltwater there is is diffused throughout a considerable mass of material and its usefulness is greatly diminished. By the end of August when this particular snow patch had disappeared, the floor of the nivation hollow was seen to be covered with angular stones and coarse mud, and the ground was much drier than when the snow patch was present. Intensive freeze-thaw action was thus limited to the area of the snow patch and its period of duration; and the rounded form of nivation hollow is contingent on the tendency of snow patches to become roughly circular by ablation (thus developing the

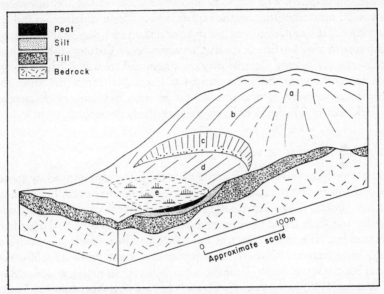

Fig. 25-3 Block diagram of a simple nivation hollow near Knob Lake, Quebec
(E. P. Henderson, *J. Geol.*, 1956, University of Chicago Press)
  a numerous rocky outcrops on hill top
  b slopes, 10-15 degrees, in drift
  c backwall, slope 30-40 degrees, in drift, with boulder accumulations
  at its foot
  d gently sloping (*c.* 5 degrees) floor in drift
  e swampy flat on solifluction deposits

least periphery per unit area) in any area where geological or existing topographical features do not play a dominating part.

Since evacuation of debris is necessary for the continued development of any nivation hollow, the circular type develops best on gently sloping rather than flat ground. The backslope forms an arcuate scar which gradually recedes, thus widening the floor and increasing the snow-holding capacity. In some areas of thick drift deposits near Knob Lake, Quebec, simple and compound nivation hollows of this sort are well portrayed (E. P. Henderson, 1956). A block diagram (Fig. 25-3) demonstrates the salient features. Back-slopes, entirely in drift, reach heights of 20 m in places, with slopes usually of the order of 30-35 degrees; the diameter may be 120 m for simple hollows or up to 800 m in the compound varieties where two or more hollows have coalesced or where mass movements have complicated the initial form. Floors slope outwards at 3 to 12 degrees and may be covered with hummocky solifluction material. The presence of thick unconsolidated and unresistant drift is the most important factor explaining both their distribution and the excellence of their development, for none occurs on rock slopes; and although some are orientated to face north-east, reflecting snow drifting with dominant south-westerly storms, the majority are orientated to face north or north-west in accord with the direction of last ice movement in the area. The direction of ice movement controlled the position and shape of the patches of thick drift in which the hollows are exclusively developed.

DEPOSITION FEATURES RELATED TO NIVATION

Downslope from snow patches accumulate the materials removed by snow-melt run-off and by solifluction. Solifluction terraces may occur below snow patches (Botch, 1946) representing successive layers of soliflual material moving out from under the base of the snow and thickening as the gradient becomes less. The smaller terraces and layers may be seasonal. The scarps of some large terraces, however, may represent not the limits of solifluction layers but the actual limits of the snow patch at various stages, as deduced by E. Watson (1966) in the case of a nivation hollow near Aberystwyth, Wales. Figure 25-4 shows the series of terraces marking the exit of Cwm Du. The uppermost one is 20 m high and its face slopes at up to 23 degrees. The terraces are built of solifluction debris, which is thought to have moved down under the snow bank as Botch suggested. The evidence that the scarps mark successive snow-patch limits is two-fold: first, the present stream trench always cuts the scarps at their lowest points (where one would expect melt-water to leave a snow patch), and secondly, the scarps are not concentric but appear to enclose a series of lobes whose axes have progressively changed. This is more easily attributed to a snow patch whose headward growth was more rapid on one side than the other (in this case, the lee side, with respect to westerly winds).

Botch claimed that a distinction might be drawn between the depositional effects of gently sloping and steeply inclined snow patches. In the former case, most debris broken off by freeze-thaw action would find its way to the base of the snow and be removed therefrom by solifluction, thus producing solifluction terraces and platforms below the snow patch. In the case of steeply sloping snow patches, a major part of the debris falling on to the

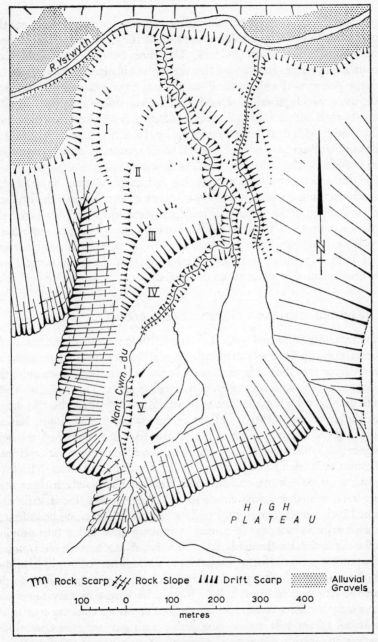

Fig. 25-4 Cwm Du, near Aberystwyth, central Wales, showing enclosing
rock walls and a series of solifluction fans I to IV. V is a protalus in
the innermost recess (E. Watson, *Biul. Peryglac.*, 1966)

snow from the frost-shattered backwall would tend to slide down over the snow surface to build up a ridge at the foot of the snow patch. Such ridges were referred to by Russell (1933); C. H. Behre (1933) used the ambiguous term 'nivation ridges' and stressed that they were only incidentally connected with the presence of a snow bank since it was frost action above the level of the snow which produced the debris. The ridges were renamed 'protalus ramparts' by K. Bryan (1934). They have often been confused with moraines. R. P. Sharp (1942) describes examples on San Francisco mountain, Arizona; some ramparts here may nourish or even be the sources of stone streams since there is evidence that some ramparts are slowly moving downslope.

The distinction between gently sloping and steeply sloping snow patches is clearly an arbitrary one, but Watson's work at least suggests the limiting angle and supports the validity of the distinction. In Cwm Du where solifluction deposits thickly mantle the floor of the hollow, the maximum snow slope is likely to have been less than 13 degrees, whereas in neighbouring Cwm Tinwen, which possesses a protalus rampart, the snow surface may have been inclined at about twice this angle.

THE TRANSITION FROM SNOW PATCH TO CIRQUE GLACIER

It has often been suggested—W. B. Wright (1914) was one of the first to do so—that nivation processes may enlarge and deepen initial depressions to such a degree that the thickening snow banks in them begin to acquire the characteristics of névé and eventually small glaciers. Theoretically, there should therefore be a graded series of landforms reflecting this transition, from the shallowest nivation hollow to the mature glacial cirque, but although isolated instances of transitional forms have been described, the field evidence is still inadequate. Some writers have used the term 'nivation cirque' (for example, Russell 1933, Watson 1966) to describe large nivation hollows resembling glacial cirques in form and situation but in which there may never have existed any true cirque glacier. Russell held that if striae and polished rock were found in such hollows, glacial action could be assumed, but since striae at least may be caused by moving snow rather than moving ice, the distinction on this basis is hardly valid. A much more fundamental distinction concerns the form of the floor of the hollow. A true glacial cirque will often possess a basin-shaped floor, with a reversed slope leading to the lip of the cirque; a nivation hollow formed by semi-stationary snow will always lack this characteristic, for removal of debris by solifluction or other non-glacial means will not be compatible with any reversed floor slope. However, in practice, it may not be possible to demonstrate conclusively the presence or absence of a rock basin in the floor, owing to subsequent deposition, and rock basins may be absent in some cirques thought to be glacial.

The distinction between snow patch and glacier is no easier to make. Certainly, the presence of ice in the lower layers of a snow patch is not a satisfactory criterion, for the ice can form here by freezing of meltwater. Lewis (1939) argued that movement was the best distinguishing feature: 'when . . . a composite mass of snow and ice begins to move over the rock floor, a new set of sculpturing agents is introduced' (p. 160). Yet slow mass movement of some snow patches is well established and there is no evidence that these are in the process of becoming miniature glaciers. Costin *et al.* (1964) thought that in the Mount Twynam area, the relationship was actually reversed, the processes of nivation (including snow sliding) following on from earlier glaciation and not leading up to glaciation.

Watson (1966) in west Wales attempted another approach, by computing the probable maximum thicknesses of snow in 'nivation cirques' now vacated by snow. Assuming the snow surface at its maximum to have extended from the top of the backwall to either the protalus rampart (Cwm Tinwen) or the outer scarps of solifluction terraces (Cwm Du), Watson shows that the greatest thickness of snow (measured vertically) was probably 38–45 m in the former case, and 75 m in the second. Since névé begins to form ice when subjected to a pressure of 30 m or more of overlying névé (Chapter 3), some glacial ice was certainly present in Cwm Du (though perhaps for only short periods) but probably not to any extent in Cwm Tinwen.

The distinction between snow patch and glacier, or between nivation hollow and glacial cirque, cannot be rigidly formulated. Transitional forms do occur, though they have not yet been widely recognized or adequately investigated, and nivation may well be responsible for the initiation of some cirque glaciers.

Nivation: Summary and Conclusion

The term nivation covers a variety of processes associated with a patchy snow cover, which give rise to the formation of distinctive erosional and depositional features. The typical erosion form is the nivation hollow which develops most rapidly in unconsolidated rocks such as glacial deposits, and which varies in form with the initial relief and geological structure of the region. Benching or terracing of hillslopes may also result from nivation. Depositional features include protalus ramparts at the foot of relatively steep snow slopes, and solifluction terraces below more gently sloping snow patches.

The principal erosive process involved in nivation is freeze-thaw action. The effectiveness of this depends on rock type and the amplitude and frequency of atmospheric freeze-thaw cycles. The thickness of the snow cover is a most important consideration, as is also the existence of permafrost in the subsoil, for it is likely that thick snow patches on unfrozen ground play a largely protective rôle. In any case, the process is most effective and potent around the edges of the snow patch, and the action is extended areally by changes in the size of the snow patch. The snow itself acts mainly by supplying meltwater, which will be unable to soak away into the subsoil if permafrost is present. Chemical weathering beneath the snow is probably not very significant except in areas of calcareous rocks.

Debris resulting from freeze-thaw action must be removed from the site of the snow patch if erosion of the ground surface is to result. Removal can only be effected on sloping ground, apart from possible slight evacuation of material in solution by downward percolation in unfrozen ground. Solifluction and meltwater run-off have both been considered as agents of transport, with the former appearing to be the more important in colder regions, and since neither can move material uphill, the floors of nivation hollows slope continuously outward.

Snow patches on ground of sufficient slope (yet not so steep as to cause avalanching) may move by slow basal sliding and/or internal creep. While this may give rise to scratched and abraded rock surfaces, it is probably not a significant process of erosion until great thicknesses of névé have accumulated and the formation of glacier ice in the basal layers has begun. Thus snow patches may develop into small glaciers, and some nivation hollows may evolve into embryo cirques.

Since the snow itself has little direct part to play in nivation apart from providing meltwater and the negative aspect of insulating the ground surface from atmospheric temperature fluctuations, a number of workers have advocated abandoning the term nivation altogether. Paterson (1951) stated that 'the phenomenon of nivation . . . is essentially a process concerned with the growth and melting of ice crystals within rock interstices, and as such has little to do with snow itself except in that the snow bank provides part of the water from which the ice crystals are developed' (p. 52), and M. Boyé (1952) asked: 'since snow neither attacks nor sweeps away, why speak of erosion by snow?' (p. 31). The validity of Boyé's criticism is somewhat doubtful in view of the evidence that slow-moving snow can in fact act as an agent of erosion, and even more so when moving rapidly as an avalanche. The term nivation remains a useful one to encompass a variety of processes connected with the presence of semi-stationary snow, provided that the limited part played directly by the snow is fully appreciated.

Avalanche Activity

Avalanches are one of the most spectacular and dangerous forms of snow activity. Their geomorphological rôle has, however, not been very extensively studied. Nevertheless some valuable work has been done recently in this field, and features that are the direct result of avalanche activity have been recognized. In mountainous areas with heavy snowfall, avalanches are very common, and it has been estimated that several tens of thousands take place in Switzerland each year. If the avalanches occur repeatedly in the same positions for many years in succession, their possible influence on the landscape cannot be ignored.

There are several different types of avalanche. Some of these can produce particular landforms, but not all of them significantly affect the landscape. Snow surface avalanches in particular do not affect the ground. In their upper courses at least, the new snow slides above the plane separating it from the older snow beneath and it is not in contact with the ground beneath. An avalanche will take place when the weight of snow exceeds the resistance of friction. The avalanche is affected by four factors: 1) the internal cohesion of the snow; 2) the thickness and density of the snow layer; 3) the character of the underlying material, whether it be old snow or rock; and 4) the slope. The first factor affects the type of movement: the snow may either slip or break up into blocks. The second factor determines the mass of the avalanche. The third factor determines the frictional effect and the fourth factor determines the effect of gravity. There is a relationship between the slope and the thickness of the snow layer required to cause an avalanche. If the slope is 50 degrees, an avalanche can theoretically occur with a layer only 5 cm thick; for a slope of 30 degrees, the critical thickness is 15 cm, and for a slope of 22 degrees, the thickness must increase to 40-50 cm before avalanching occurs. Normally avalanches do not occur on slopes less than 22 degrees although exceptionally when all the factors are favourable they may occur on slopes as low as 6 degrees.

The smoother the slope beneath the snow, the more readily an avalanche can occur. A dense vegetation cover, especially of forest, plays a large part in preventing avalanche activity. Once the vegetation has been removed and the rock smoothed by a series of avalanches or other processes, then avalanches can occur with greater ease. The avalanching process is, therefore, self-generating to a certain extent.

Avalanche activity depends on relief and surface characteristics, but climate must also be considered. Avalanches will be numerous and large where heavy snowfalls are common and temperature changes rapid, particularly those that cause the snow to melt or rain to fall on snow. The snow can absorb a large amount of rain or meltwater and this increases its weight and avalanches become much more probable. They also tend to occur more readily in areas where the tree-line is low for climatic reasons, because forests are effective in preventing avalanche activity. Statistics concerning avalanche activity are collected in Switzerland. These show an interesting relationship between elevation and the number of avalanches. The tree-line in the area is about 1500 m, and the permanent snow level is about 3000 m. The number of avalanches recorded in 1909 was as follows:

under 1500 m	394	4 per cent
1500–2000 m	2632	28
2000–2500 m	3806	41
2500–3000 m	2210	24
over 3000 m	326	3

Most of the avalanches take place in the spring, when the snow cover reaches its maximum thickness and the temperature is liable to sudden rises.

The types of avalanche include powdery avalanches, slab avalanches and wet avalanches. The wet type of avalanche has been called 'slush avalanches' by some authors. There are also ice avalanches, falling from unstable and overhanging ice masses. The latter can form a particular type of glacier that has already been mentioned (see p. 66). All the main types of avalanche can be either clean or dirty. The clean avalanches consist only of clean snow or ice, while the dirty avalanches incorporate rock material in the moving mass. The latter will have the most important geomorphological effect. The geomorphological effects of these different types of avalanche have been studied quantitatively and in detail by A. Rapp (1959, 1960) in northern Sweden and by A. Jahn (1967) in Spitsbergen.

The effects of powdery avalanches are largely secondary. They are nearly always clean and consist of the movement downslope of newly fallen powdery snow of very low density. Their main effect is the result of their very rapid velocity of about 200 to 300 km/hour. This rapid movement creates a blast of air that can cause minor damage and trigger off other avalanches. At times they also create instability in the underlying snow, generating an avalanche that reaches to the ground and thus becomes a dirty avalanche.

The slab type of avalanche, which consists of large blocks of consolidated snow, is much more likely to become a ground avalanche and hence a dirty one. It can cause considerable erosion. The blocks of snow become rounded as they move down the slope and in this form they can move far from the mountain slope. They can carry debris a considerable distance, even extending a short way up the opposite valley slope. They may dam up small streams in the valley, and floods occur when the wet snow dam suddenly gives way. Their effect is, therefore, both direct and indirect.

The wet or slush avalanches have probably the most important direct geomorphological effects. It has been reckoned that this type of avalanche can carry about ten times as much material, under certain circumstances, as the other types, according to Rapp's measurements in northern Sweden.

The effects of slab avalanches in Kärkevagge have been considered by Rapp. He describes one avalanche that was set off when a block of snow 50 m long and 2-3 m thick fell from the top of a cirque wall. The snow was hard and heavy and broke into pieces as it fell down the very steep slope, which reached 90 degrees at the top. The snow fell on to rather less steep slopes of 50-60 degrees and continued down the slope as an avalanche. The falling snow detached several large blocks of rock about 0·5-1 m in size. One of these types of avalanche, which fell in July, transported about 5-10 m³ of rock debris. This avalanche consisted of old firn snow and removed loose debris liberated by frost action from the rock wall. These relatively small avalanches of hard snow provide a means whereby rock surfaces can be cleaned of superficial debris and the avalanche can at times break off rocks loosened by frost.

The depositional forms resulting from this type of avalanche consist of debris that is roughly sorted according to size. The larger rocks roll farther down the slope than the smaller ones, but otherwise the debris is completely mixed with fine material mingled with coarse. The form of the deposits is usually in the shape of a broad tongue. The smaller avalanches do not extend very far down the slope, but the larger ones sometimes reach the valley floor or beyond. Some of the boulders that come to rest on or in the avalanche material move farther down the slope as they melt out. Avalanche debris can often be recognized by its loose packing as it melts out of the snow, and by the spread of finer material on larger boulders that were originally at the bottom of the avalanche. The boulders and cobbles in the melted-out avalanche debris show no preferred orientation, although the elongated boulders that have slid over the snow are often orientated with their long axes down-slope.

The moving boulders often cause some erosion of the ground over which they move. Examples of this type of erosion were seen in Austerdal, Norway. Boulders carried by avalanches had moved very fast down the steep valley walls and had been broken up on impact with boulders already lying on the valley floors. These boulders were also broken by the impact into many angular fragments. The dirty snow avalanches occur mainly where deep snow drifts can accumulate during the winter, They tend to recur in the same places year after year, and are particularly frequent on lee slopes. They occur mainly in the spring as the snow melts. Dirty avalanches are common in late spring when the thaw has exposed much bare ground. Rapp estimated the total amount of material moved annually by dirty snow avalanches to be of the order of 1·4 tons/m², moving an average distance of 100 m on a 30-degree slope. This gives a total of 1050 tons/m vertically.

Slush avalanches have been found to be much more effective agents of denudation than dirty snow avalanches. They are particularly important in years of rapid snow melt. The operation of slush avalanches or slushers has been described in Greenland and arctic North America. Slush avalanches are sometimes initiated by the melting of snow in a stream bed. Water may be held up behind an ice dam for some time, and then suddenly escapes down the valley when the dam gives way. An example of this kind of slush avalanche occurred on 12 June 1956 at Kärkevagge. This avalanche carried 200 m³ of rock material and extended for 350 m, reaching gently sloping ground. A still larger one occurred in 1958, carrying 300 m³ of debris. These avalanches were released from an iced-up waterfall and as they fell, strips of turf were torn from the ground, exposing the bedrock. The carrying capacity of such avalanches is indicated by the fact that a boulder 5 by 3 by 2 m was carried about 120 m down a slope of only 5 degrees. The boulders plucked from the ground first glide over the ground and then they may move by jumping; they may then form hollows where they land. Finally they roll and make smaller holes, closer together. Sometimes the holes from which the boulders were plucked can also be seen. Slush avalanches can be released on gentler slopes than normal avalanches. The saturation of the snow, which increases in weight as a result, is partly the cause. Layers of ice on the lower side of the snow patch may provide glide planes to initiate the avalanche. Sometimes slush avalanches start as a snow slab avalanche that moves into a stream valley. The snow then mixes with the water and is converted into a slush avalanche. Three major slush avalanches occurred during the period from 1952 to 1960 in this part of northern Sweden, and it is estimated that these three carried a total of at least 700 m³ of rock debris.

There seems to be a difference between the slush avalanches of the Arctic regions and the dirty ground avalanches of the Alps. Some of the particularly disastrous Alpine avalanches may, however, have had the characteristics of slush avalanches. The Alpine ones differ from the Arctic type in that the Alpine ones are released as large slab avalanches on steep slopes. They do not originate directly in stream courses. They tend to move down steep ravines and the material they carry is deposited on alluvial cones. Relief factors, therefore, probably cause the major differences between the Alpine and Arctic types of avalanche. In the Alps, there are steep incised gullies, but in the Arctic areas, more open slopes are typical.

The avalanche boulder tongues described by Rapp (1959) in Lapland are a good example of a feature that can be specifically related to avalanche activity. These constructional features also provide evidence of the capacity of avalanches to erode the slopes across which they are moving. The avalanche boulder tongue must first be differentiated from other features resulting from mass movement and stream flow, including talus cones and alluvial cones. Another rather similar form is that produced by a rockslide tongue. The rockslide tongue, however, lacks the long, straight path leading down to it. This long straight path is characteristic of the avalanche boulder tongue. The normal avalanche boulder tongue also has a markedly concave longitudinal profile. The end of the tongue may reach far out into the valley or even extend a short distance up the opposite valley slope. The surface of the tongue is flattened out and in this it differs from the talus and alluvial cones, which have convex surfaces.

There are also smaller features characteristic of avalanche boulder tongues, such as avalanche debris tails. The tails consist of small straight ridges of debris on the distal side of large boulders. They are about the same height and width as the boulders, which may attain dimensions of 1 m, behind which they have formed. The tail is often about 5-10 m long, diminishing in size towards the distal end. All the tails are parallel to each other and are elongated in the direction of the tongue. There is also sometimes a small accumulation on the proximal side of the boulder. The features form in the same way as a sand shadow round an obstacle, or, on a larger scale, crag and tail. They are an accumulation and not an erosional form.

The whole boulder tongue probably takes quite a long time to form and is not the result of one single avalanche. Two types of avalanche boulder tongues are distinguished. The first is the road-bank tongue and the second is the fan tongue. Most of both types are asymmetrical in outline transverse to their elongation. The tongues consist of angular debris that is only very poorly sorted to the extent that the larger boulders tend to occur round the margin. The features form most effectively above the tree-line and in those areas where much loose rocky sediment is available on the hillsides.

One example of the road-bank type of avalanche boulder tongue was 330 m long from the mouth of the chute and 70-80 m broad, ending in a lake. It attained a thickness of 5 m at a point 150 m from its end. The material consisted of angular boulders up to 2 m in size. The long profile decreased from 28 degrees to 22 degrees and finally flattened to 15 degrees near its end. The asymmetrical transverse profile is thought to be associated with snow banks that collect alongside the tongue. The snow becomes thicker on the lee side of the tongue, while the windward side is swept bare. Snow will accumulate in the chute above the tongue and hence promote avalanche activity. When the avalanche runs down the chute, it will tend to erode the exposed part. The material will accumulate around the bare part of the fan as the snow bank on the lee side will tend to divert the avalanche to the bare windward side. In this way, the flattish top of the feature and its asymmetry can be accounted for. In transverse profile the lee side is steeper as the material slumps when the snow bank melts.

The fan type of avalanche boulder tongue sometimes extends on to the flat ground in the valley floor. The fan starts in a broad shallow avalanche track about 200 m wide, less narrow than the chutes that lead down to the road-bank type. One fan-type avalanche boulder tongue was 650 m long and sloped at 35 degrees on the lower part of the mountain wall. The slope decreased to 30 degrees above the wall, a slope that is less steep than the angle of rest of boulders which cover the hillside where no avalanche has removed them. The upper part of the tongue consisted of an alluvial cone. This cone has been built up by streams and its eroded margins indicate that avalanches scour a wider area than that eroded by the streams. The true avalanche tongue reaches a further 300 m beyond the bottom of the alluvial fan. This indicates that avalanches carry material much farther than streams or rock-falls. At its lower limit, the tongue rises about 0·5 m above the grass surrounding it, but its general thickness appears to be about 2 m. The extent of individual avalanches is indicated by arcs of boulders on the tongue. A study of the preferred orientation of the stones on the surface of the boulder tongue showed that they are nearly all aligned parallel to the elongation of the tongue, but they show a slight divergence towards the lateral margins (Fig. 25-5). The fan type of boulder tongue is probably the work of very rare large avalanches that have the capacity to erode and carry large quantities of debris. It seems probable that this type of feature only occurs above the tree-line. They are widespread under suitable conditions in the northern parts of Scandinavia, where aerial photographs show that the hillsides are furrowed with the parallel features formed as a result of avalanche activity.

So far the depositional aspects of avalanche activity have been stressed, but it is important to recognize that the deposition must be accompanied by an equal erosion of material from the upper slopes. This activity influences the slope development processes in the areas where avalanche activity is a significant morphological agent. The road-bank tongues lead up to narrow, fairly deeply cut chutes, which are very straight. These chutes are also used by rock-falls and streams. Avalanches are, however, the most efficient mechanism whereby rock and debris are carried down these chutes. The avalanche removes loose material lying on the surface across which it is moving as well as abrading this surface. Weathering is thus allowed to proceed effectively. Not all rock types are equally susceptible to avalanche activity. Amphibolite

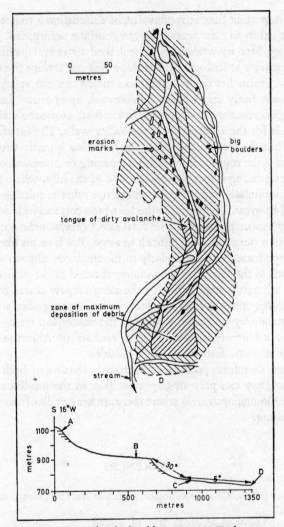

erosion
marks

big
boulders

tongue of dirty avalanche

zone of maximum
deposition of debris

stream

S 16°W

A

1100

B

900

30°

C

5°

D

700

0   500   1000   1350

metres

Fig. **25-5** An avalanche boulder tongue in Kärkevagge,
Sweden (A. Rapp, *Geogr. Annr.*, 1956)

appears to react more obviously to avalanche erosion than the mica-schist
of Kärkevagge, where avalanche chutes are not common.

The effectiveness of avalanches in modifying the slopes depends partly on
the slope gradient. The optimum slope appears to be about 30 to 35 degrees.
On steeper slopes, although some modification does occur, it is not so marked.
On slopes gentler than 30-35 degrees rock falls lose their effectiveness. One

of the most important functions of avalanche denudation is to keep the slope bare of rock debris so that weathering can continue unimpeded. Where the slope is swept bare by avalanches it will tend to retreat parallel, with a growing concavity at its base. A talus slope will not replace the rock slope. The fact that avalanches erode wide tracks means that the valley walls will tend to remain fairly smooth and undissected, apart from the relatively shallow chute formation. In areas such as Lapland, avalanche activity could be responsible for the asymmetry of the valley walls. The east-facing walls are steeper than the west-facing ones. This variation is partly structural, but also in part it is the result of avalanches steepening or maintaining the steepness of the east-facing slopes. On the lee side of the hills, where large snow drifts can accumulate, and on cols, where the relief is suitable, avalanche activity will be great. The slopes in the lee of the cols and hillsides will tend to retreat by avalanche erosion. The total effect of avalanche erosion in the denudation of a steep hillside is difficult to assess. But it seems likely that this aspect of slope formation, particularly in mountainous districts with heavy snowfall, such as the Alps, has been underestimated in the past. Avalanches in the Alps are partly responsible for the steep narrow chutes that are also used by summer storm waters and melt water. These gullies are probably kept clear mainly by avalanche activity in the winter and spring. The avalanche chutes in Norway have been called 'rasskars' by Ahlmann (1919) and they are characteristic features of many hillsides.

Avalanches, therefore, produce characteristic features of both deposition and erosion. They can play an important part in the modification of the landscape in mountainous areas where there are heavy falls of snow and rapid summer melting.

REFERENCES

H. W. AHLMANN (1919), 'Geomorphological studies in Norway', *Geogr. Annlr* 1, 1-148 and 193-252

A. ALLIX (1924), 'Avalanches', *Geogr. Rev.* 14, 519-60

A. ALLIX (1954), *L'action morphologique de la glace et celle des coulées de neige* (Mél Bénévent, Gap), 11-17

C. H. BEHRE (1933), 'Talus behaviour above timber in the Rocky Mountains', *J. Geol.* 41, 622-35

S. G. BOTCH (1946), 'Les névés et l'érosion par la neige dans la partie nord de l'Oural', *Bull. Soc. géogr. U.S.S.R.* 78, 207-34 (original in Russian; translated by C.E.D.P., Paris)

I. BOWMAN (1916), *The Andes of Southern Peru* (*Am. Geogr. Soc. Spec. Publ.* 1)

M. BOYÉ (1952), 'Névés et érosion glaciaire', *Revue Géomorph. dyn.* 3, 20-36

K. BRYAN (1934), 'Geomorphic processes at high altitudes', *Geogr. Rev.* 24, 655-6

A. J. BULL (1940), 'Cold conditions and landforms in the South Downs', *Proc. Geol. Ass.* 51, 63-70

T. C. CHAMBERLIN and R. T. CHAMBERLIN (1911), 'Certain phases of glacial erosion', *J. Geol.* 19, 193-216

F. A. COOK (1962), 'Simple transverse nivation hollows at Resolute, N.W.T.', *Geogr. Bull.* 18, 79-85

J. CORBEL (1958), 'Climats et morphologie dans la Cordillière Canadienne', *Revue can. Géogr.* 12, 15-48

A. B. COSTIN, J. N. JENNINGS, H. P. BLACK, and B. G. THOM (1964), 'Snow action on Mount Twynam, Snowy Mountains, Australia', *J. Glaciol.* 5, 219-28

S. DAVEAU (1958), 'Cône central d'éboulis de l'aiguille Rousse', *Revue Géogr. alp.* 46, 423-8

R. DETTERMANN, A. BOWSHER, and T. DUTRO (1958), 'Glaciation on the Arctic slope of the Brooks Range', *Arctic* 11, 43-61

D. L. DINELEY (1954), 'Investigations in Vest-Spitsbergen', *J. Glaciol.* 2, 379-83

J. L. DYSON (1937), 'Snowslide striations', *J. Geol.* 45, 549-57

W. E. EKBLAW (1918), 'The importance of nivation as an erosive factor, and of soil flow as a transporting agency, in northern Greenland', *Proc. natn. Acad. Sci. U.S.A.* 4, 288-93

R. F. FLINT (1957), *Glacial and Pleistocene geology* (New York)

G. E. GROOM (1959), 'Niche glaciers in Bünsow Land, Vest-Spitsbergen', *J. Glaciol.* 3, 369-76

E. P. HENDERSON (1956), 'Large nivation hollows near Knob Lake, Quebec', *J. Geol.* 64, 607-16

A. JAHN (1967), 'Some features of mass movement on Spitsbergen slopes', *Geogr. Annlr* 49, 213-25

W. V. LEWIS (1936), 'Nivation, river grading and shoreline development in south-east Iceland', *Geogr. J.* 88, 431-47

W. V. LEWIS (1939), 'Snow-patch erosion in Iceland', *Geogr. J.* 94, 153-61

F. E. MATTHES (1900), 'Glacial sculpture of the Bighorn Mountains, Wyoming', *U.S. geol. Surv. 21st A. Rep.* (1899-1900), 167-90

F. E. MATTHES (1938), 'Avalanche sculpture in the Sierra Nevada of California', *Bull. Un. géod. géophys. int.* 23

L. H. MCCABE (1939), 'Nivation and corrie erosion in West Spitsbergen', *Geogr. J.* 94, 447-65

R. L. NICHOLS (1963), 'Miniature nivation cirques near Marble Point, McMurdo Sound, Antarctica', *J. Glaciol.* 4, 477-9

T. T. PATERSON (1951), 'Physiographic studies in north-west Greenland', *Meddr Grönland* 151, 4, 60 p.

A. RAPP (1959), 'Avalanche boulder tongues in Lapland, a description of little known forms of periglacial accumulations', *Geogr. Annlr* 41, 34-48

A. RAPP (1960), 'Recent development of mountain slopes in Kärkevagge and surroundings, north Scandinavia', *Geogr. Annlr* 42, 65-200 (esp. 122-47)

W. A. ROCKIE (1951), 'Snowdrift erosion in the Palouse', *Geogr. Rev.* 41, 457-63

R. J. RUSSELL (1933), 'Alpine landforms of western United States', *Bull. geol. Soc. Am.* **44**, 927–49

R. P. SHARP (1942), 'Multiple Pleistocene glaciation on San Francisco Mountain, Arizona', *J. Geol.* **50**, 481–503

J. TRICART and A. CAILLEUX (1962), *Le modelé glaciaire et nival* (Paris), 198–207

C. TROLL (1944), 'Strukturböden, Solifluktion, und Frostklimate der Erde', *Geol. Rdsch.* **34**, 545–694 (English translation, *Snow Ice Permafrost Res. Establ.*)

R. S. WATERS (1962), 'Altiplanation terraces and slope development in Vest-Spitsbergen and south-west England', *Biul. Peryglac.* **11**, 89–101

E. WATSON (1966), 'Two nivation cirques near Aberystwyth, Wales', *Biul. Peryglac.* **15**, 79–101

J. E. WILLIAMS (1949), 'Chemical weathering at low temperatures', *Geogr. Rev.* **39**, 129–35

W. B. WRIGHT (1914), *The Quaternary Ice Age* (London, 1st Ed.), 6

Chapter 26

Periglacial Wind Action

The winds blew . . . during a Dust Age over outwash fans and plains . . . exposed and dried during the intervals between successive inundations. (J. K. CHARLESWORTH, 1957)

Wind action under cold dry conditions is responsible for the most widespread and bulky of all periglacial deposits, namely loess, though its eolian origin was not universally accepted in the past. The loess associated with regions which lay just outside the embrace of the last glaciation suggested further searches for other forms resulting from periglacial wind action, for loess represents but the finer fraction of wind-transported debris, and the strength of wind action that it demanded ought to be independently supported by wind erosion forms. Deposits of coarser disseminated wind-blown sand were identified in the 1940s, fossilized dune forms have been described and analysed since the early part of this century, while the effects of sand-blasting are now well documented in the form of ventifacts and wind-worn rock surfaces. All these features are not only of interest in themselves, reflecting a particular aspect of periglacial activity, but yield valuable evidence on the climatic conditions prevailing at certain times in the Pleistocene.

Wind Erosion

Wind erosion is manifested in the occurrence of faceted, fluted, and grooved surfaces on bedrock, but more commonly on pebbles and boulders projecting up from the ground surface. Ventifacts related to strong wind action of Pleistocene time are found today lying exposed on surfaces now no longer subject to such attack, or else buried by contemporaneous or later eolian deposits. Examples of buried ventifacts, sometimes in numbers sufficient to justify their description as 'desert pavement', are represented by the Steinsohle beneath the loess in north Germany, or by the wind-blasted stones beneath loess described by G. Johnsson (1958) in the Lund region of southern Sweden. In the United States, C. K. Wentworth and R. I. Dickey (1935) summarized the then known ventifact localities, and W. E. Powers (1936) provided some additional occurrences. Powers, for instance, recorded beautifully and unmistakably wind-grooved boulders lying on the sandy plain of former Lake

Wisconsin; and similar ventifacts and grooved boulders of South Park, Colorado, enabled him to establish that the dominant direction of the winds here responsible was from the north-west, mainly in pre-Wisconsinan glacial periods.

Ventifact formation generally requires strong winds, open topography that does not hinder their action, minimal vegetative ground cover, and an adequate supply of suitable sand grains for cutting. R. P. Sharp (1949) discusses some of the problems involved in determining former wind directions from ventifacts, with reference to the Big Horn region of Wyoming where large areas of ground are littered with ventifacts ranging from less than 2-3 cm in size to boulders 2 m in length. The smaller ventifacts may display up to twenty wind-cut faces and have clearly moved during sand-blasting; but the boulders more than 0·3 m in size and with but a single wind-cut facet are thought to be *in situ* and provide reliable indications of wind-direction. If wind-cut grooves on the boulders are to be used for this purpose, it must be remembered that the wind will itself be deflected by a large boulder; therefore, only grooves following the true dip of a wind-cut facet should be so used. The most probable causes of the shifting of the smaller stones during wind attack are either the wind itself or frost action (A. Cailleux, 1942). Sharp regards frost-heave or solifluction as most likely and deduces from this that the cutting of the ventifacts occurred in cold periods. Unmoved boulders from some thirty localities in the Big Horn region show winds blowing from directions between N.10°W. and N.60°W., essentially the same as today and reflecting the dominant effect of the mountain barriers.

Disseminated Eolian Deposits

SAND

In discussing Pleistocene eolian deposits, it is useful to draw a rough distinction between silt (grain-size 0·004-0·06 mm) and sand (0·06-1·00 mm). Most work on sand deposits has in fact concerned the grain-size range 0·25-1·00 mm (medium-coarse); and R. A. Bagnold (1941) has shown that sand movement by saltation is most effective within this range. Wind-blown sand is found in many glacial deposits and sometimes forms a thin separate mantle over the ground surface. J. Pelisek (1963) notes how such a mantle, with grain sizes dominantly in the range 0·25-0·50 mm, is common in the Bohemian lowlands and Carpathians, sometimes several metres thick. Wind speeds involved in transporting such sand deposits must have been 30 km per hour or more. Wind-blown sand grains may be distinguished by their 'frosted' or matt surfaces according to A. Cailleux (1942), though more recently this method of distinction has been criticized. Sand grains will be rounded if they have been subjected to a long period of saltation, otherwise they will be sub-angular. Cailleux's study of the Pleistocene coarse sand deposits of Europe is the most comprehensive yet undertaken. From about 3000 samples of various Quaternary deposits, the sand in the range 0·4–1·0 mm was separated, and the percentage of grains possessing a rounded and 'frosted' surface

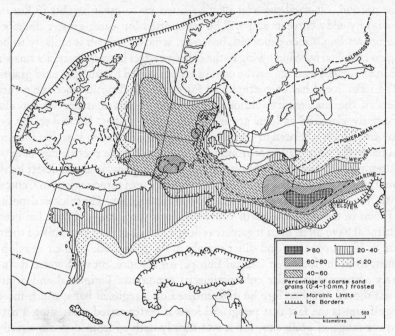

Fig. 26-1 Distribution of frosted (wind-worn) sand grains in the Quaternary deposits of Europe (*A. Cailleux, Memoir* 46, 1942, Geological Society of France)

was recorded. Figure 26-1 shows the general distribution, and does not refer to any particular phase of the Pleistocene. In Central Poland, the percentage of wind-blown grains rises towards 100, and there is clearly revealed a belt of country lying just outside the Pomeranian ice limit in which there occurs a very high frequency of wind-blown grains. Within the belt, ventifacts are relatively abundant, and beyond it lies the main zone of deposition of finer wind-blown silt or loess. Cailleux argues for strongest wind action where the proportion of frosted grains is highest, and simultaneously, a severe periglacial climate, since the larger ventifacts in this area show signs of having been overturned by frost heaving. Such conditions probably prevailed in the areas marginal to the Scandinavian ice in each of the older glacial periods, but the evidence of wind action is clearest (because best preserved) in the area bordering the last ice to reach the southern edge of the Baltic.

SILT

Whereas sand is moved by wind primarily by the process of saltation near the ground, silt-sized particles may be lifted up to great heights (3 km or more) by eddy currents and will be transported for much greater distances before settling. Once deposited, however, wind-blown silt is unlikely to be again removed by wind owing to the cohesive effect between particles smaller than about 0·06 mm, unless the deposit is bombarded by saltating sand grains. The term 'loess' has been employed for eolian silt deposits since the earlier part of the last century, although it is now known that not all deposits formerly classified as loess are of purely eolian origin. Moreover, by no means all loess is related to periglacial conditions, for much loess has formed in relatively warm climates around the margins of some present-day deserts. Attempts have been made to distinguish between periglacial and desert loess, but the distinction is not a simple one for the loess of some regions in Central Asia, for example, can be said to fall into both categories. The thickest deposits of loess in the world in north China (where a thickness of 600 m has been claimed) have often been regarded as desert loess, but the great bulk of them are of Pleistocene age and differ mainly in the magnitude of their development from the loesses of Central Europe, for instance, most of which definitely accumulated under cold periglacial conditions. Periglacial and desert loess should not therefore be differentiated on a regional basis; much more significant is the fact that periglacial loess is better sorted, possessing a narrower range of grain size, for it was mostly derived from deposits already sorted and laid down by glacial meltwater. C. Troll (1944) argues that periglacial loess is also derived from frost-pulverized debris, for the ultimate effect of comminution by frost action is to produce material of grain size 0·01-0·2 mm, and this grain size range is also the commonest range for periglacial eolian sand and loess. He adopts the extreme view that all loess demands a cold climate—without frost, no loess.

Composition of loess Figure 26-2 shows typical graphs of grain-size distribution for samples of Kansan loess (A. Swineford and J. C. Frye, 1945). Silt-sized particles are dominant—normally, between 70 and 85 per cent by weight lie in the range 0·007 to 0·003 mm—but there is also a clay fraction with particles smaller than 0·002 mm which has only become amenable to analysis with the introduction of X-ray diffraction techniques (see, for example, J. C. Frye *et al.*, 1962). Some fine or very fine sand is also usually present. The grains are angular or sub-angular, but occasionally rounded.

The chemical composition of loess is somewhat variable, especially in regard to its carbonate content which may range from nil to over 40 per cent by weight. A low $CaCO_3$ content reflects subsequent leaching. The dominant mineral in loess is quartz (50-80 per cent by weight); felspar may account for 20-25 per cent. The grains are bound together by clay minerals of which montmorillonite is often the most important—it constitutes up to 70 per cent of all clay minerals in Mississippi valley loess. Heavy mineral suites are

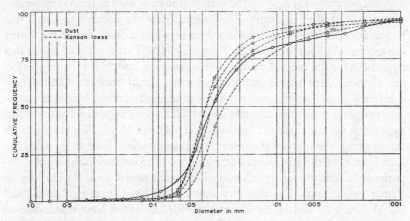

Fig. 26-2 Grain-size distribution of samples of Kansan loess and recent wind-blown dust (A. Swineford and J. C. Frye, *Am. J. Sci.*, 1945)

often distinctive and may afford a means of identifying and correlating particular loesses: the same applies to the clay mineral suites (J. C. Frye *et al.*, 1962).

The buff colour of loess exposures is well known, though in the un-weathered state (which is rare), the colour is grey. True loess is unstratified, for deposition was slow and each addition was an extremely thin layer. It is characterized by a vertical structure composed of numerous tubes representing the casts of former roots, for as the loess thickened, so the deeper parts of plant roots slowly decayed. The fine grain size and tubular structure result in a very high porosity—40 per cent is common and 65 per cent not unusual —so that the surface of loess remains dry and is relatively immune to dissection by water erosion. Wherever it is cut into by through-flowing streams however, the loess is capable, because of its cohesive nature and vertical structure, of maintaining precipitous faces, such as appear in the well-known loess bluffs of the Mississippi.

Loess may dovetail with glacial deposits (see p. 578) and with eolian sand (as in southern Poland: E. M. Dowgiallo, 1965). Fossil soils in all stages of development may be interbedded with it and appear in exposures as darker humic layers. These palaeosols are taken to mark more humid climatic intervals separating the dry loess substages, and are invaluable in providing radiocarbon dates to clarify the stratigraphic record. Thus the stratigraphy of the Wisconsinan loesses of Illinois as far back as about 40,000 years is now controlled by more than twenty radiocarbon dates.

Theories of the origin of loess   The earliest studies of loess in Europe suggested a lacustrine or alluvial origin. The originator of the eolian hypothesis was F. von Richthofen (1877), who based his arguments on studies of the China loesses; he was supported by R. Pumpelly (1879) who adopted the eolian hypothesis for the loess of central North America, and added the suggestion that the loess here was largely derived from glacial outwash. Controversy continued, however, and many continued to regard certain loesses as water-laid, especially those containing stratified silts. Recent opponents of the eolian hypothesis have included R. J. Russell (1944) and H. N. Fisk (1951), both investigating the loess of the lower Mississippi valley. Russell regarded the loess here as of fluvial-colluvial origin, derived from floodplain alluvial deposits by eluviation of the clays, and later attempted to apply the idea to Rhine valley loess. The majority opinion does not support Russell, and M. M. Leighton and H. B. Willman (1950) and others have marshalled an impressive array of evidence testifying to the eolian nature of Mississippi loess, even in the lower valley. However, the dispute has been of value, for it has stressed the possibility that certain formations labelled loess may not be eolian. H. Aumen and others (1965) reached this conclusion for some layers in the 'limon' deposits of Provence which show definite signs of water sorting.

A perennial problem is that of distinguishing between primary (eolian) loess and secondary loess resulting from hill wash or slipping of the primary loess. Inevitably one may be confused with the other and mapping of primary loess distributions becomes very difficult. Thickness measurements will be unreliable—indeed, the immense thicknesses sometimes quoted for loess in China certainly include much material which is not primary loess.

The classical theory explaining the origin of periglacial loess (see, for example, T. C. Chamberlin, 1897), which is now generally and widely accepted, envisages the main sources of wind-blown silt as glacial outwash plains. Across these plains flowed braided and heavily silt-laden meltwater streams, which flooded in spring and summer, and shrank in the later part of the year to expose broad expanses of unvegetated sediment. The sediment was replenished each year and thus provided rich sources of fine silt. As the sediments dried out each autumn, strong or gusty winds winnowed out the finer silts and rock-flour, raising great dust clouds which are today a common phenomenon on the outwash plains of Alaska, Iceland, or Spitsbergen. Some of the dust may also have come from till plains possessing little or no vegetation under a periglacial climate. The dust was eventually deposited in regions away from the ice where vegetation of grass or even forest could bind the wind-blown silt together.

The evidence generally supporting this theory is abundant and persuasive. The eolian nature of loess is confirmed by the way the deposit mantles varying bedrock formations indifferently, and by its independence of altitude (in Europe, for instance, it sweeps from near sea-level up to 500 m and over). Moreover, loess is unstratified and in places merges into or is interbedded with coarser deposits of definitely eolian character, including ventifacts. Its deposition in dry form is evident, for when subsequently wetted and loaded, it subsides or shrinks, showing that the original loess was neither soaked by water nor deposited by it (V. A. Obruchev, 1945). Its thickness diminishes away from the source regions which can often be identified by mineralogical analyses.

Further valuable insight into conditions of deposition is given by the fauna. Mollusca may occur in huge numbers—the Peoria loess of the Great Plains of the United States sometimes contains 180,000 shells per cubic metre. The shells are often small and delicate, so fragile that emplacement by running water is inconceivable. They lived on the surface, died, and were gradually entombed by the dust. The mollusca are of comparatively few species and are terrestrial forms, the only aquatic species being those associated with streams that dry up periodically. The gastropods throw further light on the climatic conditions prevailing at the time the loess was accumulating. In the Peoria loess, A. N. Leonard and J. C. Frye (1954) found no species which could resist long periods of drought or prolonged high temperatures; most required a moist organic cover on the ground surface and were suited by cooler conditions than now prevail in the Great Plains in summer. Near the main rivers, forest probably flourished and was instrumental in aiding accumulation of thick loess here (30 m or more), while a few kilometres away, mollusca and the remains of prairie animals suggest grassland with occasional woodland patches. Altogether, Leonard and Frye think conditions during the loess deposition were more favourable in this area to plant and animal life than they are today, and the idea of a barren dusty landscape in the loess accumulation areas here must be dismissed. Moreover, such favourable conditions must have lasted throughout the whole period of Peoria loess deposition, and have existed over thousands of square kilometres, from north Texas to South Dakota, and east into Ohio though there were certainly not uniform conditions over the whole of this vast area. As analyses of the loess point to a source in the outwash of valleys draining from glacier ice to the north or in the Rockies, and the mollusca were suited to a cool environment, loess deposition evidently took place in glacial rather than interglacial periods. This tallies with the stratigraphical relationships of the loess to the drift sheets: the Peoria loess, for instance, is interbedded with the Woodfordian till* in Illinois.

The picture that emerges may be summarized thus. At times of glaciation, outwash plains and to a lesser extent till plains pulverized by frost action and bare of vegetation were swept by strong, gusty, cold winds, often blowing off the ice. Dust clouds wafted across the barren tundra zone marginal to the ice, and the dust was deposited beyond, where vegetation cover existed to arrest its movement and helped to bind it together.

DISTRIBUTION OF LOESS

Continental Europe and U.S.S.R.   The greater part of the loess in this region lies beyond the maximum limit of glaciation. Loess mantles thousands of square kilometres of lowland in eastern Europe and European Russia, but is seldom thick enough to develop its own topography. The maximum thickness is about 80 m in Rumania. The vast outwash spreads of the Bug, Dnieper, Don, and Volga appear to have been the main sources. Eastward, loess reaches into Central Asia to the foothills of the Pamirs and Tien Shan, and north-eastward into Siberia.

An important loess belt in Central Europe stretches from Cracow past Breslau and Leipzig towards Köln (where it was first described in the 1840s), and lies between the limits of the Weichsel and Elster glaciations, along the northern flank of the Mittelgebirge; glacial outwash in the Urstromtäler provided the main source of silt. Important extensions of this belt into unglaciated country run up the Rhine Rift Valley, and westward into Belgium and northern France.

The main source of silt for loess in these western areas was probably the floor of the North Sea or Channel, exposed during the low sea-levels of glacial times. The low sea-level also had the effect of extending the European land-mass, thus enhancing the continentality of the climate in central Europe where dry and severely cold conditions accompanied loess deposition. Loess is rare in north-west Europe (except Iceland), for this area was largely ice-covered in the last glaciation, and its relatively maritime climate in the late- and post-glacial probably precluded much wind action. Some loess is known in Sweden, though; A. Falk (1955) has mapped loess up to 70 cm thick in the Mora area (latitude 61°), and the larger deposits at Brattforsheden in Värm-

* Late Wisconsinan; radiocarbon years 22,000–12,500 B.P.

land have been known for some time because of their interesting relationships with neighbouring sand-dune systems (see p. 582). Most Swedish loess is thought to date from about 9000 B.P. (late Yoldia or early Ancylus time).

In summary, the principal features of loess in Europe are its concentration outside the limits of the Weichsel glaciation, its association with glacial outwash which provided the source areas, its more extensive occurrence in eastern rather than western Europe and on lowland rather than upland, and the clear indications of its formation under a dry continental climate during glacial periods.

Britain Loess is rare in Britain apart from small patches in south-eastern England (for a full discussion of these see J. B. Dalrymple, 1960). The brick-earth of the London Basin, southern East Anglia, and Kent, comprises a silt loam of various origins, but much of it is probably a true wind-blown loess (S. W. Wooldridge and D. J. Smetham, 1931). The laminated variant seen in some sections probably represents deposition in small patches of standing water. Brickearth occurs up to elevations of 200 m on the Chilterns near Luton, where the inclusion of Palaeolithic implements shows its recent nature here. It is surprising that so few eolian deposits of substantial thickness have yet been found in and around the Thames valley, in which great masses of glacial outwash were laid down in the Saale and older glaciations. Perhaps the more maritime climate of Britain and the more rapid spread of vegetation were not so conducive to wind action as in continental Europe, and certainly hindered the drying-out of the surface of the outwash. The patches of loess in Britain are essentially of local origin, in contrast to the widespread loess of central and eastern Europe where powerful wind action operated over vast regions.

At Pegwell Bay, Kent, W. S. Pitcher and others (1954) have demonstrated the existence of a true eolian loess, derived from local Thanet Beds. Its age was later shown by M. P. Kerney (1965) to be the equivalent of the Older Coversand of Holland (Weichselian maximum); an overlying weak soil gave a radiocarbon date of about 13,000 years. In south-western England, also beyond the limits of glaciation, loess has only recently been recognized and more will certainly be revealed with further investigations of the superficial deposits. D. E. Coombe and L. C. Frost (1956) have examined the soils on the Lizard peninsula. These contain silt, probably derived from the Carnmenellis granite in part. In various samples, 43–57 per cent of the silt grains fall within the size range 0·01–0·05 mm. This in itself does not prove an eolian origin, but the wide scatter of silt in the region and its lack of addition today strongly suggest late Pleistocene wind action and the trapping of dust by vegetation. The silt overlies the early Pleistocene deposits of

Crousa Common. On the Durham coast, C. T. Trechmann (1919) has claimed an interglacial age for loess 3-4 m thick banked against a north-facing slope. Its mineral composition was found to be remarkably similar to Rhine valley loess near Bonn. Only one eolian deposit is so far known from Scotland (R. W. Galloway, 1961), a silt 1 m thick near Kinross resting on and derived from local fluvioglacial material.

North America   There is an enormous literature on the loess deposits of North America, to which some references have already been made. A useful recent summary of Pleistocene eolian action in the United States is given by H. T. U. Smith (1964), and the distribution of loess is displayed in the map compiled by J. Thorp and others (1952). Stretching from the Rockies to Ohio, and from the lower Mississippi to beyond the Canadian border, most of the loess now seen is of Wisconsinan (Weichselian) age, for the older loess is often either denuded, buried, or decomposed too far to be easily recognizable· As in Europe, glacial outwash provided the chief sources, with the addition of some from bare till plains or non-glacial alluvium. Its distribution in relation to source areas reflects a dominantly westerly but variable air circulation, very similar to that of today, though probably stronger.

The best known occurrences of loess in the world lie in the Mississippi basin. In an area of half-a-million km², half of which is glaciated, are found most of the phenomena associated with loess deposition (Leighton and Willman, 1950). Of particular interest is the Iowan loess (equivalent to the Morton loess of J. C. Frye *et al.*, 1962) and its unique relationship with the Iowan drift border. At the latter, the loess rises suddenly above the drift plain, looking rather like a moraine series, and attains thicknesses of 15 m or more; it is also rather sandy, with some dune-like forms. Away from the drift border on the older Kansan drift, it gradually becomes finer in grain size and thinner. The loess is similar to the Iowan drift both chemically and mineralogically, ventifacts occur on the Iowan drift beneath a scanty loess cover, and beyond the drift border, the loess mantles hills and valley slopes alike. An eolian origin is impossible to deny.

Figure 26-3 shows the variations in thickness of all Wisconsinan loess in Illinois, and demonstrates clearly the relationships of the loess to the Wisconsinan ice limit and to the major valleys carrying outwash. The loess is thickest along these valleys, especially where they are widest; Leonard and Frye (1954) claim that a forested border along the valley edges was important in trapping great thicknesses of loess here. Away from the valleys, both thickness and average particle size of the loess diminish exponentially with distance, again confirming the eolian origin (G. D. Smith, 1942). At the same time, the carbonate content also decreases, in inverse proportion to the thickness of the

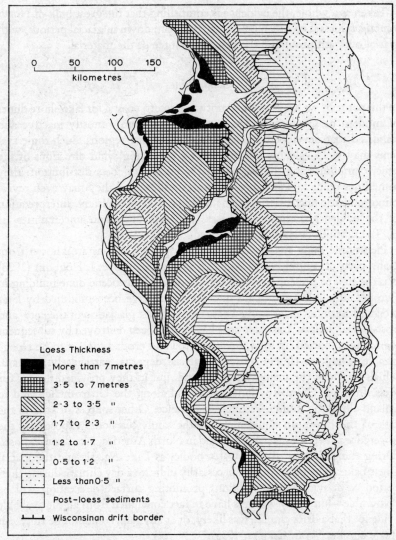

Fig. 26-3 Variations in the thickness of the Wisconsinan loess in Illinois. Note the sudden changes at the Wisconsinan drift border (R. F. Flint, *Glacial and Pleistocene Geology*, John Wiley & Sons, New York, 1957)

loess. Carbonate content at the time of deposition was over 40 per cent at the valley bluffs where loess accummulated rapidly, but in the zones of thinner loess, deposition was so slow that the loess was mostly leached of lime even when deposition had just ceased.

Taken as a whole, the evidence is compelling that the great bulk of North American loess is of eolian origin, and was laid down in glacial periods, with glacial outwash providing the principal source of the material.

Sand Dunes

In this context, we are concerned not with desert, coastal, or lake-shore dunes of modern or Quaternary age, but with dunes now mostly inactive and stabilized that developed in Pleistocene periglacial regions. Such dune systems may yield valuable information on prevailing wind directions of the time, complimenting the evidence of ventifacts or loess distribution; they sometimes point to conditions of greater aridity, and they may even reveal minor climatic fluctuations if dune building was intermittent. Interpretation of the evidence is, however, beset with difficulties and uncertainties, as H. T. U. Smith (1949) has emphasized.

Pleistocene dune systems of former periglacial regions are known from many parts of central North America and from Europe. I. Högbom (1923) was the first to present a unified picture of late Pleistocene dune building in Europe, although some of his interpretations have been modified by later work. Most dunes so far studied date from the last glaciation or the Late- and Post-glacial periods, for older ones have either been destroyed by subsequent ice advance, buried by loess, or weathered and eroded too far to be recognizable. They are closely related to glacial deposits, particularly outwash, which provided the main sources of sand, just as they provided the silt for loess. In north Germany, dunes are found along the Urstromtäler, and in Poland along the middle Vistula, for instance. Other sources of sand in the case of the European dunes included the sandy sea floors exposed by the lowered sea-levels of glacial times, and in North America, lake floors exposed during the shrinking of such water bodies as Lake Agassiz and the glacial Great Lakes. The dunes do not necessarily indicate a dry climate during their period of activity—any conditions promoting surface desiccation will be favourable—but many workers have in fact concluded from all the available evidence that a drier climate was likely, or at any rate, could not be ruled out (for example, W. S. Cooper, 1935).

The forms of Pleistocene dunes vary greatly. The only major dune type not represented is the barchan, which modern studies of active desert dunes indicate requires complete absence of vegetation or other obstacles for most perfect development. In Pleistocene periglacial regions, vegetation appears to have been present to a limited degree, increasing its hold as the climate ameliorated but while the dunes were still active. Thus, complex dune forms occur. Their basic components are the longitudinal dune, and the U-shaped ('parabolic' or 'transverse') dune. The latter, with its horns pointing upwind, is not to be mistaken for the barchan. Examples of ancient parabolic dunes are given by F. Gullentops (1957). Determinations of former wind direction from dune morphology alone can be dangerous, especially if weathering has subdued the forms. Gross errors are possible if longitudinal and transverse

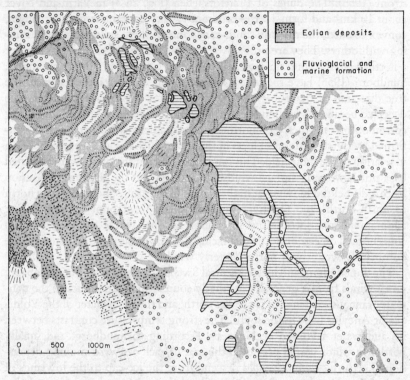

Fig. 26-4 Map of the eolian deposits in the northern part of Brattforsheden, Värmland, Sweden. The small dots indicate the loess (F. Hjulström, *Geogr. Annlr.* 1955)

dunes are confused (e.g. F. Enquist, 1932, who argued that certain Swedish dunes were longitudinal). In Europe, most Pleistocene dunes are of the transverse variety.

Dune-sand bedding may be used to establish former wind directions with fair reliability if there are steeply dipping (30-35 degrees) beds, for the true dip of these is always to the leeward in both U-shaped dunes and barchans. The method is less satisfactory if the dips are variable and less than 25 degrees. One further point should be raised in connection with determinations of wind direction: the dune will record the direction of the effective sand-moving winds, which may or may not be that of the prevailing winds. Powerful storms from one quarter may shift far more sand in several hours than gentle prevailing winds from another quarter may achieve in several days or weeks.

An example of a Late-Glacial dune-field in Sweden is described by F. Hjulström (1955). The dunes of Brattforsheden (Fig. 26-4) in Värmland cover about 10 km² and form a series of about twenty concentric curving ridges, convex towards the east and south-east, some of them subdividing and linking with others. They are probably transverse dunes and their forms suggest a north-westerly wind during their period of activity. F. Hjulström and A. Sundborg (1955) have suggested, however, that this does not represent the true direction of the isobaric gradient wind, for the surface winds may be deflected because of ground friction; thus, north-westerly surface winds may correspond to northerly or north-easterly gradient winds, and this is confirmed by the disposition of loess in the area, which has an eccentric position in relation to the dune field. This can be explained if the dunes were activated by a north-westerly surface wind but the loess was the result of deposition from a dust cloud blowing at higher levels in the atmosphere from the north or north-north-east. The hypothesis may help to account for some other anomalous loess distributions in central Europe, where the loess is often concentrated on east or north-east facing slopes, yet the dune systems indicate dominantly westerly or north-westerly winds.

The dune systems and loess of central Sweden demand strong and probably gusty winds from a generally northerly source; their age indicates the contemporaneous existence of ice to the north, and the open water of the Yoldia Sea or Ancylus Lake to the south. The strong temperature contrasts between ice and water favoured cyclogenesis, according to Sundborg, and squally northerly winds would prevail in the rear of each eastward-moving depression. The surface winds would descend from the ice-cap to the lowlands of central Sweden, the resulting föhn effect helping to produce a low relative humidity to dry the surfaces of fluvioglacial deposits.

Conclusion

Periglacial wind action is associated with cold dry climatic conditions in regions where strong winds blow off nearby glacier ice, frost-shattered till plains, and glacial outwash spreads. Wind erosion is manifested in the occurrence of grooved and polished bedrock surfaces and ventifacts. The latter if large and undisturbed can be used to infer former wind directions.

Disseminated eolian deposits consist of sand (0·06-1·00 mm) and silt (0·004-0·06 mm). Wind-blown sand grains are distinguishable by their roundness and frosted appearance, and occur in greatest abundance just beyond the last major Pleistocene ice limits. Silt-sized particles were blown in dust clouds farther from the ice and were redeposited as loess, the bulkiest of all periglacial accumulations. The eolian origin of loess is now firmly established, but within this broad category, it may be possible to differentiate between periglacial loess, with a narrow range of grain size since it was mostly derived from already-sorted fluvioglacial deposits, and desert loess of warmer climates. Periglacial loess is characterized by a dominance of grain sizes in the range 0·007-0·03 mm. It is unstratified, and its texture, faunal content, variations in thickness, and relations with contemporaneous glacial deposits testify to its eolian origin and its association with nearby glacier ice. It mantles huge areas in eastern Europe, European Russia, and North America, while smaller occurrences are being steadily revealed in Britain.

Fossilized dune systems formed under Pleistocene periglacial conditions are known in many parts of central North America and Europe. The commonest dune forms are parabolic (transverse) with horns pointing upwind, and longitudinal, and it is important to distinguish between the two types. The dune systems may then be used in some instances to reconstruct former wind directions, as in the case of some wind-eroded bedrock surfaces and large ventifacts, thus helping to add to our knowledge of Pleistocene meteorology.

REFERENCES

H. AUMEN et al. (1965), 'Petrographie des limons de Provence', Bull. Ass. fr. Étude Quat. 2, 35-49

R. A. BAGNOLD (1941), The physics of blown sand and desert dunes (London), 265 p.

K. BRYAN (1945), 'Glacial versus desert origin of loess', Am. J. Sci. 243, 245-8

A. CAILLEUX (1942), 'Les actions éoliennes périglaciaires en Europe', Mém. Soc. géol. Fr. 46, 1-166

T. C. CHAMBERLIN (1897), 'Supplementary hypothesis respecting the origin of the loess of the Mississippi valley', J. Geol. 5, 795-802

D. E. COOMBE and L. C. FROST (1956), 'The nature and origin of the soils over the Cornish serpentine', J. Ecol. 44, 605-15

W. S. COOPER (1935), 'The history of the upper Mississippi River in Late Wisconsin and Post-glacial time', Bull. Minn. geol. Surv. 26, 72-108

J. B. DALRYMPLE (1960), 'The nature, origin, age, and correlation of some of the brickearths and associated soils of south-eastern England', Unpubl. Ph.D. thesis, Univ. of London

E. M. DOWGIALLO (1965), 'Mutual relation between loess and dune accumulation in southern Poland', Geogr. Polonica 6, 105-15

F. ENQVIST (1932), 'The relation between dune-form and wind direction', *Geol. För. Stockh. Förh.* **54**, 19-59

A. FALK (1955), 'Preliminary mapping of some localities of wind-blown silt in Dalarna', *Geogr. Annlr* **37**, 112-17

H. N. FISK (1951), 'Loess and Quaternary geology of the lower Mississippi valley', *J. Geol.* **59**, 333-56

J. C. FRYE, H. D. GLASS, and H. B. WILLMAN (1962), 'Stratigraphy and mineralogy of the Wisconsinan loesses of Illinois', *Circ. Ill. St. geol. Surv.* **334**

R. W. GALLOWAY (1961), 'Periglacial phenomena in Scotland', *Geogr. Annlr* **43**, 348-53

F. GULLENTOPS (1957), 'Quelques phénomènes géomorphologiques depuis le Pleni-Würm', *Bull. Soc. belge Géol. Paléont. Hydrol.* **66**, 86-95

F. HJULSTRÖM (1955), 'The problem of the geographic location of wind-blown silt: an attempt of explanation', *Geogr. Annlr* **37**, 86-93

I. HÖGBOM (1923), 'Ancient inland dunes of northern and middle Europe', *Geogr. Annlr* **5**, 113-242

G. JOHNSSON (1958), 'Periglacial wind and frost erosion at Klågerup, S.W. Scania', *Geogr. Annlr* **40**, 232-43. See also by the same author a further article in *Geogr. Annlr* **44**, 378-404

M. P. KERNEY (1965), 'Weichselian deposits in the Isle of Thanet, east Kent', *Proc. Geol. Ass.* **76**, 269-74

M. M. LEIGHTON and H. B. WILLMAN (1950), 'Loess formations of the Mississippi valley', *J. Geol.* **58**, 599-623

A. B. LEONARD and J. C. FRYE (1954), 'Ecological conditions accompanying loess deposition in the Great Plains region', *J. Geol.* **62**, 399-404

V. A. OBRUCHEV (1945), 'Loess types and their origin', *Am. J. Sci.* **243**, 256-62

J. PELISEK (1963), 'Pleistozäne Dünensande in der tschechoslowakischen sozialistischen Republik', *Eiszeitalter Gegenw.* **14**, 216-23

W. S. PITCHER et al. (1954), 'The loess of Pegwell Bay, Kent, and its associated frost soils', *Geol. Mag.* **91**, 308-14

W. E. POWERS (1936), 'The evidences of wind abrasion', *J. Geol.* **44**, 214-19

R. PUMPELLY (1879), 'The relation of secular rock disintegration to loess, glacial drift, and rock basins', *Am. J. Sci.* **17**, 133-44

F. VON RICHTHOFEN (1877), *China* (Berlin), **1**, 758 p.

R. J. RUSSELL (1944), 'Lower Mississippi valley loess', *Bull. geol. Soc. Am.* **55**, 1-40

R. P. SHARP (1949), 'Pleistocene ventifacts east of the Big Horn Mountains, Wyoming', *J. Geol.* **57**, 175-95

G. D. SMITH (1942), 'Illinois loess: variations in its properties and distribution', *Bull. Ill. agric. Exp. Stn.* **490**

H. T. U. SMITH (1949), 'Physical effects of Pleistocene climatic changes in non-glaciated areas', *Bull. geol. Soc. Am.* **60**, 1485-516

H. T. U. SMITH (1964), 'Periglacial eolian phenomena in the United States', *Rep. 6th Conf. int. Ass. quatern. Res.* (*Warsaw, 1961*), Lodz (1964), **4**, 177-86

A. SUNDBORG (1955), 'Meteorological and climatological conditions for the genesis of aeolian sediments', *Geogr. Annlr* **37**, 94-111

A. SWINEFORD and J. C. FRYE (1945), 'A mechanical analysis of wind-blown dust compared with analyses of loess', *Am. J. Sci.* **243**, 249-55

J. THORP *et al.* (1952), *Pleistocene eolian deposits of the United States, Alaska, and parts of Canada* (map, scale 1/2,500,000, publ. *Am. geol. Soc.*)

C. T. TRECHMANN (1919), 'On a deposit of interglacial loess and some transported preglacial freshwater clays on the Durham coast', *Q.J. geol. Soc. Lond.* **75**, 173-203

C. TROLL (1944), 'Strukturböden, Solifluktion, und Frostklimate der Erde', *Geol. Rdsch.* **34**, 545-694 (English translation, *Snow Ice Permafrost Res. Establ.*)

C. K. WENTWORTH and R. I. DICKEY (1935), 'Ventifact localities in the United States', *J. Geol.* **43**, 97-104

S. W. WOOLDRIDGE and D. J. SMETHAM (1931), 'The glacial drifts of Essex and Hertfordshire, and their bearing upon the agricultural and historical geography of the region', *Geogr. J.* **78**, 243-69

H. E. WRIGHT (1946), 'Sand grains and periglacial climate: a discussion', *J. Geol.* **54**, 200-5

Index

Note: Place names in England are generally classified under their county, and in the United States of America under their state; in all other parts of the world, including Scotland, Wales and Ireland, places are classified under the name of the country. This applies also to glaciers, except those in the European Alps which will be found under Alps. Lake names are classified under Lake. Page numbers printed in bold refer to figures in the text.